High Throughput
Bioanalytical Sample Preparation

Methods and Automation Strategies

PROGRESS IN PHARMACEUTICAL AND BIOMEDICAL ANALYSIS

PROGRESS IN PHARMACEUTICAL AND BIOMEDICAL ANALYSIS 5

High Throughput Bioanalytical Sample Preparation

Methods and Automation Strategies

David A. Wells

Sample Prep Solutions Company
St. Paul, Minnesota
U.S.A.

2003

ELSEVIER
Amsterdam – Boston – London – New York – Oxford – Paris – San Diego
San Francisco – Singapore – Sydney – Tokyo

ELSEVIER B.V.	ELSEVIER Inc.	**ELSEVIER Ltd**	ELSEVIER Ltd
Radarweg 29	525 B Street	**The Boulevard**	84 Theobalds Road
P.O. Box 211, 1000 AE	Suite 1900, San Diego	**Langford Lane, Kidlington,**	London WC1X 8RR
Amsterdam, The Netherlands	CA 92101-4495, USA	**Oxford OX5 1GB, UK**	UK

First edition 2003
Second impression 2005

Library of Congress Cataloging in Publication Data
A catalog record is available from the Library of Congress.

British Library Cataloguing in Publication Data
A catalogue record is available from the British Library.

ISBN: 0 444 51029 X

Working together to grow
libraries in developing countries

www.elsevier.com | www.bookaid.org | www.sabre.org

ELSEVIER   BOOK AID
International   Sabre Foundation

∞ The paper used in this publication meets the requirements of ANSI/NISO Z39.48-1992 (Permanence of Paper).
Printed in The Netherlands.

Preface

It has been exciting to be involved with the development and implementation of high throughput techniques for sample preparation used for drug analysis in the pharmaceutical industry. My work as an independent consultant and educator in this field has allowed me to work with scientists worldwide having a broad range of expertise. It became evident during my assignments that a single source of information that reviewed the utilization of high throughput sample preparation techniques was not available. This book was written to fulfill this need for my students and colleagues.

The text begins with an introductory overview of the role of bioanalysis in pharmaceutical drug development, focused on the particular activities that are performed within each stage of the research process. A fundamental understanding of the strategies for sample preparation is reinforced next, along with essential concepts in extraction chemistry. In order to gain a mastery of knowledge about the available tools needed to perform high throughput sample preparation techniques, several chapters introduce and discuss microplates, accessory products and automation devices. Particular strategies for efficient use of automation within a bioanalytical laboratory are also presented. The subject material then reviews four common sample preparation techniques: protein precipitation, liquid-liquid extraction, solid-phase extraction and various on-line sample preparation approaches. Each technique is discussed with reference to its fundamental principles and strategies for method development and automation. The book concludes with information on recent advances in sample preparation.

The important objectives that can be accomplished when the strategies presented in this book are followed include:

(a) Improved efficiency in moving discovery compounds to preclinical status with robust analytical methods
(b) Return on investment in automation for sample preparation
(c) Improved knowledge and expertise of staff

It is my sincere desire that the reader finds this book a valuable resource for information on high throughput sample preparation methods for bioanalysis and recommends it to staff and colleagues.

The need for this book was mentioned previously but this project would never have been completed without inspiration and the support of many individuals. Early in my professional career I was inspired by two outstanding professors who instilled within me the exhilaration of investigative science and the important role of teaching students and nurturing their professional growth. I am indebted to Dr. George A. Digenis at the University of Kentucky, College of Pharmacy, and Dr. Robert E. Lamb at Ohio Northern University, Department of Chemistry for their inspiration and encouragement. As I became an educator in my own career, my inspiration continued from the students with whom I worked and who learned high throughput sample preparation techniques from me. The progress of several of them, in particular, into recognized experts has been rewarding. I have also learned from, and been motivated by, established scientists and colleagues who worked with me to develop leading edge products for solid-phase extraction and implement high throughput drug sample preparation techniques in their laboratories.

The production of this book was greatly assisted by the efforts of two individuals in particular. Teresa Wells efficiently managed my large database of 1,800 literature references in sample preparation collected over the years so that I could quickly retrieve published information for use within each chapter. She also tolerated with great patience the long hours and months that were spent assembling the considerable amount of information that went into this book ("Is it done yet?"). I was astonished that an associate, Patt Threinen, volunteered to perform the arduous task of editing this manuscript by carefully reading each chapter. I owe her profound thanks for her understanding, corrections and comments of the many writes and rewrites passed back and forth. Her extensive experience in chemistry, biology, chromatography and sales added an invaluable perspective. This text is more readable because of her indefatigable attention to detail.

My colleagues also assisted in this effort by reviewing selected chapters pertinent to their expertise; their discussions and suggestions shaped the subject material into a useful and comprehensive information resource. They are each mentioned within the chapter to which they contributed. In particular, I am grateful for the collaboration with Jing-Tao Wu who coauthored the on-line sample preparation techniques chapter. The staff at Elsevier Science was also very supportive in the production of this book.

David A. Wells

Contents

Chapter 1
Role of Bioanalysis in Pharmaceutical Drug Development

Chapter 2
Fundamental Strategies for Bioanalytical Sample Preparation

Chapter 3
High Throughput Tools for Bioanalysis: Microplates

Chapter 4
High Throughput Tools for Bioanalysis: Accessory Products

Chapter 5
Automation Tools and Strategies for Bioanalysis

Chapter 6
Protein Precipitation: High Throughput Techniques and Strategies for Method Development

Chapter 7
Protein Precipitation: Automation Strategies

Chapter 9
Liquid-Liquid Extraction: Strategies for Method Development and Optimization

Chapter 10
Liquid-Liquid Extraction: Automation Strategies

Chapter 11
Solid-Phase Extraction: High Throughput Techniques

Chapter 12
Solid-Phase Extraction: Strategies for Method Development and Optimization

Chapter 13
Solid-Phase Extraction: Automation Strategies

Capillary Electrophoresis

List of Acronyms

ACE	Automated Cartridge Exchange
ACN	Acetonitrile
ADME	Absorption, Distribution, Metabolism and Elimination
ADS	Alkyl-Diol Silica
AGP	α1-Acid Glycoprotein
AIDS	Acquired Immune Deficiency Syndrome
ALPS	Automated Laboratory Plate Sealer
ANSI	American National Standards Institute
APCI	Atmospheric Pressure Chemical Ionization
API	Atmospheric Pressure Ionization
API-ES	Atmospheric Pressure Ionization–Electrospray
AVS	Automated Vacuum System
C1	Methyl bonded silica
C18	Octadecyl bonded silica
C2	Ethyl bonded silica
C8	Octyl bonded silica
CAT	Catecholamine
CBA	Carboxylic acid bonded silica
CE	Capillary Electrophoresis
CEC	Capillary Electrochromatography
CE-MS	Capillary Electrophoresis–Mass Spectrometry
CFR	Code of Federal Regulations
CI	Confidence Interval
CSV	Comma Separated Value
CV	Coefficient of Variation
CYP	Cytochrome P450 drug metabolizing enzymes
DMF	Dimethyl formamide
DMSO	Dimethyl sulfoxide
ECD	Electrochemical Detection
EDTA	Ethylene diamine tetra-acetic acid
ELISA	Enzyme-Linked Immunosorbent Assay
EPD	Effective Pore Diameter
ESI	Electrospray Ionization
EtOAc	Ethyl acetate
EVA	Ethyl vinyl acetate
FDA	Food and Drug Administration

FTIM	First Time in Man
GFF	Glycine-L-phenylalanine-L-phenylalanine
GLP	Good Laboratory Practices
GMP	Good Manufacturing Practices
HD	High Density
HIV	Human Immunodeficiency Virus
HT	High Throughput
HTS	High Throughput Screening
i.d.	Internal Diameter
IAE	Immunoaffinity Extraction
IAM	Immobilized Artificial Membrane
IND	Investigational New Drug
IRB	Institutional Review Board
IS	Internal Standard
ISLAR	International Symposium on Laboratory Automation and Robotics
ISRP	Internal Surface Reversed Phase
IT	Ion Trap
LC	Liquid Chromatography
LC-MS	Liquid Chromatography–Mass Spectrometry
LC-MS/MS	Liquid Chromatography–Tandem Mass Spectrometry
LC-NMR	Liquid Chromatography–Nuclear Magnetic Resonance
LC-UV	Liquid Chromatography–Ultraviolet
LLE	Liquid-Liquid Extraction
LLOQ	Lower Limit of Quantitation
LOD	Limit of Detection
LOQ	Limit of Quantitation
LRIG	Laboratory Robotics Interest Group
LSD	Lysergic acid diethylamide
m/z	Mass-to-charge ratio
MALDI-TOF-MS	Matrix-Assisted Laser Desorption Ionization Time-of-Flight Mass Spectrometry
MDCK	Madin-Darby Canine Kidney
MEK	Methyl ethyl ketone
MIP	Molecularly Imprinted Polymer
MipTec-ICAR	International Conference on Microplate Technology, Laboratory Automation and Robotics
mPC-CE-MS	Membrane Preconcentration–Capillary Electrophoresis Mass Spectrometry
MRM	Multiple Reaction Monitoring
MS	Mass Spectrometry

MS/MS	Mass Spectrometry/Mass Spectrometry or Tandem Mass Spectrometry
MTBE	Methyl tert-butyl ether
MTX	Methotrexate
MW	Molecular Weight
MWCO	Molecular Weight Cutoff
NADP+	Nicotinamide adenine dinucleotide phosphate
NDA	New Drug Application
NMP	N-Methyl-2-pyrrolidinone; N-Methylpyrrolidone
NMR	Nuclear Magnetic Resonance
NSAID	Non-Steroidal Anti-Inflammatory Drug
ODS	Octadecyl silane
PAI	Pre-Approval Inspection
PBA	Phenylboronic acid
PCR	Polymerase Chain Reaction
PcSFC	Packed Column Supercritical Fluid Chromatography
PEG	Polyethylene glycol
PPT	Protein Precipitation
PSDB	Poly(styrene divinyl benzene)
PTFE	Polytetrafluoroethylene
PV	Performance Verification
PVDF	Polyvinylidene fluoride
PVM	Packard Vacuum Manifold
QC	Quality Control
Q-TOF	Quadrupole Orthogonal Acceleration Time-of-Flight
RAM	Restricted Access Media
RoMA	Robotic Manipulator Arm
RSP	Robotic Sample Processor
SBS	Society for Biomolecular Screening
SD	Standard Density
SDB	Poly(styrene divinyl benzene)
SDVB	Poly(styrene divinyl benzene)
SERM	Selected Estrogen Receptor Modulators
SIM	Selected Ion Monitoring
SOP	Standard Operating Procedure
SPA	Scintillation Proximity Assay
SPE	Solid-Phase Extraction
SPME	Solid-Phase Micro Extraction
SPS	Semipermeable Surface
SRM	Selected Reaction Monitoring
SS-LLE	Solid-Supported Liquid-Liquid Extraction

TCA	Tricyclic Antidepressants
TDM	Therapeutic Drug Monitoring
TEA	Triethylamine
TFA	Trifluoroacetic acid
TFC	Turbulent Flow Chromatography
THF	Tetrahydrofuran
TOF	Time-of-Flight
UV	Ultraviolet

Chapter 1

Role of Bioanalysis in Pharmaceutical Drug Development

Abstract

Bioanalysis is the quantitative determination of drugs and their metabolites in biological fluids. This technique is used very early in the drug development process to provide support to drug discovery programs on the metabolic fate and pharmacokinetics of chemicals in living cells and in animals. Its use continues throughout the preclinical and clinical drug development phases, into post-marketing support and may sometimes extend into clinical therapeutic drug monitoring. The role of bioanalysis in pharmaceutical drug development is discussed, with focus on the particular activities that are performed within each stage of the development process and on the variety of sample preparation matrices encountered. Recent developments and industry trends for rapid sample throughput and data generation are introduced, together with examples of how these high throughput needs are being met in bioanalysis.

1.1 Overview of the Drug Development Process

1.1.1 Introduction

The discovery and development of safe and effective new medicines is a long and complex process. Pharmaceutical companies typically invest 9–15 years of research and hundreds of millions of dollars into this effort; a low rate of success has historically been achieved. The drug development process itself requires the interaction and cooperation of scientists and medical professionals from many diverse disciplines. Some of these disciplines include medicinal chemistry, pharmacology, drug metabolism and pharmacokinetics, toxicology, analytical chemistry, pharmaceutics, statistics, laboratory automation, information technology, and medical and regulatory affairs. A progression of research activities and regulatory filings must operate in parallel, often under severe time constraints. The success of a drug development program depends upon a number of favorable selections, such as targeting a therapeutic area in which an identified drug compound offers outstanding efficacy, identifying the

1

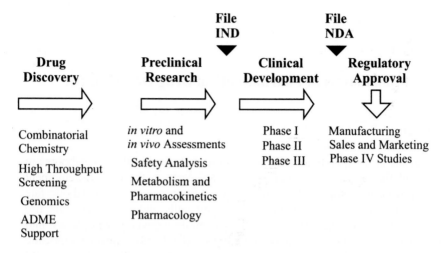

Figure 1.1. Schematic diagram of the overall drug development process and the major activities performed within each of the four major divisions.

optimal chemical structure of the drug molecule that yields the most favorable absorption, distribution, metabolism and elimination profiles, demonstrating safety, satisfying regulatory needs, as well as cost effective manufacturing and extensive sales support in the marketplace. A schematic diagram of the overall drug development process is outlined in Figure 1.1. Appropriate times for the regulatory filings are indicated.

1.1.2 Drug Discovery

Traditionally, drugs have been identified using one of three major strategies:

1. A focused and systematic approach is made to synthesize compounds that interact optimally with a target receptor or pharmacological model whose 3-dimensional structure has been elucidated, *i.e.*, rational drug synthesis. An example is the HIV-1 protease enzyme (Human Immunodeficiency Virus) whose function is important for maturation and assembly of infectious viral particles for the disease AIDS (Acquired Immune Deficiency Syndrome). The 3-dimensional structure of the HIV-1 protease enzyme is known and antiviral agents have been designed that specifically bind to this key protein and inhibit its function.

2. The active ingredients from natural plants, soil extracts and microorganisms (fungi, viruses and molds) are isolated, purified and screened for activity using various pharmacological models. This approach led to identification of paclitaxel (Taxol®), a drug used to treat various forms of cancer.

3. Accidental discovery or serendipity occurs when a drug molecule is found to work for a different target than the one for which it was originally synthesized. For example, in the search for novel drugs to treat cardiac arrhythmias, researchers discovered that imiquimod (Aldara®) was a novel immunomodulator that boosted the body's immune system; a new class of antiviral agents was discovered.

Today, however, the pharmaceutical research process is looking at new and improved ways to develop drugs, in response to several important scientific advances that have recently occurred. These advances include the identification of new and more specific drug targets (as a result of maturation in genomics and proteomics); successes with tissue growth outside of the living organism; development of faster, more sensitive and more selective analytical systems (mass spectrometry); higher throughput (as a result of robotics and laboratory automation); proliferation in synthesis techniques (combinatorial chemistry); and advances in computing and information systems (bioinformatics). In parallel with these scientific advances, business factors have changed with the consolidation of drug companies and the intense pressure to get drugs to market faster than ever before. The current focus of drug discovery research is on rapid data generation and analysis to identify promising candidates very early in the development cycle. An optimal lead candidate is selected for further evaluation.

Combinatorial chemistry techniques allow the synthesis of compounds faster than ever before, and these greater numbers of compounds are quickly evaluated for potency and pharmacological activity using high throughput screening (HTS) techniques. HTS involves performing various microplate based immunoassays with synthesized compounds or compounds from natural product isolation. Examples of assay types used in HTS are scintillation proximity assay (SPA), enzyme linked immunosorbent assay (ELISA), fluorescent intensity, chemiluminescence, absorbance/colorimetry and bioluminescence assays [1]. These HTS tests simulate a specific biological receptor or target function and a qualitative decision ("hit" or "miss") is generated [2].

Advances in genomics have increased the understanding of certain diseases at the molecular level, *i.e.*, the effect of a gene sequence on a particular illness. The role of the protein encoded by the gene is also studied. Proteins have generally been shown to be good drug targets. The effect of a drug on a protein, and thus on a specific biochemical pathway, forms the basis for a high throughput screening test. These HTS tests are usually conducted by scientists in pharmacology research groups.

Once hits are identified, chemists perform an iterative process to synthesize and screen smaller, more focused libraries for lead optimization in an effort to improve compound activity toward a specific target. Using automated techniques, ultra high throughput can be obtained by the most advanced laboratories and tens of thousands of compounds can be screened in one day. In parallel studies, information is learned on a drug molecule's absorption, distribution (including an estimate of protein binding), metabolism and elimination by sampling from dosed laboratory animals (called *in vivo* testing) and from working cells and/or tissues removed from a living organism (called *in vitro* testing since the cells are outside a living animal). These important tests are collectively referred to as ADME characteristics (Absorption, Distribution, Metabolism and Elimination).

A candidate compound that will potentially meet an important medical need receives an exhaustive review addressing all the key issues concerning its further development. Evaluation of the available data, competitive therapies, expected therapeutic benefit, market opportunities and financial considerations all contribute to the final decision to grant development status to a particular compound. A multifunctional project team is assembled to guide the development efforts into the next phase—preclinical development. By this period in the process, a patent application has been filed to prevent other companies from marketing the same compound and protect the company's investment in the research and development costs.

1.1.3 Preclinical Development

The preclinical development process largely consists of a safety analysis (toxicity testing) and continued study into a drug candidate's metabolism and pharmacology. Both *in vitro* and *in vivo* tests are conducted; many species of animals will be used because a drug may behave differently in one species than in another. An early assessment of dosing schedules in animal species can be determined, although human dosage regimens are not determined until the subsequent clinical trials in the next development phase (see Section 1.1.4).

Toxicology tests in preclinical development examine acute toxicity at escalating doses and short term toxicity (defined as 2 weeks to 3 months), as well as the potential of the drug candidate to cause genetic toxicity. Today's research efforts attempt to utilize as few animals as possible and so more *in vitro* tests are conducted. The use of metabonomics for toxicity testing is making an impact on both drug discovery (to select a lead compound) and preclinical development (to examine safety biomarkers and mechanisms). Metabonomics is a technology that explores the potential of combining state of the art high resolution NMR (Nuclear Magnetic Resonance) spectroscopy with multivariate statistical techniques. Specifically, this technique involves the elucidation of changes in metabolic patterns associated with drug toxicity based on the measurement of component profiles in biofluids (*i.e.*, urine). NMR pattern recognition technology associates target organ toxicity with specific NMR spectral patterns and identifies novel surrogate markers of toxicity [3].

Also in preclinical development, the pharmacokinetic profile of a drug candidate is learned. Pharmacokinetics is a specific, detailed analysis which refers to the kinetics (*i.e.*, time course profile) of drug absorption, distribution and elimination. The metabolites from the drug are identified in this stage. Definitive metabolism studies of drug absorption, tissue distribution, metabolism and elimination are based on the administration of radiolabeled drug to animals. It is important that the radionuclide is introduced at a position in the chemical structure that is stable to points of metabolism and conditions of acid and base hydrolysis.

Pharmacology testing contains two major aspects—*in vivo* (animal models) and *in vitro* (receptor binding) explorations. Comparisons are made among other drugs in the particular collection under evaluation, as well as among established drugs and/or competitive drugs already on the market. More informative and/or predictive biomarkers are also identified and monitored from these studies.

Detailed information about the drug candidate is developed at the proper time in preclinical development, such as the intended route of administration and the proposed method of manufacturing. In order to supply enough of the drug to meet the demands of toxicology, metabolism and pharmacology, the medicinal chemistry and analytical groups work together to determine the source of raw materials, develop the necessary manufacturing process and establish the purity of the drug product. The exact synthesis scheme and methodology needed to produce the drug are recorded in detailed reports. The pharmaceutics research group develops and evaluates formulations for the drug candidate. These formulations are assessed *in vivo* by the drug metabolism group. Quality and stability are the goals for this dosage form development effort.

In the United States, after the active and inactive ingredients of a formulation containing the candidate compound have been identified and developed, a detailed summary called an Investigational New Drug Application (IND) is prepared. This document contains reports of all the data known to date on a drug candidate's toxicology, metabolism, pharmacology, synthesis, manufacturing and formulation. It also contains the proposed clinical protocol for the first safety study in man.

All of the information contained in an IND application is submitted to the United States Food and Drug Administration (FDA). Typically, thousands of pages of documents comprise this IND. The FDA reviews the information submitted and makes a decision whether or not the drug has efficacy and appears safe for study in human volunteers. The IND becomes effective if the FDA does not disapprove the application within 30 days. The drug sponsor is then approved to begin clinical studies in humans. When questions arise, the FDA responds to the IND application with a series of inquiries to be answered and a dialogue begins between the drug sponsor and the FDA.

1.1.4 Clinical Development

1.1.4.1 Introduction

Clinical trials are used to judge the safety and efficacy of new drug therapies in humans. Drug development is comprised of four clinical phases: Phase I, II, III and IV (Table 1.1). Each phase constitutes an important juncture, or decision point, in the drug's development cycle. A drug can be terminated at any phase for any valid reason. Should the drug continue its development, the return on investment is expected to be high so that the company developing the drug can realize a substantial and often sustained profit for a period of time while the drug is still covered under patent.

1.1.4.2 Phase I

Phase I safety studies constitute the "first time in man." The objective is to establish a safe dosage range that is tolerated by the human body. These studies involve a small number of healthy male volunteers (usually 20–80) and may last a few months; females are not used at this stage because of the unknown effects of any new drug on a developing fetus. Biological samples are taken

Table 1.1
Objectives of the four phases in clinical drug development and typical numbers of volunteers or patients involved

Phase I	Phase II	Phase III	Phase IV
Establish safe dosing range and assess pharmaco-kinetics; also called First time in man (FTIM)	Demonstrate efficacy, identify side effects and assess pharmaco-kinetics	Gain data on safety and effectiveness in a larger population of patients; assess pharmacokinetics	Expand on approved claims or demonstrate new claims; examine special drug-drug interactions; assess pharmacokinetics
20–80 male volunteers	200–800 patients	1,000–5,000 patients	A few thousand to several thousand patients

from these volunteers to assess the drug's pharmacokinetic characteristics. During a Phase I study, information about a drug's safety and pharmacokinetics is obtained so that well controlled studies in Phase II can be developed.

Note that Institutional Review Boards (IRB) are in place at hospitals and research institutions across the country to make sure that the rights and welfare of people participating in clinical trials are maintained. IRBs ensure that participants in clinical studies are fully informed and give their written permission before the studies begin. IRBs are monitored by the FDA.

1.1.4.3 Phase II

Phase II studies are designed to demonstrate efficacy, *i.e.*, evidence that the drug is effective in humans to treat the intended disease or condition. A Phase II controlled clinical study can take from several months to two years and uses from 200 to 800 volunteer patients. These studies are closely monitored for side effects as well as efficacy. Animal studies may continue in parallel to determine the drug's safety.

A meeting is held between the drug sponsor and the FDA at the end of Phase II studies. Results to date are reviewed and discussion about the plan for Phase III studies is held. Additional data that may be needed to support the drug's development are outlined at this time and all information requirements are clarified. A month prior to this meeting, the drug sponsor submits the protocols for the Phase III studies to the FDA for its review. Additional information is provided on data supporting the claim of the new drug, its proposed labeling, its chemistry and results from animal studies. Note that procedures exist that can expedite the development, evaluation and marketing of new drug therapies intended to treat patients with life threatening illnesses. Such procedures may be activated when no satisfactory alternative therapies exist. During Phase I or Phase II clinical studies, these procedures (also called "Subpart E" for Section 312 of the US Code of Federal Regulations) may be put into action [4].

The company developing the drug must then consider many factors before further development is undertaken, such as the cost of manufacturing the drug (which may or may not involve new equipment purchases or changes in existing facilities), the estimated time and cost to gain final FDA approval, the competition the drug may face in the market, its sales potential and projected sales growth. The return on the company's investment is estimated. It has been observed in recent years for a major pharmaceutical company that if the return on investment on a single drug is not 100 million dollars (US) or more, the company may choose not to develop the drug further; instead, licensing the drug to a smaller company is one of several options.

1.1.4.4 Phase III

After evidence establishing the effectiveness of the drug candidate has been obtained in Phase II clinical studies, and the "End of Phase II" meeting with the FDA has shown a favorable outcome, Phase III studies can begin. These studies are large scale controlled efficacy studies and the objective is to gain more data on the effectiveness and safety of the drug in a larger population of patients. A special population may be used, *e.g.*, those having an additional disease or organic deficiency such as renal or liver failure. The drug is often compared with another drug used to treat the same condition. Drug interaction studies are conducted as well as bioavailability studies in the presence and absence of food. A Phase III study is a clinical trial in which the patients are assigned randomly to the experimental group or the control group.

From 1,000 to 5,000 volunteer patients are typically used in a Phase III study; this aspect of drug development can last from 2 to 3 years. Data obtained are needed to develop the detailed physician labeling that will be provided with the new drug. These data also extrapolate the results to the general population and identify the side effect profile and the frequency of each side effect. In parallel, various toxicology, carcinogenicity and metabolic studies are conducted in animals. The cumulative results from all of these studies are used to establish statements of efficacy and safety of the new drug.

As Phase III progresses, many commercial considerations are put into action. These matters include pricing, registration, large scale manufacturing and plans for market launch. The plan for marketing the drug is developed and additional clinical trials may be started to satisfy new labeling indications or to expand current indications that define exactly which conditions the drug is intended to treat. Note that once a drug is approved, physicians are able to prescribe its use to treat other conditions for which they feel the drug might have a beneficial effect; this use is known as "off label drug use."

A Treatment IND is a special case in which the FDA may decide to make a promising new drug available to desperately ill patients as early as possible in the drug's development [5]. In order for a Treatment IND to be instituted, there must be significant evidence of drug efficacy, the drug must treat a serious or life threatening disease (where death may occur in months if no treatment is received), and/or there is no alternative treatment available for these intended patients. Treatment INDs, when they occur, are typically made available to patients during Phase III studies before marketing of the drug begins. Any patient who receives the drug under a Treatment IND cannot participate in the definitive Phase III studies.

Another means by which promising and unique experimental agents can be made available to patients is called "Parallel Track." This policy was developed in response to the AIDS illness and allows patients with AIDS who cannot participate in controlled clinical trials to receive the promising investigational drug [6].

1.1.4.5 New Drug Application (NDA)

The New Drug Application (NDA) is the formal summary of the results of all animal and human studies, in conjunction with detailed plans for marketing and manufacturing the drug. Also, information is provided about the drug's chemistry, analysis, specifications and proposed labeling. The NDA is filed with the FDA by the drug sponsor who wishes to sell the new pharmaceutical entity in the United States.

Before final approval may be granted, the FDA conducts a PreApproval Inspection (PAI) of the manufacturer's facilities because it is very important that methods used to manufacture the drug and maintain its quality are sufficient to preserve the drug's identity, strength and purity. This inspection evaluates the manufacturer's compliance with Good Manufacturing Practices (GMP), verifies the accuracy of information submitted in the NDA, and evaluates manufacturing controls for the preapproval batches of drug formulation that were specified in the NDA. A collection of samples may be taken for analysis by other laboratories to confirm drug purity, strength, *etc.*

Once the FDA receives the NDA, it undergoes a completeness review to ensure that sufficient information has been submitted to justify the filing. If deficiencies in the required information exist, then a "refuse to file" letter may be issued to the drug sponsor. This completeness review must be finished within 60 days of filing the NDA. When a drug application is considered complete by the FDA, there is no formal time requirement in which that NDA must be acted upon. The speed of review typically depends on how unique the drug is and on the workload of the agency at the time. Typically such a review can take 2–3 years, although "fast track" status for a novel drug can allow for a shorter time for complete review.

After the NDA has been thoroughly evaluated, communication takes place with the drug sponsor about medical and scientific issues that may arise. The FDA will tell the applicant when more data is needed, when conclusions made in reports are not justified by the data, and when changes need to be made in the application. At the end of this review period, one of the following actions may occur: (1) The NDA may be "not approvable" and deficiencies in the application are clearly noted; (2) the NDA may be "approvable" after minor deficiencies are corrected, after labeling changes are made, and/or after studies are conducted that will investigate particular clinical issues; or (3) the NDA may be approved with no corrective action or delay necessary. When a director within the FDA having the sufficient authority signs an approved letter, the drug product can be legally marketed on that day in the United States. Typically, however, the precise approval date is not expected in advance and sufficient time is needed by the drug company to prepare manufacturing for the product launch.

1.1.5 Manufacturing and Sale

Plans for a drug's manufacturing are under way in parallel with efforts to complete studies needed for the NDA. Large quantities of product need to be synthesized, the formulation must be made consistently, and product packaging must be finalized. Also, quality control tests must be put into place to ensure reliable and consistent manufacturing of finished product as well as confirm drug stability in the finished dosage form. Should impurities or degradants be discovered, immediate efforts are made to identify the source of the impurity or degradation and eliminate it from the finished product. Manufacturing supplies the wholesalers with packaged drug product so that the drug can be purchased and used by pharmacies in response to receiving written prescriptions from physicians.

The product launch announces the new drug to physicians and other medical professionals. This introduction provides education about the new drug's characteristics, indications and labeling. Various marketing and advertising programs are devised and executed.

Phase IV clinical trials are those studies conducted after a product launch to expand on approved claims, study the drug in a particular patient population, as well as extend the product line with new formulations. A clinical study after the drug is sold may be conducted to evaluate a new dosage regimen for a drug, *e.g.*, fexofenadine (Allegra®) is an antihistamine sold by Aventis (Bridgewater, NJ USA). The original clinical studies indicated that a dosage of 60 mg, given every 12 h, was adequate to control symptoms of allergies and rhinitis. Their product launch was made with this strength and dosage regimen. In response to competition from a once a day allergy drug, Aventis conducted Phase IV clinical trials (after the product was on the market) with different dosages and obtained the necessary data to show that a 180 mg version of Allegra could be taken once a day and relieve allergy symptoms with similar efficacy as 60 mg taken twice a day. Aventis then filed the clinical and regulatory documentation, and obtained approval to market a new dosage form of their drug. Another example of a post-marketing Phase IV clinical study is the investigation of whether or not sertraline (Zoloft®), an antidepressant drug, could be taken by patients with unstable ischemic heart disease. Results suggested that it is a safe and effective treatment for depression in patients with recent myocardial infarction or unstable angina [7].

In order to further ensure continued drug product quality, the FDA requires the submission of Annual Reports for each drug product. Annual Reports include information pertaining to adverse reaction data and records of production, quality control and distribution. For some drug products, the FDA requires affirmative post-marketing monitoring or additional studies to evaluate long term effects. A drug company also closely monitors all adverse drug experiences collected after the sale of a drug and reports them to the FDA.

The FDA has the authority to withdraw a drug from the market at any time in response to unusual or rare occurrences of life threatening side effects or toxicity noted in the post-marketing surveillance program. A conclusion that a drug should no longer be marketed is based on the nature and frequency of the adverse effects and how the drug compares with other treatments. Some drugs that were withdrawn from the market between 1997 and 2000 include the following: Rezulin® (troglitazone), Propulsid® (cisapride), Raxar® (grepafloxacin) and Trovan® (trovafloxacin), Duract® (bromfenac), Redux® (desfenfluramine), Posicor® (mibefradil), Seldane® (terfenadine), Hismanal® (astemizole), Pondimin® (fenfluramine) and Lotronex® (alosetron). These drugs were all removed for one of the following reasons: liver toxicity, cardiac arrhythmias, drug interactions or heart damage (cardiac valve disease); the exception was alosetron which caused ischemic colitis.

In rare cases, a drug may be returned to the market after withdrawal but only when very strict and limiting measures for its continued use are put into place (*e.g.*, Propulsid). Propulsid is still available under a special investigational use designation, which means that the drug is available to people with severely debilitating conditions for which the benefits of taking the drug clearly outweigh the risks. Certain eligibility criteria must be met by each patient and additional physician office visits and paperwork are required. These limitations are put into place to assure that Propulsid will only be given to those people whose particular medical condition warrants its use.

A natural thought when a drug is taken off the market is, "How did the drug make it through clinical trials successfully?" Most often, the withdrawal occurs because of adverse effects that were not seen before marketing the drug. A rare side effect that may occur in 0.01% of the population may not be scientifically validated until the statistical population of patients taking the drug is large enough. Other times, hints of the problem may be noted through a retrospective review of data from clinical studies, but not the serious events that eventually lead to the withdrawal. Sometimes, there simply may not be any indication at all. Also, a serious side effect may only be noted when an approved drug is used in a different manner than the clinical studies were designed to investigate. Many complex factors go into the drug approval process; ultimately, the decision for a new drug approval is a balance of risks versus benefits.

1.2 Industry Trends

1.2.1 Introduction

Advances in many different disciplines have occurred to change the way drug discovery is performed today compared with even five years ago. These advances include sequencing of the human genome; identification of more drug targets through proteomics; advances in the fields of combinatorial chemistry, high throughput screening, and mass spectrometry; and improvements in laboratory automation and throughput in bioanalysis. The end result of these process improvements is that compounds can now be synthesized faster than ever before. These greater numbers of compounds are quickly evaluated for pharmacological and metabolic activity using high throughput automated techniques, with the ultimate goal of bringing a drug product to market in a shorter timeframe.

Some background material is provided next for the reader to gain a better understanding of four key industry trends: (a) combinatorial chemistry; (b) advances in automation for combinatorial chemistry, high throughput screening and bioanalysis; (c) LC-MS/MS analytical detection techniques; and (d) newer bioanalytical dosing regimens (n-in-1 dosing) made possible by the advances in detection.

1.2.2 Combinatorial Chemistry

A key component of satisfying the high throughput capability and demands of drug discovery has been the implementation of combinatorial chemistry techniques to synthesize, purify and confirm the identity of a large number of compounds displaying wide chemical diversity within a class. In place of traditional serial compound synthesis, libraries of compounds are created in 96-well plates by interconnecting a set or sets of small reactive molecules, called building blocks, in many different permutations [8–10]. Today, as many as 2,000 compounds can be synthesized in a week. Although 96-well plates serve as the most common format for reaction vessels, 24- and 48-well plates are also used by medicinal chemists.

These combinatorial chemistry and parallel synthesis strategies are used to produce a large number of compounds which are then subjected to high throughput screening to identify biological activity. Automation aids the chemist in the high throughput synthesis of these compound libraries [11], as well in the subsequent purification steps required to isolate synthesized compound from reaction starting materials, reagents and byproducts [12]. The popular strategic options for the synthesis of combinatorial libraries include solid-phase, solution-phase and liquid-phase synthesis.

Solid-phase parallel synthesis uses resins to which the starting material is attached in order to produce a large number of compounds via split and mix methods. The solid support matrix used consists of a base polymer, a linker to join the base polymer to the reactive center, and a functionalized reactive site. The immobilized reactant is then subjected to a series of chemical reactions to prepare the desired end product. The use of excess reagents drives reactions to completion. However, the need for deconvolution approaches to determine the active components within a pool has limited the utility of solid-phase synthesis. Since the synthesized compounds are attached to the solid support, this approach does offer simplified reagent removal via filtration and impurities are washed away easily during purification. The compound of interest is released from the polymer support in a final chemical release step.

Solution-phase parallel synthesis techniques are more flexible than solid-phase techniques and are often used to create focused chemical libraries. Using this approach, the reactions occur in solution and so are easily monitored by thin layer chromatography or NMR. The synthesized compound is isolated in one liquid phase; all non product species are fractionated into an immiscible liquid phase [13]. A purification step following the reaction is required and common approaches are liquid-liquid extraction, liquid chromatography, solid-phase extraction and the use of solid-phase scavengers to remove excess reagents and/or reaction impurities from crude solutions. These solid-phase scavengers (functionally modified polymers of polystyrene or bonded silica) are chosen for their inertness to the reaction products but affinity for reagents and unwanted byproducts. Scavengers are becoming more popular since they can easily be adapted to automated purification techniques via filtration [14].

The procedure for use of scavengers follows. Scavenger beads are placed into the wells of a flow-through 96-well filtration plate. A reaction block (consisting of individual wells of a flow-through 96-well plate in which the top and/or bottom of the wells can be blocked or opened to allow flow and reagent addition) is placed on top of the filtration plate (loaded with beads), so that when vacuum is applied the reaction mixture flows out of the reaction block and through the scavenger bed. A collection plate centered below the filtration plate isolates the solution.

Liquid-phase parallel synthesis combines the strategic features of solid-phase synthesis and solution-phase synthesis. This method uses a supporting polymer (*e.g.*, polyethylene glycol) that is soluble in the reaction media. Selective precipitation of this polymer can be performed for the purposes of isolation and purification. Excess reagents and byproducts are removed by simple filtration [15].

1.2.3 Automation

Automation is playing an important role in allowing researchers to meet the high throughput demands in today's research environment. A combination of robotics, liquid handling workstations and/or improved formats such as microplate sample preparation have been introduced to allow high speed analyses in combinatorial chemistry, high throughput screening and bioanalysis. An example of a typical liquid handling workstation is shown in Figure 1.2.

In combinatorial chemistry, automated workstations are available that are specifically configured for either the organic synthesis step or for the subsequent purification step. Benchtop synthesizers can perform up to 20 reactions in flasks with hands-on control. All synthesis functions (mix, heat, cool, wash, empty, cleave) have been incorporated into a single module that fits on the benchtop. A multifunctional workstation assists with the following functions: reagent preparation, reaction mapping, off-line reaction incubation, liquid-liquid extraction, compound dissolution, and compound aliquoting for

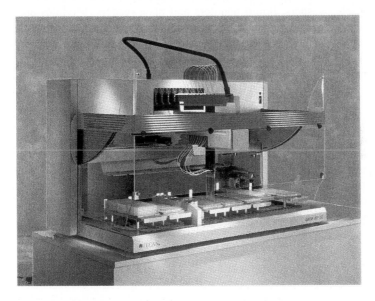

Figure 1.2. Typical example of a liquid handling workstation used to automate various sample preparation processes in drug development, the Genesis RSP. Photo reprinted with permission from Tecan.

analysis and screening. The addition of an analytical balance and vortex mixer on the workstation meet the requirements of automating synthetic chemistry conditions. A multitasking robotic workstation for synthesis features two independently controlled robotic arms that dispense reagents and solvents simultaneously, and can do so in inert environments. Equipment such as this can be upgraded to perform additional tasks and can interface with some additional components of a core system for higher throughput and higher performance.

High throughput screening utilizes robotic-feeding liquid handling workstations with higher density microplates (384-well and 1536-well formats) and plate stackers for improved productivity. Automated hit picking is a hardware and software application that automates the transfer of lead compounds from their source plates into destination plates for consolidation. The use of 96- or 384-channel disposable tip pipetting heads allows improved liquid dispensing capabilities and speeds. The demand for even greater throughput in screening procedures often requires a larger industrial process rather than a laboratory workstation approach. Independent workstation modules can be combined in an assembly line format, consisting of storage and incubation carousels, washers, liquid handlers and plate readers. Modules are simply added to put in more steps or increase capabilities. This type of system is capable of running 1,000 96-well plates (96,000 assays) per day and is compatible with 384-well plates. An ultrahigh throughput example of a fully integrated automation solution can screen 100,000 compounds in one working day [16].

Automation for bioanalysis is described in detail in Chapter 5. Briefly mentioned here, liquid handling workstations with plate grippers have greatly improved the throughput of sample preparation procedures using 96-well plates. Automation allows more samples to be processed per unit time and frees the analyst from most hands-on tasks. Using the microplate format for sample preparation allows the automation of common procedures including protein precipitation, liquid-liquid extraction, solid-phase extraction and filtration.

1.2.4 Analytical Instrumentation—LC-MS

1.2.4.1 Introduction

The preferred analytical technique in the bioanalytical research environment is liquid chromatography-mass spectrometry (LC-MS), used for qualitative and quantitative drug identification. LC-MS is preferred for its speed, sensitivity and specificity. LC is a powerful and universally accepted technique that offers chromatographic separation of individual analytes within liquid mixtures. These analytes are subjected to an ionization source and then are introduced into the mass spectrometer. The mass spectrometer separates or filters these ions based on their mass-to-charge ratio (*m/z*) and then sends them on to the detector [17]. A general scheme of this process is shown in Figure 1.3. The data generated are used to provide information about the molecular weight, structure, identity and quantity of specific components within the sample.

1.2.4.2 LC-MS Interface

The LC-MS interface is the most important element of this system. It is the point at which the liquid from the LC (operated at atmospheric pressure) meets the mass spectrometer (operated in a vacuum). Advances have occurred over the years to mate the two techniques [18, 19]. The most common ionization interface used for bioanalysis is atmospheric pressure ionization (API) which is a soft ionization process (*i.e.*, provides little fragmentation of a molecular ion). API is performed as either API-electrospray or atmospheric pressure chemical ionization (APCI).

1.2.4.2.1 Electrospray Ionization

Electrospray ionization (ESI) generates ions directly from the solution phase into the gas phase. The ions are produced by applying a strong electric field to a very fine spray of the analyte in solution. The electric field charges the surface of the liquid and forms a spray of charged droplets. The charged droplets are attracted toward a capillary sampling orifice where heated nitrogen drying gas shrinks the droplets and carries away the uncharged material. As the droplets shrink, ionized analytes escape the liquid phase through electrostatic (coulombic) forces and enter the gas phase, where they proceed into the low pressure region of the ion source and into the mass analyzer [20–22].

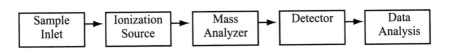

Figure 1.3. Schematic diagram of the basic components of a mass spectrometer system. The mobile phase following separation of components on a liquid chromatograph flows into the mass spectrometer for LC-MS analysis.

Analysis by electrospray requires the prior formation of ionized analytes in solution; for nonionic compounds, ions are prepared by adding acid or base modifiers to the LC mobile phase solution to promote the electrospray process [23]. Ionization thus occurs in the liquid phase with ESI.

Pure electrospray is suitable for only capillary LC and capillary electrophoresis, or conventional LC where the post-column effluent is split using a zero dead volume T-piece, reducing the flow of liquid entering the mass spectrometer. In an attempt to extend the range of solvent flow rates amenable to electrospray, modifications have been made using pneumatic and thermal assistance [23].

1.2.4.2.2 Atmospheric Pressure Chemical Ionization

APCI is similar to API-ES, but APCI nebulization occurs in a hot vaporization chamber, where a heated stream of nitrogen gas rapidly evaporates nearly all of the solvent. The vapor is ionized by a corona discharge needle [24]. The discharge produces reagent ions from the LC solvent which then ionize the sample [23]. Ionization thus occurs in the gas phase with APCI.

1.2.4.3 Mass Analyzers

The ionization source and the mass analyzer are linked since the mass analyzer requires a charged particle in order for separation to occur. The mass analyzer contains some electric or magnetic field, or combination of the two, which can manipulate the trajectory of the ion in a vacuum chamber [25]. The atmospheric pressure ionization interfaces described above are commercially available with various mass analyzers. The most popular and available mass analyzer is the quadrupole which will be described here. For information on product ion scanning, precursor ion scanning and neutral loss/gain, the reader is referred to the book chapter by Fountain [25].

The quadrupole mass analyzer is available as a single quadrupole or is configured in tandem (called a triple quadrupole) to greatly enhance its capabilities. A quadrupole mass filter typically consists of four cylindrical electrodes (rods) to which precise DC and RF voltages can be applied. Note that the tandem quadrupole mass filters are referred to as Q1 and Q3; the additional Q2 quadrupole has no filtering effect, except to use its RF voltage to guide ions through the vacuum chamber.

A mass spectrometer with a single quadrupole is capable of either full scan acquisition or selected ion monitoring (SIM) detection. In the full scan mode, the instrument detects signals over a defined mass range during a short period of time. All the signals are detected until the full mass range is covered. A mass spectrum is generated in which ion intensity (abundance) is plotted versus *m/z* ratio. Scan mode is used for qualitative analysis when the analyte mass is not known. In the SIM mode, a single stage quadrupole is used as a mass filter and monitors only a specific *m/z* ratio, allowing only that *m/z* ratio to pass through to the detector. Chemical noise is reduced and a response curve is generated for a specific ion rather than a mass spectrum. As a result, sensitivity is improved and this technique is useful for quantitative analysis of a specific analyte.

When an additional quadrupole mass analyzer is configured in tandem and coupled with a collision cell for gas reactions, several additional analytical experiments can be conducted: multiple reaction monitoring, product ion scanning, precursor ion scanning and neutral loss/gain. Tandem mass spectrometry (MS/MS) achieves unequivocal identification of a drug substance. In LC-MS/MS, the first analyzer selects a parent ion from the first quadrupole and filters out all unwanted ions. The selected ion undergoes collision in another quadrupole via bombardment with gas molecules such as nitrogen or argon. Fragmentation occurs and the other mass analyzer in the third quadrupole selects a particular product (daughter) ion. This scenario is called selected reaction monitoring (SRM) or multiple reaction monitoring (MRM). Information is provided on the chemical structure using the parent-daughter ion pair.

The key attractive features of an LC-MS/MS system are its speed, selectivity for a single MW in the presence of many other constituents and sensitivity (typically pg/mL concentrations are able to be accurately quantitated). Additional advantages are good precision and accuracy, and wide dynamic range. Note that the selectivity of LC-MS/MS reduces the need for complete chromatographic resolution of individual components, allowing a shorter analytical run time and higher throughput [26]. Typically the combined increases in selectivity and sensitivity of LC-MS/MS methods provide a >10 fold improvement in the limit of quantitation (LOQ) compared with traditional methods using ultraviolet or fluorescence detection [23]. A typical LC-MS/MS instrument as used for the analytical quantitation of drugs from biological matrices is shown in Figure 1.4.

In addition to the single quadrupole (LC-MS) and triple quadrupole (LC-MS/MS) mass analyzers already discussed, other mass analyzers include:

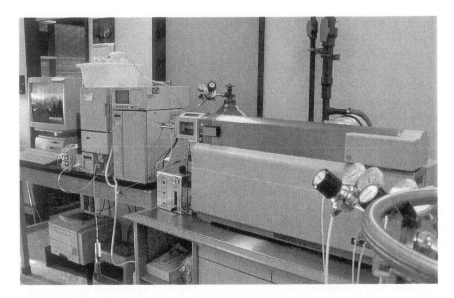

Figure 1.4. Typical example of an LC-MS/MS system used for the separation, detection and quantitation of drugs in biological matrices.

magnetic sector, time-of-flight (TOF) [27, 28], quadrupole orthogonal acceleration TOF (Q-TOF) [29] and ion trap (IT) mass analyzers [26, 30, 31]. Applications for quantitative bioanalysis are best served by LC-MS (single quadrupole) and LC-MS/MS (triple quadrupole) for added sensitivity. However, when gaining information on the metabolic route of a compound is more important than absolute sensitivity and selectivity for the parent drug, the use of ion trap and time-of-flight instruments present advantages. IT and TOF offer greater sensitivity in full scan mode than triple quadrupole MS/MS detection [32] and multiple analytes can be detected and quantified, as reported by Cai *et al.* [33] and Zhang *et al.* [34]. Another benefit of TOF-MS is its capability of accurate mass analysis which allows metabolites to be identified with greater confidence [32].

1.2.4.4 Further Reading

Additional discussion of these mass spectrometry techniques is outside the scope of this text and the reader is referred to additional resources. Mass spectrometry is introduced as a tutorial in a book chapter by Fountain [25] and the fundamentals of electrospray are presented by Gaskell [35]. A review by

Lee and Kerns [26] and a book by Lee [36] detail how LC-MS techniques are fundamentally established as a valuable tool and used in all phases of drug development. Kyranos *et al.* describe applications for LC-MS in drug discovery [37]. Hoke *et al.* describe how pharmaceutical research and development have been transformed by innovations in mass spectrometry based technologies [38]. The impact of mass spectrometry on combinatorial chemistry is described by Triolo *et al.* [39] and Süßmuth and Jung [40]. Niessen provides a review of the principles and applications of LC-MS [41]. Various reviews describe applications for LC-MS in proteomics [42, 43], analytical toxicology [44], food analysis [45], forensic and clinical toxicology [46] and forensic sciences [47]. A useful text for students of chemistry and biochemistry who wish to understand the principles of mass spectrometry while using it as a tool is the book by Johnstone and Rose [48].

LC-MS/MS techniques for drug separation, detection and quantitation have become the standard and their use continues to expand in bioanalysis. Several reviews of the capabilities of high throughput LC-MS/MS for bioanalysis, including sample preparation schemes, have been published by Brewer and Henion [49], Plumb *et al.* [23], Jemal [50], Brockman, Hiller and Cole [51], and Ackermann *et al.* [52]. Korfmacher *et al.* illustrate the important role of LC-MS/MS API techniques in drug discovery for rapid, quantitative method development, metabolite identification and multiple drug analysis [53]. Yang *et al.* describe the latest LC-MS/MS technologies for drug discovery support [54] and Rudewicz and Yang discuss the use of LC-MS/MS in a regulated environment [55]. Law and Temesi share some useful considerations in making the switch from LC-UV to ESI LC-MS techniques in support of drug discovery [56]. Rossi *et al.* describe the use of tandem-in-time MS as a quantitative bioanalytical tool [57].

Improvements continue to be made in three areas: (1) interfaces from the LC to the mass spectrometer; (2) multiple sample inlets (*e.g.*, four instead of one [55, 58, 59]); and (3) staggered parallel sample introduction schemes where one mass spectrometer inlet is shared with two or more LC columns [60–62]. These advances are described in more detail in Chapter 14, Section 14.4.4. Analysis times using these novel approaches are very fast, allowing rapid turnaround of samples to meet the needs of drug development research. Note that LC and mass spectrometry can be interfaced to NMR to create an LC-NMR-MS system. Such a configuration has been shown useful for the structural elucidation of metabolites in urine [63].

1.2.5 N-in-1 Dosing

The traditional approach of dosing one animal with one drug and collecting blood at a series of time points after administration yields valuable information on a candidate's pharmacokinetics in a living system. However, this procedure is time consuming and labor intensive, as each individual sample must undergo analysis in a serial manner. This approach cannot be used to rapidly evaluate the many hundreds of candidate compounds generated from combinatorial chemistry, for it would take too much time.

In an effort to improve throughput and reduce cost, n-in-1 dosing was devised in which multiple compounds are dosed in one animal and the selectivity of the mass spectrometer is used to individually quantitate their concentrations from the mixture. This approach is also called simultaneous multiple compound dosing or cassette dosing. The data generated by this approach yield meaningful pharmacokinetic data in a much shorter time frame, and fewer animals and fewer numbers of samples are used than in traditional methods. Examples of the n-in-1 approach for rapid pharmacokinetic screening of drug candidates using LC-MS/MS are published by Olah *et al.* [64], McLoughlin *et al.* [65], Liang *et al.* [66], Berman *et al.* [67], Cai *et al.* [68] and Bayliss and Frick [69]. A concise review of cassette dosing can be found in the book chapter by Vora, Rossi and Kindt [70].

The drug screening process is now more manageable using a fewer number of samples generated from n-in-1 dosing techniques. Each compound used for dosing is quickly evaluated and compared on a relative basis with the other dosed compounds and the most desirable candidates are selected. A typical profile of drug concentration versus time, observed in one dog after intravenous dosing with 10 drugs, is shown in Figure 1.5.

Note that the use of pooled plasma from multiple animals dosed with single unique compounds has also been demonstrated to yield a throughput advantage in the analysis of bioanalytical samples [71]. Using this approach, 10 animals may be dosed with one compound each and then all of the 1 h time point samples are pooled and subjected to sample preparation, and so on for each time point. Sample pooling in this manner eliminates the concern from cassette dosing of potential drug-drug interactions on pharmacokinetics and metabolic conversion of one compound to another compound already included in the series. The practice of sample pooling and its associated reduction in workload is described by Kuo *et al.* for six proprietary compounds from a class of antipsychotic agents [72]. Note that some variations of this approach exist,

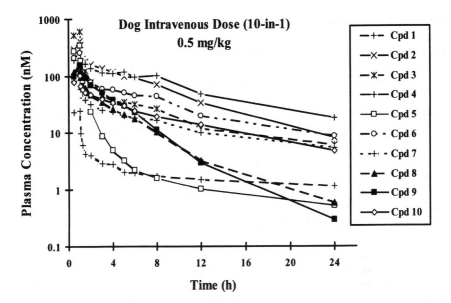

Figure 1.5. Pharmacokinetic profiles of 10 compounds that were dosed intravenously in one dog. Reprinted with permission from [73]. Copyright 2001 Elsevier Science B.V

such as the cassette accelerated rapid rat screen in which drug candidates are dosed individually (n=2 rats per compound) in batches of six compounds per set and then samples are pooled across time points to provide a smaller number of test samples for analysis [74].

Another method used to reduce the number of animals in the support of drug discovery is to serially bleed mice (removal of 10–20 μL blood) and analyze the small sample volumes using a small capillary LC column prior to MS analysis [75, 76]. Conventional techniques used to obtain a nine point pharmacokinetic curve with 4 animals per time point would require the use of 36 mice; using the serial bleeding technique, only 4 mice are used [76]. A related method to serially bleed an animal is microdialysis, in which a semi-permeable membrane is surgically implanted in a tissue of a living organism and the perfusate solution is sampled over time. Thus, little biological fluid is removed and continuous *in vivo* sampling is possible. Microdialysis works especially well for drug transport studies. An overview [77] and some applications of microdialysis [78–80] provide interesting reading.

1.3 Specific Roles for Bioanalysis in Drug Development

1.3.1 Drug Discovery—Lead Optimization

1.3.1.1 Screening in vivo for Pharmacokinetic Properties

Combinatorial chemistry and high throughput screening techniques synthesize and identify a large number of compounds that may be potential leads for continued development. Chemists synthesize and screen chemical analogs of the hits to further improve and refine a drug's activity. However, the chemist alone cannot identify the best analog for continued study because it is often the case that good leads *in vitro* are not good leads *in vivo* due to problems with absorption, metabolism or toxicity. Screening these leads is an important task in a process known as lead optimization.

Bioanalysis is the quantitative determination of drugs and their metabolites in biological fluids. Bioanalytical scientists play an integral role in the lead optimization process by performing studies to gain information on a molecule's absorption, distribution (including an estimate of protein binding), metabolism and elimination. These important ADME tests determine the likelihood of a drug candidate continuing into the preclinical phase of drug development. A summary of the experiments commonly performed to assess ADME characteristics is listed in Table 1.2.

Each sample of biological fluid from an *in vitro* or *in vivo* study is subjected to bioanalysis to determine the concentration of drug at specific time points. Bioanalysis includes the acts of sample preparation and analysis by LC-MS/MS methods. This bioanalytical technique is utilized throughout the development lifetime of all new drugs. Analysis of the drug concentration versus time data yields important pharmacokinetic information that is used in the decision making process of whether or not a new molecule should be a candidate for further development.

1.3.1.2 Screening in vitro for Pharmacokinetic Properties

The use of animals to assess ADME characteristics is a costly and time consuming process. While animals are used, especially for toxicokinetic studies to assess drug toxicity, current trends are to use *in vitro* screens which have matured in recent years and been shown to be fairly predictive. These *in vitro* methodologies use enzymes, tissues and cell cultures to allow researchers to screen for drug characteristics such as cell absorption, metabolic stability,

Table 1.2
A list of experiments that are commonly performed to assess the absorption, distribution, metabolism and elimination (ADME) characteristics of potential lead compounds in drug discovery

Parameter Examined	Typical Experiments
Absorption	Caco-2 cells, MDCK cells, PgP transport
	in vivo pharmacokinetic profiling
Distribution	*in vitro* protein binding
	in vivo tissue distribution studies
Metabolism	Metabolic stability
	−Microsomes, sub cellular fractions, hepatocytes
	P450 inhibition studies
	−Microsomes
	P450 induction studies
	−Gene chips, multiple dosing
Elimination	Quantitation of drugs and metabolites in biological fluids

drug-drug interactions, clearance, bioavailability and toxicity [81–83]. Instrumentation to accommodate cell maintenance has matured in recent years to the point where high throughput testing using these *in vitro* screens is now a viable approach to investigate the absorption and metabolism of drugs.

Note that in addition to *in vivo* and *in vitro* testing, another prediction technique is called *in silico*. This approach refers to computer modeling based on sophisticated software using information on chemical structure, receptors, enzymes and various other databases of information [84]. Another important note to mention is that *in vitro* plasma protein binding measurements (*e.g.*, equilibrium dialysis and ultrafiltration) are utilized in drug discovery in a high throughput manner but are discussed instead in Chapter 6, Section 6.6. These applications for investigating *in vitro* absorption and metabolism studies will now be described in detail: absorption, metabolic stability screening, metabolic inhibition and induction of cytochrome P450, and toxicity testing.

1.3.1.2.1 Absorption

It is important to determine whether a drug displays pharmacological activity when it is administered orally, a desirable route of administration for the general population. Therefore an estimate of absorption is desired. The approach used to thoroughly evaluate oral absorption is to assess those individual factors that contribute to drug passage through the gastrointestinal membrane, such as solubility in the lumen of the intestine, permeability across cells, and chemical stability in the stomach and small intestine. Additional criteria are evaluated, such as lipophilicity of the compound and its hydrogen bonding potential, and then an overall predictive estimate of absorption characteristics is made. While this approach is complete, it takes time to generate all the pieces of data needed.

The *in vitro* permeability of drugs through Caco-2 cells has been used as a single predictor of oral absorption in humans. Caco-2 cell monolayers are derived from human colon adenocarcinoma cells. These cells are grown on semipermeable membranes and spontaneously differentiate to form confluent monolayers that mimic intestinal absorption cells. The apical (donor) surface of the monolayer contains microvilli, as in the intestine. Permeability measurements are based on the rate of appearance of test compound in a receiving (basolateral) compartment. Bioanalysis is used to determine the concentrations of analyte in the basolateral compartment [85, 86]. The cells also express functional transport proteins and metabolic enzymes.

While the Caco-2 model is fairly predictive, it does have the limitation of requiring a 21 day culture time with frequent attention required for replenishing its nutrients. A common concern about the Caco-2 model is that it may not be truly representative of all absorption pathways in the small intestine. Another model of absorption, which reduces the tissue culture time to 3 days, is the MDCK (Madin-Darby Canine Kidney) cell line [87, 88]. This accelerated permeability model is a feasible alternative to the traditional model that provides rank ordering of compounds with improved turnaround time.

1.3.1.2.2 Metabolic Stability Screening

The metabolic stability of a new chemical entity greatly influences its pharmacokinetic profile. A drug may have high bioavailability (high absorption and low first pass metabolism) following oral dosing but extensive and rapid metabolism can reduce the time it is in the blood as an intact molecule. For example, ester groups on drugs can be cleaved by esterases in the blood and the drug is metabolized or biotransformed to a new chemical entity. Note that metabolites can also be active as well, or even more active than the parent drug. The cytochrome (CYP) P450 system in the liver is an important enzymatic pathway for the oxidative metabolism of drugs and is often the primary route for degradation, regardless of how the drug is administered. It is valuable to learn the specific P450 isoenzymes responsible for a drug's metabolism, as the information can be used to predict the fate of the drug, potential drug-drug interactions, reactive metabolites and cytotoxic mechanisms.

Many *in vitro* methodologies for assessing metabolism are used in drug discovery support. Hepatic microsomes are among the most popular systems in use. These preparations retain the activity of those enzymes that reside in the smooth endoplasmic reticulum of cells, such as the cytochrome P450 system, flavin monooxygenases and glucuronyltransferases. Cultured hepatocytes retain a broader range of enzymatic activities, including not only the reticular systems of CYP450 but also cytosolic and mitochondrial enzymes [89]. These hepatocytes are particularly useful for induction and inhibition studies where the enzymatic activities in the liver are predicted. Additionally, liver slices are sometimes used in similar metabolic screens because they retain a wide range of enzymatic activities, like hepatocytes, but more closely resemble the organ level of the liver.

Metabolic stability testing of compounds is performed in liver microsomes with a collection of subcellular materials called S9 mix in the presence and absence of enzymatic cofactors such as $NADP^+$ (Nicotinamide Adenine Dinucleotide Phosphate). Hepatocytes and liver slices are also used for this stability testing. Incubations are assembled in a microplate format. At selected time points, the metabolic reaction is stopped by the addition of cold acetonitrile and then centrifugation is performed to pellet the proteins at the bottom of the wells. The supernatants are collected following centrifugation. Substrates are analyzed by LC-MS/MS interfaced with a 96-well autosampler for high throughput operation [61, 90–96].

The use of metabolic stability screening to predict clearance is also of interest. A drug with a high clearance has a high hepatic extraction ratio and so its maximum oral bioavailability is low. Conversely, a drug with a low clearance has a low hepatic extraction ratio and its maximum oral bioavailability is high (assuming complete absorption). Since clearance provides an understanding of the ability of the body to eliminate a drug, it is often used as a pharmacokinetic screen in lead selection. The ability of hepatic microsomal stability assessments to predict *in vivo* clearance in the rat was retrospectively evaluated for 1163 compounds from 48 research programs at a pharmaceutical research company [97].

1.3.1.2.3 Metabolic Inhibition of Cytochrome P450

The assessment of a candidate drug's potential to inhibit or induce cytochrome P450 isoenzymes using *in vitro* microsomal incubations is one method for predicting possible *in vivo* drug-drug interactions. Seven isoenzymes of cytochrome P450 play dominant roles in drug metabolism: CYP1A2, 2A6, 3A4, 2C9, 2C19, 2D6 and 2E1. One of the strategies adopted to evaluate the inhibition or induction of CYP450 by a drug is to monitor its effect on the metabolism of selected compounds known as probe substrates whose specific metabolic pathway is documented to occur via a single CYP450 enzyme [98–100]. After incubation and quenching of the reaction, sample preparation is required to remove proteins and matrix interferences before analysis. Two reports from the same laboratory detail methods for high throughput CYP inhibition screening using a cassette dosing strategy [101, 102]. Characteristics and common properties of inhibitors, inducers and activators of CYP enzymes are reviewed by Hollenberg [103].

The ADME profile of a drug product can be improved when the effect of CYP P450 inhibitors and/or inducers is fully known. For example, terfenadine (Seldane®) was removed from the market (see Section 1.1.5) due to potentially life threatening drug interactions that could result in abnormalities in the electrical impulse that stimulates the heart to contract and pump blood. An active carboxy metabolite of terfenadine, named fexofenadine, was introduced as Allegra®. Fexofenadine undergoes less metabolism by CYP3A4 isoenzymes and therefore the effect of inhibitors or inducers is greatly reduced on CYP3A4 when compared with the effect with terfenadine [104]. Another example of an improved ADME profile is esomeprazole (Nexium®). Esomeprazole is the (S)-isomer of omeprazole (Prilosec®) and undergoes less metabolism by CYP2C19 isoenzymes. A beneficial result is that CYP2C19 polymorphism has less of an effect on its pharmacokinetics [104].

1.3.1.3 Toxicity Testing

Toxicity testing is an important component of screening potential lead compounds [105], as compounds often fail in the development process because of the harmful effect they may have on cells, organs or organ systems. Commonly, animals will be administered escalating drug doses and the time course profile of the drug will be determined using bioanalysis; this technique is referred to as toxicokinetics. Genotoxicity analysis is another important toxicity test that is usually performed just prior to phase I clinical trials because of the high cost involved; performing this test on all lead compounds would be prohibitively expensive. An effort has recently begun to move toxicity testing to an earlier stage in drug development to confirm viable leads more quickly. Toxicogenomics, the examination of changes in gene expression following exposure to a toxicant, offers this potential for early detection; it may also detect human specific toxicants that cause no adverse reaction in rats [106]. Cytotoxicity tests using bioluminescence are proceeding along the lines of high throughput screening by being miniaturized to 384-well formats [107].

1.3.1.4 Further Reading

A wealth of information is available in published literature concerning the role of bioanalysis in the support of drug discovery and the lead optimization process, including descriptions of the various *in vivo* and *in vitro* screening tests. A schematic diagram of this iterative course of action is shown in Figure 1.6. Both general reviews and detailed reports are recommended.

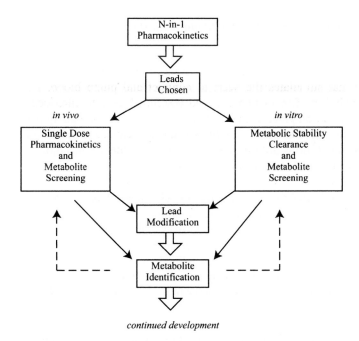

Figure 1.6. Schematic diagram of the lead optimization process and the role of metabolite screening. Reprinted with permission from [108]. Copyright 2002 John Wiley & Sons, Ltd.

Venkatesh and Lipper review the role of the development scientist in the selection and optimization of lead compounds [109]. Kennedy describes the compound selection and decision making skills necessary to manage the interface between drug discovery and drug development [110]. Perspectives on twelve months of lead optimization were provided from a chemistry team which identified the need for a balance between achieving the needed throughput and allowing time for appropriate decision making and reflection [111]. The prediction and use of pharmacokinetic properties in early drug discovery is discussed in several papers [81, 82, 112–115] (the report by Roberts [112] is highly recommended reading). The concept of a bioanalytical toolbox for performing fast turnaround in drug discovery support is a practical proposal to describe the choices available to the bioanalyst [116]. The importance of stereospecific bioanalytical monitoring in drug development is reviewed by Caldwell [117], with particular emphasis on stereospecific assays for the individual optical isomers of drugs. Some examples of enantiomeric separation and quantification in bioanalysis include the analysis of ketoprofen [118] and fluoxetine [119] in human plasma.

1.3.2 Preclinical—ADME and Metabolite Identification

1.3.2.1 ADME Studies

Once a drug candidate has received IND approval, the preclinical development phase formally begins. With a focus on accelerated timelines, the distinction between drug discovery and preclinical development is less defined than in past years. Some activities that were traditionally performed only in the preclinical phase are now begun in drug discovery with the goal of identifying optimal lead compounds as early in development as possible.

Once a drug compound reaches the preclinical development phase, its ADME characteristics are determined in many species of animals as well as in man. At this point, a defined and validated assay is used over and over for determining concentrations of drugs from *in vitro* and *in vivo* samples. It would be ideal if one analytical method could be used for every animal species, but in practice a method needs to be modified slightly and validated for each species. One method for many species of animals is described in a case study of two antibiotics [120].

While multiple compounds are encountered in drug discovery, in preclinical development fewer compounds are in use. Another differentiation between the two phases of development is that now in the preclinical stage validated methods are needed and adherence to Good Laboratory Practices (GLP) is required.

Preclinical development projects in bioanalysis commonly involve the determination of drug concentrations in biological fluids after dosing by multiple routes of administration and in multiple species of animals. Toxicokinetic studies also continue to be performed which examine the relationships among dose, pharmacokinetics and toxicity. Tissue distribution studies are also conducted, frequently with radiolabeled drug. Some common approaches to assessing tissue distribution are to excise tissues and analyze extracts by LC-MS/MS, following sample preparation; oxidize excised tissues and determine the amount of radioactivity contained in each; and perform autoradiography. This latter technique utilizes thin, whole body slices and the amount of radioactivity in each organ slice is determined.

1.3.2.2 Metabolite Identification and Characterization

Another important aspect of drug development research is the determination of the metabolic fate of lead compounds in several animal species. Many new chemical entities are terminated late in the development stage due to problems with drug metabolism. In an effort to prevent these late stage failures, it is useful to know the metabolic fate of a promising lead compound as early as possible. Typically, metabolites of lead compounds are identified in the preclinical development phase, although it is becoming more common to identify major metabolites in the discovery phase. Then, should toxic metabolites be identified (*e.g.*, some acyl glucuronides [121]), structural analogs of early drug lead candidates may be designed to block portions of the molecules that are particularly susceptible to metabolism.

Mass spectrometry coupled with liquid chromatography is an effective analytical method for metabolite profiling. The integration of data collected from the ion trap, triple quadrupole and quadrupole/time-of-flight instruments allows a comprehensive evaluation of biotransformation products. This approach is routinely used to evaluate metabolites generated from *in vitro* and *in vivo* systems. Once metabolites have been identified and characterized from definitive ADME studies, they are monitored in subsequent pharmacokinetic studies; it is common in bioanalysis to monitor concentrations of both parent drug and one or more metabolites. These parent-metabolite combinations present a useful assessment of metabolism that can be important to establish dosing regimens and assess toxicity concerns. Also note that a metabolite can present an ideal backup candidate to a drug lead, *e.g.*, desloratidine (Clarinex®)

is a metabolite of loratidine (Claritin®) that was developed and FDA approved for marketing several years after the introduction of Claritin.

Many LC-MS applications are found in the literature for metabolite characterization from *in vitro* [122–124] and *in vivo* [34, 108] experiments. An overview of the choices and decisions involved in metabolite identification and how they can be merged into a systematic approach is described by Clarke *et al.* [125]. The mass spectrometric identification of metabolites is reviewed by Oxford and Monté [126] and strategies for metabolite isolation and identification are reviewed by Wiltshire [127]. An automated high throughput approach to metabolite identification is of great practical utility. Lopez *et al.* [128] and Kim *et al.* [129] have demonstrated an automated metabolite identification approach using an ion trap mass spectrometer. Automated fraction collection of metabolites for subsequent ion trap MS, MS/MS or NMR analyses has been described by Dear *et al.* [130].

1.3.3 Clinical—Pharmacokinetics

A major effort in the clinical phase of drug development is the determination of drug and metabolite concentrations in biological fluids after drug administration to humans. The pharmacokinetic data obtained are used to support drug development in assessing the therapeutic index, drug-drug interactions, design of dosage regimes, *etc.* Since drugs today are more potent than in years past and are dosed at lower levels, very sensitive assays are required to detect the low circulating levels of drugs in biological fluids (often to pg/mL sensitivity limits). Again, LC-MS/MS is the analysis and detection technique of choice due to its high sensitivity, selectivity and speed. GLP assays are also maintained during the clinical phase of drug development.

The greatest numbers of samples to be analyzed are found in clinical drug development, where thousands of samples are often obtained from one clinical study (conducted at multiple sites). Rapid sample turnaround is sought for one or more reasons: important metabolic information from patient samples can be obtained; the next dose can be determined; or the results used to plan the next clinical study. In addition to sensitivity, selectivity and ruggedness are also important to the analytical method. Examples of the increased requirement for sensitivity as a compound moved from preclinical into the clinical phase of drug development have been detailed by Dear *et al.* [131] and Laugher *et al.* [132]. In the report by Dear *et al.*, preclinical studies required a lower limit of quantitation (LLOQ) of only 1 ng/mL in rat and dog plasma but the LLOQ needed for human clinical studies was 50 pg/mL.

1.3.4 Therapeutic Drug Monitoring

The objective of therapeutic drug monitoring (TDM) is as a guide to provide optimal drug therapy by maintaining plasma concentrations of a drug within a desired range. Some examples of drugs for which monitoring is very important are: digoxin, phenytoin, aminoglycoside antibiotics, theophylline, cyclosporine, HIV protease inhibitors, and some cardiac agents, antiepileptic drugs and antidepressant drugs. Should drug concentrations in the sampled biological fluid exceed the desired range, toxicity may result and/or the desired therapeutic effect may no longer be achieved. The knowledge of plasma drug concentrations may explain why a patient does not respond to drug therapy or why the drug causes an adverse effect. Also, for HIV infections, low drug concentrations may in some cases lead to drug resistance and treatment failure.

Most drugs can be safely taken without monitoring drug concentrations since therapeutic endpoints can effectively be evaluated by other means. For example, blood pressure determinations indicate how well the β-blocker atenolol may be working for a patient, and coagulation times accurately measure the effect of Coumadin®, an anticoagulant drug.

Drugs that are chosen for TDM have the following characteristics in common:

- The range of therapeutic and safe plasma concentrations is narrow (*i.e.*, a low therapeutic index)
- Toxicity or lack of effectiveness of the drug puts the patient at risk
- Patient compliance needs to be monitored

The therapeutic index is defined as the ratio between the maximum and minimum plasma concentrations of the drug's therapeutic range. A low therapeutic index (less than 2) means that the dose that commonly yields a sub therapeutic response is close to the dose that produces some toxicity. Most drugs have a therapeutic index greater than 2 [133]. Note that the relationship between drug concentration at the site of action and the pharmacological response observed in the patient should be considered. TDM using plasma concentration data can be justified when a correlation has been established.

Assay procedures for a drug in biological fluids are usually performed by a local or regional clinical laboratory or a national medical laboratory. Standard methods that are used for detection and quantitation of drugs are immunoassay techniques and liquid chromatography employing sensitive detection (fluorescence or mass spectrometry). Once drug concentrations and responses have been adjusted to achieve the desired therapeutic result, a regular clinical monitoring program is begun. An informative discussion of the analytical goals in therapeutic drug monitoring is provided by Bowers [134].

Additional reports provide a further perspective on therapeutic drug monitoring. The roles for chromatographic analysis in TDM have been described by Wong [135], Shihabi and McCormick [136] and Binder [137]. Some examples of TDM are reported for flecainide [138], tacrolimus [139], immunosuppressants [140, 141], an antineoplastic agent (etoposide) [142], gabapentin [143], oxcarbazepine [144] and mexiletine [145]. TDM in special populations of patients is discussed by Walson [146]. A case for performing prospective concentration to clinical response investigations during the early stages of drug development, rather than the traditional retrospective review, is proposed by Shaw, Kaplan and Brayman [147].

Acknowledgments

The author is appreciative to Danlin Wu and Mike Lee for their critical review of the manuscript, helpful discussions and contributions to this chapter. The line art illustrations were kindly provided by Woody Dells.

References

[1] G.R. Nakayama, Curr Opin. Drug Discovery Dev. 1 (1998) 85-91.

[2] G. Zlokarnik, Anal. Chem. (1999) 322A-328A.

[3] J.K. Nicholson, J. Connelly, J.C. Lindon and E. Holmes, Nature Reviews 1 (2002) 153-161.

[4] Federal Register USA (1988) October 21.

[5] Federal Register USA (1987) May 22.

[6] Federal Register USA (1990) May 21.

[7] A.H. Glassman, C.M. O'Connor, R.M. Califf, K. Swedberg, P. Schwartz, J.T. Bigger, K.R.R. Krishnan, L.T. van Zyl, J.R. Swenson, M.S. Finkel, C. Landau, P.A. Shapiro, C.J. Pepine, J. Mardekian and W.M. Harrison, J. Amer. Med. Assn. 288 (2002) 701-709.

[8] D.T. Rossi and M.J. Lovdahl, In: D.T. Rossi and M.W. Sinz, Eds., Mass Spectrometry in Drug Discovery, Marcel Dekker, New York (2002) 215-244.

[9] R.A. Fecik, K.E. Frank, E.J. Gentry, S.R. Menon, L.A. Mitscher and H. Telikepalli, Med. Res. Rev. 18 (1998) 149-185.

[10] M.A. Gallop, R.W. Barrett, W.J. Dower, S.P.A. Fodor and E.M. Gordon, J. Med. Chem. 37 (1994) 1233-1251.

[11] L.V. Hijfte, G. Marciniak and N. Froloff, J. Chromatogr. B 725 (1999) 3-15.

[12] J.F. Cargill and M. Lebl, Curr. Opin. Chem. Biol. 1 (1997) 67-71.

[13] R. Ferritto, P. Seneci, Drugs Future 23 (1998) 643-654.

[14] U.J. Nilsson, J. Chromatogr. A 885 (2000) 305-319.

[15] S.X. Peng, C. Henson, M.J. Strojnowski, A. Golebiowski and S.R. Klopfenstein, Anal. Chem. 72 (2000) 261-266.

[16] T. Winkler, U. Kettling, A. Koltermann and M. Eigen, Proc. Natl. Acad. Sci. USA 96 (1999) 1375-1378.

[17] T. Matsuo and Y. Seyama, J. Mass Spectrom. 35 (2000) 114-130.

[18] J. Abian, J. Mass Spectrom. 34 (1999) 157-168.

[19] M.P. Balogh, LC-GC 16 (1998) 135-144.

[20] R. Bakhtiar, L. Ramos and F.L.S. Tse, J. Liq. Chrom. & Rel. Technol. 25 (2002) 507-540.

[21] P. Kebarle, J. Mass Spectrom. 35 (2000) 804-817.

[22] P. Kebarle and L. Tang, Anal. Chem. 65 (1993) 972A-986A.

[23] R.S. Plumb, G.J. Dear, D.N. Mallett, D.M. Higton, S. Pleasance and R.A. Biddlecombe, Xenobiotica 31 (2001) 599-617.

[24] B. Erickson, Anal. Chem. 72 (2000) 711A-716A.

[25] S.T. Fountain, In: D.T. Rossi and M.W. Sinz, Eds., Mass Spectrometry in Drug Discovery, Marcel Dekker, New York (2002) 25-84.

[26] M.S. Lee and E.H. Kerns, Mass Spectrom. Rev. 18 (1999) 187-279.

[27] N.J. Haskins, J. Pharm. Biomed. Anal. 25 (2001) 767-773.

[28] M. Guilhaus, V. Mlynski and D. Selby, Rapid Commun. Mass Spectrom. 11 (1997) 951-962.

[29] I.V. Chernushevich, A.V. Loboda and B.A. Thomson, J. Mass Spectrom. 36 (2001) 849-865.

[30] R.E. March, J. Mass Spectrom. 32 (1997) 351-369.

[31] R.E. March, Rapid Commun. Mass Spectrom. 12 (1998) 1543-1554.

[32] D. O'Connor, Curr Opin. Drug Discovery Dev. 5 (2002) 52-58.

[33] Z. Cai, C. Han, S. Harrelson, E. Fung and A.K. Sinhababu, Rapid Commun. Mass Spectrom. 15 (2001) 546-550.

[34] N. Zhang, S.T. Fountain, H. Bi and D.T. Rossi, Anal. Chem. 72 (2000) 800-806.

[35] S.J. Gaskell, J. Mass Spectrom. 32 (1997) 677-688.

[36] M.S. Lee, LC/MS Applications in Drug Development, Marcel Dekker, New York (2002).

[37] J.N. Kyranos, H. Cai, D. Wei and W.K. Goetzinger, Curr. Opin. Biotechnol. 12 (2001) 105-111.

[38] S.H. Hoke, K.L. Morand, K.D. Greis, T.R. Baker, K.L. Harbol and R.L.M. Dobson, Int. J. Mass Spectrom. 212 (2001) 135-196.

[39] A. Triolo, M. Altamura, F. Cardinali, A. Sisto and C.A. Maggi, J. Mass Spectrom. 36 (2001) 1249-1259.

[40] R.D. Süβmuth and G. Jung, J. Chromatogr. B 725 (1999) 49-65.

[41] W.M.A. Niessen, J. Chromatogr. A 856 (1999) 179-197.

[42] J. R. Yates III, J. Mass Spectrom. 33 (1998) 1-19.

[43] F.W. McLafferty, Int. J. Mass Spectrom. 212 (2001) 81-87.

[44] H. Hoja, P. Marquet, B. Verneuil, H. Lotfi, B. Penicaut and G. Lachatre, J. Anal. Toxicol. 21 (1997) 116-126.

[45] M. Careri, A. Mangia and M. Musci, J. Chromatogr. A 794 (1998) 263-297.

[46] H.H. Maurer, J. Chromatogr. B 713 (1998) 3-25.

[47] M.J. Bogusz, J. Chromatogr. B 748 (2000) 3-19.

[48] R.A.W. Johnstone and M.E. Rose, Mass Spectrometry for Chemists and Biochemists, Second Ed., Cambridge University Press, Cambridge (1996).

[49] E. Brewer and J.D. Henion, J. Pharm. Sci. 87 (1998) 395-402.

[50] M. Jemal, Biomed. Chromatogr. 14 (2000) 422-429.

[51] A.H. Brockman, D.L. Hiller and R.O. Cole, Curr Opin. Drug Discovery Dev. 3 (2000) 432-438.

[52] B.L. Ackermann, M.J. Berna and A.T. Murphy, Curr. Topics Med. Chem. 2 (2002) 53-66.

[53] W.A. Korfmacher, K.A. Cox, M.S. Bryant, J. Veals, K. Ng, R. Watkins and C.-C. Lin, Drug Discov. Today 2 (1997) 532-537.

[54] L. Yang, N. Wu and P.J. Rudewicz, J. Chromatogr. A 926 (2001) 43-55.

[55] P.J. Rudewicz and L. Yang, Amer. Pharm. Rev. 4 (2001) 64-70.

[56] B. Law and D. Temesi, J. Chromatogr. B 748 (2000) 21-30.

[57] D.T. Rossi, K.L. Hoffman, N. Janiczek-Dolphin, H. Bockbrader and T.D. Parker, Anal. Chem. 69 (1997) 4519-4523.

[58] V. de Biasi, N. Haskins, A. Organ, R. Batemen, K. Giles and S. Jarvis, Rapid Commun. Mass Spectrom. 13 (1999) 1165-1168.

[59] M.K. Bayliss, D. Little, D.N. Mallett and R.S. Plumb, Rapid Commun. Mass Spectrom. 14 (2000) 2039-2045.

[60] Y.-Q. Xia C.E.C.A. Hop, D.Q. Liu, S.H. Vincent and S.-H.L. Chiu, Rapid Commun. Mass Spectrom. 15 (2001) 2135-2144.

[61] R. Xu, C. Nemes, K.M. Jenkins, R.A. Rourick, D.B. Kassel and C.Z.C. Liu, J. Amer. Soc. Mass. Spectrom. 13 (2002) 155-165.

[62] C.K. Van Pelt, T.N. Corso, G.A. Schultz, S. Lowes and J. Henion, Anal. Chem. 73 (2001) 582-588.

[63] G.J. Dear, J. Ayrton, R. Plumb, B.C. Sweatman, I.M. Ismail, I.J. Fraser and P.J. Mutch, Rapid Commun. Mass Spectrom. 12 (1998) 2023-2030.

[64] T.V. Olah, D.A. McLoughlin and J.D. Gilbert, Rapid Commun. Mass Spectrom. 11 (1997) 17-23.

[65] D.A. McLoughlin, T.V. Olah and J.D. Gilbert, J. Pharm. Biomed. Anal. 15 (1997) 1893-1901.

[66] L. Liang, C. Chi, M. Wright, D. Timby and S. Unger, Amer. Lab. 30 (1998) 11-14.

[67] J. Berman, K. Halm, K. Adkison and J. Shaffer, J. Med. Chem. 40 (1997) 827-829.

[68] Z. Cai, A.K. Sinhababu and S. Harrelson, Rapid Commun. Mass Spectrom. 14 (2000) 1637-1643.

[69] M.K. Bayliss and L.W. Frick, Curr. Opin. Drug Discovery Dev. 2 (1999) 20-25.

[70] J. Vora, D.T. Rossi and E.K. Kindt, In: D.T. Rossi and M.W. Sinz, Eds., Mass Spectrometry in Drug Discovery, Marcel Dekker, New York (2002) 357-376.

[71] Y. Hsieh, M.S. Bryant, J.-M. Brisson, K. Ng and W.A Korfmacher, J. Chromatogr. B 767 (2002) 353-362.

[72] B.S. Kuo, T. Van Noord, M.R. Feng and D.S. Wright, J. Pharm. Biomed. Anal. 16 (1998) 837-846.

[73] H. Zeng, J.-T. Wu and S.E. Unger, J. Pharm. Biomed. Anal. 27 (2002) 967-982.

[74] W.A. Korfmacher, K.A. Cox, K.J. Ng, J. Veals, Y. Hsieh, S. Wainhaus, L. Broske, D. Prelusky, A. Nomeir and R.E White, Rapid Commun. Mass Spectrom. 15 (2001) 335-340.

[75] I.J. Fraser, G.J. Dear, R. Plumb, M. L'Affineur, D. Fraser and A.J. Skippen, Rapid Commun. Mass Spectrom. 13 (1999) 2366-2375.

[76] K.P. Bateman, G. Castonguay, L. Xu, S. Rowland, D.A. Nicoll-Griffith, N. Kelly and C.-C. Chan, J. Chromatogr. B 754 (2001) 245-251.

[77] D.J. Weiss, R.M. Krisko and C.E. Lunte, In: D.T. Rossi and M.W. Sinz, Eds., Mass Spectrometry in Drug Discovery, Marcel Dekker, New York (2002) 377-398.

[78] M. Ye, D.T. Rossi and C.E. Lunte, J. Pharm. Biomed. Anal. 24 (2000) 273-280.

[79] C.S. Chaurasia, C.-E. Chen and C.R. Ashby Jr., J. Pharm. Biomed. Anal. 19 (1999) 413-422.

[80] P.S.H. Wong, K. Yoshioka, F. Xie and P.T. Kissinger, Rapid Commun. Mass Spectrom. 13 (1999) 407-411.

[81] D.A Smith and H. van de Waterbeemd, Curr. Opin. Chem. Biol. 3 (1999) 373-378.

[82] P.R. Chaturvedi, C.J. Decker and A. Odinecs, Curr. Opin. Chem. Biol. 5 (2001) 452-463.

[83] C.K. Atterwill and M.G. Wing, Alternatives to Laboratory Animals, 28 (2000) 857-867.

[84] W.P. Walters, M.T. Stahl and M.A. Murcko, Drug Discov. Today 3 (1998) 160-178.

[85] G.W. Caldwell, S.M. Easlick, J. Gunnet, J.A. Masucci and K. Demarest, J. Mass Spectrom. 33 (1998) 607-614.

[86] H.Z. Bu, M. Poglod, R.G. Micetich and J.K. Khan, Rapid Commun. Mass Spectrom. 14 (2000) 523-528.

[87] E. Liang, K. Chessic and M. Yazdanian, J. Pharm. Sci. 89 (2000) 336-345.

[88] J.D. Irvine, L. Takahashi, K. Lockhart, J. Cheong, J.W. Tolan, H.E. Selick and J.R. Grove, J. Pharm. Sci. 88 (1999) 28-33.

[89] D.M. Cross and M.K. Bayliss, Drug Metab. Rev. 32 (2000) 219-240.

[90] W.A. Korfmacher, C.A. Palmer, C. Nardo, K. Dunn-Meynell, D. Grotz, K. Cox, C.-C. Lin, C. Elicone, C. Liu and E. Duchoslav, Rapid Commun. Mass Spectrom. 13 (1999) 901-907.

[91] J.M. Linget and P. du Vignaud, J. Pharm. Biomed. Anal. 19 (1999) 893-901.

[92] V. Tong, T. Sheng, M.J.A. Walker and F.S. Abbott, J. Chromatogr. B 759 (2001) 259-266.

[93] J. Ayrton, R. Plumb, W.J. Leavens, D. Mallet, M. Dickins and G.J. Dear, Rapid Commun. Mass Spectrom. 12 (1998) 217-224.

[94] H.K. Lim, K.W. Chan, S. Sisenwine and J.A. Scatina, Anal. Chem. 73 (2001) 2140-2146.

[95] E. Bendriss, N. Markoglou and I.W. Wainer, J. Chromatogr. B 754 (2001) 209-215.

[96] J.J. Zheng, E.D. Lynch and S.E. Unger,J. Pharm. Biomed. Anal.,28 (2002) 279-285

[97] S.E. Clarke and P. Jeffrey, Xenobiotica 31 (2001) 591-598.

[98] C.L. Crespi, V.P Miller and B.W. Penman, Med. Chem. Res. 8 (1998) 457-471.

[99] A.A. Nomeir, C. Ruegg, M. Shoemaker, L.V. Favreau, J.R. Palamanda, P. Silber and C.-C. Lin, Drug Metab. Dispos. 29 (2001) 748-753.

[100] I. Chu, L. Favreau, T. Soares, C.-C. Lin and A.A. Nomeir, Rapid Commun. Mass Spectrom. 14 (2000) 207-214.

[101] H.-Z. Bu, L. Magis, K. Knuth and P. Teitelbaum, Rapid Commun. Mass Spectrom. 14 (2000) 1619-1624.

[102] H.-Z. Bu, L. Magis, K. Knuth and P. Teitelbaum, Rapid Commun. Mass Spectrom. 14 (2000) 1943-1948.

[103] P.F. Hollenberg, Drug Metab. Rev. 34 (2002) 17-35.

[104] A.D. Rodrigues, Presented at Symposium on Chemical and Pharmaceutical Structure Analysis (CPSA), Princeton, NJ USA (2001).

[105] M.D. Todd and R.G. Ulrich, Curr Opin. Drug Discovery Dev. 2 (1999) 58-68.

[106] A.L. Castle, M.P. Carver and D.L. Mendrick, Drug Discov. Today 7 (2002) 728-736.

[107] K. Slater, Curr. Opin. Biotechnol. 12 (2001) 70-74.

[108] P.R. Tiller and L. Romanyshyn, Rapid Commun. Mass Spectrom. 16 (2002) 1225-1231.

[109] S. Venkatesh and R.A. Lipper, J. Pharm. Sci. 89 (2000) 145-154.

[110] T. Kennedy, Drug Discov. Today 2 (1997) 436-444.

[111] S.J.F. Macdonald and P.W. Smith, Drug Discov. Today 6 (2001) 947-953.

[112] S.A. Roberts, Xenobiotica 31 (2001) 557-589.

[113] P.J. Eddershaw, A.P. Beresford and M.K. Bayliss, Drug Discov. Today 5 (2000) 409-414.

[114] G.W. Caldwell, Curr. Opin. Drug Discovery Dev. 3 (2000) 30-41.

[115] R.E. White, Annu. Rev. Pharmacol. Toxicol. 40 (2000) 133-157.

[116] C.A. James, M. Breda, E. Frigerio, J. Long and K. Munesada, Chromatographia *Supplement* 55 (2002) S41-S43.

[117] J. Caldwell, J. Chromatogr. A 719 (1996) 3-13.

[118] T.H. Eichhold, R.E. Bailey, S.L. Tanguay and S.H. Hoke II, J. Mass Spectrom. 35 (2000) 504-511.

[119] Z. Shen, S. Wang and R. Bakhtiar, Rapid Commun. Mass Spectrom. 16 (2002) 332-338.

[120] L. Bentley, J. Harden, D. Browne and H.M. Hill, Chromatographia *Supplement* 52 (2000) S49-S52.

[121] S.B. Wainhaus, R.E. White, K. Dunn-Meynell, D.E. Grotz, D.J. Weston, J. Veals and W.A. Korfmacher, Amer. Pharm. Rev. 5 (2002) 86-93.

[122] H. Zhang, J. Henion, Y. Yang and N. Spooner, Anal. Chem. 72 (2000) 3342-3348.

[123] C.E.C.A. Hop, P.R. Tiller and L. Romanyshyn, Rapid Commun. Mass Spectrom. 16 (2002) 212-219.

[124] D.Q. Liu, Y.-Q. Xia and R. Bakhtiar, Rapid Commun. Mass Spectrom. 16 (2002) 1330-1336.

[125] N.J. Clarke, D. Rindgen, W.A. Korfmacher and K.A. Cox, Anal. Chem. 73 (2001) 430A-439A.

[126] J. Oxford and S. Monte, In: R.F. Venn, Ed., Principles and Practice of Bioanalysis, Taylor & Francis, London (2000) 255-277

[127] H. Wiltshire, In: R.F. Venn, Ed., Principles and Practice of Bioanalysis, Taylor & Francis, London (2000) 302-341

[128] L.L. Lopez, X. Yu, D. Cui and M.R. Davis, Rapid Commun. Mass Spectrom. 12 (1998) 1756-1760.

[129] H.K. Lim, S. Stellingweif, S. Sisenwine and K.W. Chan, J. Chromatogr. A
      831 (1999) 227-241.
[130] G.J. Dear, R.S. Plumb, B.C. Sweatman, I.M. Ismail and J. Ayrton, Rapid
      Commun. Mass Spectrom. 13 (1999) 886-894.
[131] G.J. Dear, I.J. Fraser, D.K. Patel, J. Long and S. Pleasance, J. Chromatogr.
      A 794 (1998) 27-36.
[132] L. Laugher, R. Briggs, J. Doughty and T.A.G. Noctor, Chromatographia
      Supplement 52 (2000) S113-S119.
[133] P.L. Madan, U.S. Pharmacist 21(7) (1996) 92-105.
[134] L.D. Bowers, Clin. Chem. 44 (1998) 375-380.
[135] S.H.Y. Wong, J. Pharm. Biomed. Anal. 7 (1989) 1011-1032.
[136] Z.K. Shihabi, BioChromatography 5 (1990) 121-126.
[137] S.R. Binder, Clin. Lab. Med. 7 (1987) 335-356.
[138] T. Breindahl, J. Chromatogr. B 746 (2000) 249-254.
[139] G.L. Lensmeyer and M.A. Poquette, Ther. Drug Monit. 23 (2001)
      239-249.
[140] A. Volosov, K.L. Napoli and S.J. Soldin, Clin. Biochem. 34 (2001)
      285-290.
[141] L. Zhou, D. Tan, J. Theng, L. Lim, Y.-P. Liu and K.-W. Lam,
      J. Chromatogr. B 754 (2001) 201-207.
[142] C.-L. Chen and F.M. Uckun, J. Chromatogr. B 744 (2000) 91-98.
[143] M.M. Kushnir, J. Crossett, P.I. Brown and F.M. Urry, J. Anal. Toxicol. 23
      (1999) 1-6.
[144] H. Levert, P. Odou and H. Robert, J. Pharm. Biomed. Anal. 28 (2002)
      517-525.
[145] F. Susanto, S. Humfeld and H. Reinauer, Chromatographia 21 (1986)
      41-43.
[146] P.D. Walson, Clin. Chem. 44 (1998) 415-419.
[147] L.M. Shaw, B. Kaplan and K.L. Brayman, Clin. Chem. 44 (1998)
      381-387.

Chapter 2

Fundamental Strategies for Bioanalytical Sample Preparation

Abstract

It can be a challenge to develop bioanalytical methods that selectively separate drugs and metabolites from endogenous materials in the sample matrix. Fortunately, many different sample preparation techniques are available in order to meet the desired objectives for an assay. Successful implementation of these procedures relies on having a fundamental understanding of the strategies for sample preparation and the chemistry of the extraction process. This chapter first discusses the objectives for bioanalytical sample preparation. The many different techniques available to the bioanalyst are then introduced as a preview to the full content available in subsequent chapters of this book. Finally, some essential concepts related to extraction chemistry in sample preparation are presented: the influence of sample pH on ionization, the effects of anticoagulants and storage conditions on clot formation, and procedures for determining the matrix effect and extraction efficiency for a sample preparation method.

2.1 Importance of Sample Preparation

2.1.1 Introduction

Sample preparation is a technique used to clean up a sample before analysis and/or to concentrate a sample to improve its detection. When samples are biological fluids such as plasma, serum or urine, this technique is described as bioanalytical sample preparation. The determination of drug concentrations in biological fluids yields the data used to understand the time course of drug action, or pharmacokinetics, in animals and man and is an essential component of the drug discovery and development process.

A reliable analytical method is achieved with the successful combination of efficient sample preparation, adequate chromatographic separation and a sensitive detection technique. Liquid chromatography (LC) coupled with

41

detection by mass spectrometry (MS) is the preferred analytical technique for drug analysis. The use of tandem mass spectrometry (MS/MS) that identifies a specific parent-daughter ion pair allows unequivocal identification of a drug substance from biological samples. LC-MS/MS is widely used because it offers unmatched speed, sensitivity and specificity with good precision and wide dynamic range.

Note that although mass spectrometry allows sensitive and specific detection of analytes of interest, it benefits tremendously from a well chosen column and mobile phase to provide adequate chromatographic separation [1]. The importance of chromatography cannot be overlooked; the column separates the analytes from the much higher concentrations of endogenous materials present in samples that can potentially mask an analyte or introduce ion suppression. Some essential chromatography issues for fast bioanalysis are discussed by O'Connor [2].

Another very important component in the overall analysis is the choice of sample preparation technique which influences the cleanliness of the sample introduced into the chromatographic system. Sample preparation is necessary because most analytical instruments cannot accept the matrix directly. Three major goals for sample preparation are to:

1. Remove unwanted matrix components that can cause interferences upon analysis, improving method specificity
2. Concentrate an analyte to improve its limit of detection
3. Exchange the analyte from an environment of aqueous solvent into a high percentage organic solvent suitable for injection into the chromatographic system

Some additional goals for the sample preparation step may include removal of material that could block the tubing of the chromatographic system, solubilization of analytes to enable injection under the initial chromatographic conditions and dilution to reduce solvent strength or avoid solvent incompatibility [3].

The term sample preparation typically encompasses a wide variety of processes which include aspirating and dispensing liquids, release of drugs from the sample matrix via hydrolysis or sonication, dilution, filtration, evaporation, homogenization, mixing and sample delivery. Here, the term will primarily be used regarding the removal of endogenous compounds from the sample matrix.

Table 2.1
Objectives for bioanalytical sample preparation

1. Removal of unwanted matrix components (primarily protein) that would interfere with analyte determination
2. Concentration of analyte to meet the detection limits of the analytical instrument
3. Exchange of the solvent or solution in which the analyte resides so that it is compatible with mobile phase for injection into a chromatographic system
4. Removal of selected analyte components if the resolving power of the chromatographic column is insufficient to separate all the components completely
5. Removal of material that could block the chromatographic tubing or foul the interface to the detector
6. Dilution to reduce solvent strength or avoid solvent incompatibility
7. Solubilization of compounds to enable injection under the initial chromatographic conditions
8. Stabilization of analyte to avoid hydrolytic or enzymatic degradation

Reprinted with permission from [4]. Copyright 1989 Elsevier Science.

The different types of sample matrices encountered in bioanalysis may include the following: plasma, serum, bile, urine, tissue homogenates, perfusates, buffer, saliva, seminal fluid, dialysate solution, Caco-2 buffer and hepatocyte or microsomal incubation solution. Table 2.1 lists the many overall objectives of sample preparation for drug bioanalysis. Three major objectives from this list will now be described in more detail: removal of matrix components, concentration of analyte and solvent exchange.

2.1.2 Removal of Matrix Components

Biological samples cannot usually be injected directly into an analytical system such as LC-MS/MS because of the multitude of substances present in the sample matrix that can potentially interfere with the analysis, the chromatographic column and/or the detector. These materials include proteins, salts, endogenous macromolecules, small molecules and metabolic byproducts. If these materials were to be injected, the consequences may include the following: a rapid deterioration in the separation performance of the chromatographic column; clogged frits or lines resulting in an increased system backpressure; impaired selectivity of the sorbent in the column due to irreversible adsorption of proteins; and detector fouling that may reduce system performance and require maintenance for cleaning the source. Injection of matrix substances can also cover up and hide the drug or analyte being analyzed, making quantitation difficult (and adversely affecting the data). These materials may also coelute with the analyte of interest, falsely elevating the data. All samples can benefit from a pretreatment step before analysis in order to remove interfering components and attain a selective technique for the desired analytes.

2.1.3 Concentration of Analyte

When blood is collected from an animal or human test subject, plasma or serum is commonly isolated by centrifugation. When an anticoagulant (*e.g.*, heparin, EDTA or sodium citrate) is added to the blood immediately upon collection, plasma is obtained following centrifugation; when blood is first allowed to clot at room temperature and then centrifuged, serum is obtained. A drug of unknown concentration is contained in the isolated plasma (or serum) sample. If one milliliter (mL) of plasma is subjected to sample preparation (interferences are removed), and the drug concentration is determined in a final solvent volume of 1 mL, then no concentration has taken place. However, if this plasma were prepared for analysis and also concentrated, by having a final solvent volume of 0.1 mL instead of 1 mL, analysis of that same aliquot volume will improve the detection limit by a factor of ten ($1.0/0.1=10$). It is the goal of many, but not all, sample preparation techniques to concentrate the analyte before analysis so that the limit of detection and quantitation can be improved. A common method to concentrate analytes is to evaporate a given solvent volume to dryness and then reconstitute in a smaller volume of mobile phase compatible solvent.

2.1.4 Solvent Exchange

Pure aqueous (water based) samples cannot usually be injected into an analytical instrument because of matrix components and incompatibility with the mobile phase used in the chromatographic system (solubility and peak shape concerns). Instead, the analyte is exchanged from a 100% aqueous solution into a percentage of aqueous in organic solvent, such as methanol or acetonitrile, or into a 100% organic solvent. The final solution containing the analyte is now compatible with the mobile phase of the liquid chromatography system used for separation and detection. Solvent exchange occurs using various sample preparation techniques as described in this chapter.

2.2 General Techniques for Sample Preparation

2.2.1 Introduction

Many different sample preparation techniques are available for choosing a method to perform bioanalytical sample preparation. These techniques vary in many regards, such as simplicity, time requirements (in terms of speed and hands-on analyst time), ease of automation, extraction chemistry expertise, concentration factor and selectivity of the final extract. The particular method chosen depends on the requirements of the assay as well as the time involved to run the method. The investment in method development time is also a consideration. Fortunately, the bioanalytical chemist can choose from a range of sample preparation methodologies, as listed in Table 2.2.

Examples of two contrasts for sample preparation requirements are drug discovery and clinical development laboratories. In drug discovery, criteria of rapid sample turnaround, little time available for method development and higher limits of quantitation are acceptable. These decisive factors suggest protein precipitation as a preferred approach. However, in clinical analysis where drugs are potent and are dosed at low levels, the important criteria of ultra sensitivity, great selectivity and a rugged method point toward solid-phase extraction as the technique of choice.

Some useful perspectives and reviews of sample preparation approaches for high throughput LC-MS and LC-MS/MS analyses are described in the literature [2, 5–8] and in a book chapter by Rossi [9]. Other helpful but more general discussions of different sample preparation methodologies are found in various reviews [4, 10–14] and in a book chapter by Kataoka and Lord [3]. A review by Peng and Chiou [15] is recommended for the reader who would like to learn more about pharmacokinetics and the overall requirements for bioanalysis, including sample preparation and analytical method validation.

Table 2.2
Typical choices of sample preparation techniques useful in bioanalysis
- Dilution followed by injection
- Protein precipitation
- Filtration
- Protein removal by equilibrium dialysis or ultrafiltration
- Liquid-liquid extraction
- Solid-supported liquid-liquid extraction
- Solid-phase extraction (off-line)
- Solid-phase extraction (on-line)
- Turbulent flow chromatography
- Restricted access media
- Monolithic columns
- Immunoaffinity extraction
- *Combinations of the above*

2.2.2 Dilution Followed by Injection

Sample dilution is used to reduce the concentration of salts and endogenous materials in a sample matrix and is commonly applied to urine. Drug concentrations in urine are usually fairly high and allow this dilution without an adverse effect on sensitivity; protein concentrations in urine are negligible under normal physiological conditions. An example of a urine dilution procedure is reported for indinavir in which 1 mL urine was diluted with 650 μL acetonitrile (so that the resulting concentration of organic in the sample was equal to or less than that of the mobile phase). An aliquot of 6 μL was injected into an LC-MS/MS system [16]. Although coeluting endogenous species in urine were not seen in the selected ion monitoring mode, their presence did suppress or enhance the ionization of analytes, leading to increased variation in MS/MS responses.

Dilution is not typically used for plasma due to the high amounts of protein present and the greater effect that dilution has on sensitivity. However, dilution is very attractive for the minimal effort and time involved. One report did discuss a dilution approach for plasma in a high throughput procedure [17]. In this report, dog plasma samples were centrifuged, pipetted into wells of a microplate and then placed on an automated pipettor. A volume of 15 μL plasma was diluted with 485 μL of a solution of water/methanol/formic acid (70:30:0.1, v/v/v) containing internal standard. The samples were sealed, mixed and 5 μL were injected into an LC-MS/MS system. The dilution resulted in a slightly viscous solution with no observed precipitation. The limit of quantitation for the dilution assay (2 ng/mL) was 400 times higher than that of a more selective procedure that also concentrates the analyte (liquid-liquid extraction; 5 pg/mL LOQ). However, the advantage of the first technique was that the throughput was 50 times greater. In this case, throughput was a more important consideration than analyte sensitivity.

2.2.3 Protein Precipitation

Protein precipitation is often used as the initial sample preparation scheme in the analysis of a new drug substance since it does not require any method development. A volume of sample matrix (1 part) is diluted with a volume of organic solvent or other precipitating agent (3–4 parts), followed by vortex mixing and then centrifugation or filtration to isolate or remove the precipitated protein mass. The supernatant or filtrate is then analyzed directly. Protein precipitation dilutes the sample. When a concentration step is required, the supernatant can be isolated, evaporated to dryness and then reconstituted before analysis.

This procedure is popular because it is simple, universal, inexpensive and can be automated in microplates. However, matrix components are not efficiently removed and will be contained in the isolated supernatant or filtrate. In MS/MS detection systems, matrix contaminants have been shown to reduce the efficiency of the ionization process [1, 16, 18–26]. The observation seen is a loss in response and this phenomenon is referred to as ionization suppression. This effect can lead to decreased reproducibility and accuracy for an assay and failure to reach the desired limit of quantitation.

Protein precipitation techniques can be performed in high throughput systems using the collection plate format and several reports of these applications are available [27–31]. Procedures for performing high throughput protein precipitation are presented in Chapter 6 along with strategies for method development. The automation of protein precipitation techniques is presented in Chapter 7.

2.2.4 Filtration

Filtration is important for the removal of material or debris from a sample matrix so that the chromatographic tubing and column do not become physically blocked. For example, a precipitated protein mass can be filtered from solution and the filtrate analyzed directly [28, 32–36]. Filtration is often used to clarify raw sample matrix before another sample preparation technique (*e.g.*, turbulent flow chromatography, discussed in Section 2.2.10) as well as for the filtration of debris from the final eluate before injection. Filtration in the microplate format allows high throughput sample preparation and this approach is discussed in more detail in Chapter 6, Section 6.4.

2.2.5 Protein Removal

Equilibrium dialysis, ultrafiltration and other membrane based sample preparation methods are useful for protein removal in bioanalysis [13, 14, 37–40]. Equilibrium dialysis is a classical method to physically separate small molecular weight analytes from larger molecular weight constituents (*e.g.*, protein) in a biological sample matrix. This process occurs by diffusion through the pores of a selective, semipermeable membrane and is concentration driven. Ultrafiltration separates proteins according to molecular weight and size using centrifugal force, and is thus based on a pressure differential rather than on a concentration differential as in equilibrium dialysis. Materials below the molecular weight cutoff pass through the membrane and are contained within the ultrafiltrate; retained species are concentrated on the pressurized side of the membrane. Both equilibrium dialysis and ultrafiltration can be performed in a high throughput manner in the microplate format. Methodologies for their use are discussed in Chapter 6, Section 6.6.

2.2.6 Liquid-Liquid Extraction

Liquid-liquid extraction (LLE), also called solvent extraction, is a technique used to separate analytes from interferences in the sample matrix by partitioning the analytes between two immiscible liquids. Analytes distribute between these two liquid phases (aqueous and organic) and partition preferentially into the organic phase when the analytes (1) are unionized (uncharged) and (2) demonstrate solubility in that organic solvent. Isolation of the organic phase, followed by evaporation and reconstitution in a mobile phase compatible solvent, yields a sample ready for injection. The method provides efficient sample cleanup as well as sample enrichment.

A major advantage of LLE is that it is widely applicable for many drug compounds and is a relatively inexpensive procedure. With proper selection of organic solvent and adjustment of sample pH, very clean extracts can be obtained with good selectivity for the target analytes. Inorganic salts are insoluble in the solvents commonly used for LLE and remain behind in the aqueous phase along with proteins and water soluble endogenous components. These interferences are excluded from the chromatographic system and a cleaner sample is prepared for analysis. However, some disadvantages of LLE include its labor intensive nature with several disjointed vortex mix and centrifugation steps required. The organic solvents used are volatile and present hazards to worker safety. Also, emulsions can be formed without warning and can result in loss of sample.

Liquid-liquid extraction has been demonstrated in collection microplates to provide for high throughput sample preparation [41–52]. The use of liquid handling workstations reduces the hands-on analyst time required for this technique and offers semi-automation to an otherwise labor intensive task. Procedures for performing liquid-liquid extraction in microplates are discussed in Chapter 8. Strategies for its method development and automation are detailed in Chapters 9 and 10, respectively.

2.2.7 Solid-Supported Liquid-Liquid Extraction

The many disjointed mixing and centrifugation steps of traditional LLE can be eliminated by performing LLE in a flow-through 96-well microplate filled with inert diatomaceous earth particles. This technique is referred to as solid-supported LLE. The high surface area of the diatomaceous earth particle facilitates efficient, emulsion-free interactions between the aqueous sample and the organic solvent. Essentially, the diatomaceous earth with its treated aqueous phase behaves as the aqueous phase of a traditional liquid-liquid extraction, yet it has the characteristics of a solid support.

The procedure for use of solid-supported particles to perform liquid-liquid extraction follows. A mixture of sample (*e.g.*, plasma), internal standard, and buffer solution to adjust pH is prepared. This aqueous mixture is then added to the dry particle bed (no conditioning or pretreatment is necessary). The mixture is allowed to partition for about 3–5 min on the particle surface via gravity flow. The analyte in aqueous solution is now spread among the particles in a high surface area. A hydrophobic filter on the bottom of each well prevents the aqueous phase from breaking through into the collection vessel placed underneath. Organic solvent is then added to the wells and as it slowly flows through the particle bed via gravity the analyte partitions from the adsorbed aqueous phase into the organic solvent. This eluate is collected in wells or tubes underneath the tips of the plate. A second addition of organic solvent is performed and the combined eluates are evaporated to dryness. The residue is reconstituted in a mobile phase compatible solution and an aliquot is injected into the chromatographic system for analysis.

Solid-supported LLE in microplates is fully automatable with liquid handling workstations since each step is simply a liquid transfer or addition. The many capping and mixing steps as used for traditional LLE are unnecessary, so less hands-on time is required from the analyst. Several applications for this high throughput technique are reported in the literature [53–56]. Solid-supported liquid-liquid extraction in microplates is discussed further in Chapter 8, Section 8.3 and methods for its automation are presented in Chapter 10.

2.2.8 Solid-Phase Extraction (Off-Line)

Solid-phase extraction (SPE) is a procedure in which an analyte, contained in a liquid phase, comes in contact with a solid phase (sorbent particles in a column or disk) and is selectively adsorbed onto the surface of that solid phase. All other materials not adsorbed by chemical attraction or affinity remain in the liquid phase and go to waste. Generally a wash solution is then passed through the sorbent bed to remove any adsorbed contaminants from the sample matrix, yet retain the analyte of interest on the solid phase. Finally, an eluting solvent (usually an organic solvent such as methanol or acetonitrile that may be modified with acid or base) is added to the sorbent bed. This solvent disrupts the attraction between analyte and solid phase causing desorption, or elution, from the sorbent. Liquid processing through the sorbent bed can be accomplished via vacuum or positive displacement. Solvent exchange is followed by analysis on a chromatographic system. The term off-line refers to performing a sample preparation procedure independently of the LC-MS/MS analysis.

High throughput solid-phase extraction utilizes the 96-well microplate format. Several dozen published reports of its use for quantitative bioanalysis are available [25, 28, 45, 57–116], as well as numerous presentations which are not referenced here. Some reports have even utilized a higher density 384-well format [117–119]. Although the plates can be processed manually, liquid handling workstations are preferred; the use of 96-tip semi-automated workstations results in very high throughput. Many sorbent chemistries (>45) and formats are available in the plates to meet most needs. However, the cost per sample is relatively expensive compared with other techniques, time is required for method development, and sometimes a longer learning curve is necessary to realize its successful use. Procedures for performing solid-phase extraction in microplates are discussed in detail in Chapter 11. Strategies for method development and automation of SPE are presented in Chapters 12 and 13, respectively.

2.2.9 Solid-Phase Extraction (On-Line)

The solid-phase extraction technique is automated by the use of disposable extraction cartridges that are placed on-line with the chromatographic system. An important feature and advantage of on-line SPE, compared with off-line SPE, is direct elution of the analyte from the SPE cartridge into the LC system. The time consuming off-line steps of evaporation, reconstitution, and preparation for injection are eliminated, making on-line SPE more efficient and fully automated. Since the entire volume of eluate is analyzed, maximum sensitivity for detection is obtained.

The SPE cartridge used for on-line applications has a standard dimension of 10 x 2 mm and can withstand LC system pressures to 300 bar. A complete set of traditional sorbent chemistries (>38) is available in this format. The use of on-line SPE with LC analysis in a serial mode is straightforward; one sample is processed after the previous one has finished. With the advent of parallel and staggered parallel LC systems, throughput of on-line SPE can now be doubled; two analyses can be performed in the time taken for one analysis using serial mode.

Many applications have been reported using on-line solid-phase extraction. Its usefulness has been documented for bioanalytical applications providing drug discovery support [31, 120–122], developmental compound support [123–131] and therapeutic drug monitoring [132–145]. Some staggered parallel applications for higher throughput have also been demonstrated [145–148]. On-line SPE using disposable cartridges in this manner is discussed in more detail in Chapter 14, Section 14.3.6.

2.2.10 Turbulent Flow Chromatography

Turbulent flow liquid chromatography has been shown to eliminate the need for traditional off-line sample preparation as it allows direct injection of plasma or serum samples. This technique is performed with high flow rates and a single short column (1 x 50 mm) containing particles of a large size (30–50 μm). When a sample in a biological fluid is injected into this system, the liquid chromatograph delivers a high flow rate (~5 mL/min) of a highly aqueous solvent (typically 100% water with pH or salt additives as needed). A divert valve is set to waste for the load and wash period, which normally lasts for less than a minute. When a reversed phase sorbent is used in the extraction column, the analyte is retained via partitioning while hydrophilic components in the sample matrix, including proteins, are washed off the column and go to waste.

After the load and wash steps are completed, the flow is changed to a high organic solvent and the divert valve is switched to direct the flow into a splitter; a portion of the flow enters the mass spectrometer. Depending on the type of ion source and the ionization mode used, the flow is usually split from ~5 mL/min down to 0.3 to 1.0 mL/min. The retained analyte is eluted from the column with some limited separation and is detected by the mass spectrometer. The elution step usually takes less than one minute. Following this elution, the flow composition is changed back to high aqueous to equilibrate the column in preparation for the next sample. Again, this equilibration step can be completed in less than one minute. Therefore, the entire on-line extraction and analysis procedure can usually be accomplished in about 3 min for each sample. This single-column mode provides the distinct advantage of high throughput and simplicity.

The single-column mode of turbulent flow chromatography is the simplest and yields the shortest injection cycle. It is often the first approach tried in the development of a new assay. However, when the single-column mode does not present sufficient cleanup, an analytical column can be placed in series downstream from the first.

The dual-column mode provides the advantage of improved separation performance as well as detection sensitivity resulting from chromatographic focusing. Another advantage of the dual-column configuration mode is that the extraction and separation processes can be performed simultaneously since they are now driven by two separate LC pumps. While the sample is running on the analytical separation column the extraction column is in use with the next sample.

Many applications have been reported using turbulent flow chromatography in both the single-column mode [81, 96, 149–153] and the dual-column mode [83, 149, 154–167]. Turbulent flow chromatography is discussed further in Chapter 14; information on the single-column mode is found in Section 14.2.2 and dual-column mode in Section 14.3.3.

2.2.11 Restricted Access Media

Another technique allowing for the direct injection of plasma or serum on-line with the chromatographic system uses normal laminar flow liquid chromatography. The analytical columns used contain restricted access media (RAM). The RAM particles packed within columns are designed to prevent or restrict large macromolecules from accessing the inner adsorption sites of the bonded phase. The most popular RAM column used in bioanalysis is called internal surface reversed phase (ISRP). In this type of column, the internal particle surface (pore) is coated with a bonded reversed phase material and the outer particle surface is coated with a nonadsorptive but hydrophilic material. This dual-phase column allows effective separation of the analyte of interest from macromolecules in the sample matrix. Drugs and other small molecules enter the pores of the hydrophobic reversed phase to partition and retain while proteins and larger matrix components are excluded.

The dual-phase nature of these RAM materials allows the direct injection of the biological sample matrix onto the column; little or no pretreatment is required.

Some disadvantages with the use of RAM columns are that retention times can be long (>10 min), washing of the column is necessary between injections and required mobile phases are not always compatible with some ionization techniques used in LC-MS/MS. Dual column RAM techniques are also used. These methods utilize an analytical separation column placed in series downstream from the RAM column. A general overview of the use of restricted access materials in liquid chromatography has been published in two parts [168, 169]. Single column RAM techniques are discussed further in Chapter 14; information on the single-column mode is found in Section 14.2.3 and dual-column mode in Section 14.3.4.

2.2.12 Monolithic Columns

Monolithic columns possess a unique structure consisting of a one piece organic polymer or silica with flow-through pores. The smaller pores (mesopores, diameter about 13 nm) located on the silica skeleton provide the large surface area needed to achieve sufficient capacity. The larger pores (macropores, diameter about 2 μm) on the silica skeleton reduce flow resistance. Together, these specifications allow the use of high flow rates without generating high backpressure. In addition to the clear implications for high speed separation, this biporous structure offers the unique advantage of direct injection of the sample matrix in bioanalysis.

Several bioanalytical applications using monolithic columns have been described for drug metabolite identification and drug discovery support [170–173]. In some cases, samples were injected directly and in others pretreatment was performed using turbulent flow chromatography [174], protein precipitation with acetonitrile [175], or 96-well solid-phase extraction [101]. The use of monolithic columns is discussed further in Chapter 14, Section 14.2.4.

2.2.13 Immunoaffinity Extraction

Immunoaffinity extraction uses antibody-antigen interactions to provide a very high specificity for the molecules of interest. Antibodies are immobilized onto a pressure resistant solid support and then packed into LC columns for use in chromatography applications. These antibodies can remove a specific analyte or class of analytes from all other materials in a sample. Their recognition is based on a particular chemical structure rather than on a general reversed phase attraction such as occurs with solid-phase extraction. The great specificity of this technique can make it preferable to other approaches. The binding of antibody to antigen is a reversible process and its equilibrium can be influenced by manipulating the solution pH and aqueous/organic composition. Note that an antibody can also bind one or more analytes structurally similar to the one of interest, in a process called cross-reactivity. This phenomenon can be useful when a group of analytes is to be isolated from solution, such as for the screening of small molecule combinatorial libraries [176]. Immunoaffinity extraction can be performed off-line, but is time consuming; on-line extraction is preferred for higher throughput. This technique is also referred to as immunoaffinity solid-phase extraction and the particles as immunosorbents. Several reviews discuss advances in analytical applications of immunoaffinity chromatography [177, 178] and the characteristics and properties of immunosorbents with techniques for their successful use [179, 180]. Immunoaffinity extraction is discussed further in Chapter 14, Section 14.2.5.

2.2.14 Combinations of Techniques

Although many individual choices for sample preparation exist as previously discussed, often the combination of two techniques yields a more desirable result. The protein precipitation procedure can yield a sample that is ready for analysis but its potential to carry over matrix interferences can be problematic. The isolated supernatant can be filtered and/or centrifuged before injection but these procedures remove only particulates or proteinaceous material, not the materials causing matrix interferences. Protein precipitation is sometimes followed by liquid-liquid extraction, solid-phase extraction or an on-line technique for a more selective cleanup before analysis.

The following applications report sequential sample preparation in which an initial protein precipitation step was followed by:
1.   Turbulent flow chromatography [156]
2.   Off-line solid-phase extraction [28, 181]
3.   On-line SPE [128]
4.   On-line column switching [182, 183]
5.   Liquid-liquid extraction [44, 184–187]

Another example of sequential sample preparation is the analysis of tissue homogenates. Precipitation of proteins is performed using, *e.g.*, a mixture of zinc salt with acetonitrile [188]. The supernatant, following centrifugation and dilution with water (to decrease the organic concentration), is loaded onto an SPE bed. The SPE technique permits the more selective isolation of analyte and excludes interferences. It also permits desalting since the zinc sulfate salts are washed away.

2.3 Essential Concepts for Sample Preparation

2.3.1 Influence of Sample pH on Ionization

2.3.1.1 Introduction

Successful implementation of a sample preparation procedure relies on having a fundamental understanding of the chemistry of the extraction process. It is important to be familiar with the basic physical and chemical properties of organic solvents as well as functional groups on the analyte of interest in order to predict behavior under various extraction conditions. Exploiting strong interactions of analyte to the exclusion of matrix materials will yield selective extractions. Since most sample matrices also contain proteins, a fundamental understanding of how to exclude proteins is also valuable, as discussed in Chapter 6. One essential concept for sample preparation is the influence of sample pH on ionization. The outline of the following discussion is based on the text by Rodwell [189] and is reprinted with permission (Copyright 1999, McGraw-Hill Companies).

2.3.1.2 Dissociation Behavior of Weak Acids

Drugs, as well as endogenous chemicals, possess functional groups that are weak acids or weak bases. Commonly these groups are carboxyl or amino groups and they display a characteristic dissociation behavior as shown in equations (1) and (2).

$$R-COOH \leftrightharpoons R-COO^- + H^+ \tag{1}$$

$$R-NH_3^+ \leftrightharpoons R-NH_2 + H^+ \tag{2}$$

The protonated form of each species above is referred to as the "acid" ($R-COOH$ and $R-NH_3^+$) and the unprotonated form is called its "conjugate base" ($R-COO^-$ and $R-NH_2$). The relative strength of a weak acid and a weak base is expressed as its dissociation constant (K) which describes its tendency to ionize. The expressions for the dissociation constant for two typical weak acids are shown in equations (3) and (4).

$$K = \frac{[R\text{--}COO^-]\,[H^+]}{[R\text{--}COOH]} \tag{3}$$

$$K = \frac{[R\text{--}NH_2]\,[H^+]}{[R\text{--}NH_3^+]} \tag{4}$$

It is convenient to express the dissociation constant K as pK to avoid working with a negative exponential number, where

$$pK = -\log K \tag{5}$$

For example, acetic acid has a dissociation constant (K) of 1.76×10^{-5} and substituting this number into equation (5) yields pK = 4.75. Acetic acid is referred to as a weak acid. Stronger acids display lower pK values (*e.g.*, the monocarboxylic acid of citric acid has a pK value of 3.08, and so it is a stronger acid than acetic acid).

In equations (3) and (4) above it can be seen that when the protonated species (acid) and the unprotonated species (conjugate base) are present in equal concentration, K and $[H^+]$ are equal.

$$K = [H^+] \tag{6}$$

By taking the negative logarithm of both sides of the above equation,

$$-\log K = -\log [H^+] \tag{7}$$

Since $-\log K$ is defined as pK, and $-\log [H^+]$ defines pH, the equation becomes

$$pK = pH \tag{8}$$

The pK of an acid group is that pH at which the protonated and unprotonated species are present in equal concentrations. Expressed another way, the pK value is defined as the pH at which 50% of a drug is ionized and 50% is unionized. The pK value for an acid (also called pKa) can be experimentally determined by adding 0.5 equivalent of base per equivalent of acid. The resulting pH will be equal to the pK of the acid.

An illustration of how the titration of a weak acid with base changes the solution pH is shown in Figure 2.1 for a hypothetical weak acid having a pKa

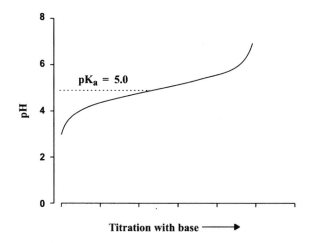

Figure 2.1. Typical example of a titration curve for a weak acid having a pKa of 5.0. The pKa is defined as the pH at which 50% of a drug is ionized.

value of 5.0. The shape of this pH change resembles a sigmoid curve because there is symmetry at the point where the drug is half ionized. At the inflection point the concentrations of the protonated and unprotonated forms of the drug are equal and the pKa of the analyte is equal to the pH of the solution.

2.3.1.3 Henderson-Hasselbach Equation

It is important to understand that many sample preparation procedures require the drug to be fully unionized for a desired reversed phase or lipophilic attraction to occur (*e.g.*, liquid-liquid extraction and solid-phase extraction). The prediction of the ionization state of a drug is essential for successful sample preparation. The Henderson-Hasselbach equation has proved to be of great predictive value to understand and describe proton equilibrium:

$$\text{For acids: pH} = \text{pKa} + \log\left(\text{[ionized]}/\text{[unionized]}\right) \qquad (9)$$

$$\text{For bases: pH} = \text{pKa} + \log\left(\text{[unionized]}/\text{[ionized]}\right) \qquad (10)$$

When the concentrations of ionized and unionized are equal, pH = pKa. When the ratio ([ionized]/[unionized]) = 10:1, pH = pKa + 1. Likewise, when the ratio ([ionized]/[unionized]) = 100:1, pH = pKa + 2. In order to promote a drug to the fully unionized state (>99% complete), the pH should be adjusted at least

2 units below or above the pKa value, depending on whether it is an acid or a base. If a 1 unit adjustment in pH is made, only a 91% conversion will take place (as calculated by equation 9). Listed below are important pH adjustment guidelines to promote the ionization of acids and bases to their fully ionized or fully unionized species.

For > 99% conversion the pH should be adjusted 2 pH units above or below the pKa value of the analyte

Acids	pH < pKa = R–COOH	unionized
	pH > pKa = R–COO⁻	ionized

Bases	pH < pKa = R–NH3+	ionized
	pH > pKa = R–NH₂	unionized

The following example shows how the Henderson-Hasselbach equation can be used to calculate the percent ionization at a known pH. Aspirin is a weak acid with a pKa of 3.5; its dissociation is shown in Figure 2.2.

What percent ionization exists in the stomach at pH 1.5? Substituting values into equation (9) yields the following:

$$1.5 = 3.5 + \log ([ionized]/[unionized]) \tag{11}$$

$$-2.0 = \log ([ionized]/[unionized]) \tag{12}$$

$$[ionized]/[unionized] = 1/100 \tag{13}$$

It is seen above that at pH 1.5 aspirin is almost completely (99%) **unionized**.

What percent ionization exists in the duodenum at pH 5.5? Substituting values into equation (9) yields the following:

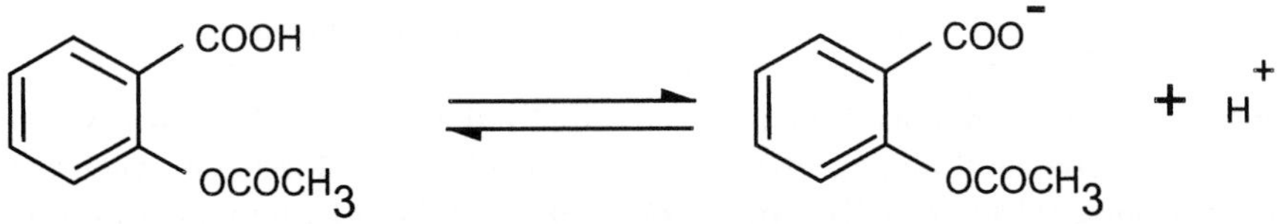

Figure 2.2. The dissociation of aspirin is shown. Aspirin is a weak acid (pKa = 3.5).

$$5.5 = 3.5 + \log ([\text{ionized}]/[\text{unionized}]) \tag{14}$$

$$2.0 = \log ([\text{ionized}]/[\text{unionized}]) \tag{15}$$

$$[\text{ionized}]/[\text{unionized}] = 100/1 \tag{16}$$

It is seen above that at pH 5.5 aspirin is almost completely (99%) **ionized**.

2.3.1.4 Prediction of pKa Values

Note that often the pKa value of a drug is not known or is not communicated to the analyst who is designing a sample preparation method. In this case, it is important to be able to *predict* pKa values. Generally, pKa values for amines range from 8.0–11.0, so the pH should be adjusted from 10.0–13.0 to promote conversion of the amine to the **unionized** form. Typical pKa values for acids may range from 3.0–5.0, so the pH should be adjusted from 1.0–3.0 to promote conversion of the acid to the **unionized** form.

Some general principles can be described to better understand the range of pKa values for acids and bases. The presence of a halogen atom near a carboxyl group strengthens the acid effect since it acts as an electron sink with the carbonyl group. Acetic acid does not possess a halogen and its pKa is 4.75. Monochloroacetic acid, having one halogen atom, is a stronger acid (pKa = 2.85) than acetic acid. Dichloroacetic acid and trichloroacetic acid are stronger yet (pKa = 1.48 and 0.70, respectively). Also, the closer a halogen atom is to a carbonyl group, the greater the effect. Aromatic amines are moderately strong acids (weak bases) because the aromatic ring creates an electron sink that reduces the negative charge on the nitrogen and facilitates dissociation of a proton. Aliphatic amines are stronger bases than aromatic amines. Therefore, the aromatic acids aniline (pKa = 4.6) and pyridine (pKa = 5.2) are weak bases and the aliphatic amines trimethylamine (pKa = 9.7) and dimethylamine (pKa = 10.7) are strong bases.

When an accurate prediction of the pKa value for an analyte is not possible, it is suggested that several different pH values be examined toward the acidic or basic end of the pH scale to bracket the expected pKa. For example, if a sample preparation method requires the analyte to be unionized and an extraction procedure failed at pH 9.0, it may be that the pKa is actually 8.5 and a pH value of 10.5 should have been used to more fully promote the unionized species. When pKa values are unknown, extractions can be performed in parallel at pH

9.0, 11.0 and 13.0, for example. At least one of these three values will likely fully convert an amine to the unionized form for the appropriate attraction to occur. Had pH 9.0 been chosen and it failed to extract unionized drug, stopping and admitting failure ("It didn't work") would not be the best course of action since the extraction chemistry was not properly considered.

2.3.1.5 Importance of Using Buffers

A solution of a weak acid and its conjugate base (or of a weak base and its conjugate acid) acts as a buffer. A buffer solution is able to effectively resist a change in pH after the addition of a strong acid or base. Water is not able to resist this change since it does not have any buffering capacity.

Solutions at a known pH are commonly used in sample pretreatment protocols to modify the pH of the biological fluid to the desired value or range. **It is highly encouraged to always add buffer solutions to plasma or other matrices so that the pH of the resulting sample can be estimated with greater confidence.**

Another important consideration is that distilled and/or deionized water is *not* pH 7.0 (neutral) by definition and this common assumption should be avoided. In fact, the pH of water can vary from day to day. Commonly it may vary in a typical range from 6.2 to 7.2. Therefore, the use of water should be avoided for dilution of matrix because the resulting pH of the mixture is not known. Even the addition of 250 µL water to 100 µL plasma has been shown to raise the pH to a value greater or lower than the expected value. When a neutral pH is desired for sample pretreatment, *always* use a buffer so that the pH of the resulting mixture is estimated with confidence.

The improper adjustment of pH is one of the most common reasons for failure of a sample preparation method to perform as expected. When preparing buffer solutions or pH modified solutions remember that some solutions containing a volatile component must be made fresh weekly (*e.g.*, 1% NH$_4$OH in methanol or in acetonitrile) or even daily, depending on storage conditions and frequency of exposure to the atmosphere. As a general rule, prepare ammonium hydroxide containing organic solutions fresh each week and prepare buffers fresh every 3 months. It is advisable to invest in a rugged pH meter and to maintain a complete inventory of complementary salts and acids (*e.g.*, phosphoric acid with potassium phosphate salts in the monobasic, dibasic and tribasic forms) so that buffers of any pH can be prepared when needed and used with confidence to manipulate sample pH.

2.3.1.6 Further Reading

It is important to be familiar with the principles of functional group chemistry and for further information the paperback book by Lemke [190] is highly recommended. A book chapter by Wiltshire is also useful to learn more about the physical and chemical properties of drugs and their extractions from biological material [191]. This Wiltshire chapter also discusses intramolecular and intermolecular forces, functional group chemistry, solubility, miscibility, ionization, pH and pK values, titration curves, buffers and strategies for solvent extraction (primarily liquid-liquid extraction).

2.3.2 Effects of Anticoagulants and Storage on Clot Formation

The choice of anticoagulant used in the preparation of plasma from whole blood may influence properties of sample flow and viscosity. Some commonly available anticoagulants in clinical use are ethylene diamine tetra acetic acid (EDTA), sodium or lithium heparin and sodium fluoride/potassium oxalate. The effect is not apparent when using freshly prepared plasma, but with repeated freeze/thaw cycles the samples containing heparin may readily produce masses of thrombin clots. These clots must be removed before sample analysis. Anticoagulants such as EDTA and sodium citrate may be generally more acceptable for use in this regard, although data are not definitive [36, 192].

These formed clots in thawed plasma are capable of plugging pipet tips and creating pipetting errors with liquid handling workstations. These errors can result in aspirating an incomplete volume via tip occlusion (air aspirated instead of sample) as well as in pipetting failures that require user intervention. One of the reliable ways to eliminate clots is to thaw the plasma, invert several times rapidly to homogenize the sample, and then centrifuge the clots to the bottom of the tube. The sampling pipet should then remove an aliquot from the middle of the tube.

Another approach to alleviate clot formation is to transfer freshly prepared plasma from Vacutainer® tubes, as generated from study samples, directly into the wells of a filtration plate for storage and freezing. This application is reported by Berna *et al.* [36], in which 20 μm polypropylene filter plates (Captiva™; Ansys Technologies, Lake Forest CA USA, now a part of Varian) were used. Seals cover the bottom of the filter plate so that plasma does not leak out the tips via gravity and start the filtration process. The top of the plate is also sealed. Prior to sample analysis the filter plate containing frozen plasma samples is allowed to thaw; the bottom cover is removed and the filter plate is placed over a clean microplate to collect the plasma filtrate. The filter plate/collection plate combination can be placed into a centrifuge to complete the filtration process, if required. Alternately, a vacuum can be applied using a vacuum collar or manifold. Once the samples have been thawed and filtered in this manner, the plasma filtrate (already in the microplate format) becomes the source plate for a sample preparation scheme. More details about this application are provided in Chapter 6, Section 6.4.5.1.

It is possible that clot formation may also be temperature dependent. If plasma is frozen at −20°C an insoluble precipitate may form upon the freeze/thaw process but storage at −80°C may dramatically prevent its formation. Also, the speed at which the sample is frozen may be important, with rapid freezing in dry ice/acetone preferred over simply placing fresh plasma samples into a freezer and allowing them to become solid over time [193].

2.3.3 Evaluation of Matrix Effect

Materials within the matrix can potentially affect the efficiency of analyte ionization in the mass spectrometer. This effect can lead to decreased reproducibility and accuracy for an assay and failure to reach the desired limit of quantitation. It is reported that the extent of ionization suppression seen is much more severe with electrospray ionization than with atmospheric pressure chemical ionization [22, 194]. Therefore, analysts need to use a post-extraction spiked matrix blank and compare the results with an analytical standard in solution to determine the influence of the matrix on the analysis (equation 17). A matrix blank is a representative biological sample that is free of the target analytes. A spiked matrix blank is a control sample that has been fortified with the target analytes at a defined, relevant level [6].

$$\text{Matrix Effect} = \frac{\text{Response of post-extracted spike}}{\text{Response of unextracted sample}} \qquad (17)$$

The absence of a matrix effect is indicated by a ratio of 1.0. Suppression of ionization results in reduced analyte response (ratio < 1.0). Total matrix suppression yields a value of zero [195]. Signal enhancement may also result (ratio > 1.0). A matrix effect may be concentration dependent. More detailed information to help the reader understand matrix effects is found in Chapter 6, Section 6.2, including information on how to minimize these effects.

A good example of the typical procedure used to measure the amount of ionization suppression or enhancement caused by the extracted matrix is found in the report by Zweigenbaum and Henion [48] in which analytes were prepared for analysis from plasma using liquid-liquid extraction. In order to determine the degree of ionization suppression, blank control human plasma was extracted. The extract was spiked with a known amount of each analyte. This post-extraction spike was then analyzed by selected reaction monitoring LC-MS and the integrated area response for each analyte was compared with the response when analyte was instead dissolved in reconstitution solvent at the same concentration as the spiked samples.

2.3.4 Determination of Recovery

Analyte recovery from a sample matrix (also called extraction efficiency) is a comparison of the analytical response from an amount of analyte added to and extracted from the sample matrix (pre-extraction spike) with that from a post-extraction spike (equation 18).

$$\% \text{ Recovery} = \frac{\text{Response of extracted spike}}{\text{Response of post-extracted spike}} \times 100 \qquad (18)$$

The efficiency for an extraction should be 100% or less. This determination is typically made at a minimum of 3 concentrations (*e.g.*, at 3 or more quality control levels) in multiple replicates (*e.g.*, 6).

A good example of the typical procedure used to measure analyte recovery is found in the same report by Zweigenbaum and Henion [48] as mentioned above. The integrated analytical response of each analyte in the pre-extraction spike of plasma was compared with the response from the post-extraction spiked plasma to determine recovery (extraction efficiency). A comparison of the ion current responses from the pre-extraction spike with the responses obtained with the analytes prepared directly in solvent (analytical standard) showed an overall analyte ion current signal reduction. These differences result from losses due to extraction efficiency and the effects of matrix suppression of ionization (Table 2.3).

The determinations of matrix effect and analyte recovery are important experiments that should always be conducted as a sample preparation method is being developed. Also, in the situation when a bioanalytical method is being optimized or undergoing troubleshooting procedures, the matrix effect as a result of ionization suppression may provide a large contribution to reduced

Table 2.3
Ion suppression and extraction efficiency of 4% isoamyl alcohol in hexane extracts

Analyte	Matrix Ion Suppression (%)[a]	Extraction Efficiency (%)[b]	Total Response from Extract (%)[a]
Raloxifene	46	34	16
4-Hydroxy-tamoxifene	57	59	34
Nafoxidine	74	60	44
Tamoxifene	81	59	48
Idoxifene	86	52	45

[a]Given as a percent of the response of analyte prepared directly in solvent
[b]Given as a percent of the response of analyte added to a blank control human plasma extract
Reprinted with permission from [48]. Copyright 2000 American Chemical Society.

performance that is often overlooked. An educational case study of ionization suppression and methods for its elimination are detailed by Nelson and Dolan [196].

Acknowledgments

The line art illustrations were kindly provided by Pat Thompson and Woody Dells.

References

[1] M. Jemal and Y.-Q. Xia, Rapid Commun. Mass Spectrom. 13 (1999) 97-106.

[2] D. O'Connor, Curr Opin. Drug Discovery Dev. 5 (2002) 52-58.

[3] H. Kataoka and H. Lord, In: J. Pawliszyn, Ed., Sampling and Sample Preparation for Field and Laboratory: Fundamentals and New Directions in Sample Preparation, Elsevier, Amsterdam (2002).

[4] R.D. McDowall, J. Chromatogr. 492 (1989) 3-58.

[5] F. Klink, LC-GC 17 (1999) 1084-1093.

[6] J. Henion, E. Brewer and G. Rule, Anal. Chem. 70 (1998) 650A-656A.

[7] J. Henion, S.J. Prosser, T.N. Corso and G.A. Schultz, Amer. Pharm. Rev. 3 (2000) 19-29.

[8] R. Bakhtiar, L. Ramos and F.L.S. Tse, J. Liq. Chrom. & Rel. Technol. 25 (2002) 507-540.

[9] D.T. Rossi, In: D.T. Rossi and M.W. Sinz, Eds., Mass Spectrometry in Drug Discovery, Marcel Dekker, New York (2002) 171-214.

[10] R.D. McDowall, E. Doyle, G.S. Murkitt and V.S. Picot, J. Pharm. Biomed. Anal. 7 (1989) 1087-1096.

[11] R.E. Majors, LC-GC 14 (1996) 754-766.

[12] D.K. Lloyd, J. Chromatogr. A 735 (1996) 29-42.

[13] M. Gilar, E.S.P. Bouvier and B.J. Compton, J. Chromatogr. A 909 (2001) 111-135.

[14] J.R. Veraart, H. Lingeman and U.A.Th. Brinkman, J. Chromatogr. A 856 (1999) 483-514.

[15] G.W. Peng and W.L. Chiou, J. Chromatogr. 531 (1990) 3-50.

[16] I. Fu, E.J. Woolf and B.K. Matuszewski, J. Pharm. Biomed. Anal. 18 (1998) 347-357.

[17] D.L. McCauley-Myers, T.H. Eichhold, R.E. Bailey, D.J. Dobrozsi, K.J. Best, J.W. Hayes II and S.H. Hoke II, J. Pharm. Biomed. Anal. 23 (2000) 825-835.

[18] R. Bonfiglio, R.C. King, T.V. Olah andk K. Merkle, Rapid Commun. Mass Spectrom. 13 (1999) 1175-1185.

[19] M. Constanzer, C. Chavez-Eng, and B. Matuszewski, J. Chromatogr. B 760 (2001) 45-53.

[20] B.K. Matuszewski, M.L. Costanzer and C.M. Chavez-Eng, Anal. Chem. 70 (1998) 882-889.

[21] D. L. Buhrman, P. I. Price and P.J. Rudewicz, J. Amer. Soc. Mass. Spectrom. 7 (1996) 1099-1105.

[22] R. King, R. Bonfiglio, C. Fernandez-Metzler, C. Miller-Stein and T. Olah, J. Amer. Soc. Mass. Spectrom. 11 (2000) 942-950.

[23] P. Sarkar, C. Polson, R. Grant, B. Incledon and V. Raguvaran, Proceedings 49th American Society for Mass Spectrometry Conference, Chicago, IL USA (2001).

[24] C. Miller-Stein, R. Bonfiglio, T.V. Olah, R.C. King, Amer. Pharm. Rev. 3 (2000) 54-61.

[25] J.J. Zheng, E.D. Lynch and S.E. Unger, J. Pharm. Biomed. Anal. 28 (2002) 279-285.

[26] K.A. Mortier, K.M. Clauwaert, W.E. Lambert, J.F. Van Bocxlaer, E.G. Van den Eeckhout, C.H. Van Peteghem and A.P. De Leenheer, Rapid Commun. Mass Spectrom. 15 (2001) 1773-1775.

[27] A.P. Watt, D. Morrison, K.L. Locker and D.C. Evans, Anal. Chem.,72 (2000) 979-984.

[28] M.C. Rouan, C. Buffet, L. Masson, F. Marfil, H. Humbert and G. Maurer, J. Chromatogr. B 754 (2001) 45-55.

[29] A.T. Murphy, M.J. Berna, J.L. Holsapple and B.L. Ackermann, Rapid Commun. Mass Spectrom. 16 (2002) 537-543.

[30] D. O'Connor, D.E. Clarke, D. Morrison and A.P. Watt, Rapid Commun. Mass Spectrom. 16 (2002) 1065-1071.

[31] J. Wang, S.-Y. Chang, C. D'Arienzo and D. Wang-Iverson, Proceedings 48th American Society for Mass Spectrometry Conference, Long Beach, CA USA (2000).

[32] M.C. Rouan, C. Buffet, F. Marfil, H Humbert and G.Maurer, J. Pharm. Biomed. Anal. 25 (2001) 995-1000.

[33] R.A. Biddlecombe and S. Pleasance, J. Chromatogr. B 734 (1999) 257-265.

[34] R.E. Walter, J.A. Cramer and F.L.S. Tse, J. Pharm. Biomed. Anal. 25 (2001) 331-337.

[35] C. De Nardi, S. Braggio, L. Ferrari and S. Fontana, J. Chromatogr. B 762 (2001) 193-201.

[36] M. Berna, A.T. Murphy, B. Wilken and B. Ackermann, Anal. Chem. 74 (2002) 1197-1201.

[37] B.M. Cordero, J.L.P. Pavon, C.G. Pinto, M.E.F. Laespada, R.C. Martinez and E.R. Gonzalo, J. Chromatogr. A 902 (2000)195-204.

[38] J.A. Jonsson and L. Mathiasson, J. Chromatogr. A 902 (2000)205-225.

[39] P.R. Haddad, P. Doble and M. Macka, J. Chromatogr. A 856 (1999) 145-177.

[40] N.C. van de Merbel, J. Chromatogr. A 856 (1999) 55-82.

[41] N. Zhang, K.L. Hoffman, W. Li and D.T Rossi, J. Pharm. Biomed. Anal. 22 (2000) 131-138.

[42] J. Ke, M. Yancey, S. Zhang, S. Lowes and J. Henion, J. Chromatogr. B 742 (2000) 369-380.

[43] L. Ramos, R Bakhtiar and F.L.S. Tse, Rapid Commun. Mass Spectrom. 14 (2000) 740-745.

[44] S. Steinborner and J. Henion, Anal. Chem. 71 (1999) 2340-2345.

[45] M. Jemal, D. Teitz, Z. Ouyang and S. Khan, J. Chromatogr. B 732 (1999) 501-508.

[46] S.H. Hoke II, J.A. Tomlinson II, R.D. Bolden, K.L. Morand, J.D. Pinkston and K.R. Wehmeyer, Anal. Chem. 73 (2001) 3083-3088.

[47] J. Zweigenbaum, K. Heinig, S. Steinborner, T. Wachs and J. Henion, Anal. Chem. 71 (1999) 2294-2300.

[48] J. Zweigenbaum and J. Henion, Anal. Chem. 72 (2000) 2446-2454.

[49] J.M. Onorato, J.D. Henion, P.M. Lefebvre and J.P. Kiplinger, Anal. Chem. 73 (2001) 119-125.

[50] Z. Shen, S. Wang and R. Bakhtiar, Rapid Commun. Mass Spectrom. 16 (2002) 332-338.

[51] N. Brignol, L.M. McMahon, S. Luo and F.L.S. Tse, Rapid Commun. Mass Spectrom. 15 (2001) 898-907.

[52] R.D. Bolden, S.H. Hoke II, T.H. Eichhold, D.L. McCauley-Myers and K.R. Wehmeyer, J. Chromatogr. B 772 (2002) 1-10.

[53] S.X. Peng, C. Henson, M.J. Strojnowski, A. Golebiowski and S.R. Klopfenstein, Anal. Chem. 72 (2000) 261-266.

[54] S.X. Peng, T.M. Branch and S.L. King, Anal. Chem. 73 (2001) 708-714.

[55] A.Q. Wang, A.L. Fisher, J. Hsieh, A.M. Cairns, J.D. Rogers and D.G. Musson, J. Pharm. Biomed. Anal. 26 (2001) 357-365.

[56] A.Q. Wang, W. Zeng, D.G. Musson, J.D. Rogers and A.L. Fisher, Rapid Commun. Mass Spectrom. 16 (2002) 975-981.

[57] J. Hempenius, J. Wieling, J.P.G. Brakenhoff, F.A. Maris and J.H.G. Jonkman, J. Chromatogr. B 714 (1998) 361-368.

[58] B. Kaye, W.J. Herron, P.V. Macrae, S. Robinson, D.A. Stopher, R.F. Venn and W. Wild, Anal. Chem. 68 (1996) 1658-1660.

[59] Y.-F. Cheng, U.D. Neue and L. Bean, J. Chromatogr. A 828 (1998) 273-281.

[60] R.S. Plumb, R.D.M. Gray and C.M. Jones, J. Chromatogr. B 694 (1997) 123-133.

[61] W.Z. Shou, M. Pelzer, T. Addison, X. Jiang and W. Naidong, J. Pharm. Biomed. Anal. 27 (2002) 143-152.

[62] M.J. Rose, S.A. Merschman, R. Eisenhandler, E.J. Woolf, K.C. Yeh, L. Lin, W. Fang, J. Hsieh, M.P. Braun, G.J. Gatto and B.K. Matuszewski, J. Pharm. Biomed. Anal. 24 (2000) 291-305.

[63] C. Souppart, M. Decherf, H. Humbert and G. Maurer, J. Chromatogr. B 762 (2001) 9-15.

[64] I.D. Davies, J.P. Allanson and R.C. Causon, J. Chromatogr. B 732 (1999) 173-184.

[65] Y.-F. Cheng, U.D. Neue and L.L. Woods, J. Chromatogr. B 729 (1999) 19-31.

[66] L.M. McMahon, S. Luo, M. Hayes and F.L.S. Tse, Rapid Commun. Mass Spectrom. 14 (2000) 1965-1971.

[67] J.S. Janiszewski, M.C. Swyden and H.G. Fouda, J. Chrom. Sci. 38 (2000) 255-258.

[68] R.J. Scott, J. Palmer, I.A. Lewis and S. Pleasance, Rapid Commun. Mass Spectrom. 13 (1999) 2305-2319.

[69] S.L. Callejas, R.A. Biddlecombe, A.E. Jones, K.B. Joyce, A.I. Pereira and S. Pleasance, J. Chromatogr. B 718 (1998) 243-250.

[70] A.C. Harrison and D.K. Walker, J. Pharm. Biomed. Anal. 16 (1998) 777-783.

[71] S.H. Hoke II, J.D. Pinkston, R.E. Bailey, S.L. Tanguay and T.H. Eichhold, Anal. Chem. 72 (2000) 4235-4241.

[72] H. Zhang and J. Henion, Anal. Chem. 71 (1999) 3955-3964.

[73] M. Jemal, M. Huang, Y. Mao, D. Whigan and A. Schuster, Rapid Commun. Mass Spectrom. 14 (2000) 1023-1028.

[74] G. Rule and J. Henion, J. Am. Soc. Mass Spectrom. 10 (1999) 1322-1327.

[75] S.X. Peng, S.L. King, D.M. Bornes, D.J. Foltz, T.R. Baker and M.G. Natchus, Anal. Chem. 72 (2000) 1913-1917.

[76] R.D. Gauw, P.J. Stoffolano, D.L. Kuhlenbeck, V.S. Patel, S.M. Garver, T.R. Baker and K.R. Wehmeyer, J. Chromatogr. B 744 (2000) 283-291.

[77] J.P. Allanson, R.A. Biddlecombe, A.E. Jones and S. Pleasance, Rapid Commun. Mass Spectrom. 10 (1996) 811-816.

[78] J. Janiszewski, R.P. Schneider, K. Hoffmaster, M. Swyden, D. Wells and H. Fouda, Rapid Commun. Mass Spectrom. 11 (1997) 1033-1037.

[79] H. Simpson, A. Berthemy, D. Buhrman, R. Burton, J. Newton, M. Kealy, D. Wells and D. Wu, Rapid Commun. Mass Spectrom. 12 (1998) 75-82.

[80] W.Z. Shou, X. Jiang, B.D. Beato and W. Naidong, Rapid Commun. Mass Spectrom. 15 (2001) 466-476.

[81] D. Zimmer, V. Pickard, W. Czembor and C. Müller, J. Chromatogr. A 854 (1999) 23-35.

[82] S. Pleasance and R.A. Biddlecombe, In: E. Reid, H.M. Hill and I.D. Wilson, Eds., Drug Development Assay Approaches, Including Molecular Imprinting and Biomarkers, The Royal Society of Chemistry, Cambridge (1998) 205-212; Methodological Surveys in Bioanalysis of Drugs, Volume 25, E. Reid, Ed.

[83] J.-T. Wu, H. Zeng, M. Qian, B.L. Brogdon and S.E. Unger, Anal. Chem. 72 (2000) 61-67.

[84] J. Ayrton, G.J. Dear, W.J. Leavens, D.N. Mallet and R.S. Plumb, J. Chromatogr. B 709 (1998) 243-254.

[85] R.C. King, C. Miller-Stein, D.J. Magiera and J. Brann, Rapid Commun. Mass Spectrom. 16 (2002) 43-52.

[86] Y. Hsieh, J.-M. Brisson, K. Ng and W.A. Korfmacher, J. Pharm. Biomed. Anal. 27 (2002) 285-293.

[87] A. Eerkes, T. Addison and W. Naidong, J. Chromatogr. B 768 (2002) 277-284.

[88] Y.-L. Chen, G.D. Hanson, X. Jiang and W. Naidong, J. Chromatogr. B 769 (2002) 55-64.

[89] D. Schütze, B. Boss and J. Schmid, J. Chromatogr. B 748 (2000) 55-64.

[90] C.Z. Matthews, E.J. Woolf and B.K. Matuszewski, J. Chromatogr. A 949 (2002) 83-89.

[91] L. Yang, N. Wu and P.J. Rudewicz, J. Chromatogr. A 926 (2001) 43-55.

[92] C.Z. Matthews, E.J. Woolf, L. Lin, W. Fang, J. Hsieh, S. Ha, R. Simpson and B.K. Matuszewski, J. Chromatogr. B 751 (2001) 237-246.

[93] M.J. Rose, N. Agrawal, E.J. Woolf and B.K. Matuszewski, J. Pharm. Sci. 91 (2002) 405-416.

[94] R.S. Mazenko, A. Skarbek, E.J. Woolf, R.C. Simpson and B. Matuszewski, J. Liq. Chrom. & Rel. Technol. 24 (2001) 2601-2614.

[95] J. Hempenius, R.J.J.M. Steenvoorden, F.M. Lagerwerf, J. Wieling and J.H.G. Jonkman, J. Pharm. Biomed. Anal. 20 (1999) 889-898.

[96] G. Hopfgartner, C. Husser and M. Zell, Therap. Drug Monit. 24 (2002) 134-143.

[97] Z. Liu, J. Short, A. Rose, S. Ren, N. Contel, S. Grossman and S. Unger, J. Pharm. Biomed. Anal. 26 (2001) 321-330.

[98] V. Cenacchi, S. Baratte, P. Cicioni, E. Frigerio, J. Long and C. James, J. Pharm. Biomed. Anal. 22 (2000) 451-460.

[99] M. Gilar, A. Belenky and B.H. Wang, J. Chromatogr. A 921 (2001) 3-13.

[100] L. Yang, T.D. Mann, D. Little, N. Wu, R.P. Clement and P.J. Rudewicz, Anal. Chem. 73 (2001) 1740-1747.

[101] Y. Deng, J.-T. Wu, T.L. Lloyd, C.L. Chi, T.V. Olah and S.E. Unger, Rapid Commun. Mass Spectrom. 16 (2002) 1116-1123.

[102] P.T. Vallano, R.S. Mazenko, E.J. Woolf and B.K. Matuszewski, J. Chromatogr. B (2002) *in press*.

[103] T.H. Eichhold, R.E. Bailey, S.L. Tanguay and S.H. Hoke II, J. Mass Spectrom. 35 (2000) 504-511.

[104] R. Bakhtiar, L. Khemani, M. Hayes, T. Bedman and F. Tse, J. Pharm. Biomed. Anal. 28 (2002) 1183-1194.

[105] K.B. Joyce, A.E. Jones, R.J. Scott, R.A. Biddlecombe and S. Pleasance, Rapid Commun. Mass Spectrom. 12 (1998) 1899-1910.

[106] H. Yin, J. Racha, S.-Y. Li, N. Olejnik, H. Satoh and D. Moore, Xenobiotica 30 (2000) 141-154.

[107] S. Hsieh and K. Selinger, J. Chromatogr. B 772 (2002) 347-356.

[108] C.K. Hull, A.D. Penman, C.K. Smith and P.D. Martin, J. Chromatogr. B 772 (2002) 219-228.

[109] I.D. Davies, J.P. Allanson and R.C. Causon, Chromatographia 52 *Supplement* (2000) S92-S97.

[110] L.-Y. Zang, J. DeHaven, A. Yocum and G. Qiao, J. Chromatogr. B 767 (2002) 93-101.

[111] M. Larsson, U. Logren, M. Ahnoff, B. Lindmark, P. Abrahamsson, H. Svennberg and B.-A. Persson, J. Chromatogr. B 766 (2001) 47-55.

[112] R.C. Simpson, A. Skarbek and B.K. Matuszewski, J. Chromatogr. B 775 (2002) 133-142.

[113] C.S. Tamvakopoulos, J.M. Neugebauer, M. Donnelly and P.R. Griffin, J. Chromatogr. B 776 (2002) 161-168.

[114] P.H. Zoutendam, J.F. Canty, M.J. Martin and M.K. Dirr, J. Pharm. Biomed. Anal. 30 (2002) 1-11.

[115] M. Gilar and E.S.P. Bouvier, J. Chromatogr. A 890 (2000) 167-177.

[116] D. Fraier, E. Frigerio, G. Brianceschi, M. Casati, A. Benecchi and C. James, J. Pharm. Biomed. Anal. 22 (2000) 505-514.

[117] G. Rule, M. Chapple and J. Henion, Anal. Chem. 73 (2001) 439-443.

[118] R.A. Biddlecombe, C. Benevides and S. Pleasance, Rapid Commun. Mass Spectrom. 15 (2001) 33-40.

[119] D.A. Campbell, T.J. Ordway, K.T.M. Dillon, L.M. Irwin, J.R. Perkins and J. Henion, Proceedings 49th American Society for Mass Spectrometry Conference, Chicago, IL USA (2001).

[120] D.A. McLoughlin, T.V. Olah and J.D. Gilbert, J. Pharm. Biomed. Anal. 15 (1997) 1893-1901.

[121] F. Beaudry, J.C.Y. Le Blanc, M. Coutu and N.K. Brown, Rapid Commun. Mass Spectrom. 12 (1998) 1216-1222.

[122] H. Ghobarah, J.C. Flynn, J.D. Laycock and K.J. Miller, Proceedings 48th American Society for Mass Spectrometry Conference, Long Beach, CA USA (2000).

[123] A. Pruvost, I. Ragueneau, A. Ferry, P Jaillon, J.-M. Grognet and H. Benech, J. Mass Spectrom. 35 (2000) 625-633.

[124] J. Prunonosa, J. Sola, C. Peraire, F. Pla, O. Lavergne and R. Obach, J. Chromatogr. B 677 (1996) 388-392.

[125] J. Prunonosa, L. Parera, C. Peraire, F. Pla, O. Lavergne and R. Obach, J. Chromatogr. B 668 (1995) 281-290.

[126] G. Garcia-Encina, R. Farran, S. Puig, M.T. Serafini and L. Martinez, J. Chromatogr. B 670 (1995) 103-110.

[127] C. Nieto, J. Ramis, L. Conte, J.M. Fernandez and J. Forn, J. Chromatogr. B 661 (1994) 319-325.

[128] M.C. Woodward, G. Bowers, J. Chism, L. St.John-Williams and G. Smith, Proceedings 48th American Society for Mass Spectrometry Conference, Long Beach, CA USA (2000).

[129] J.Y.-K. Hsieh, L. Lin and B.K. Matuszewski, J. Liq. Chrom. & Rel. Technol. 24 (2001) 799-812.

[130] A. Kurita and N. Kaneda, J. Chromatogr. B 724 (1999) 335-344.

[131] L. Borbridge, D. Lourenco and A. Acheampong, Proceedings 48th American Society for Mass Spectrometry Conference, Long Beach, CA USA (2000).

[132] M. Yritia, P. Parra, E. Iglesias and J.M. Barbanoj, J. Chromatogr. A 870 (2000) 115-119.

[133] M. Hedenmo and B.-M. Eriksson, J. Chromatogr. A 692 (1995) 161-166.

[134] O.V. Olesen and B. Poulsen, J. Chromatogr. 622 (1993) 39-46.

[135] A. Schellen, B. Ooms, M. van Gils, O. Halmingh, E. van der Vlis, D. van de Lagemaat and E. Verheij, Rapid Commun. Mass Spectrom. 14 (2000) 230-233.

[136] A. Marchese, C. McHugh, J. Kehler and H. Bi, J. Mass Spectrom. 33 (1998) 1071-1079.

[137] J. Sola, J. Prunonosa, H. Colom, C. Peraire and R. Obach, J. Liq. Chrom. & Rel. Technol. 19 (1996) 89-99.

[138] H. Svennberg and P.-O. Lagerstrom, J. Chromatogr. B 689 (1997) 371-377.

[139] H. Toreson and B.-M. Eriksson, Chromatographia 45 (1997) 29-34.

[140] E.A. Martin, R.T. Heydon, K. Brown, J.E. Brown, C.K. Lim, I.N.H. White and L.L. Smith, Carcinogenesis 19 (1998) 1061-1069.

[141] M. Yritia, P. Parra, J.M. Fernandez and J.M. Barbanoj, J. Chromatogr. A 846 (1999) 199-205.

[142] J.A. Pascual and J. Sanagustin, J. Chromatogr. B 724 (1999) 295-302.

[143] A. Desroches, M. Vranderick, E. Federov, M. Mancini and M. Allard, Proceedings 48th American Society for Mass Spectrometry Conference, Long Beach, CA USA (2000).

[144] B. Ooms and E. Koster, Proceedings 48th American Society for Mass Spectrometry Conference, Long Beach, CA USA (2000).

[145] E.H.M. Koster, B.A. Ooms and H.A.G. Niederlander, Proceedings 50th American Society for Mass Spectrometry Conference, Orlando, FL USA (2002).

[146] C.D. James, J.A. Dunn and O. Halmingh, Proceedings 48th American Society for Mass Spectrometry Conference, Long Beach, CA USA (2000).

[147] H. Bi, R.N. Hayes, R. Castien, O. Halmingh and M. van Gils, Proceedings 48th American Society for Mass Spectrometry Conference, Long Beach, CA USA (2000).

[148] D. Tang, P. Gerry and O. Kavetskaia, Proceedings 50th American Society for Mass Spectrometry Conference, Orlando, FL USA (2002).

[149] M. Jemal, Y.-Q. Xia and D.B. Whigan, Rapid Commun. Mass Spectrom. 12 (1998) 1389-1399.

[150] N. Van Eeckhout, J.C. Perez, J. Claereboudt, R Vandeputte and C. Van Peteghem, Rapid Commun. Mass Spectrom. 14 (2000) 280-285.

[151] J. Ayrton, G.J. Dear, W.J. Leavens, D.N. Mallett and R.S. Plumb, Rapid Commun. Mass Spectrom. 11 (1997) 1953-1958.

[152] J. Ayrton, R.A. Clare, G.J. Dear, D.N. Mallett and R.S. Plumb, Rapid Commun. Mass Spectrom. 13 (1999) 1657-1662.

[153] J. Ayrton, G.J. Dear, W.J. Leavens, D.N. Mallett and R.S. Plumb, J. Chromatogr. A 828 (1998) 199-207.

[154] M. Jemal, M. Huang, X Jiang, Y. Mao and M.L. Powell, Rapid Commun. Mass Spectrom. 13 (1999) 2125-2132.

[155] L. Ramos, N. Brignol, R. Bakhtiar, T. Ray, L.M. McMahon and F.L.S. Tse, Rapid Commun. Mass Spectrom. 14 (2000) 2282-2293.

[156] N. Brignol, R. Bakhtiar, L. Dou, T. Majumdar and F.L.S. Tse, Rapid Commun. Mass Spectrom. 14 (2000) 141-149.

[157] Y. Hsieh, M.S. Bryant, G. Gruela, J.-M. Brisson and W.A Korfmacher, Rapid Commun. Mass Spectrom. 14 (2000) 1384-1390.

[158] C.R. Mallet, J.R. Mazzeo and U. Neue, Rapid Commun. Mass Spectrom. 15 (2001) 1075-1083.

[159] H.K. Lim, K.W. Chan, S. Sisenwine and J.A. Scatina, Anal. Chem. 73 (2001) 2140-2146.

[160] M. Jemal and Y.-Q. Xia, J. Pharm. Biomed. Anal. 22 (2000) 813-827.

[161] J.L. Herman, Rapid Commun. Mass Spectrom. 16 (2002) 421-426.

[162] H. Zeng, J.-T. Wu and S.E. Unger, J. Pharm. Biomed. Anal. 27 (2002) 967-982.

[163] M. Jemal, Z. Ouyang, Y.-Q. Xia and M.L. Powell, Rapid Commun. Mass Spectrom. 13 (1999) 1462-1471.

[164] C. Chassaing, J. Luckwell, P. Macrae, K. Saunders, P. Wright and R. Venn, Chromatographia 53 (2001) 122-130.

[165] D. Lachance, C. Grandmaison and L. DiDonato, Proceedings 48th American Society for Mass Spectrometry Conference, Long Beach, CA USA (2000).

[166] C. Chassaing, P. Macrae, P. Wright, A. Harper, J. Luckwell, K. Saunders and R. Venn, Presented at 11th International Symposium on Pharmaceutical and Biomedical Analysis, Basel, Switzerland (2000).

[167] M. Kollroser and C. Schober, Rapid Commun. Mass Spectrom. 16 (2002) 1266-1272.

[168] K.-S. Boos and A. Rudolphi, LC-GC 15 (1997) 602-611.

[169] A. Rudolphi and K.-S. Boos, LC-GC 15 (1997) 814-823.

[170] J.-T. Wu, H. Zeng, Y. Deng and S.E. Unger, Rapid Commun. Mass Spectrom. 15 (2001) 1113-1119.

[171] G. Dear, R. Plumb and D. Mallett, Rapid Commun. Mass Spectrom. 15 (2001) 152-158.

[172] G.J. Dear, D.N. Mallett, D.M. Higton, A.D. Roberts, S.A. Bird, H. Young, R.S. Plumb and I.M. Ismail, Chromatographia 55 (2002) 177-184.

[173] R. Plumb, G. Dear, D. Mallett and J. Ayrton, Rapid Commun. Mass Spectrom. 15 (2001) 986-993.

[174] S. Zhou, M.J. Larson, X. Jiang and W. Naidong, Proceedings 50th American Society for Mass Spectrometry Conference, Orlando, FL USA (2002).

[175] Y. Hsieh, G. Wang, Y. Wang, S. Chackalamannil, J.-M. Brisson, K. Ng and W.A Korfmacher, Rapid Commun. Mass Spectrom. 16 (2002) 944-950.

[176] R. Wieboldt, J. Zweigenbaum and J. Henion, Anal. Chem. 69 (1997) 1683-1691.

[177] D.S. Hage, J. Chromatogr. B 715 (1998) 3-28.

[178] M. de Frutos and F.E. Regnier, Anal. Chem. 65 (1993) 17A-25A.

[179] N. Delaunayi, V. Pichon and M.-C. Hennion, J. Chromatogr. B 745 (2000) 15-37.

[180] D. Stevenson, J. Chromatogr. B 745 (2000) 39-48.

[181] T. Bedman, M.J. Hayes, L. Khemani and F.L.S. Tse, Proceedings 48th American Society for Mass Spectrometry Conference, Long Beach, CA USA (2000).

[182] A. Gritsas, M. Lahaie, D. Chun, T. Flarakos, M.L.J. Reimer, F. Deschamps and R. Hambalek, Proceedings 48th American Society for Mass Spectrometry Conference, Long Beach, CA USA (2000).

[183] M. Zell, C. Husser and G. Hopfgartner, J. Mass Spectrom. 32 (1997) 23-32.

[184] E.W. Woo, R. Messmann. E.A. Sausville and W.D. Figg, J. Chromatogr. B 759 (2001) 247-257.

[185] B. Lausecker, B. Hess, G. Fischer, M. Mueller and G. Hopfgartner, J. Chromatogr. B 749 (2000) 67-83.

[186] C.-L. Cheng and C.-H Chou, J. Chromatogr. B 762 (2001) 51-58.

[187] M. Cociglio, H. Peyriere, D. Hillaire-Buys and R. Alric, J. Chromatogr. B 705 (1998) 79-85.

[188] G.L. Lensmeyer and M.A. Poquette, Ther. Drug Monit. 23 (2001) 239-249.

[189] V. Rodwell, In: R.K. Murray, D.K. Granner, P.A. Mayes and V.W. Rodwell, Eds., Harper's Biochemistry, 25th Ed., McGraw-Hill/Appleton & Lange, New York (1999) 15-25.

[190] T.L. Lemke, Review of Organic Functional Groups: Introduction to Organic Medicinal Chemistry, Third Ed., Lea & Febiger, Philadelphia (1992).

[191] H. Wiltshire, In: R.F. Venn, Ed., Principles and Practice of Bioanalysis, Taylor & Francis, London (2000) 1-27.

[192] A. Cuadrado, G. Solares, S. Gonzalez, B. Sanchez and J. A. Armijo, Meth. Find. Exptl. Clin. Pharmacol. 20 (1998) 297-300.

[193] D.S. Palmer, D. Rosborough, H. Perkins, T. Bolton, G. Rock and P.R. Ganz, Vox Sang 65 (1993) 258-270.

[194] H. Mei, Y. Hsieh, N. Juvekar, S. Wang, C. Nardo, S. Wainhaus, K. Ng and W. Korfmacher, Proceedings 49th American Society for Mass Spectrometry Conference, Chicago, IL USA (2001).

[195] G. Shi, J. Wu, Y. Li, R. Gelezlunas, K. Gallagher, T. Emm, T. Olah and S. Unger, Rapid Commun. Mass Spectrom. 16 (2002) 1092-1099.

[196] M.D. Nelson and J.W. Dolan, LC-GC 20 (2002) 24-32.

Chapter 3

High Throughput Tools for Bioanalysis: Microplates

Abstract

This chapter introduces the great diversity of commercially available microplate formats to assist the analyst in selecting and using these products to perform high throughput sample preparation techniques. The history of microplates is reviewed and a discussion follows of the varieties in well shape, well bottom shape and microplate surface geometries. The range of choices for microplates as sample containers and collection devices for bioanalysis is described. The variety in flow-through microplates is then introduced and the interesting developmental history is shared. These plates are used for techniques such as filtration, solid-phase extraction and solid-supported liquid-liquid extraction. Microplate collection devices and flow-through plates are integral components of a high throughput sample preparation method. However, they require the use of one or more accessory products for various processes (*e.g.*, for sealing, mixing/shaking, centrifugation or evaporation). These accessory products are described in the next chapter. Once a mastery of knowledge about the available formats and accessory products is gained, the reader is better prepared to choose the appropriate tools that will allow for a particular sample preparation method to be performed in a high throughput manner.

3.1 Introduction

The traditional sample containers or collection devices used in bioanalysis have been individual test tubes and vials, available in many different dimensions and made from glass or plastic. Individual pipettors have been used with these tubes or vials to transfer sample, solvent and any other liquids as part of the overall sample preparation procedure. Techniques such as protein precipitation (PPT) and liquid-liquid extraction (LLE) have also traditionally used single tubes for the extraction step; solid-phase extraction (SPE) has used individual cartridges.

75

Individual tubes or vials held a stronghold as the preferred sample container until the late 1990s. At this time, several advances occurred in parallel that brought about a change from tubes to microplates as the preferred sample and collection format. Faster analytical techniques such as liquid chromatography interfaced with tandem mass spectrometry (LC-MS/MS) became more widespread and affordable, allowing researchers to analyze samples more quickly than ever before. Liquid handling workstations were shown to be very reliable and to reduce analyst hands-on time, ensuring their vital role in the laboratory for most pipetting tasks. Within the pharmaceutical industry, drug development cycles were shortened, favoring a more rapid identification of lead candidates with a more rapid development pace. When the sample preparation step became rate limiting in this process, the need for improved efficiency and throughput became evident.

While 96-well microplates had been in use in other areas of pharmaceutical research (*e.g.*, high throughput screening) they became more commonly available in larger well volumes. They also were molded from polypropylene, a plastic that has the required solvent resistance for use with common organic solvents. Filtration and solid-phase extraction were introduced in the 96-well plate format in flow-through processes. The microplate presents many efficiencies of operation such as easier labeling, sealing and manipulation. When this format is combined with multiple probe workstations, dramatically faster pipetting throughput is achieved via parallel sample processing. The microplate architecture clearly offered the increased proficiency and productivity in sample preparation processes and analyses that were sought.

High throughput applications are based on samples being reformatted from racks of test tubes or vials into the 8-row by 12-column format. Once in this configuration, the samples are contained in a compact space and each well is uniquely identified by column number and row letter, *e.g.*, A1, A2, A3, C5, F8 or H12. The tedious and time consuming task of individually labeling tubes or vials is now eliminated. The grid layout and numbering scheme for a 96-well plate are shown in Figure 3.1.

Vendors in all areas of laboratory supply and sample preparation responded to this emerging trend for increased productivity by offering a variety of microplate and extraction plates, as will be detailed in this chapter. An earlier publication by the author provides an overview of 96-well solid-phase extraction plates and was the inspiration for this subject material which has been expanded in its scope [1]. Accessory products needed to perform a sample preparation procedure using microplates are described in Chapter 4.

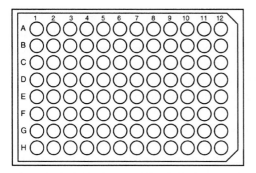

Figure 3.1. Within a microplate grid of 8-rows by 12-columns, each cell is uniquely identified by row letter and column number. This compact format replaces the use of individual tubes for sample preparation procedures.

3.2 Microplate Fundamentals

3.2.1 History

The formal history of microplate development has not been recorded in great detail; however, two references have discussed its origins [2, 3] which are briefly summarized here. The beginning of the microplate concept is generally credited to Dr. Gyola Takatsky of Hungary who, in his attempt to scale down serology tests, machined six rows of 12 connected tubes in an acrylic plastic in about 1952. Subsequently, he machined a 96-well plate in acrylic. This microplate format was improved over the years and in about 1963 Dr. John Sever at the National Institutes of Health (USA) published on an improved form for the microtiter™ system. The plates were simultaneously designed by John Liner and Nelson Cooke. A 96-well acrylic plate first became commercially available in 1965 from Cooke Laboratories (Alexandria, VA USA). In 1966 Greiner Labs (Frickenhausen, Germany) was the first company in Europe to mold these plates. In about 1968–1969, Dynatech Laboratories [now called Dynex Technologies (Alexandria, VA USA)] acquired Cooke.

Advances in plastics occurred over the years and polystyrene was found to offer greater clarity and temperature resistance than acrylic. Polystyrene then became the preferred plastic for use with aqueous samples and allowed for the manufacture of a low cost disposable plate. Polycarbonate also offered its own advantages for certain applications in serology. By about 1976, the Falcon Corning (Acton, MA USA) and Nunc (Naperville, IL USA) companies each entered the 96-well market with production molded plates; Falcon is now a part of BD Biosciences (Bedford, MA USA) and Nunc is now a part of Nalge Nunc International Corporation (Rochester, NY USA). In about 1980, plates containing filters for flow-through applications first became available. Pigmented plates (black and white) were introduced in about 1982. Today, microplates are molded by at least 15 companies worldwide.

The word "microplate" in general terms refers to the common 8-row by 12-column format of interconnected wells in a single base plate. When mentioned without any modifier or description, microplate usually is meant to describe the original design, also called a "shallow well" plate in which well volumes are about 0.35 mL. Taller plates having larger well volumes are available and are generally called "deep well" plates (Figure 3.2). Typically, a deep well plate with round wells holds a 1 mL volume and a deep well plate with square wells (not shown in illustration) holds a 2 mL volume.

Microplates are also sometimes referred to by the name "microtiter plates." However, note that the word microtiter™ used in this context is a trademarked name, first used in commerce in 1961 and registered with the U.S. Patent and Trademark Office in 1963 by Cooke Engineering. The present listed owner of this trademark name is Dynex Technologies. Another related name associated with microplates that is trademarked is MicroWell™ owned by Nalge Nunc International Corporation since 1982.

3.2.2 Applications

Many dozens of applications have utilized the 96-well plate format over the years. The same 8-row by 12-column system has been retained and an entire industry has been built around the manufacture of traditional plastic collection

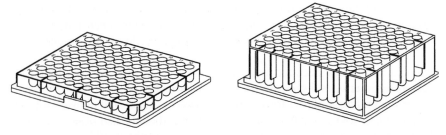

Figure 3.2. Traditional shallow well (*left*) and deep well (*right*) microplate formats differ in height and well volume.

plates as well as specialty plates using coated wells which are described in more detail by Johnson [3]. A related overview discusses the combination of membranes with micro well plates [4]. A large support industry has also developed with regard to liquid handling workstations and robotics platforms which have been designed to automate all tasks involved in plate processing, storage and analysis. Automation tools and strategies for using microplates in sample preparation procedures are described in Chapter 5.

The applications listed below generally use microplates for one of three purposes: production, storage or analysis. Some of the most common applications for microplates in clinical and drug research include:

- Clinical testing—ELISA assays, serology assays and antibiotic susceptibility assays
- Basic research—immunology, screening monoclonal antibodies, tissue culture of various normal mammalian cells and tumor cells, production or screening of ordered array genome libraries, and determination of protein and nucleic acid concentrations
- Molecular biology—DNA purification, PCR applications, and binding of peptides, nucleic acids and proteins
- Drug discovery—High throughput screening utilizing fluorescence, luminescence and other detection endpoints

Combinatorial chemistry techniques for drug synthesis have utilized the microplate configuration for the production, storage and analysis of chemical libraries. These high throughput techniques rapidly synthesize, purify and confirm the identity of a large number of compounds displaying wide chemical diversity within a class [5–8]. In place of traditional serial compound synthesis, libraries of compounds are created in 96-well plates by interconnecting a set or sets of small reactive molecules, called building blocks, in many different permutations. While polystyrene is a convenient, inexpensive and good-performing plastic for use with aqueous samples, it is not compatible with the organic solvents and harsh acids and bases used in chemical synthesis. Plate manufacturers responded to this need by offering the microplates made from polypropylene.

The volumes required by medicinal chemists can be greater than that provided by a traditional shallow well plate; deep well plates having volumes from 1–2 mL are preferred for this application. Although 96-well plates serve as a common format for reaction vessels, 24- and 48-well plates are also used by chemists when greater volumes per well are required in the same footprint.

The use of microplates in pharmaceutical bioanalysis has proliferated also. These plates are commonly utilized in the production, storage or analysis aspects of typical drug metabolism and bioanalytical applications.

Some typical uses of microplates in laboratories providing bioanalytical support to drug discovery and drug metabolism include:

- Caco-2 cell absorption assays
- Various *in vitro* tissue homogenate assays for metabolism
- Protein precipitation
- Liquid-liquid extraction
- Solid-supported liquid-liquid extraction*
- Solid-phase extraction*
- Filtration and clarification*

Flow-through microplates (*) are used for some sample preparation schemes in addition to the traditional closed well collection plates. The wells in these flow-through plates usually contain either a filter or particles in a bed or disk. The formats for these flow-through extraction plates are discussed in Section 3.6 and an introduction to their applications is found in Section 3.7.

The well density of microplates used in areas of pharmaceutical research other than bioanalysis and combinatorial chemistry has actually extended beyond 96-well. Higher density 384- and 1536-well plates are used for some applications such as high throughput screening activities in drug discovery research. A higher density well number within the same footprint requires a resulting decrease in well volume capacity. Instead of ~350 μL well volume as in 96-well (shallow) plates, a 384-well plate has a well volume of only about 120 μL (or 240 μL for a deep well version of the 384-well plate). A 1536-well plate contains a well volume of only about 12 μL. Acceptance of higher density plates would not be adopted without corresponding improvements in the accuracy and precision of automated liquid handling workstations; improved versions are able to deliver the reduced volumes required.

3.2.3 Specifications

In 1995, the Society for Biomolecular Screening (SBS) identified a clear need to establish some basic dimensional standards for microplates as their use in drug discovery research was proliferating. At that time, the general appearance of the microplate format was similar among manufacturers but subtle variations began to appear with regard to height, side indentations, surfaces and/or footprint. Problems in compatibility began to result with some plates when used with robotic and liquid handling instrumentation. The SBS started a Microplate Standards Development Committee to discuss and develop standards for submission to the American National Standards Institute (ANSI). Membership in this committee has grown to represent over 100 corporations, educational institutions and government organizations from among 15 countries. Detailed specifications and recommendations exist for many areas of the microplate design such as footprint dimensions, flange dimensions, chamfers (corner notches), plate height, side wall rigidity and well positions. Table 3.1 provides a summary of the fundamental microplate specifications as proposed by SBS. The complete individual reports of each standard should be consulted for all details.

3.3 Varieties in Microplate Well Geometry and Plate Surface

3.3.1 Well Shape and Well Bottom Shape

The 96-well microplate design has evolved over the years in response to new applications to now include many varieties in well shape and well bottom. Fundamentally, plates are molded in two well shapes—round well and square well. Also, four main types of well bottom geometries are available: flat, rounded, conical and rounded corners (Figure 3.3). Flat bottom plates are traditionally used for applications requiring optical reading; light transmission is maximal through a flat bottom. Rounded bottoms (also called U-bottom) are

Table 3.1
Fundamental microplate specifications as proposed by the Society for Biomolecular Screening

- The wells in a 96-well microplate should be arranged as 8 rows by 12 columns
- Center-to-center well spacing should be 9.0 mm (0.3543 in)
- The outside dimension of the base footprint, measured at any point along the side, shall be: Length 127.76 mm (5.0299 in), Width 85.48 mm (3.3654 in)
- The footprint must be continuous and uninterrupted around the base of the plate
- The overall plate height (shallow well design) should be 14.35 mm (0.5650 in)
- The maximum allowable projection above the top stacking surface is 0.76 mm (0.0299 in)
- The four outside corners of the plate's bottom flange shall have a corner radius to the outside of 3.18 mm (0.1252 in)

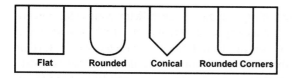

Figure 3.3. Four well bottom geometries are commonly available in microplates: Flat bottom (F-bottom), rounded bottom (U-bottom), conical (V-bottom) and rounded corners leading to flat bottom (C-bottom).

best for efficient mixing and approximate the dimensions of a test tube bottom. Conical bottoms (also called V-bottom) are good for pelleting of small particles and also for efficient retrieval of small sample volumes. Rounded corners in the bottom (sometimes called C-bottom) resulted from a modification of flat bottoms that offer improved performance in applications requiring mixing or washing in addition to light transmission.

Shallow well plates typically hold a volume of 0.3–0.4 mL, depending on the geometry of the well shape and well bottom. Deep well plates typically hold 1 mL in the round well format and 2 mL in the square well format. The need to conveniently handle volumes smaller than 2 mL but retain the versatile square well design and height of the deep well plate has resulted in manufacturers designing square well plates having approximate volumes of 0.35 mL and 1 mL, in addition to the original 2 mL volume. The overall geometry and height of the plate remain unchanged to accommodate these smaller volumes, but the well depth has been reduced, as shown in Figure 3.4. Note that square well plates generally have a conical well bottom.

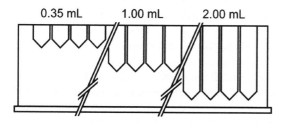

Figure 3.4. Deep well plates (square wells) can accommodate volumes of 0.35 mL, 1 and 2 mL by varying well depth, yet they retain their overall height and outer dimensions. Reprinted with permission from [9]. Copyright 1999 Advanstar Communications.

Square well plates are preferred in bioanalysis for their ability to hold a relatively larger volume in the same well height, but their use can be problematic. Wicking is known to occur, in which liquid slowly rises or creeps up the right angle edges toward the top of the plate. This phenomenon can result in the accumulation of liquid at the top of the wells and/or the promotion of evaporation; the full volume of liquid placed into the well may not be retrievable.

A greater concern with wicking from square wells is that the liquid at the top can cause well-to-well contamination or crosstalk. Also, vortex mixing using a square well is not ideal, as a round well is more preferable to ensure thorough mixing in a circular vortex mode. Note that the prevalence of wicking also depends on the surface tension and viscosity of the liquid used in the square well.

Organic solvents such as methyl tert-butyl ether, and to a lesser extent acetonitrile and methanol, are known to frequently result in wicking. Reduction of wicking can be aided by the use of a square-rounded well shape which eliminates the 90° angles that promote this phenomenon. Also, reduced wicking is observed when using a plate having a consistently low surface energy.

Until recently, the only way to achieve a 2 mL well volume in a deep well plate was to use a square well design. Nalge Nunc International has utilized a "shared wall" technology in their polypropylene DeepWell plates so that, in a round well design, a 2 mL volume can now be achieved. This shared wall design (Figure 3.5) provides optimal mixing for liquid-liquid extraction and other sample preparation and sample reconstitution procedures. This unique configuration allows the NUNC™ DeepWell plates to hold up to 50% more volume than traditional round well designs. Their low profile design also takes up 50% less space. This wide well diameter also includes the U-bottom 0.5 mL polypropylene microplates and the V-bottom 0.45 mL microplates.

Both the 1 mL and the 2 mL DeepWell plates have the standard 96-well footprint that facilitates robotic gripping and provides a sufficient area for barcode placement. The H1 and H12 cutoff corners allow for quick visual orientation. NUNC™ 96-well filter plates (1 mL) have been designed for use with NUNC DeepWell plates. These polypropylene filter plates, offered in several formats, also feature the shared wall design. Applications for filter plates are discussed in detail in Chapter 6, Section 6.4.

Figure 3.5. NUNC™ DeepWell polypropylene plates utilizing shared wall technology in the 1 mL (*left*) and 2 mL (*right*) round bottom, round well formats. Photo reprinted with permission from Axygen Scientific.

3.3.2 Microplate Surfaces

Plate designs vary not only in well shape and well bottom geometry but also with regard to the plate surface and the well surface. The top surface of the original microplate design was flat from outer edge to outer edge. Improved versions have a raised flat surface around and between the wells that does not extend to the plate edge. Another variety of plate surface (from Greiner) actually is not flat but has cavities between the wells so that any droplets splashing out will fall between adjacent wells and not into another well, reducing the potential for cross-contamination. The presence of a raised rim around circular and square wells (chimney well design) helps to reduce cross-contamination by forming a tighter seal when a cap mat or sheet is applied [9]. This chimney well design works especially well with heat sealed mats, discussed in more detail in Chapter 4, Section 4.3.4.

Surface treatments have been applied to the inside of the plastic wells to allow for use in cell culture, immunology and molecular biology applications. For example, in cell culture applications, Nunclon® MicroWell™ plates (Nalge Nunc International) have a negatively charged surface that is ideal for adhering biological cells. Antibody attachment is facilitated by attachment to polar groups treated on a polystyrene surface, as in the MaxiSorp™ surface (Nalge Nunc International). Covalent binding through an amine group can couple carboxylic acid groups of peptides, proteins, carbohydrates, and nucleic acids to polystyrene, provided by the Cova Link® NH2 Primary Amine plates (Nalge Nunc International). Further discussion of these specialty polystyrene plates is outside the scope of this text and the reader is referred to a general profile for detail [3].

3.4 Microplates as Sample Containers and Collection Devices

3.4.1 Shallow Well Plates

Almost all shallow well plates are made of polystyrene, which is not compatible with the organic solvents commonly used in pharmaceutical bioanalysis. Polypropylene plates offer the necessary chemical resistance for these applications. The use of shallow well plates as sample containers (reservoir volume of about 350 μL) is limited to small volume aqueous applications, such as those encountered in drug discovery support.

Protein precipitation can be performed in this shallow well microplate format when small volumes such as 25 μL are used. Details about protein precipitation are presented in Chapter 6, but typically 1 part plasma to 3 parts acetonitrile is used to precipitate proteins out of a biological matrix. When the total sample volume (100 μL in this example), even with the addition of a small volume of internal standard, does not exceed half the reservoir volume of the well, it may be used for vortex mixing (after plate sealing) with subsequent centrifugation. Either a U-bottom or V-bottom shallow well plate can be used successfully for this application.

When sample volumes used are extremely small, it may be possible to use a PCR plate for the protein precipitation procedure. PCR plates have a 96-well format with a conical well shape and the well volume is usually about 200 μL (although 100 μL versions are also available). These plates are available in both non-skirted and skirted varieties and with different surface and edge modifications as shown in Figure 3.6. Centrifugation of the plate requires the use of the skirted variety for structural support, unless the PCR plate is inserted into a deep well plate before centrifugation to enhance its rigidity.

Another case where polypropylene shallow well plates may be used in bioanalysis is as a collection device, but only when the eluate volumes are very small. Solid phase extraction is one application when very small elution volumes can be realized, when using small bed mass extraction plates (≤25 mg) or disk plates. When the eluate solution is compatible for direct injection into the analytical system and no further concentration is necessary, volumes of 200 μL and less can be conveniently collected in this format.

The geometry of the well and the potential for cross-contamination become very important with use of the shallow well plate, especially when placed inside a vacuum manifold. Keep in mind that when a filtration or extraction

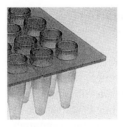

Figure 3.6. A PCR plate provides for an extremely small well volume and is useful in limited applications as a sample container or collection device. Note that different surface and skirt varieties exist. Photos reprinted with permission from Simport.

plate is positioned above a shallow well plate, the tips occupy some of the available volume. Should the tips be placed too low and become submerged in the eluate liquid during elution using vacuum, bubbling can result and lead to well-to-well contamination [9]. Also, in the case when the eluate solution needs to be dried down and then reconstituted before analysis, enough air space above the liquid level must be present for a mild to vigorous vortex action to occur upon introduction of heated nitrogen gas. Understanding these limitations, shallow well plates can sometimes be used with very small elution volumes. However, in most instances a deep well plate is better for eluate collection.

3.4.2 Deep Well Plates

Deep well plates made of polypropylene are used most often in bioanalysis as both sample containers and eluate collection devices. Clearly, the importance of the deep well microplate as a standard format is firmly supported through all stages of the analytical process from sample preparation through analysis. While this microplate design is commonly utilized beginning at the sample preparation stage, the feasibility of using it even earlier in the process for study design, sample generation and collection has been proposed by Zhang *et al.* [10]. In this report it is suggested that greater process efficiency would be obtained if a closer relationship among these tasks were developed, as they are usually performed by separate groups and the generated samples are transferred along a defined path to ultimately reach the analytical chemist. The elimination of sample transfer steps by collecting animal or human samples in the 96-well format can allow microplate sample preparation to proceed in a more rapid and straightforward manner.

The variation seen among deep well plate choices is generally in well volume (0.35 mL, 1 mL or 2 mL) and in the geometry of the well shape, well bottom and plate surface. The chimney well design having raised rims is generally preferable to provide for tighter sealing, especially with the use of heat sealable film. Sealing systems for microplates are discussed in Section 4.3. Occasionally there are other subtle differences among deep well plates such as corner notches, bottom skirt edges and flanges that may provide for ease in gripping by a robotic arm.

As another example of the variety in microplate specifications, some plates do not match the standard height of a shallow well plate or a deep well plate. One design in particular has an intermediate height taller than a shallow well but shorter than a deep well plate (Figure 3.7). It typically has a round well shape (with or without a raised rim) and a V bottom; its well volume ranges from 0.5–0.75 mL. This reduced height may be useful to yield a smaller combined height when mated with a filtration or an extraction plate, thus a smaller z-axis clearance is required for centrifugation. This intermediate height plate design can also be useful for small volume eluate collection when a shallow well plate presents too small a capacity and a deep well plate presents too large a capacity.

3.4.3 Vial Inserts and Specialty Material Plates

Although the deep well format is standardized and of great utility, situations arise when users prefer an inert glass surface instead of polypropylene. This need is met by vendors who offer a 96-position deep well polypropylene plate having glass inserts and traditional individual seal caps with PTFE (polytetrafluoroethylene) or silicone septa (Figure 3.8). Designed specifically for chromatography applications, this format is unique in that it retains the familiar

Figure 3.7. Example of a collection plate design having an intermediate height taller than a shallow well but shorter than a deep well plate; well volumes typically are from 0.5–0.75 mL. Photo reprinted with permission from Axygen Scientific.

glass vial approach to eluate collection but in the high throughput microplate format [9]. The plate with glass inserts is placed directly into an autosampler for injection. PTFE vials are also offered by some vendors in place of glass vials.

Instead of individually crimped seals on the 96 vials contained in a microplate base, a specialty cap mat such as the WebSeal™ (Chromacol Limited Corporation, Hertfordshire, United Kingdom) can be used to conveniently cap all vials at once. This mat is specifically designed to fit these vial inserts. If one of the sample vials needs to be removed for closer examination, a cutting tool allows the removal of individual vials with the sample, vial and plug as a sealed unit. Sealing systems are discussed in detail in Chapter 4, Section 4.3.

Another variation of glass inserts arranged in a microplate base is the Kombi-Screen™ (Kimble Kontes, Vineland, NJ USA). The glass inserts have a unique Push-Point™ tapered bottom design that provides sample access down to about 10 µL. A snug fit of the vials in a base plate allows their use with shaker and vortex mix units. Many needs are met by five vial sizes: 500, 750, 1000, 1500 and 2000 µL.

The use of a totally inert microplate surface is sometimes required for use with harsh solutions such as encountered in combinatorial chemistry. Plates are available made of pure glass, quartz and PTFE for these applications. Vendors

Figure 3.8. Collection plates are available having removable glass inserts with septa. Photo reprinted with permission from MicroLiter Analytical Supplies.

of these types of plates include Zinsser Analytic GmbH (Frankfurt, Germany), Sun International (Wilmington, NC USA), MicroLiter Analytical Supplies (Suwanee, GA USA) and Orochem Technologies (Westmont, IL USA).

3.4.4 Microtubes and Removable Well Plates

The familiarity of individual test tubes is presented in the microplate format by the use of microtubes, which are individual polypropylene test tubes arranged in an 8-row by 12-column rectangular grid (Figure 3.9). Microtubes fit into a polypropylene base plate or polycarbonate storage rack and retain the SBS standard 9.0 mm center-to-center spacing between wells. The volumes of these removable tubes generally range from 0.65–1.4 mL in the round well tube design and up to 2 mL in square well tubes (rounded bottoms).

Microtubes are available in many formats, such as individual tubes in bulk, tubes connected in strips of eight or twelve in bulk, and individual tubes or strips of tubes pre-assembled into storage racks. Tubes are sealed with cap strips for subsequent sample storage. In addition to a solid one piece base plate having a skirt, a grid plate is available that stands on four legs. This plate can be placed in a tray or put on a counter, in a water bath, or on a refrigerator shelf. Many manufacturers offer sequenced alphanumeric characters (A1–H12) preprinted on the bottom of each tube and assembled in a rack. Individual tubes are removed from the rack and returned since they are numbered.

Figure 3.9. Microtube varieties are available in both round well (*left*) and square well (*right*) formats. Photos reprinted with permission from Simport.

Bar coding can also be incorporated using a unique 2-dimensional hexagonal code on the tube bottom (Figure 3.10). This hexagonal code system from Micronic B.V. (Lelystad, The Netherlands) has 8.5 billion numerical combinations, ensuring unambiguous identification. Labels and markings that can potentially be lost or misinterpreted are now made obsolete.

An important precaution when using traditional microtubes is that they are not rigidly held within a storage rack and some leeway within each well is evident. If the entire rack is tilted inadvertently by hand, some tubes may actually fall out from the base since they are not firmly secured. An improvement in this original design has been made by incorporating plastic springs for gripping the microtubes. An additional lower plate keeps the microtubes straight within the well. This construction is available with both polypropylene and borosilicate glass tubes from vendors such as Trade Winds Direct (Gurnee, IL USA).

When eluates must be transferred to individual vials for injection, *e.g.*, an autosampler must be used that does not accommodate microplates, the use of microtubes is a good choice since they approximate the traditional way sample eluates have been handled. However, most users prefer deep well plates to microtubes for their rigidity and flexibility in sealing [9]. Another choice for eluate injection is the glass insert within a microplate, as discussed previously. The role for microtubes is often as a sample plate rather than a collection plate.

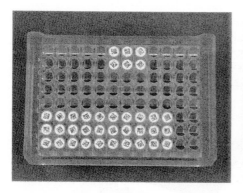 

Figure 3.10. Microtubes are available in racks with preprinted 2-dimensional bar codes on their bottoms. Photos reprinted with permission from Trade Winds Direct.

3.5 Historical Development of Flow-Through Microplates

3.5.1 Evolution from Single Columns to Microplates

Single columns or cartridges have been the traditional format in bioanalysis for performing sample preparation in a flow-through device. These columns typically contain either a filter or sorbent particles and are used for techniques such as filtration, solid-phase extraction and solid-supported liquid-liquid extraction. The columns are available in varying sizes and formats to handle a wide variety of sample volumes and sorbent loadings. Typically, bioanalytical applications use column sizes of 1, 3 and 6 mL to perform most sample preparation tasks. An industry of product support has developed around this traditional column format for sample analysis, including a variety of automated liquid handling workstations.

When solid-phase extraction is performed using individual columns, sorbent is packed between two frits (approximately 25–500 mg packing material) or is contained within a disk (about 14–20 mg). These extraction columns are used in many scientific disciplines other than pharmaceutical bioanalysis, such as environmental, food and biotechnology applications. Note that for applications in environmental and food analysis, larger sample volumes and format sizes are used. The reader is referred to Chapter 11, Section 11.1 for introductory information about the principles of solid-phase extraction theory.

As mentioned earlier in Section 3.1, several advances occurred in parallel in the pharmaceutical research setting that brought about a change from individual flow-through columns to microplates for sample preparation. Faster analytical techniques such as LC-MS/MS became more widespread and affordable, allowing researchers to analyze samples more quickly than ever before. Liquid handling workstations became more prevalent and feature-rich, and replaced most manual pipetting tasks. Within the pharmaceutical industry, drug development cycles were shortened, favoring a more rapid identification of lead candidates with a more rapid development pace. When the sample preparation step became rate limiting in this overall process, a more efficient and faster way to work with the greater number of samples was sought and a change in format toward greater productivity was developed for flow-through applications. The pharmaceutical industry responded to the challenge of higher throughput sample preparation by utilizing filtration, solid-phase extraction and solid-supported liquid-liquid extraction in a 96-well microplate format. The history of its development has been briefly reviewed by the author [1]. This improved format utilizes single blocks or plates having 96 wells that contain filters, disks or packed beds of sorbent particles arranged in the familiar 8-row by 12-column rectangular matrix. The bottom of each well contains a modified tip for the liquids to exit and they are collected in microplate wells placed underneath (or a waste tray when liquid collection is unnecessary).

3.5.2 Early Models and Prototypes of Flow-Through Plates

In the early 1990s, Qiagen Inc. (Valencia, CA USA) first demonstrated a commercial application using the flow-through microplate concept in the molecular biology market. The application was for the purification of DNA (deoxyribonucleic acid) from microbial broth prior to its sequencing. Qiagen and 3M Corporation (St. Paul, MN USA), their plate manufacturer at the time, patented this multiple well format in 1993 (USA), adding a variation (no collars around the exit tips) in 1995 (USA). A series of products was developed for molecular biology applications based on this format, including a filter-only membrane (QIAfilter™), an anion exchange sorbent enmeshed within a PTFE disk (QIAwell™) and 8-well strips containing the same materials (Figure 3.11). Since the plates are intended for use with aqueous solutions and made of polystyrene, they are not suitable for bioanalytical applications that require resistance to organic solvents. However, the 8-well strips did serve as a model for development of 96-well bioanalytical products.

A solid-phase extraction plate made of polypropylene that was suitable for bioanalytical applications was patented in 1995 (USA) by Pfizer Inc. (Sandwich, United Kingdom). This patent described a 96-well flow-through extraction plate called an assay tray, containing packed bed sorbents in a specific columnar configuration. The patent was licensed to Porvair Sciences Ltd. (Shepperton, United Kingdom) for manufacturing and the product became commercially available in 1996. Porvair Sciences subsequently bought the worldwide rights to the patent, which today is co-owned by Porvair and Pfizer.

The original plate configuration had round wells with a reservoir volume of about 0.8 mL. It was made available with column extenders, as strips of eight, positioned on the top surface of the plate to increase the reservoir volume to several milliliters (Figure 3.12). While the column extenders were important for the development of the product to handle the large sample and solvent volumes, they were found to leak occasionally and were not as robot compatible as desired. Future improvements eliminated the use of these optional extenders by replacing them with taller wells in a single molded design [1].

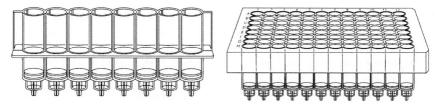

Figure 3.11. Innovators in the development of flow-through microplates for sample preparation included Qiagen and 3M with their patented disk and filtration products for molecular biology applications. *Left*: 8-well strip. *Right*: 96-well plate.

3.6 Varieties in Flow-Through Microplates

3.6.1 Introduction

Flow-through polypropylene microplates have evolved over the years as described previously but can generally be classified by the manner in which they are molded—a single block or a multiple piece block composed of a base plate having removable wells. Note that both types are also distinguished as having square wells (~2 mL volume) or round wells (~1 mL). Within the category of single molded plates, varieties are distinguished by the height of the sides of the microplate (the skirt). Most are short-skirted or tall-skirted plates. Note that a few unique varieties exist, such as the Oro-Sorb I SPE Plate

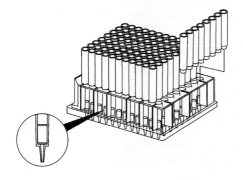

Figure 3.12. The original flow-through plate for SPE using packed particle beds was patented by Pfizer (Sandwich, United Kingdom) and used round wells with separate column extenders for handling the required sample volumes. This design was improved and is now only of historical note.

(Orochem Technologies) which has a 0.7 mL well volume, a capillary shaped exit tip and an intermediate skirt height.

An important note about short and tall skirt plates is that when placed on top of a vacuum manifold, as shown in Figure 3.13, the tips of a short skirt plate descend lower into the manifold compared with those of a tall skirt plate. Thus, compatibility with various models of vacuum manifolds needs to be confirmed before use as some manifolds are specific for only one plate design.

3.6.2 Single Molded Plates with Short Skirts

The original short skirt polypropylene flow-through microplate was the 3M Empore™ Extraction Disk Plate. 3M's manufacturing experience in molding the Qiagen plate allowed for rapid creation of a mold using polypropylene. This product was introduced in late 1996. The plate has round wells, a reservoir volume of 1.2 mL and fine tips with collars, as shown in Figure 3.14(A).

Although the reservoir volume is small, it is adequate for this design as only a thin sorbent particle-loaded disk resides in the bottom of each well. The use of 96-well microplates for SPE applications is discussed in detail in Chapter 11.

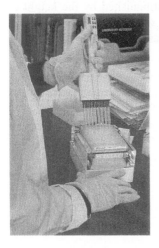

Figure 3.13. The tips from a short skirt plate sit lower within a vacuum manifold (*left*) than those from a tall skirt plate (*right*). The compatibility of a plate design needs to be evaluated before use to ensure optimal height clearance. Some manifolds are specifically designed to work with both short and tall skirt plates.

In 1997, Ansys Technologies (Lake Forest, CA USA, now a part of Varian Inc.) introduced a competing but similar 96-well plate. It also has round wells with a short skirt but a 1.3 mL volume reservoir. However, there are no collars around the exit tips and the tips are more vertical instead of tapered to a fine point. This design improves flow characteristics (*i.e.,* reduce potential for foaming and splashing under vacuum) [1]. A similar but modestly improved plate design, shown in Figure 3.14(B), is used for packed particle bed SPE by Macherey Nagel (Easton, PA USA) and for disks as well as packed particle bed SPE by Orochem Technologies.

In 2000, 3M introduced a plate with an increased reservoir volume (2.5 mL) by securely attaching essentially a one piece column extender onto the top of the plate. It now represents a large volume format in round wells with short skirt and is shown in Figure 3.14(C).

The advantages of a low sorbent bed mass disk for SPE have been well characterized: namely reduced reagent volumes, smaller elution volumes, elimination of channeling, and the ability to elute in a volume small enough for

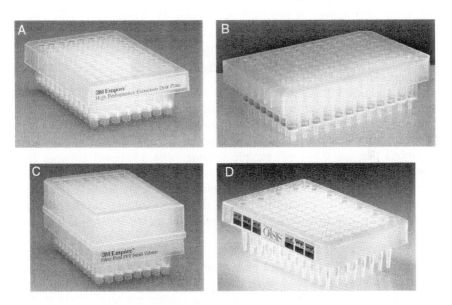

Figure 3.14. Typical examples of single mold, round well, short skirt plates for solid-phase extraction. Photos reprinted with permission from (A, C) 3M Corporation, (B) Macherey Nagel and (D) Waters Corporation.

direct injection into a chromatographic system. Manufacturers of the competing packed particle bed plate products have fashioned small bed masses of their particles in an attempt to match the performance of disks. While this effort has been somewhat successful, the geometric considerations of packing such a small particle mass into the wide diameter wells of a 96-well plate are not ideal.

In 2002, Waters Corporation (Milford, MA USA) introduced an innovative, patented plate design that retains the outer dimensions common to the microplate format but offers a redesigned inner well geometry to accommodate a 2 mg particle mass. Named the Oasis® μElution Plate, shown in Figure 3.14(D), elution volumes as small as 25 μL can be achieved using this plate design, according to the manufacturer. Note that there are some important physical considerations to observe when eluting in such a small volume, namely the collection plate geometry and the potential for adsorption to sides of the collection plate wells, depending on the choice of processing (vacuum or positive pressure).

A list of typical examples of 96-well solid-phase extraction plates in the single mold, short skirt format is provided in Chapter 11, Table 11.4. This table includes comparative information, by manufacturer, on the number and types of sorbent chemistries, sorbent bed mass range, and well geometry.

3.6.3 Single Molded Plates with Tall Skirts

A reservoir volume of just over 1 mL for the short skirt plates was acceptable for some applications which used small sample volumes (*e.g.*, bioanalytical support to drug discovery) but it was not a universal solution. Clinical applications that use 0.5–1 mL sample volume, with additional volumes for internal standard and pH adjustment buffer, often exceed the volume available in the short skirt round well extraction plates. The plate manufactured by Porvair according to the Pfizer patent holds this larger volume, but the original version using column extenders was problematic, as mentioned. It was time for an improvement.

This original 96-well particle bed plate from Porvair was modified—column extenders were replaced with taller wells in a single molded unit Figure 3.15(A). The wells were square for maximum volume but were round at the very bottom. The reservoir volume is about 2 mL, depending on sorbent mass contained within the wells. This plate design is often referred to as the "Porvair Plate," named for the original mold manufacturer. Orochem Technologies subsequently introduced its preferred plate design, which also has square wells but has modified exit tips. Notches were placed in the sides for ease in gripping by a robotic arm, as shown in Figure 3.15(B). A modification of this tall skirt plate was subsequently introduced by Waters Corporation. The sides of the Waters plates are made taller to align with the tops of the square wells, as shown in Figure 3.15(C), for ease in stacking.

A list of typical examples of 96-well solid-phase extraction plates in the single mold, tall skirt plate format is provided in Chapter 11, Table 11.4. This table includes comparative information, by manufacturer, on the number and types of sorbent chemistries, sorbent bed mass range, and well geometry.

3.6.4 Modular Plates with Removable Wells

In 1999, shortly after the introduction of the one piece 96-well flow-through plates, the modular plate was introduced. Its individual wells are removable from a common base plate and approximate the familiar column or tube design. As occurred with single molded plates, several designs were independently developed by different manufacturers. Generally, two modular plate formats are available in the marketplace—square well and round well designs. The column heights also differ among brands.

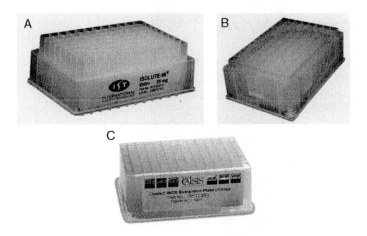

Figure 3.15. Typical examples of single mold, square well, tall skirt plates for solid-phase extraction. Photos reprinted with permission from (A) IST (now an Argonaut Technologies Company), (B) Orochem Technologies and (C) Waters Corporation.

The modular construction was introduced for several reasons:

- Provide familiarity to users in switching from columns to microplate
- Ease the method transfer process by matching sorbent mass and chemistry in each product
- Offer efficiency in cost by using only the number of wells desired or needed, eliminating waste
- Present a convenient, customizable format for method development by mixing wells containing different sorbent chemistries or sorbent mass loadings

Modular plates are available pre-assembled in a specific single sorbent and bed mass, as well as in packages of individual columns for manual insertion into a base plate. A few pre-assembled method development kits are also offered. While each vendor provides individual columns (wells), Orochem Technologies also offers wells in strips of 8 for added convenience. Photos of modular plate designs are shown in Figure 3.16. A list of typical examples of 96-well SPE plates in the modular format is provided in Chapter 11, Table 11.5. This table includes comparative information on the number and types of sorbent chemistries, sorbent bed mass range, and well geometry.

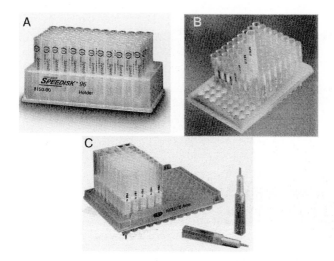

Figure 3.16. Typical examples of modular plates for SPE: (A) Speedisk® 96, (B) VersaPlate™ and (C) ISOLUTE® Array®. Photos reprinted with permission from Mallinckrodt Baker, Varian and IST (Argonaut Technologies), respectively.

3.6.5 Other Formats

The degree of protein binding is an important characteristic that is always determined for a new drug entity or candidate compound. Two common methods of determining drug protein binding involve equilibrium dialysis and ultrafiltration. Until recently, these processes were performed individually in a very time consuming and laborious manner. Equilibrium dialysis is now amenable to high throughput in a 96-well plate, such as the Equilibrium Dialyzer-96™ (Harvard Apparatus, Holliston, MA USA), shown in Figure 3.17(A). The unbound analyte is dialyzed through a membrane until its concentration across the membrane is at equilibrium; the free (unbound) fraction is measured by liquid chromatography. Note that the key discriminator of materials passing through the membrane is the molecular weight cutoff (MWCO). Each well in this system contains a separate 5- or 10-kD MWCO membrane to form two compartments. Since membranes are not shared between adjacent wells, the possibility for well-to-well cross-contamination is eliminated.

Ultrafiltration separates proteins according to molecular weight and size using centrifugal force and is based on a pressure differential rather than a concentration differential as in equilibrium dialysis. Materials below the molecular weight cutoff pass through the membrane and are contained within the ultrafiltrate; retained species are concentrated on the pressurized side of the membrane. Sample preparation using an ultrafiltration method is typically followed by direct injection of the ultrafiltrate. Ultrafiltration is also amenable to high throughput in a 96-well plate, such as the Microcon®-96 well plate

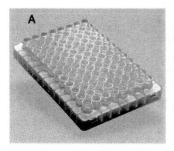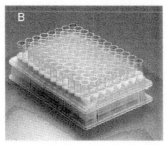

Figure 3.17. (A) Example of an equilibrium dialysis plate containing a selective, semipermeable membrane in the middle of each well. Photo reprinted with permission from Harvard Apparatus. (B) Example of an ultrafiltration plate containing a size exclusion membrane within each well. Photo reprinted with permission from Millipore.

assembly (Millipore, Bedford, MA USA) as shown in Figure 3.17(B). This multiwell ultrafiltration assembly contains a size exclusion membrane to separate free drug from protein-bound drug. Whatman (Clifton, NJ USA) also offers a 96-well ultrafiltration plate. High throughput equilibrium dialysis and ultrafiltration are discussed in more detail in Chapter 6, Section 6.6.

3.7 Applications for Flow-Through Microplates

3.7.1 Filtration Plates

Protein precipitation is a rapid and popular sample preparation technique used in bioanalysis and is discussed thoroughly in Chapter 6. In the traditional procedure, proteins are removed from the sample matrix by the addition of a precipitating agent such as acetonitrile, methanol or ethanol. Proteins are pelleted at the well bottom using centrifugation. An aliquot of the diluted supernatant is analyzed by LC-MS/MS or another technique. The throughput of this procedure can be improved by using filtration microplates. These plates contain a fine frit or membrane at the bottom of each well that filters the precipitated proteinaceous mass. Liquids are processed through the plate using vacuum or centrifugation. Many additional uses exist for filter plates, such as eluate clarification, and a more detailed discussion of these products in bioanalysis is detailed in Chapter 6, Section 6.4.

A variety of filter materials are available in plates (*e.g.*, polyvinylidene fluoride, PTFE, nylon, glass fiber, polypropylene, polyethylene) in a range of porosities (*e.g.*, 0.2–20 µm). Well volumes in filter plates range from about 0.8–2.5 mL. Filter plates are available in a single piece mold with either small or large volumes, as well as the modular format. Unfritted plates can be purchased (*e.g.*, Nalge Nunc International) for users to insert their own filter media. Also, many manufacturers custom make filter plates according to user requirements, *e.g.*, Nalge Nunc International, Innovative Microplate (Chicopee, MA USA) and Orochem Technologies.

Some typical examples of suppliers of filtration plates are listed in Table 3.2. While the information in this table is meant to be instructive, it should not be considered exhaustive. Thus, individual suppliers should always be contacted to obtain more complete and current information about their filtration products. Note that many companies offer their own manifold that is optimized for their filtration product. However, since plates are designed to industry standard footprints they should be compatible with most existing vacuum manifolds.

Table 3.2
Typical examples of vendors of flow-through polypropylene microplates used in high throughput sample preparation methods

Vendor	Solid Phase Extraction Plates				Filtration Plates	Solid-Supported LLE Plates	Vacuum Manifold (except*)
	Packed Particle Bed		Disk Technology				
	One Piece Microplate	Modular Plate	One Piece Microplate	Modular Plate			
Agilent Technologies	X						X
Alltech Associates	X						
Analytical Sales and Service						X	
Ansys (now a part of Varian)			X		X		X
Apogent Discoveries					X		
Chromacol Ltd.	X		X		X		X
Diazem Corporation	X						
Horizon Instrument						X	
Innovative Microplate					X		X
IST Ltd. (now Argonaut)	X	X			X	X	X
Macherey-Nagel	X				X	X	X
Mallinckrodt Baker			X	X			X*
Millipore					X		X
Nalge Nunc International					X		
Orochem Technologies	X	X	X	X	X	X	X

| Vendor | Solid Phase Extraction Plates | | | | Filtration Plates | Solid-Supported LLE Plates | Vacuum Manifold (except*) |
| | Packed Particle Bed | | Disk Technology | | | | |
	One Piece Microplate	Modular Plate	One Piece Microplate	Modular Plate			
Pall Life Sciences					X		
Phenomenex	X						X
Porvair Sciences	X		X		X		X
Supelco	X						X
3M Corporation			X		X		X
United Chemical Technologies	X						X
Varian	X	X	X		X	X	X
Waters Corporation	X						X
Whatman	X		X		X		X

*Positive pressure manifold

Vendors should always be contacted to obtain more complete and current information about their products

3.7.2 Solid-Phase Extraction Plates

Flow-through microplates have been reviewed in Section 3.6. Microplates intended for SPE applications contain sorbent media in each well. The sorbents are commonly a chemically bonded silica or polymer particle and display chemical affinity for certain analytes. Sorbents are packaged or secured in microplate wells by one of two approaches—loose particles packed in a column (held on top and bottom by frits) or particles contained in a thin disk and secured at the bottom of each well.

The selection of an SPE plate is based on many considerations as the author has previously reported [1]. First, the chemistry of the sorbent is chosen. Next, the sorbent configuration is selected (packed particles or disk). The sorbent mass, for packed particles, is then considered, as multiple bed mass loadings are available. The bed mass will influence the solvent and elution volume requirements, as well as the capacity for analyte and matrix components. In general, it is suggested to use the smallest bed mass that will provide sufficient capacity for the drug in its sample matrix. The use of smaller solvent volumes from a smaller bed mass format results in reduced total extraction times. Also, smaller elution volumes require less time for evaporation prior to reconstitution and analysis.

Some typical examples of suppliers of SPE plates are listed in Table 3.2. While the information in this table is meant to be instructive, it should not be considered exhaustive. Thus, individual suppliers should always be contacted to obtain more complete and current information about their SPE products. Details about sorbent chemistries, protocols and applications for 96-well solid-phase extraction are found in Chapter 11. Also in Chapter 11, Tables 11.4 (single molded plates) and 11.5 (modular plates) provide more detailed information about the various extraction plates and product specifications offered by different vendors.

3.7.3 Solid-Supported Liquid-Liquid Extraction Plates

Liquid-liquid extraction can be performed in a flow-through microplate using diatomaceous earth particles packed in wells. When used in this manner, the high surface area of the diatomaceous earth particle facilitates efficient, emulsion-free interactions between the aqueous sample and organic solvent. Essentially, the diatomaceous earth coated with the aqueous phase behaves similar to the aqueous phase of a traditional liquid-liquid extraction, yet it shows characteristics of a solid support. The procedure for use of solid-supported particles to perform liquid-liquid extraction (SS-LLE) is detailed in Chapter 8, Section 8.3. Some typical examples of suppliers of solid-supported LLE plates are listed in Table 3.2. While the information in this table is meant to be instructive, it should not be considered exhaustive. Thus, individual suppliers should always be contacted to obtain more complete and current information about their SS-LLE products.

3.7.4 Equilibrium Dialysis and Ultrafiltration Plates

An assessment of the *in vivo* properties of a new drug substance includes the measurement of how strongly that drug is bound to plasma proteins. The binding of drugs to plasma or serum proteins has been shown to have dramatic effects on pharmacodynamic and pharmacokinetic processes in the body. Two common methods of drug protein binding determinations involve equilibrium dialysis and ultrafiltration. A high throughput approach using the microplate format for these two techniques is commercially available. The applications for equilibrium dialysis and ultrafiltration are reviewed in detail in Chapter 6, Section 6.6.

Acknowledgments

The line art illustrations were kindly provided by Willy Lee and Pat Thompson. The author also thanks Greg Lawless for his careful review of the manuscript and helpful discussions. The assistance of the following vendor representatives who contributed information and reviewed text is appreciated: Linda Alexander (Macherey Nagel), Paul Bouise (Mallinckrodt Baker), Mike Brown (Porvair Sciences), Claire Desbrow (IST Ltd., now an Argonaut Technologies Company), Michael Early (Waters), Kim Gamble (MicroLiter Analytical Supplies), Greg Hoff (Millipore), Colin Keir (Axygen Europe), Rich Matner (3M), Asha Oroskar (Orochem Technologies), Cecile Poirier (Simport), Roger Roberts (Ansys Technologies, now a part of Varian), Jinghua Schneider (Trade Winds Direct), Daniel Schroen (Nalge Nunc International), Ron Sostek (Harvard Apparatus) and Rob Stubbs (Varian).

References

[1] D.A. Wells, LC-GC 17 (1999) 600-610.
[2] R. Manns, Presented at MipTec-ICAR, Montreux, Switzerland (1999).
[3] B. Johnson, The Scientist 13 (September 27, 1999) 16.

[4] A.A. Oroskar, In Vitro Diagnostic Technology 4 (January 1998) 40.

[5] D.T. Rossi and M.J. Lovdahl, In: D.T. Rossi and M.W. Sinz, Eds., Mass Spectrometry in Drug Discovery, Marcel Dekker, New York (2002) 215-244.

[6] R.A. Fecik, K.E. Frank, E.J. Gentry, S.R. Menon, L.A. Mitscher and H. Telikepalli, Med. Res. Rev. 18 (1998) 149-185.

[7] M.A. Gallop, R.W. Barrett, W.J. Dower, S.P.A. Fodor and E.M. Gordon, J. Med. Chem. 37 (1994) 1233-1251.

[8] E.M. Gordon, R.W. Barrett, W.J. Dower, S.P.A. Fodor and M.A. Gallop, J. Med. Chem. 37 (1994) 1385-1401.

[9] D.A. Wells, LC-GC 17 (1999) 808-822.

[10] N. Zhang, K. Rogers, K. Gajda, J.R. Kagel and D.T. Rossi, J. Pharm. Biomed. Anal. 23 (2000) 551-560.

Chapter 4

High Throughput Tools for Bioanalysis: Accessory Products

Abstract

Sample preparation techniques within the pharmaceutical industry have made a transition from individual columns or cartridges into the microplate format. These microplate collection devices and flow-through plates are integral components of a high throughput sample preparation method. However, they require the use of one or more accessory products to perform related tasks, such as: liquid pipetting, vacuum and centrifugation systems for processing liquids, sealing systems for securing samples for mixing or eluates before injection, mixing or shaking, evaporation systems for dry-down of eluates prior to reconstitution and autosamplers compatible with microplates. Many times, the source for these accessory products may be unfamiliar to the bioanalytical scientist. Assembled in this chapter are descriptions of the microplate accessory products required to equip a laboratory for high throughput operation. Typical examples of suppliers of these products are also provided. The previous chapter introduced collection and flow-through plates used for high throughput bioanalytical sample preparation. Knowledge of these many tools for bioanalysis is essential to perform high throughput sample preparation methods.

4.1 Multichannel Pipettors for Liquid Handling

A multichannel pipettor is an indispensable tool used to manually deliver liquids into microplates. Even when liquid handling workstations are used, there is often the need to make use of a multichannel pipettor for some part of the sample preparation procedure. A useful related function is the ability to repeatedly deliver the same volume with each press of the plunger. Frequent uses of a repeater pipettor are to deliver internal standard solution and reconstitution solution to multiple wells of a microplate. Many models of multichannel pipettors function as repeater pipettors. Note that some traditional single channel repeater pipettors may be converted into multichannel pipettors via an adapter, *e.g.*, the Eppendorf® Repeater™ Plus Pipette (Brinkmann

Instruments, Westbury, NY USA) can be fitted with a Plus/8™ cartridge to become an 8-channel multiple dispense pipettor.

Multichannel liquid handling units are commonly available as 8-channel and 12-channel pipettors. Many use electronic designs that require a charging stand for the battery and others are operated using mechanical air displacement. Note that 4-channel pipettors are also available, but greater throughput is gained using 8- and 12-channel pipettors since the microplate format consists of an 8-row by 12-column rectangular grid [1].

Many volume ranges exist for 8- and 12-channel pipettors, depending on manufacturer. These volume ranges include low volume (*e.g.*, 0.5–10 µL and 10–100 µL), mid volume (*e.g.*, 25–250 µL and 50–300 µL) and large volume (8-channel only; *e.g.*, 50–1200 µL). Generally, an 8-channel pipettor offers the most versatility since it accommodates the largest volume (50–1200 µL). A detailed summary table of pipettors and their volume specifications, sorted by manufacturer, can be found in another publication by the author [1].

The most common use for a manual pipettor is the reformatting of samples from test tube racks to a microplate for use with a 96-tip liquid handling workstation. Note that the 4-/8-probe workstations can perform this task. Since the center-to-center spacing of sample tubes does not match that of a microplate but varies, it is quite valuable to use a pipettor that has the ability to vary the spacing between its tips. This variable spacing feature allows samples or reagents to be aspirated from 12 or 13 mm test tubes having up to 14.15 mm center to center spacing and delivered into plates having 9.0 mm spacing; adjustments are made in increments of 0.1 mm. Another useful feature is a dispense head that rotates through 360° to provide a more comfortable working angle [1].

Reagent reservoirs are used with multichannel pipettors to allow for convenient access to the liquid source. These reservoirs are also called reservoir basins, troughs or trays. These products are commonly made of polypropylene, but are also made from polystyrene or polyvinyl chloride. Lids can control evaporation between uses and are optional. The traditional V-shaped reservoir design generally holds a liquid volume of 60–120 mL, depending on manufacturer. Smaller 25 mL reservoirs are available, and one variety has a divider that splits the reservoir into two sections. An 8-channel reservoir tray allows different liquids to be placed within each of the rows, or channels (volume capacity within each row is about 10 mL). Each channel may contain different elution solvents which can be particularly useful for performing method development.

It is also possible to use a square well collection plate as a solvent reservoir, as it contains 96 individual wells into which unique liquids can be placed [1].

The reagent reservoirs are also made in custom varieties to allow small volume pipetting as well as chilling. The pipetting of small volumes of internal standard solution, which may exist in a stock solution of only 5–10 mL, can be performed using a low dead volume reservoir tray. An example of such a product is a reusable polypropylene reservoir that contains dimples on its bottom in 9 mm centers (*e.g.*, Innovative Microplate, Chicopee, MA USA). A reservoir can also be chilled during use. The ChillTrough™ supports a disposable polypropylene trough liner and an environmentally friendly coolant sealed within the trough cavity (Figure 4.1).

4.2 Liquid Processing Options for Flow-Through Microplates

4.2.1 Vacuum Manifold Systems

A vacuum manifold system has traditionally been used with a flow-through microplate for its convenience, low cost and usability. Liquid levels can easily be viewed from above the plate to monitor flow progress. Most manifolds also

Figure 4.1. Cooling reagent reservoirs are available in which an environmentally friendly coolant is sealed within the trough cavity. Photo reprinted with permission from Trade Winds Direct.

have at least a small window on the side so that liquid exiting the tips can be observed. A manifold system typically consists of a two piece vacuum box, tubing, a small portable vacuum pump, and a bleed valve with gauge to assess vacuum levels. In place of a dedicated vacuum pump, many laboratories have access to an in-house vacuum source which is usually preferred for its silence. Although the load on the house vacuum may vary slightly over time and between days, the use of a bleed valve and the presence of a vacuum gauge ensure that vacuum strength can be monitored and adjusted as necessary.

Many designs for vacuum manifolds exist in the marketplace. The first widely used manifold that performed consistently was made by Qiagen Inc. (Valencia, CA USA) and consisted of a white Delrin® bottom with an acrylic top. This manifold was originally intended for use with aqueous liquids in molecular biology applications; as bioanalytical scientists used the manifold for their own applications, the organic solvents employed in extractions turned the clear acrylic top cloudy (although it still provided functionality). Over time, other vendors developed their improvements to this original design using materials such as anodized aluminum, PTFE (polytetrafluoroethylene), PTFE coated aluminum and specially treated polypropylene for improved solvent resistance.

Some typical examples of vacuum manifolds are shown in Figure 4.2. These manifolds are available from many sources; a representative supplier list can be found in Chapter 3, Table 3.2. Note that manifolds are also designed for use in applications other than bioanalysis, such as PCR purification and dye terminator cleanup in genomics applications, as well as reaction blocks for synthesis in combinatorial chemistry.

Most manufacturers of flow-through microplates offer a version of vacuum manifold that is optimized for use with the plate design(s) that they provide. When first introduced, these manifolds were specific only for either the short or the tall skirt design (see Chapter 3, Section 3.6 for details). However, over subsequent years as plate formats proliferated and the realization became evident that one laboratory often cannot standardize on only one plate design for all needs, better compatibility among competing plate designs has been introduced. Typical examples of vacuum manifolds that accommodate this dual compatibility of short and tall plates include those solid-phase extraction manufacturers (3M Corporation, St. Paul, MN USA; IST Ltd., Hengoed, United Kingdom, now an Argonaut Technologies Company; Supelco, Bellefonte, PA USA; Orochem Technologies, Westmont, IL USA) as well as automation vendors (Tomtec Inc., Hamden, CT USA; Beckman Coulter, Fullerton, CA; Packard Instruments, Meriden, CT USA, now PerkinElmer Life

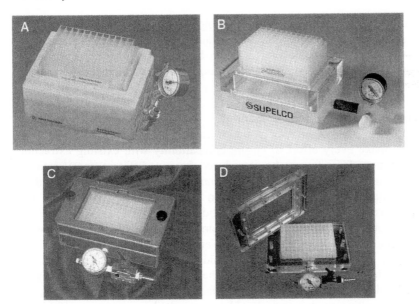

Figure 4.2. Typical examples of vacuum manifolds used with flow-through microplates include models from (A) Agilent, (B) Supelco, and (C, D) Innovative Microplate.

Sciences; Zinsser Analytic, Frankfurt, Germany; and Hamilton Company, Reno, NV USA).

When a flow-through plate is used with a vacuum manifold, liquids are either collected in a waste tray or in a receiving plate. When using a receiving plate, it is important to adjust the height of this plate within the manifold. The tips of the flow-through microplate are carefully positioned just above the collection device so that the liquid exiting the tips is fully contained within the receiving wells. Thus, the potential for cross-contamination is reduced. Shims are often required for exact positioning of the collection plate within the manifold. A unique product offered by Orochem Technologies with their manifold is a guide block. Seated on the waste tray, its spacers ensure proper seating of the exit tips from filtration and extraction plates. This block eliminates crosstalk between wells during the sample transfer process.

Note that another type of adapter is sometimes required for superior sealing of a microplate with the gasket on the top lid. When using the Array® modular plate from IST Ltd. (now Argonaut Technologies), an adapter is placed on top of the manifold at the point where the modular plate bottom meets the gasket. Not all modular plates require this adapter, however, and the manufacturer of each plate type should be contacted for vacuum sealing recommendations using specific manifolds.

The compatibility of vacuum manifolds with automated liquid handling workstations represents another point of caution. Manufacturers of vacuum manifolds make every effort to ensure compatibility with existing extraction plates but new formats can be introduced having different specifications; incompatibilities may result. An example of a plate with manifold incompatibility is the VersaPlate™ (Varian Inc., Harbor City, CA USA). The VersaPlate is taller than the traditional solid-phase extraction (SPE) plate and when used on the Quadra® 96 with the Tomtec vacuum manifold, the tips do not clear the overall plate height. The VersaPlate manufacturer solved this problem by introducing its own manifold to be used specifically on the Quadra 96. It is optimized for the z-axis height clearance. Occasionally such incompatibilities may result and the manufacturers of the extraction plates and the liquid handling workstation should be consulted to ensure manifold and plate compatibilities.

It is important for the gasket on top of a vacuum manifold to form a tight seal with the underside of the flow-through microplate. This seal must be effective simply by turning on the vacuum when using automation. The workstations often control the vacuum via on/off signals in the software. If manual intervention is required to press the plate down on the manifold lid to start the sealing process, then unattended automation using this manifold is impossible.

A unique vacuum device is the CaptiVac™ Vacuum Collar (Ansys Technologies, Lake Forest, CA USA, now a part of Varian). This transparent device is a collar that joins Captiva filtration plates or SPEC® 96-well extraction plates directly to an underlying collection plate. The unique design of the CaptiVac forms a pre aligned vacuum seal between the two components (Figure 4.3). This device also positions the outlet tips of the filtration plate to a specific distance into the collection plate wells to prevent cross-contamination of samples. No shims or adjustments are necessary as with traditional manifold units. The appealing features of this system are its compact size, simplicity of use and affordability.

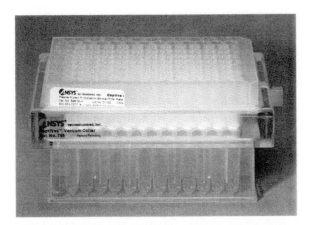

Figure 4.3. A vacuum collar is an alternative to traditional vacuum manifolds. It is positioned between a filtration or extraction plate and a deep well collection plate to form a tight vacuum seal. Photo reprinted with permission from Ansys (now Varian).

4.2.2 Centrifugation

The use of centrifugation to process liquids through extraction and filtration plates can present some advantages compared with vacuum. While vacuum strength can sometimes vary among wells within a plate, centrifugation allows a more uniform flow through all wells and at a more controlled rate. Centrifugation can be started at a low speed and, after a few minutes, increased to a greater speed; when performed in this manner, liquid flow through the wells is maintained regardless of viscosity or resistance. Centrifugation is realized to be more effective than vacuum in efficiently collecting all residual liquid from plate wells; using vacuum, some liquid is always left behind. This residual liquid remaining after vacuum can be problematic for solid-phase extraction, when residual water left from the wash step is displaced with elution solvent and is collected with the eluate. Plugged wells seen using vacuum almost always evacuate using the greater gravitational forces achievable with centrifugation.

The use of centrifugation with flow-through microplates requires the mating, or nesting, of a 96-well extraction or filtration plate with a collection plate, as shown in Figure 4.4. Usually a deep well plate is used but an intermediate height plate is also a viable option. This combination is placed into the microplate carrier of the centrifuge. Note that the height of this combination must not exceed the z-axis limitation of the microplate rotor. While most microplate rotors will be able to spin a deep well collection plate, not all of them have the added clearance for a deep well plate nested with a flow-through extraction plate. The centrifuge manufacturer should be contacted for height or z-axis limitations for a specific model. More information on centrifugation is found in Section 4.4.4.

4.2.3 Positive Displacement System

A benchtop positive displacement system for liquid processing is available from Mallinckrodt Baker (Phillipsburg, NJ USA) as the Speedisk® 96 Positive Pressure Processor. It is designed for use with Baker 96 position Speedisk columns in a base plate, matching the footprint of 96-well plates. However, it is versatile and will also work with single molded filtration and extraction plates from other manufacturers. The unit is hydraulically activated and seals the top surface of the plate against a gasket covered manifold. The 96 tips deliver nitrogen gas or compressed air at a controlled rate to process liquids through the extraction device via positive pressure displacement (Figure 4.5). A microplate is placed below the exit tips of the extraction plate to collect filtrate or eluate [1].

Positive displacement in this manner has been shown to be a very effective alternative to vacuum processing. Many users prefer the controlled flow rate from displacement and cite this feature as an important advantage, as well as its ability to completely evacuate all liquid volume from the well or column.

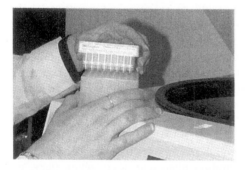

Figure 4.4. A flow-through filtration or extraction plate is nested with a deep well plate to collect filtrate or eluate, respectively, and this combination is placed into a centrifuge for processing. The centrifuge microplate rotor must have enough z-axis clearance to accommodate the total height.

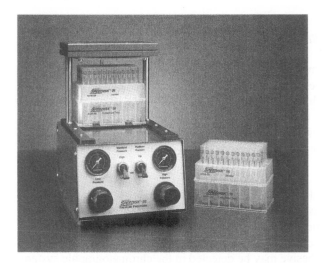

Figure 4.5. A positive displacement system for processing liquids through filtration or extraction plates provides an alternative to vacuum. Photo reprinted with permission from Mallinckrodt Baker.

4.3 Sealing Systems

4.3.1 Introduction

Several types of sealing systems for microplates are available for use in pharmaceutical research. The choice of system should be made with the exact needs of the application in mind. For example, a microplate may simply need to be covered at the lab bench for transport to another laboratory and a simple adhesive film strip will preserve sample integrity. Storage of a microplate in the freezer requires a seal having low temperature stability with seal integrity. A microplate that is to be pierced within an autoinjector requires a seal that is perforated easily to prevent coring by the injector needle. When the unused wells of an extraction microplate need to be covered to keep the vacuum focused on the wells containing liquid, a tight seal needs to be realized so that air does not enter from the atmosphere. The careful matching of sealing needs with the application in mind is very important. If these considerations are not made, unwanted and unnecessary problems may be introduced such as gumming of the injector needle with adhesive when a tape sheet is used with an autosampler instead of a pierceable cap mat.

4.3.2 Adhesive Backed Sheets

An adhesive backed film sheet is applied to a microplate and adheres to the plastic surface. The adhesive is chemically inert and is intended for use only with aqueous liquids. Therefore, its main applications are in high throughput screening, genomics and clinical applications to reduce evaporation of aqueous liquids within wells. This sheet is also available as a gas permeable plate seal for applications in cell culture. Adhesive sheets are usually effective only at room temperature.

The adhesive backed sheets are naturally considered by novices. Several concerns exist with the use of such sheets:
1. A sealed organic eluate solution may dissolve the adhesive via its vapors in a closed well
2. The chemical adhesive may be detected in the chromatographic system
3. When used with an autoinjector the adhesive can gum the injector needle after repeated injections

Adverse results from using adhesive backed sheets are often injector failure and/or chromatographic contamination. However, one acceptable application in which these adhesive film sheets are useful in bioanalysis is to seal the unused wells in a filtration or extraction microplate when used with vacuum.

An improvement over the traditional sticky adhesive film sheet is the pressure activated adhesive. Alone this film sheet is not tacky to the touch, but when firmly and evenly pressed onto a polypropylene surface it forms a strong seal. An example of this pressure activated adhesive film is called 3M Advanced Pierceable Microplate Sealing Tape (#9796) and is available from the Medical Specialties group within the 3M Corporation.

4.3.3 Cap Mats

Cap mats are single semi rigid covers that fit into each well to provide protection from the atmosphere at room temperature, shown in Figure 4.6(A). They are not intended for heating applications, although they can generally be frozen to −20°C. Cap mats are commonly made of ethyl vinyl acetate (EVA), which is not a chemically resistant plastic and so it is best used to seal aqueous liquids. However, a cap mat made from Santoprene® (a thermoplastic elastomer) does offer greater solvent resistance and more flexibility. Note that cap mats do not always fit snugly into every brand of microplate; occasionally some brand specific alignment anomalies exist, so testing them before use on a large scale is suggested.

The use of cap mats in pharmaceutical bioanalysis is commonly reserved for sealing a microplate prior to injection into a chromatographic system. In this case, a pierceable cap mat is needed, one in which the center of the well cover is precut or thin milled in a crosshatch area to facilitate piercing by an autoinjector needle. Needle coring is thus minimized and the needle exits out of the well without great resistance [1]. Pierceable cap mats are available for both round and square well plates. Strips can be removed from a perforated cap mat to match the number of samples to cover. Note that the semi rigid EVA cap mats do not generally seal well on microtubes; instead, these tubes are best sealed with microtube caps which are available in strips of 8 or 12.

Concerns over solvent resistance, the stiffness of the cap mat made from EVA, and the lack of a tight fit into wells have led to the development of cap mats made from a silicone elastomer. The underside of this sheet can be treated with PTFE for improved solvent resistance. The lack of a snug fit with the EVA cap mats has prevented their use for liquid-liquid extraction applications in bioanalysis, as solvent would exit the tops upon vigorous vortex mixing. However, the silicone mats provide a much tighter fit with most microplates and are preferred for use in this application. These mats are available in both pre scored (crosshatch design in center) and unscored versions for both round and square well plates, as shown in Figure 4.6(B). The pre scored silicone cap

 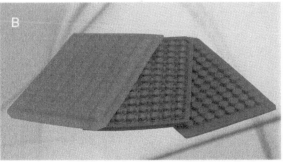

Figure 4.6. A cap mat is used to cover a microplate for storage and the pierceable version can seal a plate for use with an autoinjector. Two versions are shown: (A) semi rigid pierceable mat and (B) silicone elastomer coated with PTFE. Photos reprinted with permission from (A) MicroLiter Analytical Supplies and (B) Chromacol Ltd.

mats do offer some resealing at the injection point, useful to limit evaporation from the pierced well over time. These silicone mats (PTFE treated) are available from many suppliers such as Chromacol Ltd. (Herts, United Kingdom), MicroLiter Analytical Supplies (Suwanee, GA USA), Sun International (Wilmington, NC USA) and SSP Companies (Ballston Spa, NY USA).

Note that some designs of silicone cap mats (*e.g.*, Duo-Seal™ and Duo-Seal™ II from Ansys Technologies, now a part of Varian) can also be used to occlude the bottom tips of flow-through filter and extraction plates as well as the top surface of microplates, filter plates and extraction plates. An application for plasma sample storage and filtration illustrates this sealing approach for filter plates and is discussed in Chapter 5.4.4.

Many times it is difficult to manually apply a cap mat to the surface of a plate and achieve a good seal rapidly and reliably. Usually it requires some effort and firm pushing with the wrist. A small paint roller tool is one common solution to press on a cap mat. In the interest of worker safety and productivity, several vendors have developed a simple press to quickly and reliably apply a cap mat, shown in Figure 4.7. The plate, with cap loosely attached, is set into place and a single firm push of the handle seals the plate. Suppliers of such presses include, *e.g.*, Axygen Scientific (Renfrewshire, United Kingdom), Porvair Sciences (Shepperton, United Kingdom), Millipore (Bedford, MA USA) and Sun International.

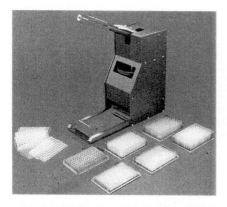

Figure 4.7. A cap mat is quickly and reliably applied with a small press. Photo reprinted with permission from Axygen Scientific.

4.3.4 Heat Applied Foil and Film Sheets

Heat sealing of microplates is performed by placing a cover sheet of film onto the surface of a microplate and applying heat evenly for several seconds using a rectangular iron or a semi-automated benchtop sealer unit. Since the bonding surface is polypropylene these seals can only be used on polypropylene plates. **Important Note: A microplate having a raised rim around each well is essential for good sealing and preventing cross-contamination between wells.** Heat sealing offers many advantages compared with other plate sealing methods:

- Secure and uniform sealing of wells
- No contaminants from adhesives
- Resealable plates for sample access
- Exceptional seal integrity at low temperatures
- Solvent resistance
- Quick application process
- Leak proof even with vigorous vortex mixing
- Cost effective

Typically, one side of the precut film is composed of a thin layer of polypropylene and the other side is aluminum foil; laminates of polypropylene and foil exist but these are tougher for an autoinjector needle to pierce. Pure polypropylene sheets are also available. Note that seals are optimized for both peelable and pierceable uses, so proper selection of product is again important to meet the needs of the application.

A foil sheet having a thin layer of polypropylene on one side is generally preferred for sealing plates before injection in bioanalytical applications (Figure 4.8). The heat quickly melts and welds the plastic on the sheet to the plastic plate surface, forming a tight seal at the raised rim. The outer, exposed surface of the cover sheet is aluminum foil which offers protection and maintains well integrity even at low temperatures. The heat sealing process is performed rapidly and the temperature of samples within the wells does not increase significantly during sealing. Temperature sensitive samples are not usually affected during the sealing procedure. Note that some aluminum foil sheets found in vendor catalogs actually have adhesive on one side (not polypropylene for heat sealing). These adhesive sheets are not appropriate for use in bioanalysis for injection or for heat sealing.

Figure 4.8. A foil sheet applied with heat presents an effective sealing option for microplates. Photo reprinted with permission from Simport.

The use of a hot iron in the laboratory to seal microplates with foil/plastic laminate sheets presents a potential workplace hazard. A safer and more practical approach is with a manual benchtop heat sealer, such as the Thermo-Sealer (ABgene North America, Rochester, NY USA), shown in Figure 4.9, the SuPEr Plate Sealer (Orochem Technologies) or the Super Sealer (Matrix Technologies Corporation, Hudson, NH USA). These instruments are compact, portable and can seal a wide range of plates including shallow well and PCR plates in addition to deep well plates. The heating plate of the sealer faces downward for safety and a built in thermal circuit prevents overheating.

Operation of the Thermo-Sealer follows. The user secures the microplate on the platform, places a sheet of film on top and manually brings the heating plate down using the handle. The plate is sealed in just a few seconds. The plate is then turned 180° and the heating plate is brought down again to complete the seal. Bioanalytical applications commonly utilize either the Easy Pierce (stable to −20°C) or Easy Pierce Strong (stable to −80°C) sheets of foil material containing a thin layer of polypropylene (ABgene). In the event that the foil sheet needs to be removed, a foil stripper tool is available to aid in manually performing this function; an edge of the foil is inserted through the center of the tool and the foil sheet is twisted away from the plate surface.

The Automated Laboratory Plate Sealer (ALPS) from ABgene is suitable for high throughput and can be integrated with robotic systems. Heat sealing can

Figure 4.9. *Left*: Thermo-Sealer for manual application of heat to a foil or film sheet placed on a microplate. *Right*: ALPS300 Plate Sealer for automated application of heat sealable films. Photos reprinted with permission from ABgene North America.

be accomplished for plates of differing heights. The unsealed plate is placed on a shuttle that is extended from the main body of the sealer, allowing full access by a robotic arm. The shuttle is drawn into the unit and the plate is sealed in about 15 s. The shuttle then slides outward, presenting the sealed plate. The ALPS can also be operated as a stand alone unit when not integrated within a robotic configuration. All parameters such as temperature and time are fully adjustable. Rolls of sealing film accommodate approximately 4,000 plates. Some additional brands of automated heat sealers in this class include Sagian™ Sealer (Beckman Coulter), Quadra Seal (Tomtec), PlateLoc™ Thermal Plate Sealer (Velocity 11, Palo Alto, CA USA) and Presto® Microplate Sealer (Zymark Corporation, Hopkinton, MA USA). Note that a microplate handling device such as the Twister™ II (Zymark Corporation) can be used with an automated plate sealer. The Twister™ II is described further in Chapter 5, Section 5.3.3.

4.3.5 Lids

When a microplate requires the use of a removable cover as in a robotics application, a lid is preferred such as Corning® Robolids (Corning Inc., Acton, MA USA). A robotic gripper arm or hand can pick the lid up and remove it without disturbing the microplate. Note that these lids are intended to be loose fitting. Polypropylene lids are preferred for their solvent resistance. When a more effective lid seal is desired, a silicone mat can be placed on top of the microplate but under the lid. A related use for a plate lid is to cover an open solvent trough when not in use such as on the Quadra 96 liquid handling workstation.

4.4 Mixing and Centrifugation of Collection Microplates

4.4.1 Introduction

Mixing and centrifugation of liquids contained within sealed microplates are two processes that are commonly performed during many different sample preparation procedures. Plasma in microplate wells, after thawing, is centrifuged to pellet thrombin clots and/or particulates before removing an aliquot. Buffer and/or internal standard are added to these same microplate wells and the liquids are thoroughly mixed before continuing the procedure. Another instance for mixing is when mobile phase compatible solution is added to a microplate, following solvent evaporation. Vortex mixing is used to reconstitute the analytes in solution prior to injection into a chromatographic system. Protein precipitation of plasma requires a centrifugation step to pellet the proteins removed from plasma after addition of a precipitating agent. Liquid-liquid extraction requires a thorough vortex mixing step for analytes to efficiently partition from the aqueous into the organic phase. Eluates contained in microplate wells are often centrifuged to pellet particulates before the injection step. Sealing systems for collection devices have been discussed in Section 4.3.

4.4.2 Mixing of Liquids within Microplate Wells

Once tightly sealed, liquids within microplate wells are commonly vortex mixed using a traditional multiple tube vortexer, also called a flat bed shaker. This compact unit, commonly available from laboratory supply houses such as VWR (Chicago, IL USA), Fisher Scientific (Pittsburgh, PA USA) and PGC Scientifics (Gaithersburg, MD USA), was originally intended for a rack of test tubes. Since the height of the clamping unit is adjustable, a microplate is easily accommodated in place of a test tube rack.

An improved approach to mixing liquids within sealed wells is represented by a pulse mixer. While a vortex mixer represents continuous action using a slow to fast setting, a multiple pulse vortex actually stops the motor for a split second; liquid falls toward the well bottom via gravity and then the mixing action is automatically restarted. This time delayed sequence helps to join the upper and lower liquid portions for more thorough and reproducible mixing. The speed of a pulse mixer is versatile; at the low end of the range, a gentle orbital motion is provided while at the high end the sample is mixed with a vigorous vortex action. A digital pulse mixer incorporates microprocessor based control into the process with tighter speed control than the analog mixer; heat can also be applied. Typical examples of pulse vortex mixers are represented by a Large Capacity Mixer (from 1 to 4 microplate capacity) and a Digital Pulse Mixer (from 1 to 8 microplate capacity, shown in Figure 4.10), both from Glas-Col (Terre Haute, IN USA).

Two important guidelines for use of sealed microplates with a vortex unit are:
1. Carefully adjust the speed to achieve the desired mixing action
2. Allow enough airspace above the liquid volume so that adequate mixing can occur within the wells

Generally, vigorous vortex mixing for about 1 min is sufficient for most reconstitution and precipitation procedures, and from 5–10 min is needed for liquid-liquid extraction procedures. Gentler mixing of solutions within microplate wells is provided by an instrument such as the Wellmix™ (Thermo Labsystems, Franklin, MA USA). The Wellmix is a four place orbital shaker with variable speed (100–1350 cycles per min) and time settings to 2 h. It displays a small fixed orbital motion and contains a sectioned rubber mat for

Figure 4.10. A pulse vortex mixer provides thorough and reproducible mixing within microplate wells. Photo reprinted with permission from Glas-Col.

secure placement of the microplates. Similar 2- and 4-plate shaker units are available from IKA Works Inc. (Wilmington, NC USA), Labnet International (Woodbridge, NJ USA) and Fisher Scientific. An orbital vortexer designed for use of 12 microplates with a robotic system is manufactured by SLR Systems (Vancouver, WA USA).

4.4.3 Evaluating Cross-Contamination in Sealed Microplates

A potential problem with mixing in microplate wells is cross-contamination. If the seal is not tight, small drops of liquid can exit through the tops of the wells and contaminate neighboring wells. It is suggested to always evaluate the chosen combination of sealing system with collection device before use.

Several approaches can be used to verify the integrity of liquids within a microplate seal. The exact solvents as will be used for a liquid-liquid extraction procedure are added to an outside row of the plate (rows A, H or both). A colored dye is added to the aqueous or organic portion so that this phase can be readily observed during the mixing procedure. Either a water soluble or an organic soluble dye can be used. The plate is sealed and the mixing procedure is performed. As mixing proceeds, the colored solutions can be observed and the speed adjusted to achieve the desired velocity of mixing within the wells.

A more complete evaluation of seal integrity can utilize a fluorescent probe (*e.g.*, fluorescamine in water or rhodamine in organic solvent). Again, the exact liquid types to be used for the procedure are added and the fluorescent probe is spiked into the mixture. However, the samples are placed into the wells in a **specific pattern** such that blank wells always fully surround each liquid filled well (Figure 4.11). The plate is sealed and the mixing procedure is performed. Following mixing, the seal is carefully removed. The seal, plate and empty wells are inspected using a handheld ultraviolet lamp. The detected fluorescence reveals where droplets have escaped or crossed wells.

The use of fluorescence can allow for the detection of much lower concentrations than that possible using a colored dye in visible light. When fluorescence is not an available option, the use of yellow and blue colored dyes in adjacent wells provides useful information. Cross-contamination may result in the formation of a green drop where blue and yellow liquids interacted.

Another useful and sensitive approach for assessing cross-contamination can be performed using radiolabeled analyte spiked in plasma. In a report using [^{3}H]-dextromethorphan, plasma aliquots spiked with tritiated analyte were placed

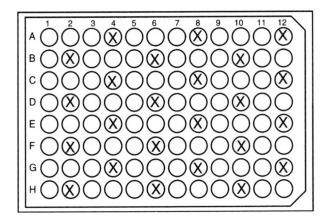

Figure 4.11. Arrangement of fluorescent probes (X) spiked into wells of a microplate for a cross-contamination test using liquid-liquid extraction. Wells surrounding the X locations are empty. Following vigorous vortex mixing, the empty wells are evaluated for presence of probe.

in alternating fashion between blank plasma samples to create a checkerboard pattern in a 96-well plate [2]. Samples were processed using a liquid-liquid extraction procedure to yield analyte contained in a back extracted acidic solution. The acidic extract from each blank sample was removed from each well and transferred to a scintillation vial, mixed with cocktail and counted for radioactivity.

Note that the procedure of removing the seal can sometimes introduce contamination from liquid that rests on the underside. When removal of the seal is performed hurriedly, droplets may splatter. **It is recommended to first centrifuge the sealed microplate following the mixing procedure and then remove the seal.** Any liquid that was on the underside of the seal after mixing will be brought to the well bottom when this technique is followed.

4.4.4 Centrifugation of Microplates

A microplate rotor is required for a centrifuge to accommodate the use of microplates. Rotors typically hold 2 or 4 plates, as shown in Figure 4.12. Note that, by definition, a microplate rotor describes shallow well plates and not necessarily the deep well plates which are more commonly used in bioanalysis. It is important to contact the vendor to make sure that deep well plates can be used with a specific microplate rotor before ordering. Likewise, an existing centrifuge having an installed microplate rotor should be evaluated for its height or z-axis clearance using a deep well plate.

A benchtop or floor model centrifuge with microplate rotor is an essential tool for performing high throughput sample preparation in microplates. Note that a centrifuge that accommodates 2 to 4 microplates must be carefully balanced by weight. A duplicate blank plate that contains the identical volume and density of solutions is used to balance the sample plate. When the same volumes are used repeatedly for a microplate procedure, the blank plate can be labeled and saved for frequent use as a balance plate. While benchtop models commonly hold 2 or 4 plates, floor models are available that hold ≥8 plates.

A sufficient centrifugation speed to pellet particulates within wells of sealed collection plates is generally from 2000–3000 rpm. A typical duration is 5–10 min. In some cases, it may be desirable to offer additional stability and rigidity to a collection plate during high speed centrifugation. When necessary, a product such as the centrifuge support plate (Porvair Sciences) can be used to mate to the underside of a collection plate before insertion into a centrifuge. Some typical examples of vendors that supply centrifuges accommodating microplates are Beckman Coulter, Jouan (Winchester, VA USA), Thermo Forma (Marietta, OH USA) and Thermo IEC (Needham Heights, MA USA). Note that flow-through extraction plates and filtration plates can also be used with centrifugation to process liquids through the plate instead of using vacuum. This application is described in Section 4.2.2.

Figure 4.12. A microplate rotor is required for the use of microplates within a centrifuge. Photo reprinted with permission from Labconco Corporation.

4.5 Solvent Evaporation Systems

4.5.1 Nitrogen and Heat

Sample preparation procedures may require the evaporation of organic solvent from microplate wells. This evaporation or dry-down procedure is performed during a solvent exchange step (replace organic solvent with an aqueous or aqueous/organic combination). Evaporation is also used to achieve concentration of analytes. Procedures such as liquid-liquid extraction and solid-phase extraction commonly utilize a dry-down step because the analyte is often isolated in a solvent that is not compatible for direct injection into a chromatographic system. A protein precipitation procedure occasionally uses a dry-down step. The volumes of analyte in pure organic solvent (or in a high percentage organic solution) that typically require a dry-down step range from 0.4–1.0 mL and occasionally as high as 1.5 mL. These solutions are evaporated to dryness and are then reconstituted in a smaller volume of a more suitable solvent for analysis [1]. An overview of sample concentration by evaporation is provided by Majors [3].

Many products are available that offer efficient and productive solutions for dry-down of eluates from 96-well collection plates. These evaporation units are compatible with all deep well plates and microtubes, as the 9.0 mm center-to-center spacing is retained. They typically deliver nitrogen gas (most units offer a heating option for the gas) uniformly through 96 stainless steel tips. The height of the tips is fully adjustable, allowing the head of the unit to be lowered as solvents evaporate. Independent controls adjust the temperature, gas flow and sometimes the duration (automatic shutoff versus continuous action). Some models of evaporation units also allow heating of the base containing the microplate or warming the plate by heated gas passing through steel needles that fit into the cavity under the plate. When dual heat sources are used they are usually independently controlled. These evaporation units have a footprint that allows them to conveniently fit inside a fume hood [1].

Evaporation units accommodate from 1 to 3 microplates. In addition to 96-well capability, units in a 384-well version are being developed. Some typical examples of 96-well evaporation units for microplates are shown in Figure 4.13. Additional models (photos not available) are from Porvair Sciences, Orochem Technologies, Glas-Col, Techne (Princeton, NJ USA) and Organomation Associates (Berlin, MA USA). Note that IST Ltd. (now Argonaut Technologies) offers both a single and double microplate version.

 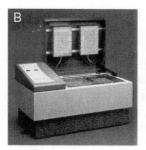 

Figure 4.13. Evaporation systems for microplates deliver heated nitrogen gas through 96needles. Photos reprinted with permission from (A) CEDRA Corporation, (B) Zymark Corporation and (C) Apricot Designs.

A practical and useful feature for an evaporation unit is the ability to remove the needle tips on top (and bottom) for cleaning. Organic solvent such as methanol, with sonication, works well to clean the tips. Occasionally, it is preferable to replace the top tips with a head accommodating disposable tips if such a feature is available. Special tips are made to offer increased resistance to solvents and non reactivity. The direction of gas flow into the wells of the plate varies amid a straight vertical, a slightly angled and a spiral path to aid in delivering a vortex flow. Some units offer interchangeable spiral needles. Usually, the position of the tips remains constant above the plate; a nice capability is for the unit to adjust this height over time by gradually raising the plate holder up toward the tips.

Safety measures in an evaporator are important, such as a pressure sensor that detects low or no gas flow and a temperature sensor that will reach a user defined maximum and then signal when that temperature is reached. A pressure relief mechanism is important to prevent the gas pressure from exceeding a defined upper limit and damaging the unit, the tubing lines and/or the user. A splashguard is a nice feature that is often positioned on the top element.

4.5.2 Centrifugal Force, Vacuum and Heat

Centrifugal concentrators use a combination of centrifugal force, vacuum and heat to speed evaporation of sample, solvent or eluate volumes. Centrifugation generates a centrifugal force of about 200xg to 500xg which prevents bumping and any physical loss of the liquid. Vacuum promotes solvent evaporation at sub ambient temperature, preventing damage to heat labile analytes and preventing oxidation. Heat is applied to counteract the cooling caused by the evaporative process under vacuum.

Centrifugal concentrators are available as single compact benchtop units and are frequently combined with separate cold trap and vacuum pump, as shown in Figure 4.14. Integrated, more compact systems are also available. The cold trap collects the vapors and fumes as they evaporate to prevent them from reaching the vacuum pump and causing corrosion over time. Systems are specifically designed for use with either aqueous, strong acid/base and/or solvent based applications. Suppliers of these types of evaporators include Labconco Corporation (Kansas City, MO USA), Thermo Savant (Holbrook, NY USA) and Jouan.

A microplate rotor is required for use with shallow well and deep well plates. A simple rotor can hold four shallow well plates stacked in pairs or two deep well plates. The use of larger chambers and higher capacity rotors can increase throughput. As an example, a rotor is available from Thermo Savant that can hold 8 deep well plates in a larger concentrator model. Tremendous throughput for research laboratories can be achieved with the Discovery SpeedVac® (Thermo Savant) that accommodates up to 48 deep well plates.

Evaporation from microplates can be slower than from traditional test tubes because the plate design makes it difficult to transfer enough heat for rapid evaporation. GeneVac Technologies (Valley Cottage, NY USA) products utilize solid aluminum heat transfer plates that fit into the bottom of a wide variety of microplates and ensure maximum heat transfer into the samples. Now, rapid evaporation rates are achieved without the risk of overheating fast

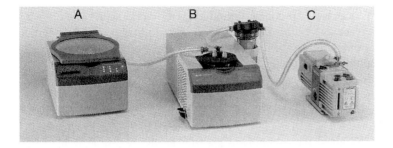

Figure 4.14. A centrifugal concentrator system includes (A) the concentrator unit, (B) a cold trap and (C) a vacuum pump. Photo reprinted with permission from Labconco.

drying samples. Some solvents and solvent mixtures (*e.g.*, acetonitrile/water 50/50, v/v) are known to bump unpredictably during standard centrifugal evaporation conditions, causing sample loss and cross-contamination. GeneVac offers a patented Dri-Pure™ system that prevents sample bumping by the combination of a gentle, ramped application of vacuum over time and increased rotational speed.

4.6 Microplate Compatible Autosamplers

A sealed microplate represents a very compact format for injection via an autosampler. The liquid samples are secured into distinct wells, sample integrity is preserved, and the collection of samples is easily handled and stored. Conveniently, samples are already uniquely labeled as A1–H12 [1]. Many different brands of microplate compatible autosamplers are available today. These autosamplers are expected to provide fast injection cycles, low carry-over, good linearity and high precision. An overview of how modern microplate autosamplers can meet the requirements for higher precision, lower carry-over and higher sample throughput is provided by Gratzfeld-Huesgen [4]. A related introductory article discusses the important issue of autosampler carry-over [5]. Some typical suppliers of these autosamplers include: Alcott Chromatography (Norcross, GA USA), Agilent Technologies (Wilmington, DE USA), Gilson Inc. (Middleton, WI USA) Hitachi Instruments (San Jose, CA USA), LEAP Technologies (Carrboro, NC USA), MDS Sciex (Norwalk, CT USA), Shimadzu Scientific Instruments (Columbia, MD USA) and Waters Corporation (Milford, MA USA).

4.7 Additional Accessory Products

The variety of accessory products introduced to work with microplates continues to increase as vendors develop ingenious solutions to solve specific needs of researchers. Many times, a product developed for a defined application in combinatorial chemistry or molecular biology may meet a related or different need in bioanalysis. Also specific solutions have been developed for the robotics platforms that sometimes are useful for semi-automated or manual applications. The range of unique microplate accessory products is diverse and the reader is encouraged to search outside the normal mainstream of suppliers in the quest for solutions to specific needs. A few examples are illustrated next.

A basic reaction block holder consists of a frame constructed of aluminum with anodized finish. A microplate is inserted into the frame which is secured with four Allen head screws (Figure 4.15). Although meant for combinatorial chemistry synthesis, the unit can be used for liquid-liquid extraction so that, once liquids are in the wells and it is sealed with a cap mat, the mat is rigidly held against the microplate. This microplate in block holder combination is then secured into a pulse vortex mixer for vigorous liquid-liquid extraction.

Heat sealable aluminum/polypropylene film provides a secure seal to retain liquid within wells of a microplate for a liquid-liquid extraction procedure. However, once the extraction is completed, the organic solvent (or aqueous solution if solvent is denser than water) needs to be aspirated and transferred to another plate. Disposable pipet tips are not always strong enough to pierce the seal. A solution to this problem is to pierce all 96 wells in one step using a piercing plate available from many suppliers such as Orochem Technologies. The plate has 96 rigid, steel tips that pierce the aluminum foil/polypropylene laminate with ease. A disposable tip can then enter each well to aspirate liquid.

Manually pipetting liquids into different wells of a microplate can be difficult and requires great concentration. A 96-well plate indexer makes it easier to perform this task. This indexer is a base upon which a microplate rests and a horizontal or vertical bar moves over the plate in a stepwise manner, row by row or column by column. The result of this system is that eyestrain is reduced and it is much easer to visually keep track of the pipetting position. These plate indexers are made of PTFE coated aluminum and are provided by suppliers such as Orochem Technologies and Horizon Instrument. A similar product, the Pipettor Guide (Diversified Biotech, Boston, MA USA) is a plastic indexer that helps align the pipettor with the microplate wells.

Figure 4.15. A reaction block holder is used as a secure chamber for a sealed plate used in liquid-liquid extraction. Photo reprinted with permission from Glas-Col.

The vigorous up and down shaking of deep well microplates is accomplished using a shaker mill such as the Model 2000 Geno/Grinder (SPEX CertiPrep, Metuchen, NJ USA). This instrument is typically used to prepare tissue for extractions of nucleic acid and protein by shaking the tissue with one or two steel balls and a buffering agent in each well of the microplate. The rate is adjustable between 500 and 2000 strokes per minute. It is feasible that this unit may be useful also for the preparation of tissue homogenates in drug metabolism prior to analysis of parent drug and metabolites. Since the tissues are in the microplate format, after the grinding procedure via extreme shaking with the steel balls, the plate can be centrifuged to pellet the mass of tissue and protein. The supernatant can be aspirated for subsequent sample filtration and/or extraction, followed by analysis.

Sonication can be provided to all the wells in a microplate by use of the Misonix Microplate Horn (Misonix Inc., Farmingdale, NY USA). Cross-contamination is eliminated with this approach since there is no direct probe intrusion into each sample. Sonication is reported by the manufacturer to be performed uniformly across all the wells in a microplate. Traditionally such a system has applications in cell disruption, high throughput screening and genomics. However, it may find a potential application in bioanalysis in certain situations for reconstitution of eluates when analytes are poorly soluble.

Some laboratories load their own particles into 96-well filter plates to meet exact needs for sample preparation methods, such as Peng *et al.* reported [6]. The MultiScreen™ Column Loader (Millipore) is a product that performs this particle loading into MultiScreen plates in a simultaneous and uniform step. Any bioanalytical processes would need the solvent resistant version of the plate called MultiScreen Resist.

The Mirror Rack from Porvair Sciences facilitates the pipetting of samples from cryovials (*e.g.*, NUNC™ vials) into the microplate format. When used with a 96-tip liquid handling head, the Mirror Rack enables sample transfer in less than a minute without the need to change pipette tips. The result of using this product is a considerable increase in productivity. The two halves of the Mirror Rack are color coded; 48-wells are transferred in two separate steps using unique pipette tips.

Acknowledgments

The line art illustration was kindly provided by Woody Dells. The author also thanks Greg Lawless for his careful review of the manuscript and helpful discussions. The assistance of the following vendor representatives who contributed information and reviewed text is appreciated: Paul Bouise (Mallinckrodt Baker), Mike Brown (Porvair Sciences), Chris Bugge (CEDRA Corporation), Claire Desbrow (IST Ltd., now Argonaut Technologies), Kim Gamble (MicroLiter Analytical Supplies), Shirley Hogenkamp (Labconco Corporation), Jim Jacso (Glas–Col), Colin Keir (Axygen Europe), Brian King (Chromacol Ltd.), Ron Majors (Agilent Technologies), Eugene Margolin (Innovative Microplate), Asha Oroskar (Orochem Technologies), Cecile Poirier (Simport), Roger Roberts (Ansys Technologies, now a part of Varian), Jinghua Schneider (Trade Winds Direct), Chris Simmons (ABgene North America), Jenny Sprung (Labconco Corporation), Lynda Thomas (Zymark Corporation), An Trinh (Supelco), Chester Wang (Horizon Instrument) and Felix Yiu (Apricot Designs).

References

[1] D.A. Wells, LC-GC 17 (1999) 808-822.
[2] R.D. Bolden, S.H. Hoke II, T.H. Eichhold, D.L. McCauley-Myers and K.R. Wehmeyer, J. Chromatogr. B 772 (2002) 1-10.
[3] R.E. Majors, LC-GC 17 (1999) 500-506.
[4] A. Gratzfeld-Huesgen, LC-GC Europe 14 (2001) 2-6.
[5] J. Dolan, LC-GC Europe 14 (2001) 148-154.
[6] S.X. Peng, T.M. Branch and S.L. King, Anal. Chem. 73 (2001) 708-714.

Chapter 5

Automation Tools and Strategies for Bioanalysis

Abstract

Bioanalytical laboratories typically process from tens to hundreds of thousands of samples per year in the course of their work. The automation of sample preparation is an important goal for laboratories to meet the high throughput demands required in pharmaceutical research and development. The choices for automation differ in complexity according to the required task. Processes can be introduced to make a manual procedure less tedious, to replace most or all steps in a manual procedure, to perform a specific application comprised of multiple linked steps or to perform a procedure around-the-clock. The size of automated instruments can vary as well, from a small benchtop instrument to a large instrument placed on a tabletop to a room-sized ultra high throughput configuration composed of individual modules linked together. The scientist needs to make informed choices about the type of automation needed to perform the identified tasks. This chapter introduces the roles for automation in bioanalysis, reviews strategies for selecting automation and provides an overview of the different types of instruments available to automate bioanalytical applications. These instruments are discussed as three distinct groups: (1) task-specific devices, (2) liquid handling workstations operating in a semi-automated or fully-automated mode, and (3) application-specific workstations or systems. The chapter concludes with a discussion of strategies for using automation in a bioanalytical setting.

5.1 Roles for Automation in Bioanalysis

Bioanalysis (the quantitative determination of drugs and their metabolites in biological fluids) is used very early in the drug development process to provide support to drug discovery programs on the metabolic fate of chemicals in animals and in living cells. Its use continues throughout the preclinical and clinical drug development phases. The detection technique of choice, liquid chromatography interfaced with tandem mass spectrometry (LC-MS/MS), allows more samples to be analyzed per unit time than ever before. As a result of this capability for rapid analysis, and the greater number of compounds able to be synthesized by combinatorial chemistry techniques, faster turnaround became expected for both pharmacological and bioanalytical evaluations. The industry demanded faster sample analysis as the race to market became more aggressive. Sample preparation was targeted as a rate limiting step in the overall procedure for bioanalysis.

The pharmaceutical industry has responded to the need for greater sample throughout in many ways. On the detection side, faster chromatographic techniques are utilized, *e.g.*, fast gradients with short columns and the greater use of monolithic columns (see Chapter 14, Section 14.2.4). On the sample preparation side, faster on-line approaches were devised, *e.g.*, parallel extractions with multiple columns (see Chapter 14, Section 14.4). Faster off-line approaches were also developed, *e.g.*, more productive formats (96-well plate) and more efficient automation solutions for these off-line sample preparation methods. The throughput of sample preparation techniques for drug bioanalysis has increased dramatically in many approaches to keep pace with the speed of detection.

The term "off-line" refers to performing the sample preparation procedure independently of the LC-MS/MS analysis. The term "on-line" refers to a serial or staggered serial approach of an instrument first preparing the sample and then analyzing it in sequence. While the first sample is being analyzed the second sample can be undergoing sample preparation. On-line sample preparation techniques are discussed in detail in Chapter 14.

While improving overall throughput and laboratory productivity are important goals for introducing automated processes into bioanalytical sample preparation methodologies, other motivating factors exist as well. Automation can reduce the hands-on time required by an analyst and allows that scientist to perform other tasks in the laboratory. Unattended operation of a process may permit multiple or overnight runs, further increasing throughput. Removing an individual from hazardous (*e.g.*, radioactive or toxic chemical usage) and/or mundane tasks can maintain worker health and safety. The automation of most processes has been shown to bring a degree of reproducibility and quality to the results that cannot be realized among different workers each performing the method manually. In addition, compliance with GLP (Good Laboratory Practices) is considerably improved as a result of the remarkable process consistency that automation provides. This commonality in the approach lends itself to easier method duplication or transfer, both among analysts and laboratories. Automation can often facilitate troubleshooting by removing the manual component processes. It can introduce process documentation into a method and also allow for less skilled workers to perform a complex task. An important overall goal for implementing automation into a laboratory workflow is greater employee job satisfaction [1].

5.2 Strategies for Selecting Automation

5.2.1 Defining How Automation will be Used and Supported

A laboratory must properly assess its needs for automation and define how it will be used and supported in the workplace. A series of questions can be addressed to help define these needs and aid in the definition of the equipment that will perform best for the task. The focus of this discussion is on the off-line processes for sample preparation.

How will the automation be used? This question is one of the first to be asked. Will the automation be used for performing the off-line techniques for protein precipitation (PPT), liquid-liquid extraction (LLE) and solid-phase extraction (SPE) in microplates? Will it be used for tubes in PPT and LLE methods but in microplates for SPE? Perhaps it will only be used for solid-phase extraction or for a particular project. Sometimes a sponsor may fund the study and define the exact equipment specifications. Defining the performance needs of the automation is discussed in Section 5.2.2.

Who will be using the automation? If one analyst will be using the automation, the sample preparation task to be performed, along with the skill of the analyst, will largely influence the choice of equipment. If several analysts will use the automation, then ease of use plays a larger role, especially when analysts display varying skills and expertise in the use of computer software and/or instrumentation.

Which applications will be automated? If the automation will be used for one technique or application only, perhaps productivity is the ultimate goal. However, if flexibility is the objective, the adaptability of the instrument to perform multiple tasks is an important criterion in the decision-making process. Also, upgradeability may be important. One configuration can be purchased (or leased, in some cases) and over time its capabilities can be upgraded.

Where will the instrument be located and what type of access is desired? Multiple instruments can be scattered among individual laboratories and supported by those researchers. The user group is small. On the other hand, one or more instruments can be located in a common area for walk up and shared access by a large user group. In the latter case, strict scheduling of instrument time is critical. Users should walk up to the instrument with all their supplies, perform the sample preparation task and then release the instrument for use by another individual. When a dry-down step is required off-line, the automation should remain available. When the sample has been dried down, the user returns to the instrument and performs the reconstitution step.

How friendly and flexible should the software be? The software itself can be a deciding factor in selecting among automation choices. Software that is straightforward with little flexibility can be understood and programmed by most analysts very quickly (fast learning curve), while software with great flexibility and many options can be daunting to some analysts, taking weeks to master. One concern with writing a sample preparation software program is the ease of inadvertently changing a default variable that can affect pipetting performance. It is often observed by the author that users may change settings without fully understanding the full effect on pipetting performance. Program defaults such as system air gaps, aspirate and dispense speeds and syringe sizes should be contained several layers deep so that only experienced users or custodians are able to change them.

How much support has management committed? When a mandate is made to utilize automation for all processes, including all pipetting tasks, enough units need to be available for each user to access in a reasonable time period. Also, Standard Operating Procedures (SOP) must be developed and followed to ensure accuracy and precision of pipetting with proper attention to maintenance and service issues.

Who will maintain the instrument? A support person or custodian is often chosen for each instrument. That person maintains supplies, runs calibration and other performance verifications, schedules service and may even act as a trainer to other users. Some laboratories have a full-time automation support specialist to maintain all instruments. **Important Note: In no case should the instruments be allowed to be used without a designated support individual.**

How much space is available? A benchtop area commonly comprises all the space that is available for instrument placement. While this space is often adequate, note that proper exhaust lines need to be placed near the instrument to remove organic solvent vapors from the work environment. Typically the best ways to accomplish vapor removal are to place the unit inside a custom-built vented enclosure or to place the unit inside a fume hood (if the dimensions permit). When multiple components are configured and linked together, a tabletop area needs to be available. Again, proper venting of organic vapors must be evaluated. Keep in mind any special electrical requirements for the equipment as the space is planned. Also be aware of computer network connections and sources for compressed air and house vacuum.

What are the productivity goals for the automation? The number of samples processed per hour is one aspect to examine with regard to automation productivity. Another aspect is the hands-on time required by the analyst. A semi-automated method that requires 15 min hands-on time may be preferable to a fully automated system that might cost more but perform at similar productivity. The cost to automate that "last step" to reach full automation can often be reached but for a price not commensurate with its value [2]. When the productivity numbers are not near the goal desired, it is usually necessary to look at automation that processes more samples in parallel, *e.g.*, move toward 96-tip pipetting when throughput needs are not met by 4- or 8-tip instruments.

What is the allowable budget? Typically, the available funds strongly influence the capability that automation can provide and the rate at which it can be implemented. A conservative approach is that one instrument is purchased and then with continued use the justification can be made for purchase of another. Also seen is the purchase of a 4- or 8-tip liquid handling workstation and then a 96-tip workstation is purchased after time or vice versa. Having a 4- or 8-tip workstation available to reformat samples from tubes to a microplate for use on a 96-tip sample preparation workstation is ideal. Other times, two or more 96-tip workstations are purchased and sample reformatting is relegated to a manual task.

What are the anticipated and supported needs for ongoing training? Just as training in the principles and operation of mass spectrometers is supported, a high priority should be placed on continued training and education in automation. Actually, a validated system assumes that only individuals that have been properly trained will be using it. Therefore, training is more than a good idea, it is often a requirement. Note that training records of system operators are often audited. Additional ongoing needs may involve the status of CFR Part 11 compliance and the budget and resources required to perform instrument validation and software upgrade to meet these requirements. Issues of software compatibility with the network already in the laboratory also require ongoing support.

Conferences in automation are a great source of continuing education and allow for awareness of applications by other users. Vendor courses are also available at their site or in-house (when there are enough participants) for in-depth training on specific automation platforms. Occasionally, companies will use independent training professionals in laboratory automation; in this case, an expert who understands the science behind the sample preparation techniques also teaches and optimizes the automation.

5.2.2 Defining Performance Needs

An accurate definition of the performance needs for an automated instrument is very important, along with a definition of how the automation will be used and supported (as discussed in Section 5.2.1). The focus for this discussion is again on off-line processes for sample preparation; on-line processes are discussed in Chapter 14.

As an example, let's propose that automation is needed in our laboratory to perform off-line sample preparation for PPT, LLE and SPE—all in microplates. The laboratory staff can be brought together and a brainstorming session conducted. The question can be asked, "What do we need a liquid handling system to accomplish for us in our bioanalytical sample preparation applications?" The group's responses are recorded. While all the desired needs for automation may never be fully realized, allow them all to be listed during this session. Remember that in brainstorming, all suggestions are compiled without discussion or criticism. Later, the responses can be reviewed and then the most important criteria can be prioritized. The list is reviewed together with knowledge of the budget and an understanding of how the instrumentation will be used. At this point, a few automation solutions can be proposed and the vendors are contacted for more information.

The following sample responses from a brainstorming session are typical. Prioritization takes place at a subsequent discussion.

1. Proven technology within the industry with a large installed customer base and network
2. Extensive programming skills not required for analyst
3. Usable on a walk up basis with multiple users per day
4. Able to perform automated methods development
5. Software easy to use and run on computers compatible with corporate directives
6. Able to prepare standards and quality control (QC) samples

7. Sample tracking capability via bar codes
8. Compatibility with LIMS system
9. Versatile; add-ons and accessories are available as new needs develop
10. Compatible with deep and shallow well microplates as well as current sample tubes (*e.g.*, 2 mL cryovials)
11. Can pipet all liquid samples, including internal standard
12. Disposable tips option to eliminate carry-over concerns
13. Able to hold 500 µL volume in disposable tips
14. Report generation capability
15. Liquid sensing capability
16. Little hands-on analyst time
17. Small volume accuracy down to 10 µL
18. Refrigeration of plasma samples on instrument deck
19. Preventive maintenance program clearly established by vendor
20. Perform protein precipitation procedures in plate format
21. Perform liquid-liquid extraction procedures in plate format
22. Rugged for constant usage each workday
23. Vendor able to validate software and instrument with proper scripts and documentation in order to meet regulations
24. Able to calibrate on a periodic basis
25. Quick service response times; expect repair in 24–48 h
26. Training available from vendor or an outside firm
27. IQ/OQ procedures defined
28. Able to be enclosed to vent solvent vapors

5.2.3 Selecting a Vendor

Many questions should be asked about the vendor before an automation purchase. The following list offers some pertinent suggestions.

1. Vendor response time in providing field service to this region, both in terms of hardware and software support
2. Commitment and response time for telephone support
3. Depth of technical service provided (backup personnel in territory)
4. Time vendor will spend training staff upon installation
5. Availability of additional training via classroom instruction
6. Preventive maintenance schedule
7. Software validation by manufacturer
8. Hardware validation by manufacturer
9. Performance verification performed by vendor for a fee
10. Extent of documentation supplied

11. Warranty terms
12. Service contract after initial warranty expires
13. Installed customer base and customer referrals, *i.e.,* reputation
14. Length of time vendor has supported the intended application

The quality of vendor support networks can vary considerably. Their relative importance depends on how complex a system is purchased and how much reliance is made on the vendor to install, operate, maintain and continually develop the system. The real cost of each automated system should consider all associated warranties, training, documentation, technical and applications support, continuing service and the upgrade policy.

The total capabilities of the system to be purchased need to be understood so that it is known what additional add-on options are available and which would be beneficial in obtaining. Sometimes, options are withheld from purchase for reasons of cost control. However, it is suggested that the initial purchase order should include as many additional racks and options as can be anticipated.

It is difficult to determine system reliability prior to purchase and use of the equipment in support of daily laboratory activities. Ask the manufacturer to supply references of users who may give a candid review of their experiences. Some companies are willing to offer the use of a demonstration unit for a hands-on evaluation prior to making a purchase. The process of acquiring and installing the demonstration unit will often provide insight into the quality and availability of the vendor's technical and service support for this system. As each individual step of the automated process is evaluated, keep in mind that the system will only be as reliable as the weakest link in the process. For more complicated or customized systems, it is wise to agree on performance specifications as part of the purchase agreement. This performance expectation shared up front serves as a good and fair source of protection for both the customer and the manufacturer [1].

5.2.4 Obtaining Customer Referrals

It is wise to contact other users of the automation chosen, before purchase, to ask for an honest assessment of their experiences to date. Phone calls should be made to these other bioanalytical laboratories and they can be asked for their experiences in using the particular equipment. In the case of a newly introduced piece of equipment, these same questions can be used to get a general idea of the vendor's offerings in the areas of hardware, software and after sale support.

Some suggestions for questions to ask owners of the equipment to be purchased include the following:

1. What do they like about the instrument?
2. What do they dislike about it?
3. How much time does it take to process a microplate?
4. Have they seen carry-over issues?
5. How often is the instrument "down" or in need of service?
6. What other instrument did they seriously consider?
7. Why did they decide to purchase this instrument over another?
8. If they had to make the decision again, would it be the same?
9. Would they buy products from this vendor again?
10. Would they buy a second instrument of this same type?
11. Is there a network or user group established?
12. How easy is the instrument to use?
13. How intuitive is the software to program a method?
14. How does their laboratory handle responsibility for the instrument and training of additional users?
15. How responsive has the manufacturer been to requests for service or support?
16. What are their **most** and **least** favorite features of this instrument?
17. What is the total hands-on time required for a method?
18. What accuracy and precision numbers have they seen when performing performance verifications?
19. What schedule do they follow for cleaning the instrument?
20. How long did it take for users to learn the software and start programming their own methods?
21. What additional components or options would you recommend be purchased with this system?

Section 5.2 has described in general terms the strategies and processes for selecting automation for use in a bioanalytical laboratory. Clearly, it is not easy to choose the right equipment without defining the exact needs, budget and support plan, and investing the necessary amount of time in discussions and product research.

The next few sections provide an overview of the types of instrumentation available for automating bioanalytical applications. The goal of these sections is to educate the reader in typical examples of the choices available. While the information is meant to be instructive, it should not be considered extensive and the individual manufacturers should always be contacted to obtain more complete and current information about their products.

5.3 Task-Specific Automation

5.3.1 Introduction

The typical idea of automation is often that of a large instrument having the flexibility to perform many different tasks. It is perceived as very expensive and its operation as very complex. However, this scenario is not always accurate. Currently, components that are task-specific are available to perform repetitive motions very reliably. A lower investment in price allows their acquisition and use. This lower cost can be particularly important for smaller laboratories that are looking to add automation to their workflow. In recent years, the approach of using low budget instrumentation to automate the most labor intensive sample preparation steps (*i.e.*, "bottlenecks") has shown to be most effective.

Two of the most common task-specific pieces of automation used for high throughput applications are a benchtop pipetting unit for dispensing liquids and a microplate-handling device to move, load and unload plates from a defined radius in its work area. Also, those laboratories that still use tubes may benefit from a capping/uncapping unit that can prepare tubes for use on a liquid handling workstation that performs sample reformatting (tubes to microplate).

5.3.2 Dispensing Liquids

The dispensing of liquids into microplates is a single task that can be automated and speed the sample preparation process dramatically, compared with manual pipetting. A compact stand-alone unit can deliver liquids into 96-well plates with very good accuracy and precision. Many units now also offer the capability to use 384-well plates. In its most basic form, this unit consists of a horizontal stage that moves source and target plates under a single dispensing head. An operator stands at the instrument to place microplates and activate the dispensing process. This type of unit can be useful for sample preparation by adding organic solvent to a microplate to perform protein precipitation or liquid-liquid extraction procedures. The cost of these units is usually much less than that of a full-featured liquid handling workstation and can represent a point of entry for automation into a laboratory.

Some typical examples of these small footprint liquid dispensing units having 96-tip capability are the Personal Pipettor™ (Apricot Designs, Monrovia, CA USA), the EDR384S/96S (Labcyte, Union City, CA USA) and the Hydra-96™ (Robbins Scientific, Sunnyvale, CA USA, now a part of Apogent Discoveries). The Personal Pipettor can be configured with 1, 3 or 4 labware stations on the small horizontal deck; the EDR384S/96S accommodates 3 stations; and the Hydra96 has 1 station (Figure 5.1). These dispensing units can perform protein precipitation in microplates and also liquid-liquid extraction. A variety of tools can be placed on the deck. By adding a vacuum manifold to the Personal Pipettor and EDR384S/96S, these instruments can perform filtration and solid-phase extraction applications as well. The Hydra-96 is a one position unit that has fixed washable needles and dispenses volumes as low as 100 nL (several syringe sizes are available). The Hydra line represents a series of instruments with additional features (automated wash, automated plate positioning, automated plate stacking).

The 96-tip units offer great throughput and are preferred; however, alternate units having only 8- or 12-tips (probes) are also available in the market. Some

Figure 5.1. Examples of small footprint 96-tip liquid dispensing units. *Left:* Hydra-96. *Right:* EDR384S/96S. Photos reprinted with permission from Robbins Scientific and Labcyte, respectively.

typical examples of 8- or 12-channel small footprint dispensing units include the MultiDrop Deep Well (Thermo Labsystems, Franklin, MA USA), µFill™ (Bio-Tek® Instruments, Winooski, VT USA) and the FlexDrop Precision Reagent Dispenser (PerkinElmer Life Sciences, Boston, MA USA).

The size of the stage (*i.e.,* number of stations) differentiates many models from each other; the small footprint units typically offer a 1- , 3- or 4-position stage. Increasing the stage size adds to the functionality of the instrument but increases the overall dimensions and results in additional cost. Optional features that are found in some of these instruments may include a stacker, access to multiple reagent vessels, use of disposable tips, aspiration and evacuation of liquids and liquid level sensing.

An example of a more feature-rich dispensing unit with great flexibility is the RapidPlate™ (Zymark Corporation, Hopkinton, MA USA) which has a six-position turntable for use with a fixed 96-tip dispensing head. While most dispensing instruments in this 96-tip class utilize a fixed position dispensing head, the Multimek® (Beckman Coulter, Fullerton, CA USA) moves its dispensing head over a six-position work deck that can be filled with plates, tips, reservoirs and a wash station. The RapidPlate and the Multimek are at the upper range in cost for 96-tip dispensers although they are very feature rich and integrate superbly with additional components.

5.3.3 Moving Microplates

A versatile piece of automation that adds great functionality to a liquid handling system is a microplate handling device. This unit is essentially a track-mounted robotic arm (up to a 345° axis of rotation) capable of handling microplates in landscape and portrait orientations. Its most basic function is to feed plates onto the deck of an instrument (*e.g.,* liquid handler, plate reader, plate storage) and remove them.

The most widely used universal microplate handler on the market is the Twister™ (Zymark). The newer Twister™ II Advanced Capabilities Microplate Handler offers greater capacity (up to 400 plates), faster transport and the ability to access the X-Y index positions of almost any liquid handling deck. Unique features of the Twister II (Figure 5.2) are its telescoping arm and rotating wrist joint which allow great compatibility with over 70 other automated systems. CLARA™ integration software (Zymark) supports the Twister products.

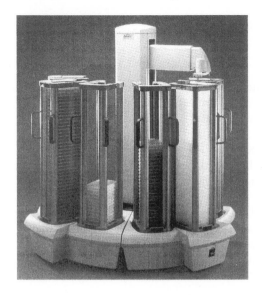

Figure 5.2. Example of a universal microplate handler, the Twister™ II. Photo reprinted with permission from Zymark.

Another model of a benchtop microplate-handling device is the MICROLAB® SWAP plate-handling robot (Hamilton Company, Reno, NV USA). The Hamilton MICROLAB Vector software supports the SWAP device and allows its integration with liquid handling workstations such as the MICROLAB STAR and the MICROLAB 4000. Many other microplate handlers are available in the market, such as the Assist from Thermo Labsystems.

5.3.4 Tube Capping and Uncapping

The capping and uncapping procedure using test tubes is a tedious, hands-on task. It is repeated multiple times when performing a liquid-liquid extraction procedure. An aid to automating this action is to use a capping/uncapping tool just prior to placing the opened tubes onto the deck of a liquid handling workstation set up to accommodate a method. An example of this task-specific tool is a four gripper capping/decapping system (Horizon Instrument Inc., King of Prussia, PA USA). This type of unit can fit inside a fume hood. By pushing a button the operator can uncap a batch of 48 tubes within 5 min. After the liquid handling procedure using the tubes is completed, the same caps or new caps are placed onto the original tubes, greatly reducing the risk of contamination between tubes and exposure of the operator.

5.4 Liquid Handling Workstations

5.4.1 Introduction

The key distinguishing feature of liquid handling workstations, also known as automated liquid handling systems, is movement of a probe in Cartesian axes (X-Y-Z) across a deck or work surface that contains items such as tubes, microplates, solvent troughs, disposable tips and other types of devices needed to perform a given task. This probe is part of a pipetting system commonly composed of 1-, 4-, 8- or 96-channels. One manufacturer also offers 6-, 12- and 48-channel units. Independent gastight syringes driven by stepper motors are used for pipetting liquids. Common pipetting tasks performed include addition of reagents, solutions to adjust pH, internal standard solution, dilutions and liquid transfers (*e.g.*, from tube to tube, plate to plate, tube to plate, tube or plate to autosampler vials) and reconstitution. These workstations wash and rinse fixed probes and work with disposable tips as well. They are controlled by computer software and the operator directs the actions by writing a program consisting of a sequence of steps.

Many different X-Y-Z liquid handling systems are available on the market, differing in deck size, the number of probes (also called tips in this text), speed and pipetting features and capabilities. The standard specification for well-to-well spacing in a microplate is 9.0 mm; when the liquid handling workstation can adjust the spacing of the probes wider than 9.0 mm, then it can be used to reformat samples from tubes into microplates. The entry level units can perform sample preparation procedures using microplates but do so in a semi-automated mode, *i.e.*, for SPE a manual step is required to remove the waste tray and insert a clean collection plate for the elution step.

As models of instruments were improved and expanded, the functionality and usefulness of an integrated gripper arm became clearly evident. Labware movement around the deck and into external devices is enabled in such workstations. The presence of a gripper arm on a workstation permits it to operate in a fully automated mode (disassembly and reassembly of a vacuum manifold) which is important to obtain maximum laboratory productivity.

The following sections provide an overview of the types of liquid handling workstation products available for automating bioanalytical applications. These systems are grouped for discussion first by number of tips or probes (1-, 4-, 8- tip and 96-tip) and then, if applicable, by semi- versus fully-automated capabilities. The goal of these sections is to educate the reader in typical examples of the choices available. While the information is meant to be instructive, it should not be considered exhaustive and individual manufacturers should always be contacted to obtain more complete and current information about their products.

5.4.2 Semi-Automated Solutions—Single Tip (Probe)

Liquid handling workstations consisting of one probe for liquid pipetting functions represent an entry level and simplified approach to automation. While the capabilities and throughput cannot match those of the more feature rich models having greater tip numbers, the low cost makes them a good value. The learning curve for operation is short. Smaller laboratories that cannot afford the top of the line models can often find a good performing unit in the single tip workstation category. The deck size of these units can vary; at the minimum, they usually hold four microplates and can easily fit inside a fume hood. Larger models can accommodate up to 38 microplates on the deck surface.

A typical example of an ultra compact single tip X-Y-Z liquid handling workstation is represented by the QuikPrep (Orochem Technologies Inc., Westmont, IL USA), shown in Figure 5.3. This model utilizes a fixed probe and can transfer volumes up to 1 mL. Accessories such as a vacuum manifold or a heating/cooling reactor block for some chemistry applications can be placed on the deck. The QuikPrep can be upgraded to a 4-tip model and can also handle disposable tips (20 µL and 200 µL).

Larger deck sizes and liquid level detection capability are features of workstations by Hamilton Company and Tecan US (Research Triangle Park, NC USA). Hamilton offers the single probe MICROLAB 4000 (18 microplate capacity), while the 4200 has a 38-microplate capacity. Note that the MICROLAB series can be upgraded to meet a laboratory's growing needs. A single probe model can be purchased initially and dilutor modules and a multiple probe head can be added later. The 4000 can be upgraded to a 4- or 48-tip head, while the 4200 can replace its single probe with a 4-, 6-, 8-, 12- or 96-tip head.

The Hamilton MPH-8+1 represents a special configuration of the MICROLAB 4200 series designed for flexible hit picking automation. The MPH-8+1 has eight low volume probes for multichannel pipetting and a longer probe for single well pipetting to pick random hits. This design allows both individual well and multichannel pipetting with the same fixed probe instrument.

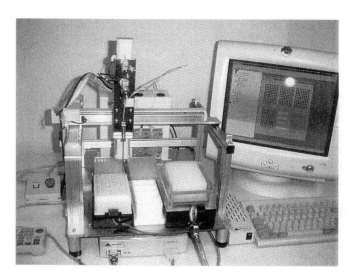

Figure 5.3. Example of a single tip liquid handling workstation, the QuikPrep. Photo reprinted with permission from Orochem Technologies

Tecan offers a single probe model in the MiniPrep™ Series. MiniPrep 75 has a 12-microplate deck capacity with liquid level detection capability; 8-tips (probes) are available, as well as a two-arm configuration (1-tip/1-tip or 8-tip/8-tip). Some other models of compact single tip liquid handling workstations are manufactured by J-KEM Scientific (St. Louis, MO USA), Thermo Orion (Beverly, MA USA) and Nichiryo America Inc. (Flanders, NJ USA).

5.4.3 Semi-Automated Solutions— 4- or 8-Tips (Probes)

5.4.3.1 MultiPROBE II

Liquid handling workstations having 4- or 8-probes (one manufacturer offers 6 or 12 probes) are rich in features and can perform pipetting activities for a wide range of disciplines. The MultiPROBE® II (4-probe) (Packard Instruments, Meriden, CT USA, now a part of PerkinElmer Life Sciences) and the MultiPROBE II HT (8-probe; see Figure 5.4) are described here as typical examples of workstations in this performance class. Two other models in this category include the Genesis™ (Tecan) and the MICROLAB (Hamilton Company); their distinguishing features will be noted in this discussion.

Figure 5.4. Example of an 8-probe liquid handling workstation which can perform a wide variety of pipetting tasks for bioanalytical sample preparation and drug discovery applications, the MultiPROBE II EX HT. Photo reprinted with permission from PerkinElmer.

The MultiPROBE II is offered in two deck sizes—16- and 32-microplate capacity. The larger version is noted as EX for EXpanded deck size, in which left and right expansion modules are utilized. The probes operate independently and offer great flexibility in liquid processing. The VersaTip® PLUS adapters enable the use of eight fixed washable probes (stainless steel probes coated with PTFE) or eight probes with disposable tips without any change in hardware; most other brands designate four fixed and four with disposable tips. Both fixed and disposable tips can be used in the same assay. Fixed probes are washed within procedures using a peristaltic pump and wash station. Disposable tip sizes used are 20, 200 and 1000 µL. A low volume fixed probe is available to dispense down to 500 nL. Liquid handling using the fixed probe mode uses positive displacement in conjunction with system liquid (deionized water) while the disposable tip mode employs air displacement to move liquid through the tip.

Varispan™ offers variable spacing between sampling probes from 9 to 20 mm (4-probe) or 9 to 40 mm (8-probe). The unit can access a wide variety of plates, tubes, vials and reagent vessels since tip spacing is adjustable. Positional reproducibility is better than 1 mm which allows the use of 384-well plates in addition to 96-well plates. Accusense™ offers patented, liquid level sensing on each probe down to a 50 μL volume. Liquids such as methanol and dimethyl sulfoxide (DMSO) can be detected, in addition to low ionic and non-ionic solutions. Note that the organic solvent methyl tert-butyl ether (MTBE) cannot be sensed, important when performing LLE methods.

Pipetting performance is optimized via the use of performance files. These files contain information such as aspirate and dispense speeds (specific to syringe volumes), air gaps, liquid delays, waste and blowout volumes, speeds for entering and exiting liquids and limits for liquid submersion level. Default performance files are provided for liquids and solvents and the user can modify them to work with unique solvents or to fine-tune pipetting accuracy. Syringe sizes can be interchanged to offer a range of pipetting volumes.

Computer control of the workstation is performed using WinPREP™ software. Standard templates are provided for common assays and tasks, *e.g.*, reagent addition, single or multiple liquid transfers, serial dilutions, plate replication, sample pooling, mixing, hit picking, and reformatting from tubes to plates. Custom procedures can be created by linking basic pipetting actions such as aspirate, dispense, flush, wash, get tip, drop tip, *etc.* The software has great flexibility and can read from external text or CSV (comma separated value) files, accepts protocol extensions using custom scripts and can link to external executable programs.

Adapter support tiles securely hold a variety of labware items onto the deck of the liquid handling workstation. Software manipulation of labware offers drag-and-drop capability. Custom labware can be added by defining its physical specifications in the software and securing it to the deck using the appropriate support tile. Rotation of labware is permitted 90° increments. A typical arrangement of labware items on the deck of a liquid handling workstation is shown in Figure 5.5. Various modules can be integrated with the instrument, such as a PlateStak™ microplate exchange system, a labware shaker, a thermal cycler, heating and cooling tiles; robotic systems can be integrated, as well. The addition of these modules allows the MultiPROBE to perform a variety of complete applications, *e.g.*, cytochrome P450 microsomal assays and PCR assays. Ancillary devices can be located adjacent to the deck, if not placed directly onto the deck.

Well maps are built into each labware item that accommodates samples. The well map defines the exact order that samples will be aspirated. The orientation

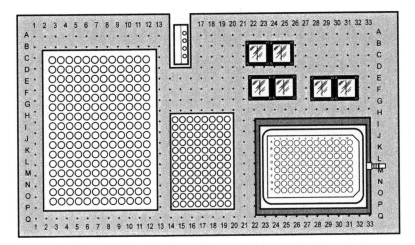

Figure 5.5. Labware items are placed onto the deck of a liquid handling workstation in any arrangement and orientation desired, as long as the support tiles line up with the notches in the deck. Shown to the right of position A13 is a flush/wash station for the sampling probes, although it can be placed anywhere on the deck. Also shown for illustration are a test tube rack, a 96-well microplate, solvent troughs, and a solid-phase extraction plate on top of a vacuum manifold.

can be defined by row or by column, and the direction can be up or down, left or right. Also, the capability exists to do cherry picking and specify any order of wells from the first to the last position. This well map feature is very convenient when using partial SPE plates.

A vacuum pump or house vacuum can be connected to a manifold on the deck of the instrument. Integrated vacuum control software provides plug and play use and includes all functions required for full automation. This system provides for vacuum build up, high and low vacuum settings and venting. Applications such as solid-phase extraction, microplate filtration and nucleic acid extraction assays can be automated on the deck of the instrument. Use of a vacuum manifold for an assay requires manual intervention to remove the waste tray just before final collection of eluate or filtrate; addition of a gripper arm (discussed in Section 5.4.4) allows full automation of this task. Note that any manifold can be used on the deck of the MultiPROBE in a semi-automated mode.

The "check well" procedure of the MultiPROBE II is very useful for sample preparation using 96-well solid-phase extraction plates. When this procedure is turned on, the probes enter a well that should be empty from a previous step (*e.g.*, after a vacuum step) and check for the presence of liquid using its Accusense™ liquid level sensing technology. If a well is found to contain liquid, then that well can be flagged for full aspiration and transfer or it can be avoided from future pipetting steps (so that it will not overflow and contaminate surrounding wells).

Improved throughput in reagent addition can be gained with the use of a six-way valve having lines directly connected to a solvent bottle instead of an open solvent reservoir on the deck of the instrument. Software activates and turns the valve to the desired position in order to use a solvent for a reagent addition step. Use of this approach is beneficial as it provides for convenient off the deck placement for multiple reagents, provides a larger reservoir volume than can be held in a solvent trough (150 mL maximum), and allows better control of the environment in which the solvents are contained. Use of the six-way valve provides faster throughput compared with traditional reagent addition using syringes with volumes of 500 µL and larger.

5.4.3.2 Genesis RSP

The GENESIS RSP (Robotic Sample Processor) from Tecan is a 4- or 8-probe liquid handling workstation that is available in a choice of deck sizes and configurations. A photograph of this instrument is found in Chapter 1, Figure 1.2. Liquid handling operations are controlled by the Gemini™ software (Tecan), which in addition to the features mentioned above in regard to the WinPrep software supports 21 CFR part 11 compliance. The Genesis and Genesis Freedom series are liquid-column driven, have three available deck sizes (highly configurable; 28, 42, and 66 microplates), and can use washable fixed or disposable tips. Tecan's Te-VacS SPE manifold comes in models appropriate for bioanalytical or molecular biology applications, including a model resistant to organic solvents (*e.g.*, ethyl acetate). As many as 8 Te-VacS modules have been configured onto a single instrument. A check for clogged wells can also be incorporated into a protocol, as in the dedicated Tecan system for SPE based viral RNA isolation which has been validated for clinical use in Europe. The addition of a <u>R</u>obotic <u>M</u>anipulator <u>A</u>rm (RoMA) offers full automation capability and allows integration with plate washers, readers and a wide range of other devices. GENESIS instruments can form the foundation for an application-specific workstation approach, as discussed in Section 5.5.

5.4.3.3 MICROLAB STAR

The MICROLAB STAR (Hamilton Company) is a flexible pipetting workstation with 4-, 8- 12- or 16-probes that display variable spacing. Unique to the STAR is a patented CO-RE technology (Compression induced O-Ring Expansion). CO-RE represents advancement in tip coupling that allows positive grasping, sealing and releasing between the channel and the tip. Compatibility is demonstrated with both disposable tips and steel probes. Dual mode (capacitive and pressure based) liquid level detection provides sensing for both conductive and non-conductive liquids. An optional internal gripper called iSWAP is available to manipulate plates on the instrument deck or transfer them to adjacent microplate compatible instruments. The MICROLAB AVS vacuum system, discussed in Section 5.4.4.2, can be added to the deck of the STAR. Another useful option is a temperature controlled microplate carrier that can maintain temperatures on the instrument deck from $-22°C$ to $+60°C$ using a Peltier system.

5.4.4 Fully Automated Solutions—4- to 8-Tips (Probes)

5.4.4.1 Gripper Integration Platform

The addition of a gripper arm enables full automation capability in 4- and 8-probe liquid handling workstations, such as the MultiPROBE II EX (HT) used as a typical example in this discussion. PerkinElmer calls this feature the Gripper Integration Platform. It provides a secondary arm with integrated gripper tool capable of picking and placing microplates and selected vacuum manifold components around the deck of the MultiPROBE II EX. Extended travel is enabled beyond the right expansion module so that integration with labware can be made, *e.g.*, mixers, incubators, heaters, plate hotels, readers, shakers and washers. This secondary gripper arm offers 5 axes of motion (X, Y, Z, θ, G) with rotation to 345°; therefore, it can move plates from a landscape to a portrait orientation. This particular arm grasps from either the outside or inside of components.

The addition of a gripper to a liquid handling workstation allows it to complete the full automation of certain sample preparation processes such as solid-phase extraction and filtration using microplates. Without a gripper, an operator must manually remove the top of a vacuum manifold, take out the waste tray, insert a collection plate, and reassemble the manifold; however, the gripper takes care of these actions to complete the automation process (Figure 5.6).

5.4.4.2 Vacuum Manifold and Software Control

The vacuum manifold used with a gripper arm must meet exact specifications. Most manufacturers supply their own manifold for use with their gripper arm. A requirement of the vacuum manifold is that it must be self-sealing when a plate is placed on top, so the gasket material must perform every time the vacuum is applied. Also, in order to provide flexibility among manufacturers, a manifold has to be able to accommodate varieties in heights of flow-through microplates and collection plates. Note that the waste tray also needs to be removed by the gripper. A special waste tray is supplied by the manufacturer with appropriate notches for retrieval.

PerkinElmer offers their manifold as a universal manifold that fits most SPE plates, filtration plates and collection plates. Additional versions of the manifold afford a gridded assembly as used for certain plates manufactured by Millipore Corporation (Bedford, MA USA) and a manifold called "microfilter" that is optimized for plates having short drip directors.

The software options for controlling vacuum can vary among manufacturers and are included in the differentiating features. The PerkinElmer vacuum control system has a separate control box that allows two levels of vacuum (high and low settings up to 25 in Hg or 850 mbar); two regulators are used as well as liquid traps. Either house vacuum or a vacuum pump can be utilized.

Figure 5.6. A vacuum manifold placed on the deck of a liquid handling workstation allows the use of 96-well filtration and extraction plates. A gripper arm can be used to assemble and disassemble the manifold for unattended operation. Photo reprinted with permission from PerkinElmer.

The software allows immediate access to vacuum (no delay) and can alternate between high and low settings for different time durations. A quick burst of vacuum for 5 s can be followed by low vacuum for 30 s, followed by another high burst and a longer sustained low vacuum. At the end of the cycle, a manual prompt can be activated (if desired) which is beneficial during method development when trying to optimize the vacuum parameters.

The Hamilton Company offers a different approach to vacuum control for 96-well SPE and filtration plates using its AVS module (Automated Vacuum System), shown in Figure 5.7. The vacuum manifold isolates the conditioning steps from the elution steps by providing two separate vacuum chambers. The automated, moveable carrier positions the plate over the elution or conditioning chamber. The conditioning chamber is directly plumbed to the waste management system with automated waste evacuation. The AVS isolation insert protects each nozzle from cross-contamination during the conditioning and sample flow steps. The adjustable collection plate support accessories accommodate a range of microplates. There is no waste tray to remove. One 96-well plate can be processed per vacuum station. An additional AVS unit can be placed on the deck to double capacity.

Figure 5.7. The MICROLAB SWAP shuttles plates into and out of the Automated Vacuum System placed on the deck of a MICROLAB liquid handling workstation.

The control module for the AVS includes the vacuum, waste pumps and a vacuum sensor to control the applied vacuum for up to two vacuum manifolds. The vacuum control module is software controlled and allows user-defined vacuum levels from 0–24 in Hg (0–810 mbar). The vacuum level is monitored and maintained using a closed loop control system. This system adjusts internal valves to maintain the user defined parameters throughout the duration of the vacuum cycle. The MICROLAB AVS is compatible with the following Hamilton Company Models: STAR, MPH, MPH-96/48 and the 4000 Single Probe. Note that the MICROLAB SWAP can be used with the AVS to shuttle plates into and out of the vacuum station.

5.4.4.3 Additional Systems

5.4.4.3.1 Gilson SPE 215

The Gilson SPE 215 for automated extraction and filtration procedures is based on its Multiple Probe 215 Liquid Handler (Gilson Inc., Middleton, WI USA). The SPE 215 is an X-Y-Z liquid handling workstation with 4- or 8-probes (Figure 5.8). It has a specialized Z-arm with an integrated sealing foot for providing positive pressure elution. An advantage of this approach is that the same pressure is applied across the seal foot, unlike vacuum manifolds which

Figure 5.8. The SPE 215 presents a unique approach to microplate processing by moving a microplate from a position above a waste tray to a position over a collection plate for elution. The 8 probes deliver solvents to the plate using positive pressure.

demonstrate a pressure drop as the well clears. The SPE 215 works with a variety of solid phase extraction formats, including 96-well SPE plates and cartridges (1 mL and 3 mL). This ability to accept both microplates and individual cartridges is unique among liquid handling workstations. Note that the use of 1 mL cartridges without "tabs" or "wings" fit within the 9 mm well-to-well spacing specification for microplates and approximate the 8-row by 12-column format. An integral gas manifold pushes solvents and samples through the SPE sorbent bed or disk, enabling an efficient elution using positive displacement. Dual channel probes supply liquid and gas separately.

The SPE 215 is a flexible system capable of supporting a wide range of racks for samples and reagents. The unit accepts any of Gilson's large selection of SPE racks and custom configured solutions can be created. An optional Solvent Delivery Station is available for selection of up to 8 solvents from individual reservoirs. The Gilson 735 Sampler Software controls the instrument. When the optional 889 Multiple Injection Module is added, the SPE 215 System injects up to eight samples in parallel or in series into chromatographic systems.

5.4.4.3.2 Biomek 2000

The Biomek 2000 Laboratory Automation Workstation (Beckman Coulter) is a highly versatile system that incorporates pipetting, diluting, dispensing, plate heating and cooling, plate to plate transfers and high capacity operations. The unique approach of the Biomek 2000 is that the dispensing head can swap pipette tools, both single tip and 8-tip varieties (Figure 5.9). Note that when using the 8-tip pipette tool, the maximum volume is 200 µL. Wash steps for probes are reduced by using disposable tips for most operations. Patented liquid level tracking is performed. A gripper arm is accessible by swapping out a pipette tool. Components sit on the deck of the Biomek on labware holders or tiles.

When a vacuum manifold is configured on the deck, filtration and solid-phase extraction in microplates can be performed. The gripper arm easily disassembles and assembles the Biomek vacuum manifold. The Beckman Coulter BioWorks™ software uses graphics based menus and prompts. Biomek Stacker Carousels can be added to supply more microplates to the system. Up to three Stack Areas can be used with a Biomek Side Loader. An ORCA® robotic arm (Beckman Coulter) on a track can also be added to this configuration using these side loaders. Two excellent applications describing use of the Biomek 2000 as a versatile liquid handling unit to perform protein precipitation in microplates are reported by Watt *et al.* [3] and Locker *et al.* [4].

Figure 5.9. The Biomek 2000 liquid handling workstation uses interchangeable 8-tip and single tip pipette heads with a gripper arm for movement of components on the deck. A configuration for filtration using a microplate on a vacuum manifold is shown.

5.4.4.3.3 Speedy

The Speedy (Zinsser Analytic GmbH, Frankfurt, Germany) is a 4-probe liquid handling workstation with a vacuum manifold that can be assembled and disassembled with a gripper tool. The probe spacing is variable from 9 to 38 mm and the instrument has liquid level sensing. A unique feature of the Speedy is its ability to use both vacuum and positive pressure (nitrogen or compressed air) within a procedure. Positive pressure is applied via the probes for processing viscous samples through an extraction or filtration plate, such as during the load step with plasma. High throughput can be accomplished using two vacuum stations on the deck. Its software is named WinLissy (Zinsser Analytic).

5.4.4.3.4 Genesis

The Tecan Genesis robotic workstation adds the Robotic Manipulator Arm (RoMA) to the 4- and 8-tip dispensing probes to create a fully automated application for filtration and extraction. The Genesis Freedom Series allows addition of a third robotic arm to create additional capabilities. Three types of robotic gripper arms can be mixed and matched—the standard RoMA, which allows 270° rotation, an extended length RoMA which allows access to modules placed below the worktable, and a pick-and-place arm which is limited to smaller objects like tubes and labware but moves them much more rapidly. Among the below deck modules is an automated centrifuge capable of performing SPE applications such as gel filtration that are not amenable to processing by vacuum.

5.4.4.3.5 MICROLAB AT Plus 2

The Hamilton MICROLAB AT Plus 2 uses unique disposable positive displacement pipette tips. These tips are well suited to pipetting blood, avoiding clots and, in addition, they are drip-free with volatile solvents; an air gap is not utilized as in traditional designs. This 12-probe system has been adapted to perform automated plasma protein precipitation and filtration in conjunction with the Ansys Captiva™ filtration system (Ansys Technologies, Lake Forest, CA USA, now a part of Varian Inc.). A full rack of 96 samples can be processed from original barcoded tubes to isolated 96-well microplate filtrates in 20 minutes. With its semi-automated vacuum manifold, it can also perform a variety of 96-well microplate SPE and filtration processes.

5.4.5 Semi-Automated Solution—96-Tips

The Quadra® 96 Model 320 (Tomtec) is a liquid handling system using 96 individual polypropylene tips for rapid throughput pipetting operations (Figure 5.10). The tips hold a maximum volume of 450 µL, which is the largest volume of all the semi-automated pipettors. It aspirates liquids from microplates or open vessels (solvent reservoir). Each pipette tip is connected to its own independent piston. Pistons are moved up for liquid aspirating and down for liquid dispensing via a microprocessor controlled stepper motor. As each piston moves up or down, it displaces an air column that causes liquid to be aspirated or dispensed. A double O-ring seals each piston to its pipette tip. The pipette tip head remains fixed in location. A patented design uses hollow pistons to allow pressurized air to blow out through the tips.

The Quadra 96 has a six-position deck that is stepper motor driven to move left, right, forward and backward. All of the six positions or stages on the deck can be elevated to a user adjustable height. Elevating the stage is necessary because the tips, once in position, remain stationary. Liquids must be moved up to touch the tips for aspirating and, likewise, plates must be moved up for dispensing. The labware holding the tips is moved up for inserting tips into the head as well as for their removal.

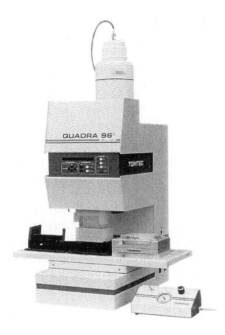

Figure 5.10. The Quadra 96 is a typical example of a 96-tip liquid handling workstation configured for microplate solid-phase extraction using a vacuum manifold. Photo reprinted with permission from Tomtec.

The Quadra 96 is conveniently controlled by computer software via a computer. The unit is operational from the front panel buttons, but it is easier to use the computer software. This instrument has shown great versatility for performing PPT, LLE and SPE for bioanalytical applications and these uses are discussed in more detail in Chapters 7, 10 and 13, respectively. Note that an operator is required to be at the instrument to swap reagent reservoirs (when more are required than the deck can hold), turn vacuum on and off, swap used tips for new tips, and disassemble and reassemble the vacuum manifold. This 96-tip liquid handling workstation is considered to perform semi-automated procedures for bioanalytical sample preparation.

Additional models of the Quadra may be of interest for some applications. The Quadra 96 Model 230 has a smaller 3-position shuttle but uses the same large volume tips (1–450 µL range) as the 6-position shuttle Model 320. Two 384-well versions of the Quadra 384™ are available as 3- and 6-position shuttles having a tip volume range of 0.5–60 µL.

5.4.6 Fully Automated Solutions – 96-Tips

5.4.6.1 Sciclone ALH

The Sciclone Advanced Liquid Handler (ALH) Workstation (Zymark), shown in Figure 5.11, simplifies a broad range of liquid handling operations that previously required the use of several devices. The instrument includes a 20-position deck and the ability to perform multiple dispensing operations. The Sciclone ALH equipped with a high volume head uses disposable pipette tips (1–200 µL dispensing volume); bulk reagent dispensers are also accessible. As a 96-tip liquid handling workstation, the Sciclone ALH can perform protein precipitation in collection microplates and all reagent additions for LLE.

The positive pressure filtration accessory can be added to provide automation of flow-through microplate applications. This system (Figure 5.12) consists of a pressure manifold and a liquid waste collection locator. The manifold is fitted with a precision gasket to seal the edges of a standard solid-phase extraction plate or a filtration plate. The system provides user controllable positive pressure to the plates, resulting in high quality separations and purifications.

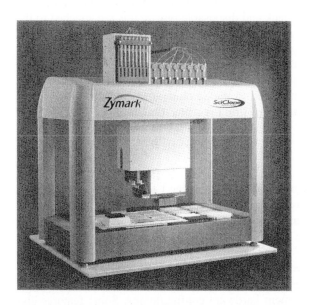

Figure 5.11. Example of a fully automated 96-tip liquid handling workstation, the Sciclone ALH. Photo reprinted with permission from Zymark.

Figure 5.12. The positive pressure filtration accessory for the Sciclone can be used to provide automation capability for 96-well filtration and solid-phase extraction applications. Photo reprinted with permission from Zymark.

The method for performing a solid-phase extraction procedure on the Sciclone ALH follows. A 96-well SPE plate is positioned on top of the liquid waste collection locater. The bulk reagent dispenser distributes the conditioning reagent across the 96-well plate. The pressure manifold is picked up by the Sciclone head and located on top of the SPE plate; the required air pressure is applied (up to 15 psi). This same process is repeated for a second conditioning solvent. In order to load the sample into the SPE plate, the pressure manifold is automatically parked on the deck and disposable pipette tips are attached to the head. The sample is aspirated from the sample plate and then loaded into the SPE plate. After ejecting the tips, the manifold is picked up and used to apply pressure to the plate. Wash steps are performed using the same procedure described for the conditioning steps. When the plate is ready for elution, a collection microplate is transported to the Sciclone deck and the SPE plate is placed on top of it. The pressure manifold is seated on top of the SPE plate and the pressure is activated.

5.4.6.2 Cyberlab

Cyberlab plate handling workstations (Gilson Inc.) are versatile full-featured liquid handling workstations similar to the Sciclone ALH previously discussed. Three different deck sizes are available, depending on the number of plates to be processed and the items to be integrated onto the deck. The largest is the C-400, which is capable of processing up to 350 plates or more, although the total number plates will decrease as ancillary items are added to the deck. A key advantage of the Cyberlab series is application-specific customization of deck components and instrument hardware configurations. Some features and operational details of this instrument follow.

A 96-channel pipette head allows dispensing of up to 200 µL of reagent into each well of a 96-well plate in one step (or into a 384-well plate in four steps). Pipette heads are interchangeable. Cyberlab has removed one of the more time consuming and repetitive maintenance tasks of the 96-channel pump by creating an O-ring greasing station. The software keeps track of tip changes, and at every user defined number of changes the greasing station is visited.

The tip installer is a pneumatic device on the deck that precisely and consistently loads tips onto the 96-channel pump. It is also used to pierce the foil that is sometimes used in heat sealing a microplate. Tips can be reused or discarded. The 8-channel pump is used to pipette and mix up to 200 µL of solutions on a column-by-column basis. Unlike the 96-channel pump, it does not require a separate tip installer. However, it cannot reuse tips.

The plate gripper is used to move plates, lids, tip boxes and other accessories on the deck. The gripper is designed so that it can remove and replace stacked plates. A barcode reader can be placed on the deck. This scanner is used to confirm the identities of plates before they are used. A plate sealer can also be located on the deck to apply and heat seal an aluminum foil over 96-well plates.

An incubator is used to heat up to three deep or shallow 96-well plates, as used for some ADME/Tox *in vitro* applications, and is controlled by an advanced temperature regulator. The instrument can be set to heat the plates for any amount of time.

A plate labeler incorporates an Intermec 3420 label printer with a Cyberlab robotic device to automatically print and attach barcode and alphanumeric information to plates. It can be used with a plate nest location to label plates or with a tube labeler tool to wrap the labels around a reagent tube.

Vortex mixers allow Cyberlab instruments to mix single or multiple tubes and microplates. By using the gripper arm to place and remove test tubes and microplates, human intervention is eliminated. Cyberlab has also incorporated precision weighing into its workstations. An analytical balance can be added. Items are placed on the scale by the gripper arm, eliminating human intervention.

Special accessories for vacuum block filtration applications are available for the Cyberlab Plate Handling Workstations. The unit provides all pipetting operations and plate movements, and manipulates the vacuum manifold ring and collection plates as necessary using the plate gripper tool. Commercially available plates in kits from vendors (*e.g.*, MultiScreen™ by Millipore) are compatible with the system for the automation of PCR cleanup and plasmid purifications.

5.4.6.3 MICROLAB MPH-96

The MICROLAB MPH-96 contains 96 separate aspirating and dispensing probes for simultaneous processing of entire microplates. Liquids are delivered via 96 individual flow paths using 96 individual precision syringes. The volume of liquid transfer is independently programmable for each row. Liquid level sensing is built in and an efficient wash system is utilized to prevent carry-over. The MPH-96 can be configured to handle a wide range of volumes from <1 µL to 500 µL. A smaller version of this instrument, the MPH-48, takes up less bench space and offers the same pipetting capabilities with 48 syringes and probes. By adding the MICROLAB AVS module, as discussed previously, this vacuum capability allows the use of 96-well filtration and SPE plates.

5.4.6.4 Tecan Te-MO96

The Te-MO96 (Tecan Multi-pipetting Option) can be configured as a standalone instrument or added to the deck of the Genesis and Genesis Freedom robotic processors. The Te-MO can use disposable tips and fixed/-washable manifolds in the same pipetting run, and features an efficient on-board washing station. Some SPE manifolds can be incorporated. When on the deck of the Genesis, wells can be accessed individually by the 8-tip arm as well as by the gripper arm. The standalone version can be combined with a stacker. In the case of SPE applications involving repeated washes with high volumes of solution, Tecan's USA branch offers a 96-head dispenser for use with the Te-VacS on a custom basis. This approach represents an inexpensive way to greatly increase throughput for certain applications.

5.5 Application-Specific Workstations or Systems

5.5.1 Introduction

Some applications in drug metabolism research require capabilities beyond that provided by the typical liquid handling units as described in Section 5.4. The liquid handler becomes an integral component within an expanded system built with peripheral pieces of laboratory equipment. Software is used to interface these components and allow for their complete integration into a fully automated system. A specific set of components is assembled to perform a particular application. These systems are referred to as application-specific workstations. Upgrades are available for these systems to meet increased demands for throughput and/or functionality. Examples of workstations from Tecan will be discussed to provide an understanding of this approach.

Another tactic to design full automation around an application is to use configurable modular components that can be assembled, disassembled, configured and reconfigured to meet current and future needs. Modular components from Thermo CRS (Burlington, Ontario, Canada) built around a robotic arm will be described to illustrate this approach.

5.5.2 Expanded Capabilities Based on a Workstation

A specific application in drug discovery support, performing cell permeability studies as part of *in vitro* ADME testing, has been automated as the Cell Permeability Workstation (Tecan). This system, shown in Figure 5.13, consists of a Genesis liquid handling workstation equipped with all of the component pieces necessary to automate routine compound permeability assays such as Caco-2, MDCK (Madin-Darby Canine Kidney) cell line or Immobilized Artificial Membrane (IAM) assays. This workstation takes advantage of automation compatible 24- and 96-well insert formats commonly used for these tests. The system includes temperature controlled plate carriers, transport buffer stations, and 4-position shaking incubators. A potential upgrade for this system involves integration of an automated station for assessment of cell monolayer integrity.

The demand for higher throughput *in vitro* ADME testing has placed a burden on resources used to grow and maintain cell culture systems. Tecan has also developed the Automated Cell Maintenance System. It consists of a Genesis liquid handling instrument with a Heraeus® Cytomat® automation compatible incubator (Kendro Laboratory Products, Newtown CT USA). The system provides the necessary sterile environment for cell culture and matches it with the large plate capacity (150–185) of the Cytomat incubator in a small footprint, self-contained workstation for all cell culture applications (adherent or suspension cell). The appropriate plumbing is included in the system to handle media storage and aseptic media dispensing and removal. Precise temperature control is also provided. This system completely automates cell seeding and the routine media changes that are associated with performing cell based assays; it is also used in the support of Caco-2 and hepatocyte *in vitro* ADME assays.

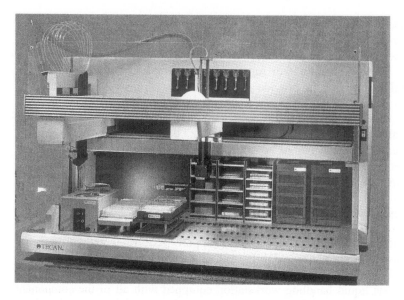

Figure 5.13. The Cell Permeability Workstation automates a series of procedures commonly used for *in vitro* ADME testing. Photo reprinted with permission from Tecan US.

Tecan offers two types of *in vitro* Drug Metabolism Workstations. The first is again based on the Genesis liquid handling workstation equipped with the appropriate capabilities to perform metabolic stability and cytochrome P450 inhibition assays. The basic Genesis is upgraded with a temperature-controlled orbital shaking platform to handle typical microsome or hepatocyte stability experiments. In addition, it is possible to upgrade the workstation by integrating a fluorimeter into the system to handle cytochrome P450 inhibition studies with commercially available substrates. This workstation provides all the equipment necessary to perform the most common types of *in vitro* drug metabolism assays.

The second configuration is a fully automated system built around a Genesis Freedom instrument. This unit is a 3-arm liquid handling system with an integrated microplate centrifuge. This configuration allows complete automation of all steps in a metabolic stability assay up to and including protein precipitation prior to sample analysis with LC-MS/MS.

Tecan also offers various upgrades to the basic ADME workstation platforms including the integration of LC injectors directly on the deck of the Genesis instruments. This allows the Genesis liquid handler to perform the assay and inject the sample directly into a chromatographic system for analysis and detection using mass spectrometric techniques. Another option is the integration of SPE blocks on the worktable of the Genesis for sample cleanup prior to the LC-MS/MS analysis.

Zymark offers an application-specific workstation concept, as well. The Mini-Staccato™ consists of the Twister II with another component such as a plate sealer or plate reader. A liquid handler such as the Sciclone ALH can be added to improve capability. The Staccato™ workstation, such as one configured for ADME/Tox applications, adds a robotic arm to the collection of the Sciclone ALH, incubators, washers and filtration units.

5.5.3 Custom Configured Capabilities Involving Robotics

Instead of taking the approach to develop an application-specific workstation, Thermo CRS has acknowledged the need for constant change in drug discovery and sample preparation methodologies. While changes occur in pharmaceutical R&D, such as assay techniques, instrument technologies, staff, facilities and research goals, one automation platform has been designed to meet current and future needs. The CRS Dimension4™ encompasses a designed modularity of software and hardware components. Any size of automated lab system can be configured, from a single tabletop workstation to an ultra high throughput production line, simply by adding components. The open architecture CRS POLARA™ 2 software controls all applications seamlessly and supports components from a wide variety of manufacturers. Once a system is configured, when the need arises it can be dismantled and/or recombined to meet new throughput or assay needs without losing the investment made to date. The key elements of this approach involve the foundation (modular tables), plate movers and software.

CRS Dimension4 modular tables (800 mm square and 800 mm x 1200 mm) provide a stable, flexible and scalable mounting platform for automated workstations. These tables can easily be connected for expansion or reconfiguration; they can also be divided. The components needed for an assay are identified and placed on the tables, *e.g.*, liquid handling workstation, microplate hotel and a robotic arm to move microplates into and out of devices. The Catalyst robot (Thermo CRS) with microplate hand represents a typical arm used.

Expansion of an application to perform ultra high throughput sample preparation or assays simply involves connecting additional modular table sections, adding a linear plate track (LPT) through the center of the tables, and adding components such as liquid handling workstations, microplate hotels, plate readers, *etc.* (Figure 5.14). In the conventional approach to robotic automation, each test plate is transported by means of a robot moving along a liner track. CRS Dimension4 changes that concept by moving test plates rapidly along a linear track to stationary robots where the plates are transferred. The CRS Flip Mover represents the high speed microplate transfer device of choice to shuttle microplates between an instrument and the LPT; average transfers are 2 s per plate. CRS PlateLevel™ and advanced motion control technologies with reliable gripping permit the Flip Mover to move microplates so quickly without spilling.

High throughput applications for bioanalysis can be configured using a liquid handling workstation with various CRS components—robotic arm, microplate

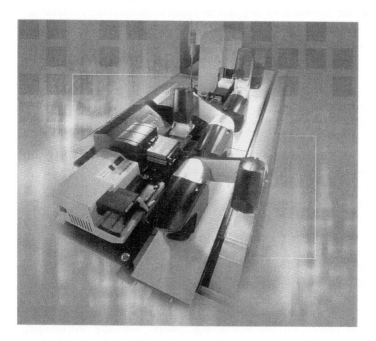

Figure 5.14. A modular automation approach moves microplates rapidly along a linear track to stationary robots where the plates are transferred. Photo reprinted with permission from Thermo CRS.

hotel, centrifuge and/or a vacuum filtration station. While a centrifuge can be used for filtration applications, the vacuum system is preferred because it is easier to automate and the centrifuge operation is time consuming. However, when a centrifuge is required, such as to spin plasma before aliquoting or to spin a collection plate containing filtrate or eluate resulting after a solid-phase extraction, it is automatable. Typical sample preparation techniques automated on robotic platforms, such as the Thermo CRS system, are protein precipitation in collection microplates, filtration of precipitated protein using 96-well filtration plates, and solid-phase extraction in 96-well SPE plates. The procedures for performing PPT and SPE are described in more detail in Chapters 6 and 11, respectively.

The CRS vacuum filtration station operates in the following manner. The robotic arm places filtration microplates on the upper manifold section and a collection microplate is placed below to capture the filtrate. Microplate guides ensure accurate positioning. A pneumatic actuator automatically raises the upper manifold section, providing robotic arm access to the collection microplate. The vacuum regulator allows adjustment of vacuum strength and a shutoff valve is built in. This CRS instrumentation allows modulation of the vacuum, a desirable feature; instead of the plate being subjected to a hard, strong vacuum immediately, the strength of the vacuum is restricted via pressure valves and begins with only a small differential across the plate. The vacuum is then ramped upward to reach the desired strength. Manipulating the vacuum pattern in this manner has been found to eliminate some common problems such as foaming of plasma samples and crosstalk between well tips.

The Zymate™ robot system (Zymark) represents the traditional concept of robotics and custom configured components, in which a central robotic arm is the focal point for a range of activities custom configured about its perimeter (Figure 5.15). The Zymate XP robot is now the third generation laboratory robot for such an automation system. The robot incorporates technology for interchangeable hands to carry various tubes and containers, and to perform functions such as gripping, weighing, pipetting, vortexing, centrifuging and evaporating. Tactile sensing capability is included in the robot hand and all robot axes have optical encoders to verify successful completion of functions. Bioanalytical applications have been demonstrated using the Zymate (and its predecessor versions) for protein precipitation, solid-supported liquid-liquid extraction using tubes filled with diatomaceous earth, and solid-phase extraction in microplates.

Figure 5.15. Example of a robotic arm that can be the focal point for a range of activities custom configured about its perimeter, the Zymate XP. Photo reprinted with permission from Zymark.

5.6 Strategies for Using Automation in a Bioanalytical Setting

5.6.1 Introduction

The previous Sections 5.3, 5.4 and 5.5 have provided an overview of the many varieties of automated instruments and an introduction to their capabilities. It is now appropriate to discuss some strategies for their use in performing off-line sample preparation methods for bioanalysis. First, some general considerations for automation are presented so that the strengths of the automated equipment are exploited for maximum productivity. The importance of regular performance verifications is then introduced along with proper practices with regard to documentation and record keeping. While it is easy to continue the manual pipetting aspect of a sample preparation procedure, such as preparing standard curves and QC samples and/or adding internal standard solution, the ease of allowing automation to accomplish these tasks is discussed.

Performing a routine sample preparation procedure is one of the main uses for automation. Strategies are discussed for effectively carrying out automated tasks for PPT, LLE and SPE. Comparisons and contrasts are made among 4-/8-tip and 96-tip workstations for these routine procedures. Also, some effective strategies are discussed for how to best accomplish method development using automation. Note that most automated instruments are not ideal for performing method development by using variable solvents and volumes; a few brands can do it well and some others can be generally adapted, as this section will describe.

5.6.2 General Considerations

5.6.2.1 Allow Automation to Eliminate Manual Pipetting Entirely

Allowing the liquid handling automation to accomplish all pipetting tasks can eliminate the introduction of human pipetting errors. The accuracy of pipetting is much higher when performed by a liquid handler compared with manual operation, especially when small volumes are aliquotted. It is often difficult, even with concentration and the use of overlaid pipetting grids, to manually pipet different plasma samples and volumes, pH adjustment buffer and internal standards into 96 unique wells of microplates for the preparation of a sample plate. However, it is much easier to pipet the same solvent and volume multiple times as when adding organic solvent for a liquid-liquid extraction procedure; here, multichannel pipettors can be used effectively. Why continue this manual approach when automation is available in the laboratory? Many varieties of workstations allow the user to specify exact well positions for each solvent delivery step. The sample plate can be prepared in this manner as well as the calibration standards and QC samples.

The accuracy and precision of the automation with regard to pipetting performance are determined at regular intervals. When maintained and used properly, automation performance is reliable. Many varieties of automation allow the user to adjust default settings for pipetting in order to allow for individual variation in the liquid type. A specific set of pipetting defaults can be used for plasma samples, organic solvents, aqueous solutions and 50/50 mixtures of organic/aqueous solutions. These settings can usually be overridden for individual steps when necessary.

5.6.2.2 Consider Modifying an Existing Practice to More Efficiently Utilize Automation

Sometimes it is relatively easy to modify an existing practice to more efficiently utilize automation. For example, if half of the SPE methods in the laboratory use C18 bonded silica chemistry and the other half use mixed phase cation chemistry, then two methods or approaches are programmed into the automation. While these methods are similar, some unique aspects exist for the mixed phase (extra conditioning step with buffer, an extra wash step, and different wash and elution solvents). Consider developing all methods using mixed phase chemistry because most analytes used with C18 can likely be made to work just as well on mixed phase. The advantage of this approach is that one method is used for all analytes. Potential error for using the wrong program is decreased when a generic method is the norm. Another example is to consider using a 1:1 ratio (v/v) of organic to aqueous for all internal standard solutions so one does not have to struggle to achieve the required volume reproducibility if 100% organic solvent is used.

5.6.2.3 Design Procedures with Flexibility in Mind

Procedures can be developed using variables to add flexibility to a program. The number of samples can be set up as a variable that is input at the start of a program. The dispense volume can be adjusted for one or more steps, *e.g.*, for the internal standard volume. Some instruments (*e.g.*, MultiPROBE II) allow for the variable dilution of sample volumes by reading an external text file. In this scenario, sample volumes can be diluted 1:1, 1:2, 1:4, *etc.* with blank matrix during the transfer step from sample tube into the sample microplate. An air gap is used to separate the two matrix volumes within a pipet tip and they are dispensed together into the same microplate well. The external text file specifies the volumes and locations for the blank matrix and samples.

Once a procedure is worked out and is defined for specific labware items, *e.g.*, a 2 mL square well SPE plate on a Porvair manifold, flexibility can be added to the program execution by changing the labware definitions to that of a short skirt disk plate on a 3M manifold; then the program is saved under a different name. When volumes for conditioning, washing and elution are set up as variables, simply load the program (specifying the correct plate and manifold combination) and run.

5.6.2.4 Automation Will Not Improve Poor Extraction Chemistry

If the extraction chemistry is not optimal, the recovery and purification of an extraction procedure will not improve once the procedure is automated. The method may, in some cases, display improved precision, but poor recovery and sample cleanup are almost always maintained. Only assays that demonstrate successful extraction chemistry should be automated. Note that automation **does not** eliminate the need for method development.

5.6.2.5 Analyze the Laboratory Workflow Pattern to Maximize Productivity

Automated instruments can be placed within different laboratories or the instruments can all be located in a common area for walk up and shared access. Either approach can work, depending on the users involved, the general workflow pattern, and who has the responsibility of being the instrument custodian. The goal of a location analysis is to maximize productivity by eliminating time traveling to and from a location. Also, it must be ensured that the work environment around the instrument has enough space for storing supplies, solvents and waste materials. Placing these instruments close to the mass spectrometers is often desirable, so that the samples, when prepared, can be sealed and easily taken to an autosampler for injection into a chromatographic system.

5.6.2.6 Scheduling Concerns

A common occurrence in using shared instruments is that everyone in the group may want to use the instrument at the same time each day. Scheduling conflicts can arise, causing frustration among users and delays in getting samples prepared and on the autosampler for analysis in the afternoon. In this case, it is best to encourage staggered time usage for the automated instrument and ensure that users do not over extend their allotted time. A common situation to avoid is that a user may occupy the instrument while waiting for a dry-down step to finish; the evaporation unit is a separate piece of equipment. The automation should be made available to others immediately after finishing a procedure. Once the user has the samples dried down and is ready for reconstitution, the waiting queue for the liquid handler can be re-entered.

5.6.3 Performance Verification and Record Keeping

5.6.3.1 Importance of Implementing a Regular Performance Verification Schedule

Performance verification (PV) is a measurement of the accuracy and precision that a liquid handling instrument displays with regard to its pipetting ability. It is essential to determine and document the accuracy and precision of the liquid handling functions so that confidence in its ability is ensured. Periodic confirmation of this data is required by the FDA (Food and Drug Administration, United States) to document system performance.

Performance verifications should be conducted on each instrument at a defined frequency as established in the SOPs, *e.g.*, every month or every 2 or 3 months. Questions always arise when reviewing data, but when PVs are routinely scheduled and conducted for each instrument, the pipetting abilities of the equipment are usually not subject to further inquiry (unless some pattern of failure associated with certain tips is seen). This report of periodic assessment is recorded and filed in the logbook for each instrument to meet documentation guidelines and allow for user access to this information. A good practice is to also post the latest PV results on or next to the instrument.

5.6.3.2 PV Determinations and Procedures

Performance verification is usually determined for three volumes within a defined range (low, mid and high), and at a minimum of six replicates per volume. This scenario is performed, at the minimum, for two solutions displaying the range of viscosities to be encountered, such as a 100% aqueous and a 100% organic solution. When internal standards are dispensed in 50/50 aqueous/organic mixtures, usually no test is needed for the 50/50 solution separately because the viscosity of a 50/50 solution is intermediate between that of 100% aqueous and 100% organic solution; it can be added, of course, when a strong preference exists on the part of the analyst or laboratory manager. The volume range used is typically a function of the tip size. A 50 µL disposable tip may be tested at 10, 25 and 50 µL for the low, mid and high volumes. A fixed tip used with a 500 µL syringe size may be tested at volumes of 50, 250 and 500 µL. The low volume tested should not be higher than the lowest volume normally used for pipetting; these volumes chosen as low, mid and high should bracket the usual volumes.

The general procedure for conducting a PV follows. Using gloves, empty tubes are carefully weighed and recorded to four decimal places. The tubes are put into place on the instrument to receive liquid. In the case of a 96-tip pipettor, microtubes can be used (which are individually removable from the base). A 4-/8-tip pipettor can utilize microtubes or glass test tubes placed in a rack. A volume of liquid is dispensed into each tared tube. These tubes are then weighed again and the difference in weight is recorded.

Statistical calculations are performed to determine values for the mean, standard deviation and %CV (coefficient of variation). These data are compared with acceptable numbers as specified in the departmental SOPs. In the event that one or more pipetting determinations fall outside of the acceptable specifications, the test is run again. Repeated failure usually signifies a mechanical adjustment issue or other problem; perhaps a correction needs to be made in the default settings for pipetting and system variables.

Accuracy in solvent delivery is a function of the reagent type (biological matrix, aqueous, organic or aqueous/organic mixture), the speed of pipetting, and the physical dimensions of the delivery hardware (tips or probes). A gravimetric determination using specific gravity calculations is an accepted way to generate accuracy data. Note that spectrophotometric determinations with dye solutions are also done but require calibration curves for measurement of pipetting accuracy.

PerkinElmer has published a liquid handling application note describing the use of the "Gravimetric Performance Evaluation Option" for the MultiPROBE II (HT) [5]. The combination of a Mettler-Toledo analytical balance #SAG285/L (Mettler-Toledo Autochem, Vernon Hills, IL USA) with a humidity chamber is placed directly on the deck of the MultiPROBE II using special mounting hardware. The balance pan is connected to a balance control touchpad and the MultiPROBE computer by communication cables. This application is reported to meet FDA guidelines and is usable with any combination of tips, performance files and volumes. Two related applications present calibration and performance testing of liquid handling workstations in a GLP laboratory [6, 7].

Sometimes it is necessary to quickly assess pipetting precision among multiple tips, as in a 96-tip liquid handling workstation or in an 8-tip workstation. Perhaps one specific tip or piston head is being examined and precision, relative to the others, needs to be investigated. A convenient approach to assess pipetting precision is to use a dilute solution of a fluorescent dye in water (*e.g.*, 0.07M D-6145, **1**) or in methanol (*e.g.*, 0.14M Rhodamine, **2**). The solution containing dye is placed in a reagent reservoir and a fixed volume is aspirated. Dispensing is performed into a shallow well plate suitable for use in

1   **2**

a fluorescence reader. The plate is then placed into the reader and the fluorescence intensity per well is measured. An examination of the signal intensity per well position can pinpoint pipetting deficiencies per tip or piston head. Note that this procedure measures pipetting precision but not accuracy (unless a calibration curve is included). An example of data obtained from a pipetting exercise using fluorescent dye in water is shown in Figure 5.16. In the top graph for Figure 5.16, the precision among all 96 tips was similar. Any outliers are easily noticed, as shown in the bottom graph.

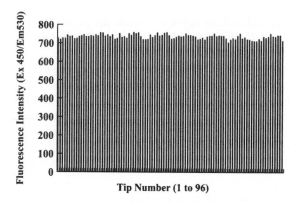

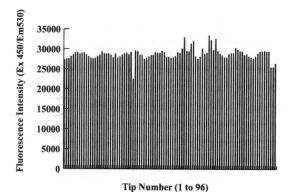

Figure 5.16. Fluorescence measurements from all 96 tips in a determination of tip-to-tip precision upon pipetting an aliquot (50 µL) from an aqueous solution of D-6145 (**1**). *Top*: Good precision shown among all 96 tips (0.0015M solution). *Bottom*: Several tips are demonstrated as outliers (0.07M solution).

Table 5.1
Volume parameters used in a test of pipetting precision on the Quadra 96 liquid handling workstation

QC Vol. Tested (μL)	First Air Gap (μL)	Water (μL)	Second Air Gap (μL)	Methyl Orange (Vol., Conc.)	Final Well Vol. (μL)	Final Dye Conc. (%)
10	100	190	15	10 μL, 0.1%[a]	200	0.005
25	100	175	15	25 μL, 0.04%	200	0.005
50	100	150	15	50 μL, 0.02%	200	0.005
100	100	100	15	100 μL, 0.01%	200	0.005

Tomtec recommends a particular testing procedure to measure pipetting head performance on the 96-tip Quadra 96. The protocol involves aspirating an initial air gap followed by a variable volume of deionized water, a second air gap, and then a variable volume of aqueous methyl orange dye solution (10, 25, 50 and 100 μL of a 0.1%, 0.04%, 0.02% and 0.01% concentration, respectively) . The volumes of water, air gap and dye solution used for this test are listed in Table 5.1. The volume within each tip is then dispensed into a microtiter plate. The volume within each well is mixed four times. The plate is sealed and shaken for 30 s using an external orbital shaker unit. The seal is removed and bubbles are carefully broken without removing dye solution from the well (centrifugation is effective to disperse the bubbles). The plate is then read in a spectrophotometer at 492 nm with simultaneous reference at 620 nm. The optical density of each well is reported for the four volumes tested. Statistical numbers are generated with regard to %CV.

5.6.3.3 PV Considerations

Performance verifications require some important considerations. It is essential to conduct them using the same pipetting features as performed for actual samples. For example, with the MultiPROBE II, use the corresponding pipetting mode (blowout mode for disposable tips or liquid displacement mode for fixed tips), performance files and system settings (including system and transport air gaps) for the PV test as are used for pipetting plasma, organic solvents, aqueous solutions and 50/50 organic/aqueous solutions. Likewise, for the Quadra 96, use the same air gap volumes, tip equilibration scheme and dispense/aspirate speeds for the performance verification test as used for actual sample matrices and solvents.

The PV results are specific for the equipment used. A MultiPROBE II uses a given syringe size, which is interchangeable. If a 500 μL syringe is used for the

PV testing, and at some point it is exchanged for a 100 μL syringe, the previous results are invalid and the PV tests should be run again using the 100 μL syringe.

Multiple replicates **per tip** should actually be used, as the data will show whether or not a specific tip is outside of specification. By increasing the number of replicates per tip in addition to per volume, the number of samples now becomes quite large. However, using automated methodology to perform the test, as previously described for the MultiPROBE, reduces the hands-on nature of the evaluation dramatically. Another approach to reduce hands-on time is to use the Prelude™ workstation (Zymark) to automatically tare test tubes and record the final weight once liquid has been dispensed into the tubes. An analytical balance is contained on the Prelude and tubes are moved in and out of racks by the gripper tool.

5.6.3.4 Instrument Maintenance and Cleaning

Instrument maintenance is a very important responsibility. A logbook that documents dates for each cleaning procedure, maintenance visit and service call should be kept for each instrument. Programs for both cleaning and preventive maintenance should be strictly followed.

Note that with the Quadra 96 the seals must be replaced at least annually (every six months is better) because the O-rings used in the piston head are rubber and moving parts on rubber will cause wear. Using a 100 μL volume, the %CV may change from a norm of 0.9–1.5% to >3% when the O-rings are worn. Tomtec offers a GLP service contract that entitles the user to receive a new piston tip head every six months. This piston head is tested at the factory and verified for pipetting performance using the methyl orange dye test. This head is then shipped to the customer for installation. Once the head is installed into the customer's unit, its pipetting performance can be reconfirmed. Also offered by Tomtec is a GLP User Guide so that laboratories can perform their own installation and operation qualifications.

PerkinElmer recommends a regular cleaning procedure for the tips and transfer lines in the MultiPROBE II. The purpose of this task is to cleanse the system of tiny amounts of accumulated biological matrix as well as to prevent the growth of bacteria, crystals or precipitates within the tubing, valves and syringes. Users generally clean the system tubing with a Lysol® IC solution that is diluted with water (1:200, v/v). PerkinElmer recommends against using bleach to clean or flush the system tubing. Bleach can damage the valves, sampling tips and/or adapters and can leach ions from stainless steel to cause pitting and promote rusting. **It is a good practice to change the entire system tubing in the liquid handling workstation once a year**.

The frequency of this cleaning depends on the instrument's usage. Some laboratories clean monthly and others every three months. PerkinElmer recommends a quarterly maintenance cycle or after 200 h of operation, whichever comes first. In addition to pipetting performance and system decontamination as mentioned previously, some other tests to be run as part of a maintenance program include a tip pickup test, a liquid level sensing test, a tip alignment test and a random move test.

5.6.4 Preparation of Standards and Quality Control Samples

5.6.4.1 Using a PyTechnology Robot

The preparation of standards and quality control samples has traditionally been performed using manual pipetting. However, automation can execute this task reliably. Short, Lloyd and Lynch have described the use of a Zymark PyTechnology™ robot to perform serial dilutions for standards and QC samples within a drug discovery support group [8]. A custom program was written that queried the operator for information to build the standard curve, such as the concentrations (ng/mL) from highest to lowest, number of aliquots to prepare for each concentration, volume of each aliquot (µL) and the extra volume to add to the original tubes. The robot used this information to recursively calculate the volumes necessary to perform a stepwise serial dilution in the supplied sample matrix. The system fine tuned the target volumes for each level using the feedback from the analytical balance. A starting table generated by the robot indicated required volumes.

The operator places test tubes in the rack starting with the highest standard concentration (example: 3 mL spiked at 25 ng/mL). The starting table indicates how much spiked matrix and blank matrix is required (example: 10.3 mL of blank matrix). The high concentration spike is followed by empty tubes in sequence. These tubes are marked with the appropriate concentration. The robot places the next tube in sequence in the balance and then makes the neces-sary dilutions using the Standard Curve Work Table as a reference. An aliquot of the high spiked serum sample is pipetted into the tube, along with an aliquot of blank serum to create the desired concentration. Upon completion of the full procedure, the robot prepares the Standard Curve Work Results Table. This table contains the concentrations requested, the actual volumes used for dilutions, the actual concentration performed, and the percentage error. The table shows how accurately the robot has prepared the dilutions.

5.6.4.2 Using a Liquid Handling Workstation

Calibration standards and QC samples are usually placed into tubes and arranged within the sample tube rack so that they are delivered to a microplate in the same manner as the samples. However, a feature of the multiple probe liquid handling workstations is that the calibration standards and QC samples can actually be prepared by the instrument, rather than prepared off-line manually. Descriptions of this procedure for preparing standards and QC samples with a Biomek 2000 multiple probe liquid handler (Beckman Coulter) have been reported by Watt *et al.* [3] and Locker *et al.* [4].

Calibration standards, quality controls and internal standard were all diluted from a 1 mg/mL top standard to the desired concentrations by the Biomek 2000 instrument and placed in a 96-well plate, the Dilutions Plate as shown in Figure 5.17 (*Top*). A full range of concentrations was prepared for the calibration standards; internal standard required only a 1:10 dilution and QC samples were diluted to yield three final concentrations. The Dilutions Plate was then used to prepare the Sample Plate that accommodated duplicate 12-point calibration curves, 6 sets of quality control samples at three concentrations, and 54 unknowns. The layout of the Sample Plate for this particular application is shown in Figure 5.17 (*Bottom*). Good agreement was found in dilutions prepared by the automated versus manual methods. Note that 1-, 4- and 8-tip liquid handling workstations that can specify the exact well maps or well positions can selectively omit the addition of internal standard into wells intended for the double blank, recovery wells and matrix effect wells.

Although these plates prepared for use on the Biomek 2000 are required to be microplates, an analogous procedure can be used to prepare these standards in test tubes for use on a liquid handling workstation that can reformat from tubes into plates. Using this approach, the test tubes prepared from the PyTechnology robot can be placed on the deck of a liquid handling workstation (*e.g.*, MultiPROBE II) and those tubes can be used as the source tubes for preparation of the sample plate.

5.6.5 Sample Pretreatment

A very important sample pretreatment step when using automation is the centrifugation or filtration of the sample before aliquoting. This procedure is

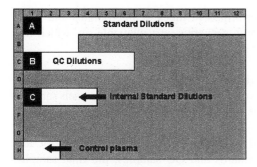

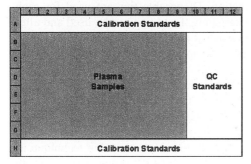

Figure 5.17. *Top:* Microplate layout of a dilution plate. *Bottom*: Microplate layout of a sample plate. Both plates are prepared by a liquid handling workstation prior to performing an automated sample preparation method. Reprinted with permission from [4]. Copyright 2001 Elsevier Science.

strongly recommended for biological samples such as plasma to avoid plugging of pipet tips, especially when the samples are old or when plasma samples have been prepared with heparin and have experienced repeated freeze/thaw cycles. It is possible to have partial plugging of tips which is difficult to visually detect; in this case, partial occlusion can allow the aspiration of air instead of sample, reducing the volume of sample actually pipetted.

Once the sample has been aliquotted into a well or vial, some additional common pretreatment procedures performed in bioanalysis are to add internal standard directly to the sample matrix and mix, as well as to add a solution to adjust the overall sample pH, again followed by mixing. While it has been common to add sample, internal standard and pH adjustment solution manually into tubes or a microplate, automation can also perform this task reliably and reduce the amount of hands-on analyst time considerably. Depending on whether or not the procedure is performed in tubes or microplates and how much automation is desired, several scenarios are available in which to accomplish this sample pretreatment scheme.

A 4-/8-probe liquid handling workstation that reformats from tubes into microplates can accept the sample volumes in tubes and transfer each sample aliquot into a microplate. Internal standard can be added from a reagent trough into each microplate well, and the solution mixed. A buffer can be added to that well and the contents again mixed. After performing this sequence for all samples, and for the calibration standards and QC samples, the final sample plate is prepared and ready for extraction. Disposable tips are used for each transfer that is mixed with plasma. Alternately, fixed tips can be used and after each contact with sample the tips are washed and rinsed with water, and then dipped into a trough of organic solvent for a more thorough wash/rinse cycle.

Alternately, that same 4-/8-probe liquid handling workstation can prepare the samples with internal standard and buffer in test tubes, and then transfer the final prepared mixture from tube to microplate well to assemble the sample plate. The only difference is that the test tube is the receiving vessel instead of a microplate well, until the end of the procedure. Note that this technique as described in a microplate works nicely for PPT and LLE procedures, in which the sample plate becomes the extraction plate. For an SPE procedure, the sample mixture can remain in tubes or in a microplate; the total volume is then transferred into an SPE Plate. Of course, when using a 96-tip liquid handling workstation, in which test tubes are not allowable, the microplate format should be used to prepare the sample plate.

5.6.6. Routine Sample Preparation Procedures

A 96-tip liquid handling workstation such as the Tomtec Quadra 96, Gilson Cyberlab or Zymark Sciclone ALH is ideal for performing a given extraction method rapidly and routinely. Its 96-tips aspirate and deliver solvents. An entire procedure can usually be performed in about 5 min for PPT and 15 min for SPE, while LLE requires more time because of the several pause, seal, mix and centrifugation steps required (approximately 45 min).

An important commitment extending the quoted 5, 15 or 45 min extraction times is the additional time it takes to reformat samples from tubes into a microplate or microtubes. Samples **must** be in a 96 well plate for use on the Quadra 96 and other 96-tip workstations. Preparation of this plate can be done manually but automation is recommended to avoid errors.

A 4-/8-probe liquid handling workstation such as the MultiPROBE II or equivalent is ideal to reformat samples from tubes into a microplate for use on the Quadra 96, Cyberlab or Sciclone. A template program can be designed for the MultiPROBE II to perform this sample reformatting for a given extraction method. The MultiPROBE II can accommodate different sample tube sizes and will transfer an aliquot of sample matrix from tube to plate. It can also add a pH adjustment buffer to the wells in the plate, and add internal standard solution. Disposable tips are used for internal standard and matrix transfers, and fixed tips for pH adjustment solution transfers. Sample mixing within the plate wells can be performed after the final dispense step "post-dispense mix" or as a "pre-aspirate mix" immediately prior to transfer. Variables can be set up within the sample reformatting program to ask for user input on number of samples, aliquot volume of sample matrix, volume of internal standard solution, and volume of pH adjustment buffer. An advanced technique allowable on the MultiPROBE II is to perform variable dilution of samples by reading an external text file.

While the MultiPROBE II is operating, no user intervention is required. The analyst can be optimizing the LC/MS/MS instrument parameters. Once the reformatting procedure is completed, the sample plate can be taken to the 96-tip workstation and the extraction performed. Automation strategies are discussed in more detail for PPT, LLE and SPE in Chapters 7, 10 and 13, respectively.

A 96-tip or a 4-/8-probe liquid handling workstation can be used for the sample reconstitution step following evaporation. The procedure is a simple reagent transfer step. Speed will vary between the two systems, of course. The 4-/8-tip workstation that can use well maps can selectively deliver solutions to specific wells, useful with partial plates. There is no need to manually reconstitute by pressing a repeater pipettor 96 times when a liquid handling workstation is available.

Clearly, in terms of productivity for performing routine extractions, the combination of a 4-/8-tprobe workstation and a 96-tip workstation is ideal. It is recognized, however, that not all laboratories may have these two pieces of equipment available to perform routine procedures. If one piece of automation is desired that will do both primary tasks—sample reformatting and routine extraction—then a 4-/8-probe workstation is the first choice. Although slower than a 96-tip workstation, it will perform both sample reformatting and extraction for PPT, LLE and SPE. When speed is the greater determining factor rather than flexibility, a 96-tip workstation is the first choice. The sample plate will need to be prepared manually, but in a laboratory that is adopting automation for the first time, manual preparation for this task does not introduce any change in procedure. As laboratories grow and purchase more automation, the trend is for one 4-/8-probe workstation to reformat samples continuously to serve two or more 96-tip workstations.

The use of automated liquid handling workstations, as reported in the literature, encompasses many sample preparation techniques. The reader can learn how they have been used, in a general sense, by reviewing the reported applications. Simpson *et al.* [9] reported the first semi-automated applications for 96-well SPE for bioanalysis using the 4-/8-probe MultiPROBE. The first report using the Quadra 96 is the publication by Janiszewski *et al.* [10]. These two reports are highly recommended reading.

A summary of selected applications describing the use of semi- and fully-automated liquid handling workstations to perform bioanalytical sample preparation techniques is provided in Table 5.2. A better appreciation can be gained of the strengths of certain instruments by reading the discussion section within these reports. For example, a Hamilton MICROLAB AT Plus 2 work station was reported to be ideal for performing LLE since it uses a positive displacement delivery technique. As discussed in this report [11], the Hamilton MICROLAB uses plunger-in-tip positive displacement pipette tips that permit the transfer of volatile organic solvents without dripping. Also, the design has no seals which can be eroded over time by the organic solvents typically used for LLE. In addition to practical information, these published reports often detail the time savings realized using the automation technique and may compare the automated results with those from a previous manual operation.

5.6.7 Method Development Procedures

The method development experiments that are commonly performed involve the selective addition of solutions and/or organic solvents, often with varying volumes. A PPT experiment may involve examining different ratios of sample to precipitating agent as well as different precipitating agents. A LLE experiment may be similar, examining different ratios of sample to organic solvent as well as different organic solvents. SPE method development can be more complex, requiring evaluation of different sorbent chemistries, different sample pH adjustment solutions, a series of gradually increasing percentages of organic in water as wash solvents, several elution solvent compositions as well as varying volumes for the elution step.

Table 5.2
Selected applications reporting the use of liquid handling workstations to perform bioanalytical sample preparation techniques in the microplate format

Category	Instrument Name (Mfr.)	Reformat Samples	PPT	LLE	SPE
4-/8-Probe					
	Biomek 2000 (Beckman Coulter)		[3, 4] [12–15]		[16, 17]
	Genesis (Tecan)	[18, 19]	[20]		[21–24]
	MICROLAB (Hamilton)	[25, 26]		[11, 27]	
	SPE 215 (Gilson)				[28]
	MultiPROBE (Packard, now PerkinElmer)	[29–38]	[39]	[40–43]	[5, 9] [41–53]
96-Tip	Multimek (Beckman Coulter)			[54]	
	Personal Pipettor (Apricot Designs)		[19, 55]	[56]	[57]
	Quadra 96 (Tomtec)		[29–31] [58–67]	[25] [68–78]	[10, 26] [31–38] [61, 67] [79–91]
Robotic	Zymate (Zymark)		[92–94]	[95–97]	[97–111]

PPT = Protein precipitation
LLE = Liquid-liquid extraction
SPE = Solid-phase extraction

Among these three sample preparation techniques, the common features are delivering unique sources of solvent and varying volumes to specific wells of a microplate. A good choice for method development is a 4-/8-probe liquid handling workstation for its maximum flexibility and capacity in handling multiple reagent sources, ability to deliver specific reagents to specific wells of a microplate via use of well maps, and ability to selectively vary the volumes delivered to specific wells. These instruments can read from an external text file that specifies source location, destination and volumes to be used. SPE method development procedures in which two sorbent chemistries are examined side-by-side can simply utilize two extractions in parallel on the deck of the multiple probe liquid handler. Some instruments may not allow for two unattended extractions to be configured on the deck; in this case, the second can be run after the first has finished.

The one shortcoming of using a 4-/8-probe liquid handling workstation for method development purposes is the time involved to perform the total experiment, typically from 30–90 min, depending on the technique (PPT, LLE or SPE), the number of steps to perform and the volumes used. However, to its advantage, the multiple probe liquid handlers can reformat samples so a rack of sample tubes can be placed on the instrument and the experiment run from that point forward. Note that most 96-tip liquid handling workstations can only deliver the **same volume** to all wells of a destination plate; different volumes within different tips are not an option.

A 96-tip liquid handling workstation can be used for limited method development (different solutions or solvents delivered to specific regions of the plate) when the reagent reservoir or trough is set up in rectangular grids or coordinates. For example, normally all 96 tips will enter a common reagent reservoir and dispense that one solvent to all wells of the destination plate. However, when that reservoir is divided into 4, 6 or 8 rectangular sections, specific solutions can be placed in those separate areas; upon aspirating liquids, those solutions are delivered to the corresponding locations in the destination plate. There is not as much flexibility in specifying the exact positions to receive specific solutions in the destination plate, but this approach can be useful. The key to this process is using custom reservoir troughs. The ultimate custom reservoir is a deep well plate containing 96 individual wells since solutions can be placed into any arrangement desired.

Considering the proposed ideal scenario of having both a 4-/8-tip and one or more 96-tip liquid handling workstations in the bioanalytical laboratory, this arrangement holds true when considering method development as well as routine extraction needs. The 96-tip workstation is faster, preparing multiple plates per day. The 4-/8-tip workstation is used to reformat samples from tubes into microplates; these instruments tend to have more idle time and can be better utilized by being setup to execute method development experiments on a regular basis, *e.g.*, each morning. Whether the method development programs are for PPT, LLE or SPE does not matter because the instrument has the flexibility to accommodate all these extraction procedures. Should the situation arise when more capacity is necessary and the 96-tip workstation is being used every hour of the workday, the 4-/8-tip workstation can always step in and perform the same work, although in a longer time period (*e.g.*, 60 min instead of 15 min). Automation strategies for method development are discussed in more detail for PPT, LLE and SPE in Chapters 7, 10 and 13, respectively.

5.6.8 Using Partial SPE Plates

The use of partial SPE plates with automated workstations is a commonly discussed issue. When all 96 wells of a disk or particle-loaded plate are processed on a 96-tip workstation, all wells receive solvents and then the sorbent bed is allowed to dry. What effect does this wetting have on the performance of the wells? In one regard, the wells will be cleansed or rinsed from the solvents and this process can only help. However, the opposing argument is that the capacity of the sorbent bed may be reduced and/or the sorbent chains will not be in a ready state and fully able to attract analyte on the second usage. In addition to performance arguments, cost arguments exist as well. The SPE plates can be expensive and discarding unused wells is considered a waste of money.

When the use of partial plates is very important to an analyst and it is the opinion that those unused wells must remain dry, a 4-/8-probe workstation should be used since destination wells for an extraction can be easily and quickly specified. It may be possible, in some cases, to remove tips from a 96-tip workstation to prevent some wells from receiving solvent, but the removal of too many tips may unbalance the tip rack and cause difficulty in loading.

5.7 Resources for Further Information

Many published references further describe the role for automation in bioanalysis. The automation of sample preparation for pharmaceutical and clinical analysis is described in a book chapter by Wells and Lloyd [1]. The attributes of automated sample preparation and an overview of solutions specifically for solid-phase extraction are discussed by Smith and Lloyd [112, 2]. Trends and techniques in automating solid-phase extraction using cartridges, 96-well plates and on-line formats have been reviewed by Rossi and Zhang [113]. An article and book chapter by Jordan discuss automation of SPE in cartridge format [114, 115]. The validation and calibration of automated workstations is discussed in a book chapter by Tomlinson [116]. Two useful reports present how to improve the quality and speed of automated sample transfers [117] and understanding the basics of liquid handling on a MultiPROBE II [118].

Additional references focus on a review of specific techniques and/or aspects of automation integration and planning. The integration of automation and LC-MS for drug discovery bioanalysis is discussed in an article and a book chapter by Rossi [119, 120]. A perspective on the implementation, calibration and compliance issues of automated sample preparation in a contract research organization is provided by Kagel *et al.* [121]. An overview of the use of robotics within sample preparation and other analytical processes is described by Luque de Castro and Velasco-Arjona [122]. Perspectives on succeeding with bioanalytical automation in a matrix environment are presented [97].

Many conferences, journals and news sources focus specifically on automation. Four annual scientific conferences prominently feature laboratory robotics and automation: ISLAR (International Symposium on Laboratory Automation and Robotics), Lab Automation and its European complement EuroLab Automation, MipTec-ICAR (International Conference on Microplate Technology, Laboratory Automation and Robotics) and the Society for Biomolecular Screening (SBS) Conference. Automation-specific news and literature can be found in the *Journal of the Association for Laboratory Automation* and in *Laboratory Automation News*, both sponsored by the Association for Laboratory Automation. A special interest group exists called the Laboratory Robotics Interest Group (LRIG). This group has regional chapters and a discussion group is hosted on the Internet.

The automation vendors themselves can be contacted for application notes and technical notes that describe the use of their products for bioanalysis. While it is common for manufacturers of mass spectrometry equipment to hold User Group meetings at regional and national conferences, a few automation manufacturers have begun a similar user forum or have introduced a means to network with their users. Now that the interest for automation news and applications is sustained and automation is recognized to be fundamental for maintaining laboratory operations, additional means of information sharing are expected to develop.

Acknowledgments

The author is appreciative to Danlin Wu for his critical review of the manuscript, helpful discussions and contributions to this chapter. The line art illustration was kindly provided by Willy Lee. The assistance of the following vendor representatives who contributed information and reviewed text is appreciated: Jeremy Bredwell (Zymark), Dave Hansen (Hamilton Company), Gordy Hunter (Gilson), Hansjörg Haas and Christy Jacobs (Thermo CRS), Lynn Jordan (Zymark), Asha Oroskar (Orochem Technologies), Greg Porter (Tecan), Robert Speziale (Tomtec), Shelly Staat (Packard Instruments, now a part of PerkinElmer Life Sciences), Lynda Thomas (Zymark) and Shirley Welsh (Beckman Coulter).

References

[1] D.A. Wells and T.L. Lloyd, In: J. Pawliszyn, Ed., Sampling and Sample Preparation for Field and Laboratory: Fundamentals and New Directions in Sample Preparation, Elsevier, Amsterdam (2002).

[2] G.A. Smith and T.A Lloyd, LC-GC 16 (1998) 21-27.

[3] A.P. Watt, D. Morrison, K.L. Locker and D.C. Evans, Anal. Chem. 72 (2000) 979-984.

[4] K.L. Locker, D. Morrison and A.P. Watt, J. Chromatogr. B 750 (2001) 13-23.

[5] P. Ahrweiler, P. Li and A. Bilimoria, Application Note LHA-021, PerkinElmer Life Sciences, Boston, MA USA (2001).

[6] J.R. Aliant and G. Smith, Proceedings International Symposium on Laboratory Automation and Robotics, Boston, MA USA (2001).

[7] J. Bruner, L. Birkemo, K. Jordan, G. Smith and J. Ormand, Proceedings International Symposium on Laboratory Automation and Robotics, Boston, MA USA (2000).

[8] J. Short, T. Lloyd and E. Lynch, Proceedings International Symposium on Laboratory Automation and Robotics, Boston, MA USA (2000).

[9] H. Simpson, A. Berthemy, D. Buhrman, R. Burton, J. Newton, M. Kealy, D. Wells and D. Wu, Rapid Commun. Mass Spectrom. 12 (1998) 75-82.

[10] J. Janiszewski, R.P. Schneider, K. Hoffmaster, M. Swyden, D. Wells and H. Fouda, Rapid Commun. Mass Spectrom. 11 (1997) 1033-1037.

[11] R.D. Bolden, S.H. Hoke II, T.H. Eichhold, D.L. McCauley-Myers and K.R. Wehmeyer, J. Chromatogr. B 772 (2002) 1-10.

[12] M. Lam, J. Shen, J.-L. Tseng and B. Subramanyam, Proceedings 48th American Society for Mass Spectrometry Conference, Long Beach, CA USA (2000).

[13] A. Gritsas, M. Lahaie, D. Chun, T. Flarakos, M.L.J. Reimer, F. Deschamps and R. Hambalek, Proceedings 48th American Society for Mass Spectrometry Conference, Long Beach, CA USA (2000).

[14] K.P. Bateman, G. Castonguay, L. Xu, S. Rowland, D.A. Nicoll-Griffith, N. Kelly and C.-C. Chan, J. Chromatogr. B 754 (2001) 245-251.

[15] D. O'Connor, D.E. Clarke, D. Morrison and A.P. Watt, Rapid Commun. Mass Spectrom. 16 (2002) 1065-1071.

[16] T.H. Eichhold, R.E. Bailey, S.L. Tanguay and S.H. Hoke II, J. Mass Spectrom. 35 (2000) 504-511.

[17] S.H. Hoke II, J.D. Pinkston, R.E. Bailey, S.L. Tanguay and T.H. Eichhold, Anal. Chem. 72 (2000) 4235-4241.

[18] C. De Nardi, S. Braggio, L. Ferrari and S. Fontana, J. Chromatogr. B 762 (2001) 193-201.

[19] R. Xu, C. Nemes, K.M. Jenkins, R.A. Rourick, D.B. Kassel and C.Z.C. Liu, J. Amer. Soc. Mass. Spectrom. 13 (2002) 155-165.

[20] J. Wang, S.-Y. Chang, C. D'Arienzo and D. Wang-Iverson, Proceedings 48th American Society for Mass Spectrometry Conference, Long Beach, CA USA (2000).

[21] D. Schütze, B. Boss and J. Schmid, J. Chromatogr. B 748 (2000) 55-64.

[22] A.C. Harrison and D.K. Walker, J. Pharm. Biomed. Anal. 16 (1998) 777-783.

[23] M. Larsson, U. Logren, M. Ahnoff, B. Lindmark, P. Abrahamsson, H. Svennberg and B.-A. Persson, J. Chromatogr. B 766 (2001) 47-55.

[24] C.K. Hull, A.D. Penman, C.K. Smith and P.D. Martin, J. Chromatogr. B 772 (2002) 219-228.

[25] S.H. Hoke II, J.A. Tomlinson II, R.D. Bolden, K.L. Morand, J.D. Pinkston and K.R. Wehmeyer, Anal. Chem. 73 (2001) 3083-3088.

[26] R.D. Gauw, P.J. Stoffolano, D.L. Kuhlenbeck, V.S. Patel, S.M. Garver, T.R. Baker and K.R. Wehmeyer, J. Chromatogr. B 744 (2000) 283-291.

[27] A. Shen, H. Mai, M.S. Chang, H.T. Skriba, Q.C. Ji, R. Wieboldt, D. Daszkowski and T. El-Shourbagy, Proceedings International Symposium on Laboratory Automation and Robotics, Boston, MA USA (2000).

[28] C. Linget, B. Guieu and C. Delmotte, Proceedings Pittsburgh Conference on Analytical Chemistry and Spectroscopy (Pittcon), New Orleans, LA USA (2002).

[29] W.Z. Shou, H.-Z. Bu, T. Addison, X. Jiang and W. Naidong, J. Pharm. Biomed. Anal. 29 (2002) 83-94.

[30] M. Berna, A.T. Murphy, B. Wilken and B. Ackermann, Anal. Chem. 74 (2002) 1197-1201.

[31] M.C. Rouan, C. Buffet, L. Masson, F. Marfil, H. Humbert and G. Maurer, J. Chromatogr. B 754 (2001) 45-55.

[32] W.Z. Shou, M. Pelzer, T. Addison, X. Jiang and W. Naidong, J. Pharm. Biomed. Anal. 27 (2002) 143-152.

[33] C.Z. Matthews, E.J. Woolf, L. Lin, W. Fang, J. Hsieh, S. Ha, R. Simpson and B.K. Matuszewski, J. Chromatogr. B 751 (2001) 237-246.
[34] M.J. Rose, S.A. Merschman, R. Eisenhandler, E.J. Woolf, K.C. Yeh, L. Lin, W. Fang, J. Hsieh, M.P. Braun, G.J. Gatto and B.K. Matuszewski, J. Pharm. Biomed. Anal. 24 (2000) 291-305.
[35] C. Marquez, C. D'Arienzo and J. Gale, Proceedings 46th American Society for Mass Spectrometry Conference, Orlando, FL USA (1998).
[36] C.Z. Matthews, E.J. Woolf and B.K. Matuszewski, J. Chromatogr. A 949 (2002) 83-89.
[37] A. Eerkes, T. Addison and W. Naidong, J. Chromatogr. B 768 (2002) 277-284.
[38] Y.-L. Chen, G.D. Hanson, X. Jiang and W. Naidong, J. Chromatogr. B 769 (2002) 55-64.
[39] R.A. Biddlecombe and S. Pleasance, J. Chromatogr. B 734 (1999) 257-265.
[40] D.S. Teitz, S. Khan, M.L. Powell and M. Jemal, J. Biochem. Biophys Meth. 45 (2000) 193-204.
[41] M. Jemal, D. Teitz, Z. Ouyang and S. Khan, J. Chromatogr. B 732 (1999) 501-508.
[42] W. Naidong, W. Shou, Y.-L. Chen and X. Jiang, J. Chromatogr. B 754 (2001) 387-399.
[43] D. Zimmer, V. Pickard, W. Czembor and C. Müller, J. Chromatogr. A 854 (1999) 23-35.
[44] C.S. Tamvakopoulos, J.M. Neugebauer, M. Donnelly and P.R. Griffin, J. Chromatogr. B 776 (2002) 161-168.
[45] I.D. Davies, J.P. Allanson and R.C. Causon, Chromatographia 52 (2000) S92-S97.
[46] I.D. Davies, J.P. Allanson and R.C. Causon, J. Chromatogr. B 732 (1999) 173-184.
[47] R. Bakhtiar, L. Khemani, M. Hayes, T. Bedman and F. Tse, J. Pharm. Biomed. Anal. 28 (2002) 1183-1194.
[48] J. Hempenius, R.J.J.M. Steenvoorden, F.M. Lagerwerf, J. Wieling and J.H.G. Jonkman, J. Pharm. Biomed. Anal. 20 (1999) 889-898.
[49] M. Jemal, M. Huang, Y. Mao, D. Whigan and A. Schuster, Rapid Commun. Mass Spectrom. 14 (2000) 1023-1028.
[50] Z. Liu, J. Short, A. Rose, S. Ren, N. Contel, S. Grossman and S. Unger, J. Pharm. Biomed. Anal. 26 (2001) 321-330.
[51] W.Z. Shou, X. Jiang, B.D. Beato and W. Naidong, Rapid Commun. Mass Spectrom. 15 (2001) 466-476.
[52] G. Hopfgartner, C. Husser and M. Zell, Therap. Drug Monit. 24 (2002) 134-143.

[53] J. Hempenius, J. Wieling, J.P.G. Brakenhoff, F.A. Maris and J.H.G. Jonkman, J. Chromatogr. B 714 (1998) 361-368.

[54] D.L. McCauley-Myers, T.H. Eichhold, R.E. Bailey, D.J. Dobrozsi, K.J. Best, J.W. Hayes II and S.H. Hoke II, J. Pharm. Biomed. Anal. 23 (2000) 825-835.

[55] D.L. Hiller, A.H. Brockman, L. Goulet, S. Ahmed, R.O. Cole and T. Covey, Rapid Commun. Mass Spectrom. 14 (2000) 2034-2038.

[56] J. Ke, M. Yancey, S. Zhang, S. Lowes and J. Henion, J. Chromatogr. B 742 (2000) 369-380.

[57] G. Rule, M. Chapple and J. Henion, Anal. Chem. 73 (2001) 439-443.

[58] Y. Hsieh, M.S. Bryant, J.-M. Brisson, K. Ng and W.A Korfmacher, J. Chromatogr. B 767 (2002) 353-362.

[59] R. Bakhtiar, J. Lohne, L. Ramos, L. Khemani, M. Hayes and F. Tse, J. Chromatogr. B 768 (2002) 325-340.

[60] S.L. King, D.J. Foltz, T.R. Baker and S.X Peng, Proceedings 48th American Society for Mass Spectrometry Conference, Long Beach, CA USA (2000).

[61] Y. Hsieh, J.-M. Brisson, K. Ng and W.A. Korfmacher, J. Pharm. Biomed. Anal. 27 (2002) 285-293.

[62] N. Sadagopan, L. Cohen, B. Roberts, W. Collard and C. Omer, J. Chromatogr. B 759 (2001) 277-284.

[63] W.A. Korfmacher, K.A. Cox, K.J. Ng, J. Veals, Y. Hsieh, S. Wainhaus, L. Broske, D. Prelusky, A. Nomeir and R.E White, Rapid Commun. Mass Spectrom. 15 (2001) 335-340.

[64] L. Ramos, N. Brignol, R. Bakhtiar, T. Ray, L.M. McMahon and F.L.S. Tse, Rapid Commun. Mass Spectrom. 14 (2000) 2282-2293.

[65] Y. Hsieh, G. Wang, Y. Wang, S. Chackalamannil, J.-M. Brisson, K. Ng and W.A Korfmacher, Rapid Commun. Mass Spectrom. 16 (2002) 944-950.

[66] Y. Zhang, J. Hollembaek and O. Kavatskaia, Proceedings 49th American Society for Mass Spectrometry Conference, Chicago, IL USA (2001).

[67] T. Bedman, M.J. Hayes, L. Khemani and F.L.S. Tse, Proceedings 48th American Society for Mass Spectrometry Conference, Long Beach, CA USA (2000).

[68] H. Zhang and J. Henion, J. Chromatogr. B 757 (2001) 151-159.

[69] N. Zhang, K.L. Hoffman, W. Li and D.T Rossi, J. Pharm. Biomed. Anal. 22 (2000) 131-138.

[70] L. Ramos, R Bakhtiar and F.L.S. Tse, Rapid Commun. Mass Spectrom. 14 (2000) 740-745.

[71] S. Steinborner and J. Henion, Anal. Chem. 71 (1999) 2340-2345.

[72] N. Zhang, S.T. Fountain, H. Bi and D.T. Rossi, Anal. Chem. 72 (2000) 800-806.

[73] N. Brignol, L.M. McMahon, S. Luo and F.L.S. Tse, Rapid Commun. Mass Spectrom. 15 (2001) 898-907.

[74] K. Heinig and J. Henion, J. Chromatogr. B 732 (1999) 445-458.

[75] J. Zweigenbaum, K. Heinig, S. Steinborner, T. Wachs and J. Henion, Anal. Chem. 71 (1999) 2294-2300.

[76] J. Zweigenbaum and J. Henion, Anal. Chem. 72 (2000) 2446-2454.

[77] J.M. Onorato, J.D. Henion, P.M. Lefebvre and J.P. Kiplinger, Anal. Chem. 73 (2001) 119-125.

[78] Z. Shen, S. Wang and R. Bakhtiar, Rapid Commun. Mass Spectrom. 16 (2002) 332-338.

[79] R.C. King, C. Miller-Stein, D.J. Magiera and J. Brann, Rapid Commun. Mass Spectrom. 16 (2002) 43-52.

[80] L. Yang, N. Wu and P.J. Rudewicz, J. Chromatogr. A 926 (2001) 43-55.

[81] A.E. Niggebrugge, E. Tessier, R. Guilbaud, L. DiDonato and R. Masse, Proceedings 48th American Society for Mass Spectrometry Conference, Long Beach, CA USA (2000).

[82] G. Rule and J. Henion, J. Am. Soc. Mass Spectrom. 10 (1999) 1322-1327.

[83] L.M. McMahon, S. Luo, M. Hayes and F.L.S. Tse, Rapid Commun. Mass Spectrom. 14 (2000) 1965-1971.

[84] W. Li, D.T. Rossi and S.T. Fountain, J. Pharm. Biomed. Anal. 24 (2000) 325-333.

[85] J.-T. Wu, H. Zeng, Y. Deng and S.E. Unger, Rapid Commun. Mass Spectrom. 15 (2001) 1113-1119.

[86] H. Zhang and J. Henion, Anal. Chem. 71 (1999) 3955-3964.

[87] S.X. Peng, S.L. King, D.M. Bornes, D.J. Foltz, T.R. Baker and M.G. Natchus, Anal. Chem. 72 (2000) 1913-1917.

[88] J.S. Janiszewski, M.C. Swyden and H.G. Fouda, J. Chrom. Sci. 38 (2000) 255-258.

[89] L. Yang, T.D. Mann, D. Little, N. Wu, R.P. Clement and P.J. Rudewicz, Anal. Chem. 73 (2001) 1740-1747.

[90] H. Yin, J. Racha, S.-Y. Li, N. Olejnik, H. Satoh and D. Moore, Xenobiotica 30 (2000) 141-154.

[91] P.H. Zoutendam, J.F. Canty, M.J. Martin and M.K. Dirr, J. Pharm. Biomed. Anal. 30 (2002) 1-11.

[92] M.C. Woodward, G. Bowers, J. Chism, L. St.John-Williams and G. Smith, Proceedings 48th American Society for Mass Spectrometry Conference, Long Beach, CA USA (2000).

[93] J.Y.-K. Hsieh, C. Lin and B.K. Matuszewski, J. Chromatogr. B 661 (1994) 307-312.

[94] S. Depee, Proceedings International Symposium on Laboratory Automation and Robotics, Boston, MA USA (2001).

[95] J. Hempenius, J. Wieling, J.H.G. Jonkman, O.E. de Noord, P.M.J. Coenegracht and D.A. Doornbos, J. Pharm. Biomed. Anal. 8 (1990) 313-320.

[96] J.J. Tomlinson, J.R Alianti and G.A Smith, Proceedings International Symposium on Laboratory Automation and Robotics, Boston, MA USA (1994)

[97] G.A. Smith, R.A. Biddlecombe, J. Bruner and J. Ormand, Proceedings International Symposium on Laboratory Automation and Robotics, Boston, MA USA (1998)

[98] S. Pleasance and R.A. Biddlecombe, In: E. Reid, H.M. Hill and I.D. Wilson, Eds., Methodological Surveys in Bioanalysis of Drugs, Volume 25: Drug Development Assay Approaches, Including Molecular Imprinting and Biomarkers (1998) 205-212.

[99] K.B. Joyce, A.E. Jones, R.J. Scott, R.A. Biddlecombe and S. Pleasance, Rapid Commun. Mass Spectrom. 12 (1998) 1899-1910.

[100] T.D. Parker III, D.S. Wright and D.T. Rossi, Anal. Chem. 68 (1996) 2437-2441.

[101] J.Y.-K. Hsieh, C. Lin, B.K. Matuszewski and M.R. Dobrinska, J. Pharm. Biomed. Anal. 12 (1994) 1555-1562.

[102] R.A. Biddlecombe, C. Benevides and S. Pleasance, Rapid Commun. Mass Spectrom. 15 (2001) 33-40.

[103] R.J. Scott, J. Palmer, I.A. Lewis and S. Pleasance, Rapid Commun. Mass Spectrom. 13 (1999) 2305-2319.

[104] T.L. Lloyd, S.K. Gupta, A.E. Gooding and J.R. Alianti, J. Chromatogr. B 678 (1996) 261-267.

[105] Y. Deng, J.-T. Wu, T.L. Lloyd, C.L. Chi, T.V. Olah and S.E. Unger, Rapid Commun. Mass Spectrom. 16 (2002) 1116-1123.

[106] T.L. Lloyd, T.B. Perschy, A.E. Gooding and J.J. Tomlinson, Biomed. Chromatogr. 6 (1992) 311-316.

[107] S.L. Callejas, R.A. Biddlecombe, A.E. Jones, K.B. Joyce, A.I. Pereira and S. Pleasance, J. Chromatogr. B 718 (1998) 243-250.

[108] R.A. Biddlecombe and S. Pleasance, Proceedings International Symposium on Laboratory Automation and Robotics, Boston, MA USA (1996) 445-454.

[109] J. Ayrton, G.J. Dear, W.J. Leavens, D.N. Mallet and R.S. Plumb, J. Chromatogr. B 709 (1998) 243-254.

[110] J.J. Zheng, E.D. Lynch and S.E. Unger, J. Pharm. Biomed. Anal. 28 (2002) 279-285.

[111] S. Hsieh and K. Selinger, J. Chromatogr. B 772 (2002) 347-356.

[112] G.A. Smith and T.L. Lloyd, LC-GC Europe 11 (1998) 21-27.

[113] D.T. Rossi and N. Zhang, J. Chromatogr. A 885 (2000) 97-113.

[114] L. Jordan, LC-GC 11 (1993) 634-638.

[115] L. Jordan, In: N.J.K. Simpson, Ed., Solid-Phase Extraction: Principles, Techniques and Applications, Marcel Dekker, New York (2000) 381-410.

[116] J.J. Tomlinson, In: C.M. Riley and T.W. Rosanske, Eds., Progress in Pharmaceutical and Biomedical Analysis, Volume 3: Development and Validation of Analytical Methods, Pergamon, Oxford (1996) 185-208.

[117] G. Smith, J. Bruner, C. Buckner and B. Biddlecombe, Proceedings International Symposium on Laboratory Automation and Robotics, Boston, MA USA (2001).

[118] J. Ormand, J. Bruner, L. Birkemo and M. Emptage, Proceedings International Symposium on Laboratory Automation and Robotics, Boston, MA USA (2000).

[119] D.T. Rossi, J. Autom. Methods Management 24 (2002) 1-7.

[120] D.T. Rossi, In: D.T. Rossi and M.W. Sinz, Eds., Mass Spectrometry in Drug Discovery, Marcel Dekker, New York (2002) 171-214.

[121] J.R. Kagel, W. Donati, L.E. Elvebak and J.A. Jersey, Amer. Lab. 33 (2001) 20-23.

[122] M.D. Luque de Castro and A. Velasco-Arjona, Anal. Chim. Acta 384 (1999) 117-125.

Chapter 6

Protein Precipitation: High Throughput Techniques and Strategies for Method Development

Abstract

This chapter introduces the techniques used to remove proteins from a biological sample matrix prior to analysis. While not as selective as other sample preparation methods, the simplicity and universality of this approach have broad appeal for many applications, particularly in support of drug discovery. High throughput procedures for removing proteins from the matrix are presented. These protocols include precipitation of protein in wells of collection microplates, filtration of precipitated protein in filter microplates, and membrane separation techniques in microplates. Efficient method development and optimization strategies for protein precipitation are also discussed. Automation strategies for performing these high throughput techniques with liquid handling workstations are presented in Chapter 7.

6.1 Understanding the Technique

6.1.1 Fundamental Principles

Sample matrices for bioanalysis almost always contain some amount of protein, along with other endogenous macromolecules, small molecules, metabolic byproducts, salts and possibly coadministered drugs. These components must be removed from the sample before analysis in order to attain a selective technique for the desired analytes. However, achieving that selectivity often requires the use of extraction procedures [*e.g.*, liquid-liquid extraction (LLE) or solid-phase extraction (SPE)] that are sometimes perceived as complex (requiring optimal solvent and pH conditions) or too time consuming to develop and/or perform.

It is most important to remove the protein from a biological sample because that protein, when injected into a chromatographic system, will precipitate upon contact with the organic solvents and buffer salts commonly used in

199

mobile phases. The precipitated mass of protein builds up within the column inlet. The result is reduction of column lifetime and an increase in system backpressure. When protein is carried through the analytical system it may reach the mass spectrometer and foul the interface, requiring cleaning. A common approach for removing protein from the injected sample, one that is amenable to high throughput applications in microplates, is precipitation using organic solvents, ionic salts and/or inorganic acids. The precipitated mass can be separated by either centrifugation or filtration; analysis of the supernatant or filtrate, respectively, is then performed.

Another approach to protein removal is physical separation by diffusion through a selective, semipermeable membrane using equilibrium dialysis or ultrafiltration methods. A dialysis membrane positioned within a well or chamber allows passage of small molecular weight drugs and other materials in solution that are not bound to protein. The protein macromolecules and drug/-protein complexes in solution remain on the other side of the membrane. Equilibrium dialysis is concentration driven while ultrafiltration is pressure driven (using centrifugal force). Both dialysis methods are now amenable to high throughput by using the microplate format. In this chapter, these various high throughput approaches for performing protein precipitation (PPT) will each be detailed after first discussing some experimental and procedural aspects of the technique.

6.1.2 Precipitating Agents

Proteins play an important role in the transport and storage of drug substances. Most biological matrices contain protein to varying extents. Protein binding phenomena are known to influence drug-drug interactions in the clinical setting. Among the various plasma proteins, serum albumin is the most widely studied and is regarded as the most important carrier for drugs. Gamma globulins represent another plasma protein. The presence of these materials for bioanalysis, however, is problematic and they must be removed before chromatographic separation and detection.

6.1.2.1 Acids Used as Precipitating Agents

Proteins are known to precipitate from solution when subjected to strong acids, organic solvents and certain salts of heavy metal cations. The following acids (in typically used concentrations) protonate basic sites on the protein to change its conformation, subsequently forming insoluble salts at a pH below their isoelectric point: trichloroacetic acid (TCA; 10%, w/v), perchloric acid

(6%, w/v) and metaphosphoric acid (5% w/v). Recall that the isoelectric point is the pH at which a protein molecule has a neutral charge; at a pH higher than its isoelectric point, the protein will act as a base, and at a pH lower than its isoelectric point, it will act as an acid. A protein has its minimum viscosity at its isoelectric point and can be coagulated more easily at this value. These aforementioned acids are very effective protein precipitants but they are generally regarded as too harsh for many labile analytes when injected directly into LC-MS/MS systems [liquid chromatography (LC) with tandem mass spectrometry (MS/MS)]. However, their great efficiency at low volume ratios makes them an attractive choice. Two examples of sample preparation methods that used acid precipitation prior to analysis by mass spectrometry (MS) include a microsomal incubation that was quenched with an equal volume of 0.3M TCA [1] and a 100 µL plasma precipitation that used 50 µL cold aqueous 10% TCA [2]. The use of a cold solution stops further enzymatic activity for *in vitro* metabolic reactions by reducing the temperature quickly below the reaction temperature of 37°C. Direct injection of supernatant solutions from 10% TCA precipitation into LC-UV systems has also been reported [3].

6.1.2.2 Organic Solvents Used as Precipitating Agents

Organic solvents are known to lower the dielectric constant of protein solutions and increase the protein-protein interactions. Solvents such as methanol, ethanol, acetonitrile (ACN) and acetone are reported to be slightly less effective than acids for their degree of protein precipitation [4] but are preferred in bioanalysis because the conditions are very mild and analyte degradation is avoided. These organic solvents are also more compatible with the LC mobile phases commonly used and can be injected directly. After vortex mixing, followed by pelleting the protein at the bottom of the tube or well, the supernatant is suitable for direct injection. Note, however, that only small volumes of acetonitrile supernatant (*e.g.*, 1–5 µL) can be injected using this approach. The high concentration of acetonitrile in the injection volume may upset the chromatographic separation and cause peak distortion when the mobile phase contains a low percentage of the organic component.

Among choices for organic solvents used as precipitating agents, acetonitrile and methanol are most often utilized for bioanalytical sample preparation; each has its advantages and drawbacks in certain situations. For example, using acetonitrile can result in poor recovery [5] and also late eluting peaks. Methanol is not as efficient as acetonitrile when used in the same ratios but it does tend to produce a white, flocculent precipitate that facilitates better mixing; the supernatant appears clearer, rather than cloudy or slightly dark in contrast to acetonitrile. The supernatant isolated from methanol precipitation has been reported to result in a lower frequency of plugged membrane filters compared with that from acetonitrile precipitation [6, 7].

6.1.2.3 Quantities of Precipitating Agents

The ratio of precipitating agent to sample matrix (*e.g.*, plasma) is important for efficient removal of proteins. Acids achieve efficient removal (>98%) of proteins at very low ratios—1 part plasma to 0.2 part trichloroacetic acid or 0.4 part perchloric acid. However, the commonly used organic solvents require greater volumes relative to plasma, as detailed in Table 6.1. A ratio of 1 part plasma to 1 part acetonitrile yields 97.2% efficacy for protein removal, while a ratio of 1:3 is maximal, yielding 99.8% efficacy. However, methanol yields only 73.4% efficacy at a 1:1 ratio but 98.9% at a 1:3 ratio. A 1:4 ratio using methanol is needed to reach 99.2% efficacy [4]. The best results for precipitating proteins from plasma are obtained using a ratio of 1 part plasma to 3 parts acetonitrile or 1 part plasma to 4 parts methanol. Blanchard experimentally determined that a ratio of 1 volume plasma requires at least 3 volumes of acetonitrile to obtain efficient protein precipitation; the Lowry assay was used for measuring protein [4]. Another report confirmed the 1:3 ratio as optimal [8]. However, it is frequently seen that published applications report the use of a smaller 1:2 ratio [9–15].

Table 6.1
The relative efficacy of some precipitants given as percentage of plasma proteins precipitated

Precipitant	pH of Supernatant	0.6 vol	1 vol	2 vol	3 vol	4 vol
Acetonitrile	8.5–9.5	45.8	97.2	99.7	99.8	99.8
Acetone	9.0–10	33.6	96.2	99.4	99.2	99.1
Methanol	8.5–9.5	32.2	73.4	98.7	98.9	99.2
Ethanol	9.0–10	41.7	91.4	98.3	99.1	99.3
10% TCA	1.4–2.0	99.6	99.5	99.8	99.8	99.8
10% $HClO_4$	<1.5	98.9	99.1	99.1	99.1	99.0
5% HPO_4	1.6–2.7	98.1	98.3	98.4	98.2	98.1

Reprinted with permission from [4]. Copyright 1981 Elsevier Science.

6.1.2.4 Other Protein Precipitating Agents

Hydrated salt ions, such as saturated aqueous ammonium sulfate, reduce the available water molecules for protein-aqueous interactions and decrease protein solubility. Heavy metal cation salts of zinc and copper coordinate with negatively charged carboxyl groups to affect the protein net charge and solvation properties. These metal salts are effective precipitating agents but are not as popular for bioanalysis since nonvolatile salts are not compatible with liquid chromatography mass spectrometry techniques. A typical metal salt used for precipitation is zinc sulfate heptahydrate (10%, w/v) prepared in 0.5N sodium hydroxide. Note that a combination of a zinc salt with an organic solvent has been shown to be a very effective agent to both lyse cells and precipitate proteins when using whole blood; a finer precipitate is formed using this combination compared with acetonitrile alone. The composition of this precipitant is a mixture of acetonitrile, water and zinc sulfate heptahydrate solution (350 mg/mL water) in a volume ratio of 500/500/50 [16].

Since zinc and copper salts should not be injected directly, their use is best reserved as a sample pretreatment step for another technique. For example, proteins from tissue homogenates (*e.g.*, mouse or rat brain) can be precipitated using this organic/water/salt mixture and the supernatant after centrifugation can be loaded directly onto solid-phase extraction media for further cleanup. The subsequent SPE procedure will remove these salts, as well as isolate and concentrate the analyte in a more selective approach. Note, however, that the percentage of organic in the sample should not exceed 10% for the SPE load step and an aqueous dilution of the extract obtained following protein precipitation is usually necessary.

Some additional combinations of precipitating agents have been used for various applications. A dilute acid solution is commonly added to help maintain compatibility with the mobile phase used for LC-MS/MS analysis. A mixture of acetonitrile and ethanol (9:1) with 0.1% formic acid in a 1:5 ratio of sample to solvent has been used [17]. Similarly, the same solvent combination but with 0.1% acetic acid in the same ratio has been reported [18]. Methanol containing 1% acetic acid [7], as well as acetonitrile containing 1% acetic acid [19] in a 1:1 ratio has also been utilized. Another variety of acid added to acetonitrile is 0.1% trifluoroacetic acid, used in a 1:2 sample to solvent ratio [13]. One report used a combination of perchloric acid with methanol: 25 µL HClO4 were added to 250 µL plasma and mixed, 200 µL methanol were added and mixed, followed by centrifugation [20].

6.1.3 Procedures and Usage

Protein precipitation is often used as the initial sample preparation scheme in the analysis of a new drug substance since it does not require much method development time. A guideline is simply followed for dilution of a given volume of matrix with a given volume of organic solvent (or other precipitating agent), followed by vortex mixing and then centrifugation or filtration to remove the protein. The supernatant or filtrate is then analyzed directly. When a concentration step is required, the supernatant can be isolated from the pellet, evaporated to dryness and then reconstituted before analysis. An alternate concentration scheme might use liquid-liquid extraction, solid-phase extraction or an on-line technique for more selective cleanup before analysis (Figure 6.1).

Protein precipitation techniques are often performed by bioanalytical laboratories in support of drug discovery programs. In this situation, an analytical method needs to be developed quickly, the sample numbers are small

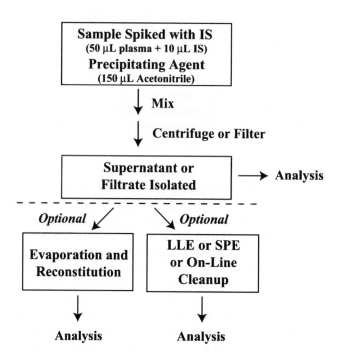

Figure 6.1. Schematic diagram of the protein precipitation technique for bioanalysis. Optional steps are shown.

(~50 or less) and manual labor is acceptable for the sample preparation step. This generic extraction approach is more important for these laboratories than achieving selectivity and sometimes sensitivity. The high resolving power of LC-MS/MS analytical methods has reduced the need for selective cleanup procedures. Typical sample matrices that are used with protein precipitation techniques are plasma, serum, tissue homogenates and *in vitro* incubation mixtures. Typically, a sensitivity of 1–10 ng/mL can be achieved from 50 µL sample volumes by protein precipitation using LC-MS/MS techniques.

Protein precipitation as a sample preparation approach can be used beyond drug discovery, particularly for bioanalytical support in Phase I clinical studies. However, as a drug candidate moves into Phase II clinical studies and beyond, the bioanalysis is usually more demanding of a more selective and sensitive (pg/mL) sample preparation technique. Attaining these goals is generally not achievable by protein precipitation alone.

6.1.4 Advantages

Protein precipitation is one of the four main drug sample preparation methodologies performed in laboratories (along with LLE, SPE and on-line techniques) because it is a simple and almost universal procedure for small drug molecules in plasma. There is usually no pH modification involved as with other techniques so the exact nature of the analyte (ionized or unionized) is not as important to the success of the method. Its speed is a real advantage and there is little method development time invested before proceeding to method validation.

This approach to sample preparation is also inexpensive and does not harm the analyte in any way. Common laboratory equipment is available to perform the work. Certain on-line systems may involve an additional hardware purchase or complex instrument configurations that allow direct injection of biological samples using column switching techniques. These direct injection schemes are attractive for their speed and added selectivity, and are reviewed in more detail in Chapter 14.

Precipitation is also an effective means to disrupt strong protein binding interactions with analytes. High recovery (>95%) of analyte can be obtained even when the extent of protein binding exceeds 99% [5]. Many times, protein precipitation is performed as a pretreatment step for another technique, and it can be easily added to the procedure with little additional effort. Very small sample volumes (*e.g.*, 20–50 µL plasma) can be efficiently eliminated of proteins using small volume microplate wells since this precipitation technique has minimal sample transfer and isolation steps.

6.1.5 Disadvantages

Satisfactory analyses have been demonstrated with this rapid sample preparation approach, but it has several disadvantages. Protein precipitation typically dilutes the sample by a factor of three or more. Therefore, it is a useful technique only when analyte concentrations are relatively high (1-10 ng/mL) and the detection limits allow adequate quantitation. However, the supernatant can be evaporated using nitrogen and heat to reverse this dilution effect. The evaporation step requires an additional transfer step (with possible transfer loss) and time for the dry-down and reconstitution procedures. The analyte volatility for newly synthesized compounds is often unknown, and lower recovery may result by introducing a dry-down step for an analyte that is labile to heat. Note that a lower recovery upon dry-down may also be due to the potential contributions of analyte adsorption to the sample container and inefficiency in resolubilizing the dried extract.

Problems can arise from LC column and mass spectrometer interface fouling since the efficiency of protein removal is not complete (only about 95–99%, see Table 6.1). Gradual accumulation of small amounts of protein and/or particulates may occur and be noticeable only after injection of a large number of samples. The completion of a full run (70–96 samples) is sometimes not possible due to this fouling, which can elevate the system backpressure and cause a high pressure shutdown error that aborts the run. Occasionally, analytes can be trapped in the pelleted protein after centrifugation, and a wash step of the protein mass may be necessary to recover more analyte [21]. In this case, after the supernatant is removed, a volume of precipitating agent is again added to the pellet (*e.g.*, acetonitrile) and the sample container is vortex mixed and centrifuged. The supernatant is isolated and combined with that from the first extraction. This supernatant can also be filtered upon isolation to remove small masses of precipitated protein [22].

The use of an additional wash step in this manner adds time to the overall procedure and dilutes the sample further (now at twice the volume of the original precipitation). Dry-down followed by reconstitution is then usually necessary to concentrate the analyte. Note that when an analyte is trapped in the protein mass in this manner, it may be worthwhile to vary the order of addition of components and/or change the sample container material in an attempt to reduce this occurrence. Other approaches to reduce the trapping occurrence are to change to a different precipitating agent and/or adjust the pH of the sample matrix before precipitation.

Protein accumulation is not the only factor that can cause problems when using protein precipitation techniques; lipid components are also undesirable. Various types of lipids can potentially accumulate in the LC column and elute slowly upon each injection, deteriorating column efficiency over time [23]. Also, when an organic solvent is used as the precipitating agent and an aliquot is injected directly, the analytes are now dissolved in a strong eluant. Injection of a volume of organic solvent (*e.g.*, acetonitrile) into a much weaker mobile phase can cause elution in the solvent front when injection is made into short LC columns (*e.g.*, 10 mm x 1 mm i.d.) [1]. More frequently, distorted peak shape results (rather than lack of retention) when injecting sample in a higher percent organic solution than contained in the mobile phase. The case of variable ionization suppression, discussed in the next section, is even more of a problem than the accumulation of matrix components and the percentage of organic in the injected sample.

6.2 Ionization Issues

6.2.1 Understanding Matrix Effects

The major precaution when using protein precipitation as a sample preparation technique is that matrix components are not efficiently removed and will be contained in the isolated supernatant or filtrate. These endogenous materials may adversely affect detection by ultraviolet (UV) and tandem MS methods but in different ways. In UV detection systems, these contaminants will add peak area to the solvent front and may reduce the analyte response at low concentrations by introducing a high background signal. In addition, coeluting materials may overlap the analyte peak of interest, making quantitation difficult. One report suggested that coelution of matrix and analyte from the LC column can be primarily attributed to column overloading with matrix components rather than identical analyte-matrix retention behavior [24]. In MS/MS detection systems, matrix contaminants have been shown to reduce the efficiency of the ionization process [10, 13, 25–33] using atmospheric pressure ionization (API) techniques. The observation seen is a loss in response and this phenomenon is referred to as ionization suppression. This effect can lead to decreased reproducibility and accuracy for an assay and failure to reach the desired limit of quantitation.

It is reported that the extent of ionization suppression seen is much more severe with electrospray ionization (ESI) than with atmospheric pressure chemical ionization (APCI) [13, 34]. In both processes, analyte in a liquid stream is converted into gas phase ions that can be sampled by the mass spectrometer. However, each technique produces the charged analyte in a different manner, as nicely described by King *et al.* [13]. In electrospray, analyte is ionized in the liquid phase inside the electrically charged droplets. The analyte ions in solution are then liberated into the gas phase. In APCI, the neutral analyte is transferred into the gas phase by vaporizing the liquid in a heated gas stream. Chemical ionization of the gas phase analyte occurs in a separate step.

The relative importance of many potential factors influencing ionization suppression is not yet fully understood. King indicated that gas phase reactions leading to the loss of net charge on the analyte is not likely the most important process involved. Rather, a more plausible explanation is that changes in the droplet solution properties caused by the presence of high concentrations of nonvolatile solutes (*e.g.*, sulfate and phosphate salts, precipitated analyte, interfering compounds) cause most of the ionization suppression in electrospray ionization of biological extracts.

Detailed analyses of the extent of ionization suppression introduced by endogenous plasma interferences using ESI tandem MS techniques have been reported by Bonfiglio *et al.* [10] and Müller *et al.* [35] and comparisons were made among protein precipitation, LLE and SPE techniques. As described in the report by Bonfiglio, an infusion pump was placed post column in order to deliver a constant amount of analyte (caffeine, phenacetin or a proprietary compound) into the LC stream entering the ion source of the MS. Blank extracts from different sample preparation techniques were injected. The presence of endogenous components as eluted from the LC column caused a variation in ESI response of the infused analyte, as shown in Figure 6.2 for protein precipitation extracts for each analyte.

The extent of ionization suppression was shown to depend on the sample preparation method [10]. LLE (using methyl tert-butyl ether extracts) and SPE (using C2, C8 and C18 extracts) showed measurable losses of ESI response compared with injections of mobile phase spiked with analyte; protein precipitation samples (acetonitrile extracts) showed even greater loss of ESI signal. This effect was seen for all three analytes. Filtering the samples showed only slightly smaller improvements. The ESI response was also documented to be compound dependent within a given sample preparation method; the more

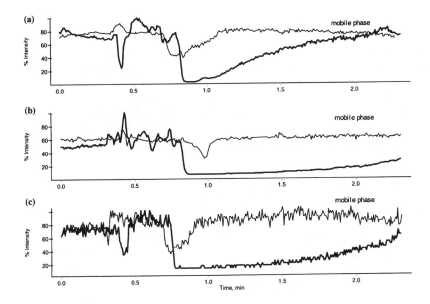

Figure 6.2. The effect of injecting blank plasma protein precipitation samples (prepared using acetonitrile) on column with post column infusion of analytes: (a) proprietary compound, (b) phenacetin and (c) caffeine. The ESI infusion chromatogram from a mobile phase injection is overlaid with an infusion chromatogram following injection of each plasma extract. The difference between the two is due to the effect of endogenous plasma components eluting from the column. Panels (a) through (c) are ESI ion infusion chromatograms for the test compounds compared with a protein precipitation blank injected on column. Selected reaction monitoring and an isocratic mobile phase were used. Reprinted with permission from [10]. Copyright 1999 John Wiley & Sons, Ltd.

polar analytes appeared to be more sensitive to loss of signal. Another important finding documented in this work [10] was that the time required for the ESI signal to return to pre-injection levels varied among compounds and sample preparation methods. Late eluting components in the samples could cause additional suppression beyond the run time of the assay. In a related report, the optimization of protein precipitation procedures was investigated with regard to ionization suppression issues, also using a constant infusion of drug into the LC stream entering the ion source of the MS [30]. It was found that ionization suppression varies with the precipitant used, and also that the suppression varies over time. Figure 6.3 shows the ionization suppression from

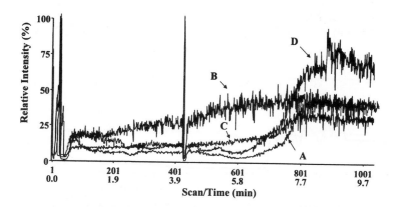

Figure 6.3. Ionization suppression is shown to vary with the precipitant solution used with plasma: (A) ethanol, (B) trichloroacetic acid 10% (w/v), (C) acetonitrile and (D) methanol. Mobile phase was methanol/water (50:50, v/v). Unpublished data [30] reprinted with permission from the authors (P. Sarkar, C. Polson, R. Grant, B. Incledon and V. Raguvaran, Eli Lilly Canada Inc.).

four different precipitant solutions used with plasma: (A) ethanol, (B) trichloroacetic acid 10% (w/v), (C) acetonitrile and (D) methanol.

Ionization suppression also varies with the mobile phase used [30]. A comparison was made of ethanol precipitated plasma extracts injected in three slightly different mobile phases: methanol/water (1:1), methanol/0.1% formic acid (1:1), and methanol/10mM ammonium formate (1:1). It was found that formic acid or ammonium formate in the mobile phase reduced the ionization suppression for an acidic drug. A comparison was also made of the extent of matrix effects using different tandem mass spectrometers [34]. It was found that matrix effects vary between sources (ESI and APCI), the gradient used (fast versus slow), and with the instrumentation. The varying degrees of matrix effects seen among instruments are proposed to be caused by differences in API source design. A case study of ion suppression in LC-MS/MS and methods used to eliminate it are detailed by Nelson and Dolan [36].

While plasma and urine as sample matrices are well studied, a related report examined the effects on electrospray ionization efficiency using microsomal incubation media [32]. In this report, a diverse set of 27 analytes was used to evaluate matrix effects from microsomal media (Tris buffer, NADPH and microsomes) after acetonitrile protein precipitation. It was found that direct flow injection MS/MS analysis, without any sample preparation or LC separation, gave an average 2.2-fold to 5-fold matrix suppression in MS response from Tris buffer and NADPH. The more polar analytes were affected the greatest, in agreement with another report [30]. A smaller matrix effect was observed when off-line SPE (Oasis® HLB, Waters Corporation, Milford MA USA) in automated 96-well plate format was used prior to analysis or when a fast gradient LC separation technique was employed.

6.2.2 Minimizing Matrix Effects

Attempts have been made to minimize matrix effects from protein precipitation extractions. The best approach to achieve resolution from biological interferences is by a more selective extraction technique. A more extensive sample cleanup results in fewer coeluting matrix components which allow less variability at lower levels and, in general, this results in a lower limit of quantitation [28, 17]. However, due to time constraints, balanced with the demand in drug discovery to evaluate a large number of compounds, the time is simply not available to develop this more selective technique.

When it is decided that the protein precipitation technique is to be used, a recommended approach is to use a stable isotope of the analyte as the internal standard (IS) so that both compounds coelute and are affected in the same manner by ESI suppression; the peak area ratios are still reliable and a smaller coefficient of variation results [10, 26, 27, 30, 37]. Shown in Table 6.2 are data comparing an isotopically labeled internal standard with an unlabeled internal standard following various protein precipitation conditions, demonstrating the benefits of improved precision using the isotope-labeled IS. Note that an investment in the synthesis of a stable isotope labeled analogue is usually undertaken only for a compound further along in the development cycle; analytes in discovery and preclinical programs rarely have stable isotopes available for use as internal standards.

Another tactic to reduce matrix effects is the use of diluted protein precipitated extracts so that less mass is injected on column. High (>6.0) values of k' [9, 27] are achieved on the analytical LC column and provide adequate separation from unretained or slightly retained matrix components. The performance of present day LC-MS/MS systems generally achieve adequate sensitivity for most applications using these diluted extracts. The combination of a more selective extraction with better chromatographic separation is also valid [26] to minimize the influence of matrix effects. In general, a greater emphasis should

Table 6.2
Comparison of the effect of isotope-labeled and unlabeled internal standards (IS) used within various precipitation conditions promoting ionization suppression

Precipitant	Approximate Retention Time (min)	% Change in Ratio for a Labeled Internal Standard	% Change in Ratio for an Unlabeled Internal Standard
Methanol	0.6	34.6	157.0
Ethanol	0.5	−42.3	28.6
Acetonitrile	0.5	−0.4	91.4
Trichloroacetic acid	6.8	10.3	−99.1
m-Phosphoric acid	4.8	1.4	14.6

% Change in Ratio = [(ratio of aqueous standard to IS)/(ratio of human plasma precipitated standard to IS)] - 1 x 100

Mobile phase was acetonitrile/water (50:50, v/v)

Analyte and internal standards eluted at a retention time corresponding to maximum suppression for the organic precipitants. A change in analyte charge resulted in a retention time shift.

Unpublished data [30] reprinted with permission from the authors (P. Sarkar, C. Polson, R. Grant, B. Incledon and V. Raguvaran, Eli Lilly Canada Inc.).

be placed on the chromatographic separation of matrix coextractables so that any compromises in speed and sensitivity are minimized. A good approach to obtain longer k' values without longer run times is to use a monolithic column for the chromatographic separation, as discussed further in Chapter 14, Section 14.2.4. Besides these tactics, another approach is to change the ionization type from ESI to APCI. APCI does not suffer ion suppression to the same degree as ESI because it uses a corona discharge that provides more complete ionization.

In related important reports, the use of rapid gradient LC-MS/MS for analysis coupled with flow diversion of the solvent front [11, 17] allowed the introduction of protein precipitated samples into the mass spectrometer without the need for frequent source cleaning. There did not appear to be significant trends in matrix effects when comparing flow rate or column length [9]. A comparison of high organic isocratic LC analysis with ballistic gradient ultrafast LC concluded that both approaches are subject to matrix interferences [17]. An investigation using lower ESI flow rates in the nanoliter per minute range (using a nanosplitting device) documented reduced ionization suppression and improved the concentration sensitivity of the assay [38].

6.3 High Throughput PPT Techniques Using Collection Microplates

6.3.1 Introduction

Protein precipitation traditionally has been performed in test tubes, but this approach is labor intensive with its required tube labeling and frequent manipulations. The manual test tube method does not meet the high throughput needs required by the emphasis on rapid pharmaceutical drug development. Since the microplate format is commonly used in autosamplers for injection into the chromatographic system, it is desirable to retain that format and perform the sample preparation procedure in a microplate, as well.

Two general approaches are common for performing protein precipitation in the high throughput microplate format:

1. Use a collection plate or microtube rack, pellet the precipitated protein at the bottom of wells by centrifugation and collect the supernatant for analysis
2. Use a filtration microplate to trap the precipitated protein on top of the filter and collect the filtrate for analysis

Each of these approaches is practical for performing high throughput protein precipitation methods. The determining factors for selection of one approach over the other include the extent of available hardware and automation accommodating the microplate format, total cost of materials, physical manipulations and the degree of transfer loss deemed acceptable. The collection plate methodologies will now be discussed in detail, followed in the next section (6.4) by filtration plate methodologies.

6.3.2 Procedure

The collection plate technique for PPT involves using wells of a microplate in place of traditional test tubes. There is no need for labeling, as each well has its own unique coordinate specified by row letter and column number (*e.g.*, A3). Both 1 mL and 2 mL microplates are used, but deep well polypropylene plates are preferred for their larger volume capacity, especially for mixing. Microtubes (*e.g.*, 1.2 mL) may also be used. Note that it is easier to seal a deep well plate once with one mat covering 96 wells (Figure 6.4) than it is to seal microtubes twelve times with 8-well cap strips (or 8 times with 12-well cap strips), and so microplates are used more often than microtubes for this procedure. A single sealing mat may be used with microtubes only if the seal is

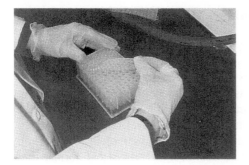

Figure 6.4. Protein precipitation is performed in a deep well collection microplate. A cover mat is applied for the mixing and centrifugation steps.

tight. Sealing systems for microplates and microtubes are discussed in detail in Chapter 4, Section 4.3. A protocol for performing protein precipitation using collection plates is listed in Table 6.3.

A microplate containing protein pellets at the bottom of each well, following centrifugation, can be placed directly on the autosampler after sealing with a pierceable mat. The autoinjector needle depth can be adjusted so that it aspirates only from the liquid above the pellet. However, depending on how

Table 6.3
Protocol for performing protein precipitation in microplate wells; typical volumes are shown

- Dispense 1 aliquot plasma into wells
- Dispense internal standard solution and briefly mix
- Dispense precipitating agent
    a) 3 aliquots of acetonitrile or
    b) 4 aliquots of methanol
- Cap or seal microplate tightly
- Vortex mix for 3 min
- Centrifuge at 3000 rpm x 10 min to pellet protein at bottom of wells
- Transfer supernatant into a clean microplate without disrupting pellet
    a) Evaporate supernatant and then reconstitute in mobile phase compatible solvent, or dilute supernatant directly with an equal volume of aqueous solution
    b) Seal microplate
    c) Vortex mix briefly
- Inject aliquot

well the pellet is compacted (which is influenced by the precipitating agent used and the force used), disruption can be caused by the autoinjector needle or from normal plate handling. Any disturbance increases the possibility of protein mass being injected and potentially plugging column frits and/or tubing. If this approach is adopted, the geometry of the well becomes very important. A well bottom that is of the tapered, V-bottom or pyramid bottom style is preferred for the ability to focus the protein mass.

A more common practice in using a collection plate for precipitation is to transfer an aliquot of the supernatant following centrifugation to a clean microplate to avoid the possibility of pellet disruption upon repeated injections and plate manipulation in the autosampler. A pipetting workstation or liquid handling workstation can be used for this transfer step. Although the transfer may be performed manually using a multichannel pipettor, it is difficult to consistently aspirate these volumes; also, the insertion depth is variable when done by hand and pellet disruption is more likely. The volume aspirated is one that does not disturb the protein pellet upon removal (determined by trial and error). The destination microplate often contains a small volume of aqueous buffer (*e.g.*, 25 mM ammonium formate or 0.1% formic acid) to make the resulting solvent composition compatible with mobile phase for subsequent injection. Alternately, the aqueous modifier can be added to the collection plate already containing the transferred supernatant. After this step, the collection plate is sealed, briefly vortex mixed, and placed into the autosampler for injection.

The potential concerns regarding this practice are that disruption of the pellet could occur upon pipetting (introducing some solid mass into the pipet tip) and that the transfer will never be 100% complete since some volume of liquid must be left behind on top of the pellet surface. With careful manipulation of the pipetting procedure and practice, however, the pellet can be effectively left behind. The use of an internal standard is appropriate to correct for transfer loss. Thorough mixing prior to centrifugation is expected to result in a homogeneous supernatant.

Note that instead of direct injection, or dilution of supernatant followed by injection, another option is evaporation followed by reconstitution. This latter approach introduces a concentration step into the procedure, often important to meet assay sensitivity needs. However, the tradeoff to achieving greater sensitivity is the additional time required for this dry-down and reconstitution step. Also important to note is that concentration of the analyte from the supernatant volume also concentrates the mass of matrix materials carried over from the precipitation, potentially introducing greater ionization suppression. Although the collection plate approach to protein precipitation is simple and can be done manually, automation is introduced when the number of samples to be extracted becomes large and the procedure is to be performed routinely. Automation is particularly useful to aspirate the supernatant at a fixed height and for its liquid level sensing capability. Automation strategies for protein precipitation are discussed in Chapter 7.

The protein precipitation procedure yields a sample that is ready for analysis, but its potential to carry over matrix interferences can be problematic. Sometimes, the isolated supernatant following precipitation is subjected to another technique to further clean up the sample. For example, applications have been reported in which an initial protein precipitation step was followed by turbulent flow chromatography [39], off-line solid-phase extraction [40, 41], on-line SPE [42], on-line column switching [43, 44] or liquid-liquid extraction [45–49]. The isolated supernatant can also be filtered and/or centrifuged before injection of an aliquot into the chromatographic system.

6.3.3 Accessory Products

The microplate format is very attractive for performing protein precipitation as it dramatically reduces the number of manipulations and eliminates individual tube labeling. Performing protein precipitation in microplates requires very few accessory products, as listed in Table 6.4. Pipettors are needed for sample transfers and reagent solvent dispensing. A collection plate or microtube sample rack is needed, and a second microplate is used when the supernatant is transferred after pelleting the protein. Microtubes are preferable by some users for manual aspiration of supernatant because the individual tubes can be removed and the pellets examined. A sealing system, used for the mix step and also for the injection step, is required. In terms of laboratory hardware, a unit

Table 6.4
Accessory products needed for performing protein precipitation in microplates

- Single channel pipettor
  (for individual samples)
- Multichannel pipettor or repeater pipettor
  (for solvent and internal standard)
- Collection microplate or microtube rack
- Sealing mat or caps (including mat capper or heat sealer)
- Mixing unit
- Centrifuge with microplate carriers

to mix liquids (multiple tube vortex mixer or equivalent) and a centrifuge that accommodates deep well microplates are needed. The consumable cost of performing this procedure includes the microplates, sealing mats, mat capper or heat sealer, and solvent acquisition and disposal expenses. Additional costs related to pipetting and automation will be incurred, of course, such as disposable pipette tips.

In drug discovery support, sample volumes of plasma and serum can be very low, in the order of 10–25 µL. A volume of 25 µL plasma is commonly precipitated with 75 µL of an organic solvent (*e.g.*, acetonitrile). The resulting total volume of supernatant isolated can be <100 µL. In the collection plate format, the key to successful use of protein precipitation with these small volumes is to choose a microplate well design that has a low reservoir volume (*e.g.*, 200–350 µL), tapered bottom and narrow well width, *i.e.*, one with a greater vertical resolution, so that enough liquid can be aspirated above the pellet. In the case when 25 µL plasma volumes are used, a suggested microplate format for this low volume application is a PCR plate or a shallow well microplate with V-bottom, rather than the larger 1 mL and 2 mL deep well microplates. PCR plates consist of 96 tapered wells having a reservoir volume of about 200 µL and are available from several manufacturers in both skirted and non-skirted varieties (see Chapter 3).

6.3.4 Applications

The simplicity and speed of protein precipitation in the microplate format make this sample preparation technique a popular procedure for high throughput bioanalytical applications. A summary of reported applications utilizing the collection plate for precipitation reactions is provided in Table 6.5. Many laboratories providing bioanalytical support to drug discovery programs have chosen this approach as their preferred sample preparation procedure prior to LC-MS/MS analysis [50, 51].

Traditionally, most reported applications for PPT in collection plates utilize direct injection of the supernatant. Dilution of the supernatant is also performed to adjust the percentage of organic prior to injection. Occasionally, a dry-down and reconstitution step is utilized when sample volumes reach 250 µL and greater. A three-fold dilution (750 µL) yields a total volume of 1 mL which almost always needs to be concentrated in order to meet the sensitivity needs for an assay. Since the microplate format is utilized, evaporation is rapidly performed using a 96-tip dry-down unit. Reconstitution of the dried extract can be quickly achieved using a 96-tip pipettor or liquid handling workstation. This

Table 6.5
Typical examples of applications using collection microplates for protein precipitation with LC-MS/MS analysis

Analytes	Sample and Precipitant Volumes; Centrifugation Conditions	Volume of Supernatant Transferred	Injection, Dilution or Evaporation and Reconstitution	Automation	Ref.
Imipramine, clomipramine and fluoxetine	Rat plasma 50μL, IS 10μL, ACN 100μL; 2000g x 15min	100μL	Dilute in 50μL 25mM Ammonium formate pH 3; inject 50μL	Biomek 2000	[11] [14]
Proprietary analytes	Monkey, rat and dog plasma 50μL, ACN with IS 100μL; 4000rpm x 10min	50μL	Dilute in 200μL 0.1% HCOOH; inject 20μL	Quadra 96 or MultiPROBE	[12]
Gleevec™ and metabolite CGP74588	Human plasma 200μL, IS in MeOH 50μL, spiking solution 50μL, ACN 250μL; 1000xg x 10min	250μL	Inject 10μL	Quadra 96	[52]
Cyclophospha-mide and metabolite	Mouse plasma 250μL, ACN 750μL; 3000rpm x 15min	500μL	Dry-down, recon in 25μL ACN/water (1:1); inject 3μL	Quadra 96	[53]
Ribavarin	Human plasma 100μL, ACN 500μL	Unspecified	Dry-down and recon in 500μL mobile phase	MultiPROBE and Quadra 96	[54]
Proprietary analytes	Plasma 25μL, ACN with IS 50μL; 4000rpm x 10min	Add 200μL 0.1% HCOOH; 100μL	Dilute in 100μL 0.1% HCOOH; inject 20μL	None	[9]
Ketoconazole	Human plasma 75μL, IS 50μL, spike 25μL, ACN 200μL; 1000xg x 10min	250μL	Inject 20μL	Quadra 96	[55]

dry-down and reconstitution step adds time to the overall assay, anywhere from 20–40 min per plate depending on how quickly the supernatant solution evaporates. However, there is no transfer loss as the same collection plate used for the dry-down step is reused for the reconstitution step and is subsequently sealed, mixed and placed on the autosampler for injection.

An example of the need for rapid sample preparation procedures in bioanalysis is the fast track status of drug entity STI571 (Gleevec™; Novartis, East Hanover, NJ USA) conferred by the Food and Drug Administration. The duration from first dose in man to completion of the New Drug Application filing was about 2.6 years. The method chosen, developed and validated to complete the pharmacokinetic studies with the speed required to meet target dates was a semi-automated protein precipitation procedure using deep well collection plates [52]. The method described used an LC-MS/MS run time of 2.5 min (injection to injection cycle). Typical batch sizes were two to four plates a day. Linearity and reproducibility of calibration curves for Gleevec in human plasma were acceptable between 4 and 10,000 ng/mL. The mean accuracy of standard calibration samples for ST1571 covering the above concentration ranged from 98.0% to 102% with %CV values from 1.66% to 5.97%. Automation in the form of a 96-tip liquid handling workstation (Quadra® 96; Tomtec Inc., Hamden, CT USA) was used to transfer supernatant (250 μL) from plates following centrifugation into clean collection plates for injection (10 μL). Filtration plates were used on occasion to filter large aggregates from the supernatant solutions upon transferring into the microplates used for injection.

A notable report regarding high sample throughput for protein precipitation as performed in a collection plate, with liquid handling by a Quadra 96 workstation, is the combination of this approach with monolithic column chromatography. The separation of methylphenidate and ritalinic acid was achieved in 15 s using a 3.5 mL/min flow rate through the monolithic column. Throughput was documented as 768 protein precipitated rat plasma samples (8 full microplates) analyzed within 3 h 45 min [56].

The combination of precipitation with an on-line purification column (that also serves as a separation column) yields a powerful analytical system for direct determination of drug concentrations in plasma samples to support pharmacokinetic studies. In this application, as reported by Hsieh *et al.* [51], ionization suppression in the ESI source was overcome by improving the chromatographic separation and lengthening analyte retention times via the use of a longer gradient separation. The supernatants isolated following precipitation were injected onto a Capcell Pak MF™ (Shiseido Co. Ltd., Tokyo, Japan) mixed phase C8 analytical column. The divert valve (post-column) was switched to waste to remove macromolecules from the injected (10 μL) plasma matrix while the LC pump delivered mobile phase A. Macromolecules such as protein passed quickly through the column due to its restricted access surface; the column retained small drug molecules via the C8 reversed phase. After 1.5 min, the column effluent was diverted from waste to the mass spectrometer for analyte detection. The LC pump then delivered mobile phase B over a time of 2.5 min to elute and separate all analytes. Following separation, the LC pump switched from mobile phase B back to mobile phase A.

In addition to plasma samples, other matrices commonly used with protein precipitation methods for sample preparation include microsomal incubations and tissue homogenates. Microsomal incubations for *in vitro* ADME (Absorption, Distribution, Metabolism and Excretion) stability studies in 96-well plates were quenched with an equal volume of 0.3M TCA and then centrifuged before transfer to a clean microplate and subsequent injection into an LC-MS/MS system [1]. In this report, TCA was preferred over the traditional acetonitrile precipitation using three times the volume (as used in [57]) because a number of compounds that were dissolved in 50/50 NADPH/acetonitrile buffer eluted in the solvent front upon injection onto short LC columns; the analytes were tested and found not to be acid labile. An *in vitro* metabolite identification study in 96-well plates quenched the microsomal reaction by the addition of two volumes of acetonitrile; centrifugation was followed by injection into an LC-MS/MS system [58].

Sample preparation of tissue homogenates is usually similar to that performed for plasma, in that three volumes of precipitating agent are added to one volume of homogenate; the microplate is sealed and vortex mixed, followed by centrifugation and analysis [53, 59]. Alternative precipitants suggested for protein precipitation in this manner include 0.5M NaOH containing 10% (w/v) zinc sulfate heptahydrate, as was used to precipitate protein from rat brain homogenate [40], as well as acetonitrile, water and zinc sulfate heptahydrate solution (350 mg/mL water) in a volume ratio of 500/500/50 [16]. Note that when using tissues, especially with use of salts which cannot be injected into an LC-MS/MS system, it is generally necessary to further clean up the supernatant isolated after precipitation by a more selective technique, such as LLE or SPE, as mentioned previously.

6.4 High Throughput PPT Techniques Using Filtration Microplates

6.4.1 Plate Specifications

Filtration microplates are used in many different applications within the pharmaceutical drug development process. A representative list of these applications is provided here.

1. Medicinal Chemistry; as a filtration plate to clarify acid, base or organic digests of plant materials
2. Medicinal Chemistry; as a support plate in which to place solid extraction media for natural product isolation
3. Combinatorial chemistry; as a support plate in which to place resin scavenger media for solution phase synthesis
4. Genomics; dye terminator removal
5. Genomics; plasmid DNA binding
6. Genomics; lysate clarification
7. Proteomics; on-membrane proteolytic digestion prior to MALDI-TOF-MS (matrix assisted laser desorption ionization time-of-flight mass spectrometry)
8. Bioanalysis; filtration of precipitated protein
9. Bioanalysis; filtration of plasma or serum before use with direct injection or traditional sample preparation techniques
10. Bioanalysis; filtration of eluates and/or reconstituted extracts before analysis
11. Bioanalysis; as a support plate in which to place extraction media (sorbent particles) for solid-supported liquid-liquid extraction, solid-phase extraction or exclusion chromatography

Note that the filtration microplate products available in the marketplace are not suitable for all of these applications. Many filter plate products that are found in distributor catalogs were designed only for general filtration applications and contain larger porosities of 10–20 μm. The decision remains with the user of whether or not the specifications of the product meet the intended analytical requirements. Also note that the plates intended for genomics filtration are usually made of polystyrene which does not offer the solvent resistance required for applications in combinatorial chemistry and bioanalysis. Considered as a whole, the variety of filter materials available in such products is great [*e.g.*, polyvinylidenefluoride (PVDF), polytetrafluoro-ethylene (PTFE), nylon, glass fiber, polypropylene and polyethylene). Ranges of porosities (generally from 0.2–20 μm) and volume reservoirs (from

0.7–2.2 mL) are available as well. An overview of the membrane filtration process, with descriptions of the various membrane filter materials and construction, is presented by Lombardi [60]. Further details about the variety of microplate formats can be found in Chapter 3.

Regarding the application for protein precipitation in bioanalysis, a filter plate product should satisfy the following needs:

- The plate construction is polypropylene (not polystyrene)
- Solvent should not flow through the filter until vacuum or positive pressure is applied
- Proteins are efficiently trapped on the surface of the filter and do not enter the filtrate (generally a filter porosity rating of 1 μm or less is desirable)
- Geometry of the plate allows convenient use with volumes from 100 μL to 1 mL
- Filter does not display irreversible or nonspecific analyte binding
- Well-to-well consistency is demonstrated across the plate
- The plate is compatible with automation
- The plate establishes strong sealing when used with a vacuum manifold
- Filtrate blanks from the plate are clean upon analysis; no contaminants are introduced from the filter and plate materials

6.4.2 Procedure

The use of filtration plates offers the advantage of eliminating the manual off-line steps of mixing and centrifuging, as used with collection microplates (discussed in Section 6.3). A comparison of these two techniques is shown in Figure 6.5. A protocol for using filtration microplates involves the addition of plasma (spiked with internal standard) and precipitant solution. These two solutions are (ideally) delivered simultaneously inside the wells of the plate. The precipitation occurs immediately and protein is trapped on the surface of the filter in each well. Vacuum or positive pressure is used to process liquid through the filter plate. A collection plate isolates the purified filtrate. Many reports in the literature state that the internal standard is contained within the precipitating solution when an organic solvent is used. However, this approach is <u>not</u> recommended. **Internal standard should be added directly to the plasma sample before the addition of acetonitrile so that the binding of the analyte can be mimicked by the internal standard.** If internal standard is premixed with a solution of acetonitrile, inaccurate quantitation can result.

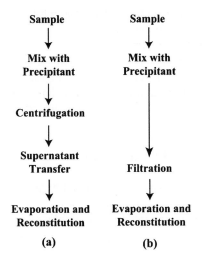

Figure 6.5. Comparison of the procedural steps required for performing protein precipitation using (a) centrifugation in collection microplates or (b) filtration in flow-through filter plates. Reprinted with permission from [61]. Copyright 2001 Elsevier Science.

Some procedural issues must be addressed in order for this system to deliver satisfactory results. One concern is the order of addition of sample and solvent to the microplate wells. When organic solvent is added first, it usually will drip through the filter unless that filter is of special construction such that it will not allow the passage of solvent under gravity conditions. For example, the Protein Crash Plate (Orochem Technologies, Westmont, IL USA) meets this requirement with its proprietary filter. Another filtration plate (Whatman Inc., Clifton, NJ USA) meets this requirement with its two layer membrane in which the top layer is a coarse filter and the bottom layer is described as oleophobic (which retains liquids without dripping until vacuum or centrifugation is applied); the bottom layer is of a fine porosity (0.7 μm).

The potential concern is that when organic solvent as precipitating agent is added first and drips through the filter upon gravity flow, a portion may pass through before the full volume of plasma sample has been added, possibly resulting in less than complete precipitation. Alternately, if plasma is added first, a concern is that it will cover the surface of the wells and, upon the addition of organic solvent or other precipitating agent, may occlude a portion of the well diameter at the surface of precipitation. Passage of the remaining liquid through the well can be obstructed, making it more difficult to pass that liquid through a smaller surface area.

The effect of various procedural approaches on performance of protein precipitation in microplates has been evaluated by Biddlecombe and Pleasance [8]. The order of reagent addition to the wells of a filtration microplate was investigated by determining the efficiency of protein precipitation for three modes of addition:

a) Plasma is delivered to the well, followed by acetonitrile in a second step
b) Plasma and acetonitrile are sequentially aspirated (separated by an air gap) and dispensed together in one step
c) Acetonitrile and plasma are sequentially aspirated (separated by an air gap) and dispensed together in one step

The preferred approach for sample and precipitating solvent addition is reported to be (c) above, the sequential aspiration of acetonitrile, air and plasma in the pipette tip of the manual pipettor or automated liquid handling system (Figure 6.6). This mixture is then forcibly dispensed into the well of the filtration microplate. Adequate mixing is achieved by this action and is followed by either vacuum or centrifugation, as summarized in Table 6.6.

The method of processing liquid through the filter plate is another concern. Although centrifugation and vacuum processing are two common options, centrifugation can, in some cases, be preferable. In order to efficiently trap proteins, the filter used generally must have a porosity of 1 μm or less (0.45 μm is a common filter media stock). This fine porosity, combined with a

Table 6.6
Protocol for performing protein precipitation in a filtration microplate

- Mate filtration plate on top of a deep well collection plate (for centrifugation) or assemble in a vacuum manifold (for vacuum)
- Aspirate precipitating reagent (3 volumes if acetonitrile, 4 volumes if methanol), an air gap, and 1 volume of sample (*e.g.*, plasma) in same pipette tip
- Forcibly dispense liquids into wells of a filtration microplate
- Process liquids through filter using centrifugation (3000 rpm x 10 min) or vacuum (~10 in Hg; increase up to 20 in Hg as necessary; 340–680 mbar)
- Seal collection plate containing filtrate
- Inject aliquot

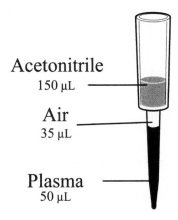

Acetonitrile
150 μL

Air
35 μL

Plasma
50 μL

Figure 6.6. The sequential aspiration of acetonitrile (3 volumes), an air gap, and then plasma (1 volume) in the same pipette tip, with forcible delivery, has been shown to be an effective way to achieve adequate mixing inside wells of a filtration plate.

hydrophobic filter stock can yield resistance to gravity flow, which is desirable as previously mentioned. This situation becomes problematic when trapping of proteinaceous mass occludes part of the membrane, reducing surface area and causing the remaining liquid to filter through a narrower area of the available diameter.

Vacuum is a practical approach to filtration but when using proteinaceous solutions it sometimes cannot achieve enough force to pull liquids through a fine porosity membrane; residual liquid can be left above the membrane or line the inside of the wells below the membrane. Indeed, incomplete collection of the entire volume was noted using the Whatman filter plate product with vacuum [8]. Centrifugation can achieve these needed higher forces and is preferred to pass the full volume of liquid through a fine porosity filter. The filter plate is mated on top of a deep well collection plate for the precipitation reaction and then is placed into a centrifuge for processing at about 3000 rpm for 10 min.

When problems such as plugging or incomplete flow are encountered, it may be sensible to evaluate the reasons for choosing a filter plate over simple protein precipitation in microplate wells. The great advantage of the filter plate is its ability for more complete automation. Another situation for use of a filter plate is when the protein pellet is very fine and does not compact tightly in a microplate well following centrifugation.

6.4.3 Accessory Products

The microplate format is very desirable for performing protein precipitation in the filtration mode since it reduces the number of manipulations dramatically and individual tube labeling is eliminated. The accessory products required for performing protein precipitation in filtration microplates are listed in Table 6.7. They are similar to those needed for protein precipitation in microplate wells with the addition of a filtration plate, vacuum source and manifold. Alternately, a centrifuge can be used in place of a vacuum manifold for liquid processing through the plate. Pipettors are needed for sample and reagent solvent delivery into the filtration microplate. A collection plate or microtube rack is needed to collect the filtrate. A sealing system to cover the microplate before injection is also necessary. The biggest additional cost of performing the filtration procedure beyond commonly available supplies and solvents is the filtration microplate itself. This filtration method presents added value to the sample preparation protocol for:

    a)   Collection of maximum volume of filtrate without transfer loss
    b)   Ease and speed of the isolation
    c)   Quality of results obtained

Table 6.8 lists typical examples of filter plates marketed specifically for protein precipitation filtration in bioanalysis. While the information is meant to be informative, it should not be considered comprehensive (some companies did not respond to requests for information), and individual vendors should always be contacted to obtain more complete and current information about their products.

Table 6.7
Accessory products needed for performing protein precipitation in filtration microplates

• Single channel pipettor (for individual samples)
• Multichannel pipettor or repeater pipettor (for internal standard and solvent)
• Filtration microplate
• Vacuum manifold
• Collection microplate
• Sealing mat system (press fit or heat sealed)
• *Optional*: Centrifuge with microplate carriers

Table 6.8
Typical examples of polypropylene filtration microplates

Vendor	Plate Name (Well Volume)	Description
Nalge Nunc International	Filter Plate (1 mL)	Proprietary construction of a coarse frit with fine glass fiber filter; unfritted plate also available
Ansys (now a part of Varian)	Captiva™ (1.2 mL)	Polypropylene (20 μm, 0.45 μm and 0.2 μm); glass fiber (10 μm)
Whatman	Unifilter Protein Precipitation Plate (2 mL)	Top layer is coarse filter and bottom layer is 0.7 μm
3M	Filter Plate PPT (in 1.2 mL and 2.5 mL versions)	Gradient density filter composed of a laminate of nonwoven polypropylene microfibers
Porvair Sciences and Varian	Protein Microlute™ and as BondElut™ Matrix PPT Plate (2 mL)	Proprietary Vyon™ frit, a combination of ultrahigh molecular weight polyethylene and high density polyethylene
Orochem Technologies	Protein Crash Plate and Filtration Plates (2 mL)	Crash plate is a proprietary system; filtration plates are polypropylene (20 μm, 0.45 μm and 0.2 μm) or glass fiber (10 μm); custom manufacturer with any filter media
Innovative Microplate	Filter Plate (0.8 mL)	PVDF, PP and nylon (0.45 μm); glass fiber (0.8 μm, 1.3 μm); custom manufacturer with any filter media
Analytical Sales & Service	Advantage-SEP Filtration Plate (1 mL)	Proprietary graded screen
IST Ltd. (now Argonaut Technologies)	ISOLUTE® Protein Precipitation Plate (2 mL) and Array® modular plate (1 mL and 2 mL)	Proprietary dual fritted system

Vendors should be contacted to obtain complete and current product information

The 3M filter plate (3M Corporation, St. Paul, MN USA) represents an example of a gradient density filter, in which multiple sheets or layers of nonwoven, blown polypropylene microfibers are laminated together. At least one of the sheets has an effective pore diameter (EPD) smaller than the EPD of at least one upstream sheet. In addition to a coarse (top) filter and a fine (bottom) filter, two layers are sandwiched in between. Also, a layer may consist of multiple identical sheets. An example of a gradient density filter construction is shown in Figure 6.7.

In terms of porosity rating, the 3M PPT Filter Plate has been tested with a range of monodisperse beads and has been determined to remove 98% of particles 10 μm and larger, and 50% of particles 2 μm and larger [62]. The fact that this composite filter gives adequate performance for protein precipitation applications is likely related to the multiple layers and gradient density construction. Generally in a single layer membrane a 1 μm filtration rating or less is preferred for adequate filtration performance following protein precipitation; otherwise breakthrough of protein occurs.

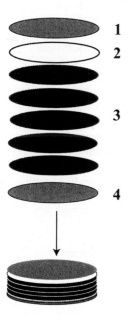

Figure 6.7. Schematic diagram of a gradient density filter comprised of a laminate of fiber materials. Four different porosities (**1**, **2**, **3** and **4**) comprise the filter construction in this example, and one porosity (**3**) is repeated in multiple layers.

6.4.4 Applications

The protein precipitation technique using filter plates has been successfully reported by many laboratories, as summarized in Table 6.9. These applications describe user experiences with filtration plates from the following vendors: Porvair Sciences [8], 3M Corporation [8, 18, 40, 63], Whatman [8, 15] and Ansys Technologies (Lake Forest, CA USA, now a part of Varian Inc.) [12]. These four brands of plates have been included in the first publications due to the novelty of the application and their early presence in the market; by no means are other plate brands to be excluded from consideration. By matching the specifications of filter plates with performance needs, many other brands are also expected to work reliably. As with all sample preparation devices, improvements are expected to be introduced over time.

The sequential aspiration of acetonitrile, an air gap, and then plasma is a viable approach to solvent and sample addition, performed either manually via pipettor or in an automated mode by a liquid handling workstation. Sufficient mixing within the well of the plate has been reported to occur and the effectiveness in precipitating proteins has been deemed adequate [8, 15, 40, 63]. Rather than adding liquids onto a dry filter, an initial wetting with water was reported to aid in enhancing the reproducibility of the results [63]. Rouan *et al.* reported the use of delivering about 200 µL of water to each filter in the plate and then applying a slight vacuum to wet the membrane thoroughly [40, 63]. The excess water went to waste, and then a clean collection plate was inserted into the vacuum manifold to continue the filtration procedure.

One technique used to avoid the dual sample/solvent delivery approach is to actually perform the mixing inside the filtration plate, just as would be done in a closed collection plate. Cap mats or seals that occlude the bottom tips of the filter plates as well as the top surfaces are used. An example of such a seal is the Duo-Seal™ 96 (Ansys Technologies). This design is unique (shown in Figure 6.8), allowing it to be inverted and used in a different manner. The seal covers the bottom of the filter plate so that plasma does not leak out and, when the sealing mat is inverted, can be used to cover the top surface. Two seals, one on bottom and the other on top of the plate, thus effectively enclose the contents of the filtration wells. An interesting note is that a fully sealed filter plate can simultaneously cover both the outlet tips from one plate and the inlet ports of a second plate when stacking two plates together.

Table 6.9
Typical examples of applications using filtration microplates for protein precipitation

Analytes	Sample and Precipitant Volumes; Order of Addition	Filtration Plate Mfr.	Inject, Dilute then Inject, or Dry-Down and Reconstitute	Automation	Analytical Technique	Ref.
Salbutamol (Albuterol)	Rabbit plasma 100μL, ACN with IS 300μL; A+P same tip	3M, Porvair and Whatman	Dry-down and recon in 100μL water; inject 25μL	MultiPROBE and Zymate	LC-MS/MS	[8]
ICL670	Human plasma 100μL, IS 10μL, ACN 300μL; A+P same tip	3M	Dilute with 200μL 1M o-phosphoric acid; inject 200μL	Manual multichannel pipettor	LC-UV	[40] [63]
Sanfetrinem	Human plasma 50μL, ACN 100μL, IS in water 100μL; A+P same tip	Whatman	Dilute with 100μL water; inject 40μL	Genesis	LC-MS/MS	[15]
Proprietary analytes	Rat plasma 100μL, IS 10μL, ACN/ethanol (9:1) with 0.1% acetic acid 500μL; sonicate 10min in deep well plate and transfer to filter plate	3M	Dry-down and recon in 100μL mobile phase; inject 10μL	Manual pipettor	LC-MS/MS	[18]
Metallo-proteinase inhibitors	Rat plasma 100μL, methanol 500μL	Ansys	Dry-down, recon and inject; no details	Quadra 96	LC-MS/MS	[6]

ACN, Acetonitrile; A+P, Acetonitrile + Plasma; IS, Internal Standard; recon, reconstitute

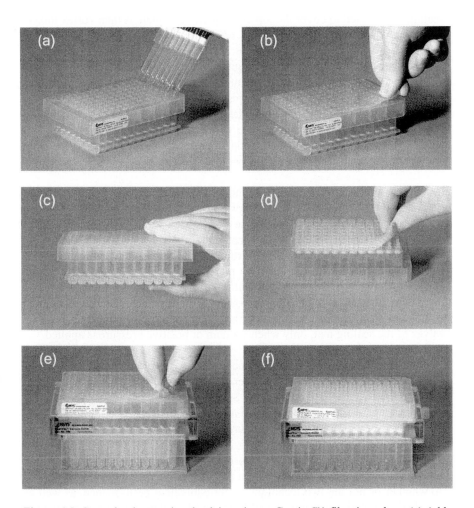

Figure 6.8. Procedural steps involved in using a Captiva™ filtration plate: (a) Add plasma and precipitant into wells which are sealed at the bottom with a mat, (b) cover top of plate with another mat, (c) mix, (d) invert plate, remove bottom seal, (e) mate with collection plate and vacuum collar, apply vacuum and (f) isolate filtered plasma in collection plate. Photos reprinted with permission from Ansys Technologies.

The procedure for plasma precipitation and filtration in a closed filtration plate follows. The bottom of the plate is sealed with a mat. Plasma and precipitating agent are added separately into the wells of the filtration plate. A second sealing mat is used to cover the top of the plate. The plate is then vortex mixed. The plate is inverted on a bench top, the bottom seal is carefully removed, a vacuum collar and a collection plate are mated and then inverted back to normal. These steps are illustrated sequentially in Figures 6.8a-f. Note that the assembly of a traditional vacuum manifold may be problematic, as liquid may drip out of the wells of the filtration plate as the plate and manifold are being assembled. A more appropriate approach in this case is to use vacuum filtration with a small vacuum collar rather than with a full manifold assembly. Another viable approach is to use centrifugation instead of vacuum, where the filter plate mated with collection plate is processed.

Instead of inverting the filter plate containing precipitated plasma and solvent, removing the bottom seal from the tips and mating with a vacuum collar and collection plate, another technique has been innovated. The newer Duo-Seal II™ design of this sealing mat actually transforms the closed bottom of the mat, originally having convex round wells, into one way plastic valves that look like small exit tips. The filter plate is sealed at the bottom in the same manner as described above, but the sealing mat never needs to be removed. The filter plate is inserted directly into a vacuum manifold or collection plate, and liquid will flow through and out of the plate tips, and through the one way valves of the sealing mat, only when vacuum or positive pressure is applied.

The time required to prepare and process a plate of samples using filtration is generally about 20 min [8, 40]. Another 20 min is needed when a dry-down and reconstitution step is used [8]. Sample throughput depends on the chromatographic run time per sample and extent of automation, but generally would be from two to four plates per day per analyst.

Protein precipitation procedures performed in tube or collection plate format have been compared with the filtration plate format. These studies investigated (a) whether or not the two techniques offer equivalent performance, as well as (b) whether materials are introduced from the filter plate that may contribute to ion suppression. In one study, a side by side evaluation was performed using an acidic compound in rat plasma [18], while another study used a highly protein bound iron chelator in human plasma [63]. Overall, the data from both studies were deemed equivalent in terms of accuracy, precision, reproducibility and linearity. Data were slightly more reproducible for the filtration method [18]. Improved reproducibility was likely the result of fewer manipulations and the elimination of a transfer step to another plate. Regarding the ionization suppression issue, it was found that there were no differences in the level of ion suppression obtained using various filtration microplates (Whatman, Porvair,

3M and Ansys Technologies) compared with the standard protein precipitation method in tubes or plate wells [8, 12].

The extraction of tissue homogenates using filtration plates can be problematic, due to the large mass of materials generated in tissue upon precipitation. Plasma is generally regarded to contain about 10% protein [30], while tissues contain a greater proportion of protein per unit mass. Blocked wells have been reported with the 3M plate when using brain tissue [40]. A more successful approach for the use of tissue homogenates is to precipitate proteins in wells of collection microplates, centrifuge, and then remove the supernatant for a subsequent, more selective sample preparation scheme, such as SPE [40, 53]. Note that it may be beneficial to filter this supernatant through a filtration microplate as it is transferred to a clean collection plate [7].

6.4.5 Additional Bioanalytical Uses for Filter Plates

6.4.5.1 Sample Filtration

Direct injection techniques, *e.g.*, on-line turbulent flow chromatography, on-line SPE and dilution of sample followed by direct injection using column switching techniques (discussed in more detail in Chapter 14), can benefit from filtration of the sample before injection [64]. This step will remove particulates and debris that can potentially foul the LC lines, column frits and/or mass spectrometer interface. The filtration plate is mated with a deep well collection plate, samples are placed into wells, and the filtrate is collected by centrifugation. Vacuum can also be used in conjunction with a manifold or a vacuum collar. A related need for sample filtration is to remove particulates from urine before analysis by NMR techniques as performed for metabonomics research in drug discovery [65].

In fact, all automated liquid handling workstations and pipetting systems can benefit from sample filtration because of the universal issues related to plasma clot formation that introduce pipetting challenges. Generally, plasma samples are frozen before analysis and this freeze/thaw process can introduce or promote clot formation because the proteins separate from the water. The clot formation prevalence and magnitude become greater with repeated freeze/thaw cycles [66]. These clots are capable of plugging pipet tips and creating pipetting errors with automated systems. These errors can result in aspirating an incomplete volume via tip occlusion by a clot (air aspirated instead of sample), as well as in pipetting failures that require user intervention. One of the reliable ways to eliminate clots is to thaw the plasma, invert several times rapidly to homogenize the sample, and then centrifuge the clots to the bottom of the tube. The sampling pipet should then remove an aliquot from the middle of the tube.

One approach to alleviate this clot formation is to transfer freshly prepared plasma into the wells of a filtration plate for storage and freezing. This can be done directly from Vacutainer® tubes generated from study samples. This application is reported by Berna et al. [12], in which 20 μm polypropylene filter plates (Captiva™; Ansys Technologies) were used. Seals cover the bottom of the filter plate so that plasma does not leak out the tips via gravity and start the filtration process. The top of the plate is also sealed. Note that the Duo-Seal 96 product is also prescored to provide the analyst with the ability to remove strips of eight seals at a time when the entire 96 wells are not required. Prior to sample analysis, the filter plate containing frozen plasma samples is allowed to thaw; the bottom cover is removed and the filter plate is placed over a clean microplate to collect the plasma filtrate. The filter plate/collection plate combination can be placed into a centrifuge to complete the filtration process, if required, or a vacuum can be applied using a vacuum collar or manifold.

Once the samples have been thawed and filtered in this manner, the plasma filtrate (already in the microplate format) becomes the source plate for a sample preparation scheme. A liquid handling workstation or multichannel pipettor can successfully aspirate aliquots from the microplate with great success (clots have been removed), as reported by a similar application [67]. As Berna described [12], additional freeze/thaw cycles did not seem to reintroduce clots to the samples, eliminating the need to refilter the plasma.

It is wise to evaluate the potential for nonspecific binding of analytes to the filter media before use of this approach. A polypropylene filter may be preferable over glass fiber since it is similar in composition to common microplates, tips and sample tubes used elsewhere within a procedure. In order to test for nonspecific binding, a series of identical standards is prepared and split into two groups, each prepared with its own standard curve. One group is processed without using the filter plate and the second group is processed through the filter plate. Next, internal standard is added and each group is analyzed under the same conditions. Differences between the two groups of standards can be evaluated by examination of analyte response, accuracy and precision when comparing drug concentrations pre- and post-filtration [12].

6.4.5.2 Eluate Filtration

Another important role for filtration microplates is to clarify the final eluate before injection into an LC-MS/MS system. This process is performed following the more selective sample preparation techniques, such as solid-phase extraction and liquid-liquid extraction. Note that the final eluate (in an organic solution) may appear clear, but the reconstitution process (after dry-down with nitrogen and heat) sometimes causes the formation of particulates. The reconstituted eluate solutions are commonly filtered to remove small particles (matrix debris and/or any sorbent particle fines from SPE media), as described in [68] which used a 96-well filter plate consisting of a 0.45 μm nylon filter with 20 μm prefilter. Presenting a clean sample for injection is important to maintain hardware lifetimes and normal LC column pressures. Those users who inject hundreds of samples a day are often adamant about filtering their eluate solutions prior to injection.

The procedure for clarifying eluate solutions follows. Eluates are commonly collected in a deep well microplate following solid-phase extraction or liquid-liquid extraction. The organic eluate solutions are usually evaporated to dryness and then reconstituted in a small volume (*e.g.*, 100–200 μL) of mobile phase compatible solvent. The reconstituted volume is aspirated using a 96-tip liquid handling workstation and dispensed into a filtration plate that is mounted above a deep well collection plate. This filter plate/collection plate combination is then centrifuged at about 3000 rpm x 5 min. Since the reconstitution volumes used are small, centrifugation is the preferred processing technique in order to recover the full eluate volume. The filter plate is discarded and the deep well plate containing filtrate is now the container for injection; it is sealed and placed into an autosampler. An alternate approach is to filter the organic eluate solution into a deep well plate and then the filtrate is evaporated and reconstituted. This alternate approach, however, is not as widely adopted because many times the reconstitution procedure (post-filtration) will introduce precipitation of materials due to a pH change or an effect of the organic solvent.

Filtration plates are available in commercial and custom varieties for the purposes of sample and eluate clarification. The bioanalytical user generally evaluates and chooses a commercial filtration plate after verification of its performance. It is important to evaluate nonspecific binding of analyte(s) to the membrane as well as the filtration efficiency. Rather than use a commercial product, many users have identified the specifications for a certain frit or membrane material and a given porosity that meets their objectives for sample preparation; arrangements are made with a vendor who accepts custom manufacturing requests (*e.g.*, Orochem Technologies, Innovative Microplate, Chicopee, MA USA or Applied Separations, Allentown, PA USA) to place these materials into the microplate format. A custom assembly in this manner may make the final price higher than another commercially available product (if unit volume goals are not met) but the value of performance and consistency often satisfies the pricing concern.

6.5 Method Development and Optimization Strategies

6.5.1 Selecting Precipitation Conditions

The most popular and successful protein precipitation approaches for pharmaceutical bioanalysis use organic solvents for the reason mentioned earlier regarding mobile phase compatibility. A method development strategy involves the evaluation of at least two precipitating agents in parallel, *e.g.*, methanol and acetonitrile, or better yet, three solvents such as methanol, acetonitrile and ethanol. Many times, one solvent will be preferable over the other for both performance and convenience; it is only through a comparative evaluation that the optimal precipitant becomes known.

An example of a three solvent evaluation is provided by King *et al.* [6]. Methanol, acetonitrile and ethanol in 1:5 ratios (plasma/solvent) were compared for their performance in precipitating proteins from plasma. Filtration plates were used for clarification. The authors reported that with acetonitrile there was insufficient mixing and "over coagulation of the denatured proteins" that led to protein breakthrough and plugging of filters. Ethanol "was the best for the separation of the analytes from the denatured proteins" and filters did not become plugged. Methanol reportedly yielded the best mixing among all three solvents; protein breakthrough was minimal, and low vacuum was effective in processing all the liquid through the filter. Here, methanol was chosen and used in a 1:5 ratio, *e.g.*, 25 µL plasma to 125 µL methanol.

The ratios of organic solvent to plasma can also be examined, *e.g.*, ratios of 1:3 versus 1:4. Generally, researchers demonstrate success with a 1:3 ratio using acetonitrile and a 1:4 ratio using methanol or ethanol. If minimal volumes are desired, it may then be appropriate to examine extraction performance using only a 1:2 ratio. An example of a plate layout for method development is shown in Figure 6.9; here, three solvents are evaluated, two ratios each, in the microplate format. Alternately, these same organic solvents can be evaluated in

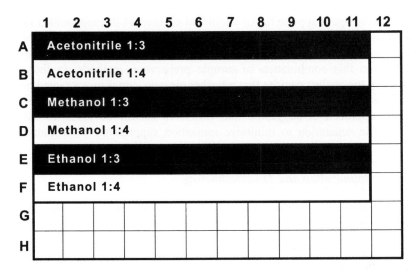

Figure 6.9. Plate layout example as arranged for a comparison of acetonitrile, methanol and ethanol as precipitating agents for protein precipitation. Two ratios of sample to solvent (1:3 and 1:4) are compared for each precipitant. Blanks are included within each set for the determination of recovery and matrix effect. Within each row, samples 1–12 can be arranged as follows: B/B = True blank (no analyte, no IS; n=1), blank plasma (no analyte; n=4), wells designated for matrix effect determination (n=2), wells designated for recovery determination (n=4) and an empty well (n=1) that can be reserved for a test solution or system check at completion of extraction.

parallel, but instead of examining two ratios of sample to solvent for each, the effect of pH adjustment can be compared with no pH adjustment. It is very important to include wells containing blank plasma in this study for examination of the matrix effect for each solvent and ratio combination. Determination of the matrix effect uses a post extraction spiked matrix blank and the results are compared with an analytical standard in neat solution to determine the influence of the matrix on the analysis (see Chapter 2, Section 2.3.3). The matrix effect can sometimes be the determining factor in choosing the final precipitation conditions.

When whole blood represents the biological sample, success has been obtained using a combination of a heavy metal cation salt plus organic solvent for both cell lysis and the precipitation of protein from the red blood cells *e.g.*, acetonitrile, water and zinc sulfate heptahydrate solution (350 mg/mL) in a volume ratio of 500/500/50 [16]. Tissue homogenates also benefit from this dual precipitating agent approach. An additional cleanup of the supernatant or filtrate is recommended for tissue homogenates, using either off-line (SPE or LLE) or on-line (turbulent flow or SPE) techniques. A more selective cleanup will result from this combination of sample preparation methods. In those instances when the analyst decides that a further cleanup step is unwarranted or adds time which is not available, certainly then the organic solvent precipitation approach is suggested and measures should be taken in the chromatographic separation to minimize ionization suppression from matrix effects.

6.5.2 Method Optimization and Troubleshooting

In the development of an analytical method that uses protein precipitation for the sample preparation, many factors other than the choice of precipitating agent must be considered. These factors may influence the success of the method, the ease of isolation of supernatant or filtrate, the compactness of the pelleted protein, analyte stability, and the overall compatibility with the analytical system upon injection. Some of these factors to consider are [30]:

- Precipitant interference with the analytical method, related to ionization suppression
- The pH stability of the analyte
- The organic and aqueous solubility of the analyte
- The metal coordination tendency of the analyte
- The lot to lot variability of plasma sources and their potential interference with analysis
- Species for which the method will be developed
- Binding of analytes to the precipitated proteins

6.5.2.1 Efficiency Optimization

The factors contributing to ionization suppression and tactics to minimize this influence have been discussed (Section 6.2). The pH stability of the analyte, if not considered, can sometimes result in cleavage or hydrolysis of a labile species if strong acid or base precipitants are used. Solubility of analyte in the solutions and solvents used is of course not to be overlooked. Note that the precipitation efficiency can differ among various lots of human plasma.

A study evaluated the efficiency of protein removal among different precipitating agents [30]; 4 lots of plasma were evaluated with 4 precipitants (acetonitrile, zinc sulfate, trichloroacetic acid and m-phosphoric acid) that

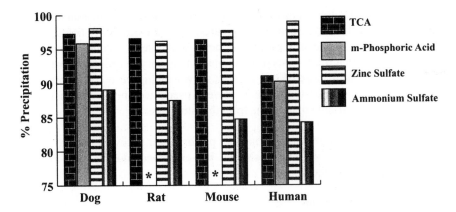

Figure 6.10. Comparison of the efficiency of protein removal from plasma of different species using acidic, ionic and salt precipitants. Two volumes of precipitant were used with one volume of plasma.
*Data for rat and mouse plasma with m-phosphoric acid could not be obtained due to cloudiness in the supernatant that did not allow for spectrophotometric measurement and quantitation.
Unpublished data [30] reprinted with permission from the authors (P. Sarkar, C. Polson, R. Grant, B. Incledon and V. Raguvaran, Eli Lilly Canada Inc.).

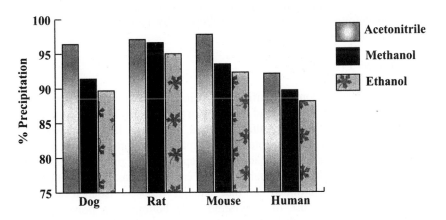

Figure 6.11. Comparison of the efficiency of protein removal from plasma of different species using three organic solvents as precipitants. Two volumes of precipitant were used with one volume of plasma. Unpublished data [30] reprinted with permission from the authors (P. Sarkar, C. Polson, R. Grant, B. Incledon and V Raguvaran, Eli Lilly Canada Inc.).

represented acidic, ionic and salt species. Two volumes of precipitant were used with one volume of plasma. The greatest lot to lot variability noted in this study was seen when m-phosphoric acid was used. Another important consideration is that these types of precipitants may vary in efficiency based on the species of plasma used [30], as also shown in Figure 6.10. Figure 6.11 shows variances due to the organic solvent used (acetonitrile, methanol or ethanol). The important conclusions from this study are:

1.   One choice of precipitant may not perform best for every species
2.   Lot to lot differences should be examined during the method development process

6.5.2.2 Binding to Protein

When the recovery of analyte as isolated from the supernatant (or filtrate) is not as high as expected (*e.g.*, <70%), it may be possible that the analyte is still bound to protein. Analyte may be trapped within the precipitated proteinaceous mass at the bottom of the sample well. In order to avoid this occurrence, a different precipitating agent may be chosen and/or a pH adjustment made to reduce the extent of binding (*e.g.*, addition of a small amount of acid to the acetonitrile precipitating solvent). Alternately, a washing of the protein mass can be performed using a second (or third) aliquot of precipitating agent. The multiple aliquots are combined and, since the volume has increased, will likely require concentration via evaporation followed by reconstitution in mobile phase compatible solvent.

6.5.2.3 Transfer Problems

Transfer loss may become significant, resulting in a lower recovery than desired. In this instance, a second washing of the pipette tip can be used to rinse adsorbed or incompletely eluted analyte. Perhaps a larger volume than desired is left behind at the protein-liquid interface upon pipetting; in this situation, an adjustment can be made in the depth that the pipette tip descends into the precipitation well. Through trial and error, the maximum depth to be penetrated can be determined so that enough mass is isolated to yield acceptable results. The use of a pyramid style bottom in a polypropylene microplate can yield more favorable results than a round well bottom plate, as the protein mass is contained in a tapered bottom, leaving more room above to isolate liquid from the pelleted protein.

6.5.2.4 Adsorption Problems

In some instances the adsorption of analyte to glass or plastic may be an issue. The examination of an alternate variety of collection plate can be tried. If adsorption to plastic is an issue, a rack containing 96 glass microtubes can be used; similarly, glass lined wells are available within deep well plates (see Chapter 3). On the other hand, if glass adsorption is an issue, samples should be loaded into a polypropylene microplate and collected in a polypropylene device. Note that the composition of the filter can present some binding potential as well when using filtration plates. Glass fiber filters are sometimes untreated and represent potential adsorption sites via residual silanols. Also, polypropylene membranes can present a similar potential to adsorb analytes, to the same extent as polypropylene plates. This attraction of analyte to filter is generally regarded to be analyte specific. An evaluation of the performance data can suggest the potential for nonspecific binding to filter, or well material, and an experiment can be carried out using analyte spikes both pre- and post-filtration.

6.5.2.5 Porosity Problems

Another important note when using filtration microplates is that not all filter varieties available in plates possess the necessary porosity to capture proteins and the required hydrophobicity to resist gravity flow from organic and aqueous solvents. Two of the most common problems seen using filter plates are: (1) breakthrough of proteins into the filtrate and (2) organic solvent that drips through prematurely. When either occurrence is seen in a filtration microplate, determine the exact specifications of the membrane in the plate being used. *What is the porosity rating of the membrane?* If the filter has a 5 µm, 10 µm or 20 µm rating, those filters adequately remove particulates from raw plasma but are not always efficient enough to remove finer sizes of precipitated proteins from treated plasma. Some protein mass may leak through the filter and contaminate the filtrate.

Occasionally it has been observed that a catastrophic failure has occurred. This failure happens in individual wells within a filtration microplate when the filter in those wells is not sealed tightly; solvent is allowed to pass along the edge of the filter, around and out to contaminate the filtrate. In this case, a manufacturing quality control problem has resulted.

When organic solvent drips through under gravity, the membrane does not present enough resistance and a more hydrophobic one is desired. One workaround to this solvent dripping issue is to first add a small volume of water (*e.g.*, 100–200 µL) to the filtration wells and pass part or all of that solution through to waste [40, 63]; a collection plate is added and then the addition of sample and organic solvent proceeds. When water is left above the filter in this instance, the water can sometimes provide a cushion of solvent on top of the filter, allowing enough time for the protein precipitation reaction to occur at and above its surface.

6.5.2.6 Incomplete Protein Removal

When concern is expressed over the amount of protein leaking through the filtration wells and being injected into the analytical system, the protein content of the supernatant or filtrate can be evaluated. The classic approach to quantitating protein content is to use the Lowry assay, as used by Blanchard [4] and Biddlecombe and Pleasance [8]. Another approach is to add a fluorescent dye to the supernatant or filtrate, one which strongly binds to protein and is fluorescent in the bound state. For example, fluorescamine (**1**) [spiro(furan-2(3H),1'(3'H)-isobenzofuran)-3,3'-dione,4-phenyl], available as F-2332 from Molecular Probes (Eugene, OR USA), is bound to primary amino groups at pH 8 and becomes fluorescent. An aliquot of fluorescamine in a pH 8 buffer can be added to the supernatant or filtrate and the intensity of fluorescence evaluated or quantitated against a standard curve as an indication of the amount of protein carried over into the analysis.

1

When coeluting matrix interferences carry over into the analysis, and loss of signal results from suppression of ionization, perhaps another ionization mode should be evaluated (*e.g.*, APCI rather than ESI). Likewise, greater emphasis can be placed on the chromatographic separation of matrix coextractables. For example, the use of rapid gradient LC-MS/MS for analysis coupled with flow diversion of the solvent front (for the first minute of the run) allowed the introduction of protein precipitated samples into the mass spectrometer without the need for frequent source cleaning [11]. This approach was found to require, however, the introduction of a makeup flow via an external pump. The external flow source was operated through a switching valve controlled by timed external contacts so that the LC flow was either directed toward the MS or toward waste, and vice versa for the makeup flow [11]. Keep in mind that extremely poor recovery and sensitivity may point toward the protein precipitation approach as not selective enough for the requirements of the particular analytical assay. In this latter case, another sample preparation technique should be evaluated, such as liquid-liquid extraction, solid-phase extraction or an on-line approach.

6.5.2.7 Filter Plugging

In the case when flow restriction is an issue, the vacuum can be left on continuously during the procedure in an effort to alleviate the blockage. Pulsing the vacuum, via a series of short on/off bursts, is also beneficial in some cases. Note that the vacuum source should be set to maximal when flow problems are evident. Alternately, when vacuum is not strong enough to allow for liquid processing through a filter, centrifugation can achieve higher g-forces and should be used. In this case, a filter plate is mated with a collection plate and this combination is placed into a centrifuge; processing is usually achieved at 3000 rpm x 10 min. When difficulty with flow persists, a longer duration can be used.

6.6 High Throughput Equilibrium Dialysis and Ultrafiltration

6.6.1 Introduction

The binding of drugs to plasma or serum proteins, *e.g.*, albumin and/or alpha 1-acid glycoprotein (AGP) [69], has been shown to have dramatic effects on pharmacodynamic processes and pharmacokinetics, *i.e.*, the time course of drug distribution in the body. It is generally recognized that only the unbound (free) drug portion is available for diffusion, transport and interaction with pharmacological targets [70–73]. When protein binding is low, the drug may fail to reach throughout the tissues in the body; when protein binding is extremely high, drug is tightly bound and the free drug concentration may be too low to reach a therapeutic concentration and exert its desired pharmacological effect. During an assessment of the *in vivo* properties of a new drug substance, the degree of protein binding is an important characteristic that is always determined. Two common methods of drug protein binding determinations involve equilibrium dialysis and ultrafiltration.

Note that while these two techniques are widely used for protein binding assays, these approaches are amenable to other applications as well. Some of the other uses for various membrane separations include desalting prior to analysis, detergent removal, protein purification, removal of excess primers from PCR products, purification of DNA sequencing reactions, receptor and ligand binding assays, protein-protein interaction studies, and the determination of protein/DNA interactions. Several reviews are available that include discussion of dialysis, ultrafiltration and other membrane based sample preparation methods for bioanalysis [74–80].

6.6.2 Equilibrium Dialysis

Equilibrium dialysis is a classical method to physically separate small molecular weight analytes from larger molecular weight constituents (*e.g.*, protein) in a biological sample matrix. This process occurs by diffusion through the pores of a selective, semipermeable membrane and is a concentration driven process. The dialysis tubing used is made of regenerated cellulose in a series of molecular weight (MW) cutoffs [MWCO] (*e.g.*, 3.5 kD, 6–8 kD, 12–14 kD) from vendors such as Spectrum Laboratories (Rancho Dominguez, CA USA). Although this approach is accurate, it requires manual manipulations to assemble a number of dialysis cells and is very time consuming.

A format for equilibrium dialysis that is amenable to higher throughput is the 96-well plate such as the Equilibrium Dialyzer-96™ (Harvard Bioscience, Holliston, MA USA), shown in Figure 6.12. The unbound analyte is dialyzed through a membrane until its concentration across the membrane is at equilibrium; the free (unbound) fraction is measured by LC, retaining the microplate format. Note that there is no discrimination of the small molecular weight materials passing through the porous membrane along with the analyte of interest; the key discriminator is the MWCO. Each well in this system contains a separate 5 or 10 kD MWCO membrane in its middle to form two compartments. Since membranes are not shared between wells, the possibility for well-to-well cross-contamination is eliminated.

Sample is added to the top of the membrane and the well is sealed with clear cap strips. An equivalent volume of buffer is added into the corresponding bottom wells and these wells are sealed with blue cap strips. While it is possible to use sample volumes as low as 75 µL, volumes from 150–200 µL are recommended in order to facilitate sample removal from the plate without compromising the membrane by accidental puncture. The dialyzer plate is

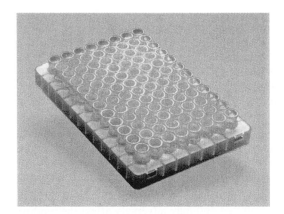

Figure 6.12. Example of an equilibrium dialysis plate containing a selective, semi-permeable membrane in the middle of each well. Photo reprinted with permission from Harvard Apparatus.

secured into a plate rotator unit which operates through a 360° circle at a uniform speed to ensure sufficient interaction among all samples across the membranes. Full equilibrium is reached in about 3 h. The plate is removed from the rotator unit, the cap strips are removed from the bottom wells, and samples are carefully transferred into a clean microplate for analysis by LC or another separation and detection technique. Note that the original plate design does not conform precisely to all standard microplate specifications but the footprint is compatible for use with multichannel pipettors and liquid handling workstations. However, the manufacturer continues to innovate and the introduction of a more standard microplate design that is readily automatable (by gripper arms) is expected to be introduced.

Another manufacturer, Spectrum Laboratories, offers a 96-well dialysis plate named Spectra/Por 96-Well Microdialyzer that can be used for equilibrium dialysis applications. This plate design accommodates about a dozen different choices of MWCO filters ranging from 0.1 to 300 kD; nominal sample volumes of 150 µL are recommended, although the minimum and maximum sample volumes are 24 µL to 560 µL, respectively. Note that these filters are sheets that are clamped into place across the plate interior surface and are not uniquely secured within each well such as the Equilibrium Dialyzer-96 design. Also note that the plate itself has 96 wells but does not conform precisely to standard microplate specifications.

6.6.3 Ultrafiltration

Ultrafiltration separates proteins according to molecular weight and size using centrifugal force and is thus based on a pressure differential rather than on a concentration differential as in equilibrium dialysis. Materials below the molecular weight cutoff pass through the membrane and are contained within the ultrafiltrate; retained species are concentrated on the pressurized side of the membrane. A key advantage using ultrafiltration is that a more rapid separation is achieved compared with equilibrium dialysis; it is also a simpler procedure to perform and lacks dilution effects. One disadvantage noted with this technique, however, is that since a portion of the aqueous phase is forced through the membrane and away from the protein solution, that protein compartment becomes more concentrated and may adversely affect the binding characteristics. Also, if the thin membrane is compromised in any way during centrifugation or handling, there can be breakage and/or leakage of protein from the upper chamber.

Traditionally, ultrafiltration is accomplished using individual filter units, such as the Centricon® (Millipore, Bedford MA USA). These units are available in tubes that hold a maximum volume of 2 mL and yield a final concentrate volume of 25–50 µL. The low binding Millipore Ultracel-YM® membrane is available in five MWCO choices: 3 kD, 10 kD, 30 kD, 50 kD and 100 kD (designated YM-3, -10, -30, -50 and -100, respectively). Sample preparation using an ultrafiltration method is typically followed by direct injection of the ultrafiltrate. An application for sameridine in plasma using ultrafiltration followed by coupled column chromatography has been described [81]. In this case, a volume of 2 mL plasma yielded 400 µL ultrafiltrate. Another application detailed the use of a fully automated ultrafiltration step (Millipore Microcon® units with a Packard MultiPROBE liquid handling workstation; Packard Instruments, Meriden, CT USA now a part of PerkinElmer Life Sciences) to replace solid-phase extraction used for the determination of zidovudine and lamivudine in human serum by LC-MS/MS [82].

Ultrafiltration in a high throughput 96-well plate format is available from Millipore as the Microcon®-96 well plate assembly. This multiwell assembly contains a size exclusion membrane (YM-10 or YM-30) to separate free drug from protein bound drug. The plastic composition of the plate is novel, comprised of 97.5% polypropylene and 2.5% PTFE. This patented composition reduces the nonspecific binding of compounds especially in the protein-free filtrate used for analysis. Microcon filter units are provided preloaded into a base plate. This assembly is operated in a centrifugal pressure mode and is compatible with standard centrifuge microplate swinging bucket rotors. Receiving plates used within the multiwell filter assembly can have U-shaped bottoms or flat bottoms. These plates are used to secure the open ends of the filter units using a single seal across the plate, rather than multiple 8-well or 12-well cap strips.

The maximum sample volume that Microcon concentrators can hold in the 96-well format depends on the plate configuration and the sample type. When using the flat bottom plate, the maximum sample volume is 350 µL for serum samples and 300 µL for dilute protein samples. When using the U-bottom plate, the maximum sample volume is 300 µL for serum samples and 250 µL for dilute protein samples. Once samples (typically 250 µL) are secured in the filter units, centrifugation is performed at 2000 x g to 3000 x g for 15 min to 60 min, depending on sample composition. Another manufacturer, Whatman, offers a 96-well ultrafiltration plate.

6.6.4 Applications

The development of a 96-well ultrafiltration plate has been a manufacturing challenge due to the difficulty in ensuring membrane integrity and preventing cross-contamination. The development of such a product has been detailed in two presentations by Whatman and applications for its use in genomics have been discussed [83, 84]. One application described the cleanup of DNA sequencing reactions to simultaneously eliminate both high and low molecular weight templates. This task was met by filling the ultrafiltration plate with Sephadex™ G-50 (Amersham Bioscience Corp., Uppsala, Sweden) DNA grade gel [83]. A related application used Sephadex G-25 beads for separation of protein/ligand complexes (high MW) from inactive molecules (low MW) for drug discovery screening [85]; however, a regular 96-well filtration plate was used here, not an ultrafiltration plate.

The Millipore Microcon-96 ultrafiltration plate was evaluated for its robustness, accuracy and precision in the protein binding determination of various compounds (radiolabeled acetaminophen, lidocaine and diazepam) in a drug discovery setting [86]. Advantages noted with this methodology included the requirement of less test material and plasma, a rapid experiment (3 h to test 48 compounds in duplicate), potential for automation and the ability to direct inject ultrafiltrate into an LC-MS/MS system. The disadvantages noted included variable ultrafiltrate volumes obtained across the plate and between assays, and the high variability of binding determined. The use of these plates was suggested for screening applications to characterize plasma protein binding as low, medium or high.

An expanded application from the same authors evaluated the Microcon-96 with the Equilibrium Dialyzer-96 system, comparing ultrafiltration with equilibrium dialysis in a high throughput format [87]. The same analytes were used. In reference to the Dialyzer-96 method, advantages noted were good reproducibility of binding both across the plate and between plates, consistent sample volume, high throughput and no significant protein leakage into the buffer compartment. With equilibrium dialysis, the protein binding is not affected by nonspecific binding due to an equilibration of free drug and analysis in both compartments. Overall, both systems offered high throughput but more variability compared with traditional methods.

Equilibrium dialysis and ultrafiltration in 96-well plates were also compared by Walsh *et al.* using the analytes nadolol, propranolol, verapamil and warfarin [88]. The equilibrium dialysis plate displayed reasonable precision and accuracy with moderate throughput and was recommended as a screening method to categorize lead compounds according to their approximate binding characteristics. A follow-up assessment using standard assays was also suggested. In contrast, the ultrafiltration plate showed widely variable precision and inconsistent accuracy. Also noted for the ultrafiltration plate was that it lacked the standard 9 mm well-to-well microplate spacing, presenting a challenge to automation.

Another report evaluated the Microcon-96 plate for the determination of the free fraction of 12 marketed drugs in plasma in combination with a Tecan Genesis™ 150 liquid handling workstation (Tecan US, Research Triangle Park, NC USA) [89]. Samples were analyzed using fast LC-MS/MS (1.5 min/sample). Data obtained with the 96-well YM-30 ultrafiltration plate were compared with data determined by individual Millipore Centrifree YM-30 devices. Overall, the 96-well method provided an accurate estimation of plasma free fraction. The ultrafiltration time (45 min) was much shorter than conventional methods (>20 h). The method was shown to be compatible with automation. One disadvantage noted was the higher variability between wells. In a related report, Microcon-96 plates were used to separate free drug from protein bound drugs and the filtrate was further analyzed using dual-column turbulent flow chromatography with MS detection [90]. This technique worked well as a predictive ADME tool.

Equilibrium dialysis is regarded as the preferred method to determine the free drug fraction. The development of a high throughput equilibrium dialysis method to support drug discovery studies is detailed in the application by Kariv, Cao and Oldenburg [91]. Plasma protein binding of three radiolabeled drugs (propranolol, paroxetine and losartan) with low, intermediate and high binding properties, respectively, were chosen for assay validation. The equilibrium dialysis was performed by rotating the plate containing samples and buffer at 37°C from 20 to 22 h. Overall, this study concluded that the data obtained using low volume equilibrium dialysis is in agreement with published values determined by traditional equilibrium dialysis techniques. The microplate design of this device allows its use in high throughput applications to match the rapid pace required by drug discovery laboratories when using automated liquid handling workstations and LC-MS/MS detection techniques.

Acknowledgments

The author is appreciative to Chris Bugge, Russell Grant and Tom Lloyd for their critical review of the manuscript, helpful discussions and contributions to this chapter. Line art illustrations were kindly provided by Richard King, Russell Grant and Woody Dells.

References

[1] R. Xu, C. Nemes, K.M. Jenkins, R.A. Rourick, D.B. Kassel and C.Z.C. Liu, J. Amer. Soc. Mass. Spectrom. 13 (2002) 155-165.

[2] B.A. Staton, E. Shobe, J. Palandra, A. Pearsall, J. Creegan, S. Walton, J. Morris and T. Heath, Proceedings 48th American Society for Mass Spectrometry Conference, Long Beach, CA USA (2000).

[3] J.S. Torano, A. Verbon and H.-J. Guchelaar, J. Chromatogr. B 734 (1999) 203-210.

[4] J. Blanchard, J. Chromatogr. 226 (1981) 455-460.

[5] M.C. Rouan, F. Marfil, P. Mangoni, R. Sechaud, H. Humbert and G. Maurer, J. Chromatogr. B 755 (2001) 203-213.

[6] S.L. King, D.J. Foltz, T.R. Baker and S.X Peng, Proceedings 48th American Society for Mass Spectrometry Conference, Long Beach, CA USA (2000).

[7] M. Lam, J. Shen, J.-L. Tseng and B. Subramanyam, Proceedings 48th American Society for Mass Spectrometry Conference, Long Beach, CA USA (2000).

[8] R.A. Biddlecombe and S. Pleasance, J. Chromatogr. B 734 (1999) 257-265.

[9] A.T. Murphy, M.J. Berna, J.L. Holsapple and B.L. Ackermann, Rapid Commun. Mass Spectrom. 16 (2002) 537-543.

[10] R. Bonfiglio, R.C. King, T.V. Olah and K. Merkle, Rapid Commun. Mass Spectrom. 13 (1999) 1175-1185.

[11] A.P. Watt, D. Morrison, K.L. Locker and D.C. Evans, Anal. Chem. 72 (2000) 979-984.

[12] M. Berna, A.T. Murphy, B. Wilken and B. Ackermann, Anal. Chem. 74 (2002) 1197-1201.

[13] R. King, R. Bonfiglio, C. Fernandez-Metzler, C. Miller-Stein and T. Olah, J. Amer. Soc. Mass. Spectrom. 11 (2000) 942-950.

[14] K.L. Locker, D. Morrison and A.P. Watt, J. Chromatogr. B 750 (2001) 13-23.

[15] C. De Nardi, S. Braggio, L. Ferrari and S. Fontana, J. Chromatogr. B 762 (2001) 193-201.

[16] G.L. Lensmeyer and M.A. Poquette, Ther. Drug Monit. 23 (2001) 239-249.

[17] P.R. Tiller and L.A. Romanyshyn, Rapid Commun. Mass Spectrom. 16 (2002) 92-98.

[18] R.E. Walter, J.A. Cramer and F.L.S. Tse, J. Pharm. Biomed. Anal. 25 (2001) 331-337.

[19] L. Ramos, R. Bakhtiar and F. Tse, Rapid Commun. Mass Spectrom. 13 (1999) 2439-2443.

[20] A. Janoly, N. Bleyzac, P. Favetta, M.C. Gagneu, Y. Bourhis, S. Coudray, I. Oger and G. Aulagner, J. Chromatogr. B (2002) in press.

[21] Th. Meyer, J. Bohler and A.W. Frahm, J. Pharm. Biomed. Anal. 24 (2001) 495-506.

[22] S.D. Garbis, A. Melse-Boonstra, C.E. West and R.B. van Breemen, Anal. Chem. 73 (2001) 5358-5364.

[23] F.M. Rubino, J. Chromatogr. B 764 (2001) 217-254.

[24] B.K. Choi, D.M. Hercules and A.I. Gusev, J. Chromatogr. A 907 (2001) 337-342.

[25] M. Constanzer, C. Chavez-Eng, and B. Matuszewski, J. Chromatogr. B 760 (2001) 45-53.

[26] B.K. Matuszewski, M.L. Costanzer and C.M. Chavez-Eng, Anal. Chem. 70 (1998) 882-889.

[27] I. Fu, E.J. Woolf and B.K. Matuszewski, J. Pharm. Biomed. Anal. 18 (1998) 347-357.

[28] D. L. Buhrman, P. I. Price and P.J. Rudewicz, J. Amer. Soc. Mass. Spectrom. 7 (1996) 1099-1105.

[29] M. Jemal and Y.-Q. Xia,Rapid Commun. Mass Spectrom. 13 (1999) 97-106.

[30] P. Sarkar, C. Polson, R. Grant, B. Incledon and V. Raguvaran, Proceedings 49th American Society for Mass Spectrometry Conference, Chicago, IL USA (2001).

[31] C. Miller-Stein, R. Bonfiglio, T.V. Olah, R.C. King, Amer. Pharm. Rev. 3 (2000) 54-61.

[32] J.J. Zheng, E.D. Lynch and S.E. Unger, J. Pharm. Biomed. Anal. 28 (2002) 279-285.

[33] J. Smeraglia, S.F. Baldrey and D. Watson, Chromatographia 55 *Supplement* (2002) S95-S99.

[34] H. Mei, Y. Hsieh, N. Juvekar, S. Wang, C. Nardo, S. Wainhaus, K. Ng and W. Korfmacher, Proceedings 49th American Society for Mass Spectrometry Conference, Chicago, IL USA (2001).

[35] C. Müller, P. Schäfer, M. Störtzel, S. Vogt and W. Weinmann, J. Chromatogr. B 773 (2002) 47-52.

[36] M.D. Nelson and J.W. Dolan, LC-GC 20 (2002) 24-32.

[37] J.V. Sancho, O.J. Pozo, F.J. Lopez and F. Hernandez, Rapid Commun. Mass Spectrom. 16 (2002) 639-645.

[38] E.T. Gangl, M. Annan, N. Spooner and P. Vouros, Anal. Chem. 73 (2001) 5635-5644.

[39] N. Brignol, R. Bakhtiar, L. Dou, T. Majumdar and F.L.S. Tse, Rapid Commun. Mass Spectrom. 14 (2000) 141-149.

[40] M.C. Rouan, C. Buffet, L. Masson, F. Marfil, H. Humbert and G. Maurer, J. Chromatogr. B 754 (2001) 45-55.

[41] T. Bedman, M.J. Hayes, L. Khemani and F.L.S. Tse, Proceedings 48th American Society for Mass Spectrometry Conference, Long Beach, CA USA (2000).

[42] M.C. Woodward, G. Bowers, J. Chism, L. St. John-Williams and G. Smith, Proceedings 48th American Society for Mass Spectrometry Conference, Long Beach, CA USA (2000).

[43] A. Gritsas, M. Lahaie, D. Chun, T. Flarakos, M.L.J. Reimer, F. Deschamps and R. Hambalek, Proceedings 48th American Society for Mass Spectrometry Conference, Long Beach, CA USA (2000).

[44] M. Zell, C. Husser and G. Hopfgartner, J. Mass Spectrom. 32 (1997) 23-32.

[45] S. Steinborner and J. Henion, Anal. Chem. 71 (1999) 2340-2345.

[46] E.W. Woo, R. Messmann. E.A. Sausville and W.D. Figg, J. Chromatogr. B 759 (2001) 247-257.

[47] B. Lausecker, B. Hess, G. Fischer, M. Mueller and G. Hopfgartner, J. Chromatogr. B 749 (2000) 67-83.

[48] C.-L. Cheng and C.-H Chou, J. Chromatogr. B 762 (2001) 51-58.

[49] M. Cociglio, H. Peyriere, D. Hillaire-Buys and R. Alric, J. Chromatogr. B 705 (1998) 79-85.

[50] W.A. Korfmacher, K.A. Cox, M.S. Bryant, J. Veals, K. Ng, R. Watkins and C.-C. Lin, Drug Discov. Today 2 (1997) 532-537.

[51] Y. Hsieh, M.S. Bryant, J.-M. Brisson, K. Ng and W.A Korfmacher, J. Chromatogr. B 767 (2002) 353-362.

[52] R. Bakhtiar, J. Lohne, L. Ramos, L. Khemani, M. Hayes and F. Tse, J. Chromatogr. B 768 (2002) 325-340.

[53] N. Sadagopan, L. Cohen, B. Roberts, W. Collard and C. Omer, J. Chromatogr. B 759 (2001) 277-284.

[54] W.Z. Shou, H.-Z. Bu, T. Addison, X. Jiang and W. Naidong, J. Pharm. Biomed. Anal. 29 (2002) 83-94.

[55] L. Ramos, N. Brignol, R. Bakhtiar, T. Ray, L.M. McMahon and F.L.S. Tse, Rapid Commun. Mass Spectrom. 14 (2000) 2282-2293.

[56] N. Barbarin, D.B. Mawhinney, R. Black and J. Henion, J. Chromatogr. B (2002) in press.

[57] D.L. Hiller, A.H. Brockman, L. Goulet, S. Ahmed, R.O. Cole and T. Covey, Rapid Commun. Mass Spectrom. 14 (2000) 2034-2038.

[58] C.E.C.A. Hop, P.R. Tiller and L. Romanyshyn, Rapid Commun. Mass Spectrom. 16 (2002) 212-219.

[59] K. Heinig and F. Bucheli, J. Chromatogr. B 769 (2002) 9-26.

[60] R. Lombardi, LC-GC 16 (1998) S47-S52.

[61] S.X. Peng, J. Chromatogr. B 764 (2001) 59-80.

[62] P. Ellefson and D.A. Wells, U.S. Pat. No. 5,472,600 (5 December 1995).

[63] M.C. Rouan, C. Buffet, F. Marfil, H Humbert and G.Maurer, J. Pharm. Biomed. Anal. 25 (2001) 995-1000.

[64] J. Shen, J.-L. Tseng, M. Lam and B. Subramanyam, Proceedings 49th American Society for Mass Spectrometry Conference, Chicago, IL USA (2001).

[65] J.K. Nicholson, J. Connelly, J.C. Lindon and E. Holmes, Nature Reviews 1 (2002) 153-161.

[66] D.S. Palmer, D. Rosborough, H. Perkins, T. Bolton, G. Rock and P.R. Ganz, Vox Sang 65 (1993) 258-270.

[67] Y. Zhang, J. Hollembaek and O. Kavatskaia, Proceedings 49th American Society for Mass Spectrometry Conference, Chicago, IL USA (2001).

[68] R.S. Mazenko, A. Skarbek, E.J. Woolf, R.C. Simpson and B. Matuszewski, J. Liq. Chrom. 24 (2001) 2601-2614.

[69] Z.H. Israili and P.G. Dayton, Drug Metab. Rev. 33 (2001) 161-235.

[70] J.C. McElnay and P.F. D'Arcy, Drugs 25 (1983) 495-513.

[71] J.H. Lin, D.M. Cochetto and D.E. Duggan, Clin. Pharmacokin. 12 (1987) 402-432.
[72] J.M.H. Kremer, J. Wilting and L.H. Janssen, Pharmacol. Rev. 40 (1988) 1-47.
[73] W.A. Craig and P.G. Welling, Clin. Pharmacokin. 2 (1977) 252-268.
[74] N.C. van de Merbel, J.J. Hageman and U.A.Th. Brinkman, J. Chromatogr. 634 (1993) 1-29.
[75] M. Gilar, E.S.P. Bouvier and B.J. Compton, J. Chromatogr. A 909 (2001) 111-135.
[76] B.M. Cordero, J.L.P. Pavon, C.G. Pinto, M.E.F. Laespada, R.C. Martinez and E.R. Gonzalo, J. Chromatogr. A 902 (2000) 195-204.
[77] J.A. Jonsson and L. Mathiasson, J. Chromatogr. A 902 (2000) 205-225.
[78] J.R. Veraart, H. Lingeman and U.A.Th. Brinkman, J. Chromatogr. A 856 (1999) 483-514.
[79] P.R. Haddad, P. Doble and M. Macka, J. Chromatogr. A 856 (1999) 145 177.
[80] N.C. van de Merbel, J. Chromatogr. A 856 (1999) 55-82.
[81] E. Eklund, C. Norsten-Hoog and T. Arvidsson, J. Chromatogr. B 708 (1998) 195-200.
[82] K.B. Kenney, S.A. Wring, R.M. Carr, G.N. Wells and J.A. Dunn, J. Pharm. Biomed. Anal. 22 (2000) 967-983.
[83] D. Tsou and O. Penezina, Presented at North American Membrane Society, Lexington, KY USA (2001).
[84] D. Tsou and O. Penezina, Proceedings of the Pittsburgh Conference (Pittcon), New Orleans, LA USA (2002).
[85] P.A. Wabnitz and J.A. Loo, Rapid Commun. Mass Spectrom. 16 (2002) 85-91.
[86] N.S. Dow, C. Curran, K. Halloran, J. Hill, V. Ramachandran, D. Shelby, D. Hartman, Proceedings of the American Association for Pharmaceutical Scientists Annual Meeting, Denver, CO USA (2001).
[87] N. Dow, K. Halloran, S. Quimson and D. Hartman, Proceedings of the Society for Biomolecular Screening Conference, Baltimore, MD USA (2001).
[88] J.P. Walsh, R.O. Angeles, K.M. Jenkins and R.A. Rourick, Proceedings 50th American Society for Mass Spectrometry Conference, Orlando, FL USA (2002).
[89] Y.Y. Lau, Y.-H. Chen, X. Cui, A. Thomas, R. White and K.-C. Cheng, Proceedings 50th American Society for Mass Spectrometry Conference, Orlando, FL USA (2002).

[90] A.E. Niggebrugge, C. MacLauchlin, D. Dai, L.A. Ford, A.T. Menendez and A.S. Chilton, Proceedings 50th American Society for Mass Spectrometry Conference, Orlando, FL USA (2002).
[91] I. Kariv, H. Cao and K.R. Oldenburg, J. Pharm. Sci. 90 (2001) 580-587.

Chapter 7

Protein Precipitation: Automation Strategies

Abstract

The use of liquid handling technology to automate the protein precipitation procedures for collection microplates and filter plates is an important step toward improving the throughput of these universal sample preparation methods. All liquid handling steps can potentially be performed by a workstation—sample transfer, addition of internal standard and precipitant solution, transfer of supernatant to a clean microplate and preparation for injection via dilution or reconstitution in a mobile phase compatible solvent. Additionally, automation can be utilized prior to the precipitation procedure in order to prepare the calibration standards and the quality control samples in the microplate format. This chapter introduces strategies and reviews procedures for the automation of the protein precipitation procedure in microplates and filter plates using a 96-tip and a 4-/8-probe liquid handling workstation.

7.1 Automation of Protein Precipitation in Microplates Using a 96-Tip Workstation

7.1.1 Strategies

7.1.1.1 Introduction

A typical example of a 96-tip liquid handling workstation is the Quadra® 96 (Tomtec Inc., Hamden, CT USA). The discussion of automation strategies here will focus on this instrument, although the general approach is similar for other 96-tip workstations and pipetting units. More information about these workstations can be found in Chapter 5, Sections 5.3 and 5.4.5.

A summary of the tasks to be performed for a protein precipitation (PPT) procedure in the microplate format is listed in Table 7.1. Note that a few manual tasks in this procedure cannot be automated by common liquid handling workstations, such as the plate sealing, vortex mix and centrifugation

255

steps, and an optional evaporation step; however, the productivity of the overall procedure using microplates is great enough to justify these manual steps. Since some hands-on time is required, the overall procedure is considered to be semi-automated.

7.1.1.2 Identify Tasks and Roles for Automation

In designing an automation strategy, the user must decide at what part of the method sequence that the liquid handler first becomes involved. Preparation of the sample plate for extraction (sample plus internal standard addition and brief mixing) can be performed off-line either manually or using a multiple probe liquid handling workstation. At this point, the prepared microplate can be placed onto the 96-tip workstation deck for rapid pipetting. The following

Table 7.1
Summary of the tasks performed for a protein precipitation procedure in the microplate format

Task	Manual	Liquid Handling Workstation
Prepare sample plate		
• Aliquot samples into microplate	X*	X*
• Deliver aliquot of internal standard		X
• Mix		X
Deliver precipitant solution		X
Seal	X	
Vortex mix	X	
Centrifuge	X	
Unseal or pierce foil	X	
Transfer Supernatant		X
Dilute Supernatant (*Optional*)		X
Evaporate Supernatant (*Optional*)	X	
Reconstitute (*if evaporation used*)		X
Vortex mix	X	
Seal for injection	X	

*Can be performed either manually or by a 4-/8-probe liquid handling workstation

discussion is based on the plasma samples being reformatted into the microplate format off-line and then the 96-tip liquid handling workstation performs all subsequent pipetting tasks.

Three main liquid transfer functions with respect to liquid handling automation are performed during a protein precipitation procedure in the microplate format. These general functions are:

1. Transfer of precipitant solution into the sample plate for precipitation
2. Transfer of a volume of supernatant from the extraction plate into a clean collection microplate following the vortex mix and centrifugation steps
3. Dilution of the transferred supernatant with an aqueous liquid –or– reconstitution of the extract following an evaporation step

7.1.1.3 Deck Layout and Labware

The deck of the Quadra 96 has six positions that will accommodate the hardware (tips), extraction plate, transfer plate and all reagent reservoirs required to perform this procedure. This deck moves to the left, right, forward and backward relative to a fixed position under a 96-tip aspirating/dispensing head. When one of the six positions is directly under the tips, the stage is raised to a programmed height to meet the tips and the aspirate or dispense step is performed. Labware components sit on either a short or a tall nest, or placeholder, in each position. The actual nest type used influences the height that the stage is raised. A typical arrangement of these components on the Quadra 96 deck, as configured for PPT, is shown in Figure 7.1.

TIPS in 1 *Tip Jig*	Precipitant Solution in 2 *Tall Nest*	(empty)
Transfer Plate (for Supernatant) in 6 *Short Nest*	Sample (Precipitation) Plate in 5 *Short Nest*	Reconstitution Solution or Aqueous Diluent in 4 *Tall Nest*

Figure 7.1. Typical deck layout of the Quadra 96 liquid handling workstation as configured for performing protein precipitation in the microplate format.

7.1.1.4 Preparation of the Sample Plate

Samples are routinely collected in screw capped tubes or microcentrifuge vials at the study site and at some point they must be reformatted into the microplate format. Putting samples into the microplate format is the key to being able to use high throughput procedures for the subsequent steps of pipetting, transfers, evaporation, reconstitution and injection into the analytical system. In order to use a 96-tip workstation, samples must be reformatted from these individual source tubes into a microplate. This sample plate can be prepared off-line either manually or using a multiple probe liquid handler. Once the samples are arranged in this 8-row by 12-column array, all subsequent pipetting tasks can be performed by the liquid handling workstation. The ideal scenario is to have a 4-/8-probe liquid handling workstation prepare the sample plate (*e.g.*, plasma samples, internal standard, calibration standards, appropriate blanks and quality control samples), although not all laboratories are able to afford both a 96-tip and a 4-/8-probe liquid handling workstation. Further details about preparing a sample plate using a 4-/8-probe liquid handling workstation are discussed in Chapter 5, Section 5.6.4.2.

7.1.1.5 Addition and Transfer of Precipitant Solution

Once the sample plate has been prepared and placed on the deck of the Quadra 96, the first action by the instrument is to aspirate the precipitant solution (*e.g.*, acetonitrile) from a reagent reservoir and dispense into the wells of the sample plate. Note that the tip capacity of the Quadra 96 is only 450 μL (425 μL when using a 25 μL air gap). Volumes larger than this amount must be delivered using multiple pipetting steps. When the total volume has been dispensed, the program is paused and the plate is removed from the deck, sealed, mixed vigorously and centrifuged to pellet the solid protein mass at the bottom of the well.

The manner in which the precipitant solution is delivered can be important. When acetonitrile is dispensed from the tips held at a point above the surface of the sample, it contacts the sample at the liquid interface and precipitation occurs immediately within the narrow diameter of the well. A protein plug may form at this interface, making it difficult for the remaining acetonitrile to reach the unprecipitated sample below. An off-line vigorous mix step usually adds the needed efficiency to this process. However, another approach to the delivery of acetonitrile within the microplate well is often preferable. Following aspiration of acetonitrile, a small air gap (*e.g.*, 15 μL) follows. This air gap at the end of the tip allows entry into the plasma without making contact with the acetonitrile. The tips descend to the well bottom and then the volume of acetonitrile is dispensed. Better mixing within the well is achieved using this approach. **Important: Do not mix the precipitated solution with the tips (aspirate/dispense repeatedly within the well), as protein clumps will stick to the polypropylene tips as they exit the mixture, potentially removing liquid and analyte with them.**

An alternate order of addition of sample and precipitant solution is also practical, and in some cases is preferable to the method discussed above. A clean microplate is prefilled with the desired volume of acetonitrile. The tips are rinsed with water in a reagent reservoir. Sample is aspirated from a source microplate using the same tips and is dispensed into the plate containing acetonitrile; its delivery is performed drop by drop (rather than all at once) above the surface of the acetonitrile in the wells. When all samples have been dispensed, an air gap still remains within each tip; the tips are lowered to the bottom of the wells. The air gap is dispensed from the bottom of the wells which provides additional mixing capability. This approach offers an alternative processing method that does not require the off-line sealing and mixing steps. Note that Tomtec also offers an oscillating shaking nest for the Quadra deck that can replace the off-line mixing component of the procedure.

The precipitation plate, following vigorous off-line vortex mixing and centrifuging, is placed back onto the deck of the Quadra 96. The next step lowers a set of clean tips into the precipitation plate to a specified depth (determined by trial and error) so that maximum volume can be aspirated without disrupting the protein pellet at the bottom of the well. This volume of supernatant is transferred into a clean collection microplate located in the next deck position. Note that the Quadra 96 does not utilize liquid level sensing technology. In this case, the tips descend to a fixed depth and aspirate a preset volume. The volume aspirated can be larger than the actual supernatant volume, in which case air will be aspirated at the end of the cycle because the tips stay at a fixed depth.

7.1.1.6 Dilution or Evaporation Followed by Reconstitution

The microplate containing the transferred supernatant can be further processed in several ways. In a few instances, an aliquot is injected directly from each well, as described in the report by Hsieh *et al.* [1]. Usually, however, the percentage of organic in the supernatant is more than can be tolerated by the chromatography system, and/or the pellet is not compact and solid. More commonly, the supernatant is removed and is either diluted with a volume of aqueous or evaporated to dryness, followed by reconstitution. Dilution, and likewise reconstitution, can easily be performed by the 96-tip liquid handling workstation using one liquid transfer step, as described in the report by Shou *et al.* [2]. The appropriate solution is simply placed into a reagent reservoir to perform this step. Once the liquid transfer is completed, the plate is removed from the deck, sealed, briefly mixed and then placed into an autosampler for injection into a chromatographic system.

Although the protein precipitation procedure in microplates is inexpensive, fairly quick to perform and does not require much variation, a drawback to this technique is that matrix components are carried over into the system. The user may discover that only a small number of injections can be tolerated on a liquid chromatographic column before high pressure buildup occurs by particulates accumulating on the inlet frit. One way to reduce this mass of injected material is to filter the supernatant as it is transferred. Instead of adding the supernatant to a collection microplate on the deck of the Quadra 96, it is added into a filter plate secured on a vacuum manifold. A collection microplate inside the manifold isolates the filtrate.

7.1.1.7 Throughput Considerations

Very fast sample turnaround is needed for bioanalytical programs that support the drug discovery process in pharmaceutical research. Sample preparation using semi-automated protein precipitation is a cost effective and reliable means to reach throughput goals. Although an analyst must remain at the instrument when methods are run, a 96-tip workstation yields rapid sample preparation times. An entire plate can be processed in under 10 min for the actual liquid handling steps; additional time is necessary to prepare the sample plate and perform the off-line manual processing tasks as listed in Table 7.1.

The following example of a high throughput protein precipitation method was previously mentioned in Chapter 6, Section 6.3.4, but is repeated here for completeness within this chapter and because it is such a dramatic illustration of the potential of PPT automation. A report of protein precipitation performed in a collection plate, with liquid handling by a Quadra 96 workstation, also utilized monolithic column chromatography for fast analytical separation. The resolution of methylphenidate and ritalinic acid was achieved in 15 s using a 3.5 mL/min flow rate through the monolithic column. Throughput, in terms of sample preparation and analysis, was documented as 768 protein precipitated rat plasma samples (8 full microplates) analyzed within 3 h 45 min [3].

7.1.2 Method Example

A typical example of a protein precipitation method that can be automated on the Quadra 96 is one used for extraction of proprietary compounds and an internal standard from rat plasma in support of drug discovery applications. Sample volumes in this case are typically low (50 µL). The complete PPT method is listed in Table 7.2.

This protein precipitation method must be translated into a series of sequential actions for the Quadra 96 to execute. A typical Quadra 96 program for performing this method is detailed in Table 7.3. Each item, or numbered line in the program, involves an action to be performed at one of the six positions that is moved under the 96-tip aspirating/dispensing head. When pipetting, each line specifies a volume of liquid, an air gap volume, and the stage height setting as a number. Note that the stage height setting for each pipetting step depends on the specific nest type used and the preference of the user. Thus, stage height settings are not included in this table as individual variations apply.

Table 7.2
Typical example of a protein precipitation procedure performed in a collection microplate

Prepare sample plate

- Aliquot 50 µL plasma from vials or tubes and transfer into 1 mL (or 0.75 mL) round well microplate
- Add IS 25 µL and briefly mix

Dispense acetonitrile 150 µL

Seal microplate

Vortex mix 5 min

Centrifuge 3000 rpm x 10 min

Transfer supernatant to clean microplate without disrupting protein pellet

Evaporate to dryness

Reconstitute in 150 µL mobile phase compatible solvent

Briefly mix well contents

Seal plate with pierceable cap mat

Table 7.3
Typical Quadra 96 program for performing protein precipitation in the microplate format. The labware positions defined for this program are provided in Figure 7.1. Stage height settings are not included as individual variations apply .

1. Load Tips at 1

2. Mix 150 µL 3 times at 2

3. Aspirate 150 µL from Precipitant Solution in 2, 25 µL Air Gap

4. Aspirate 10 µL [stage height 0] from 2, 0 µL Air Gap

5. Dispense 185 µL to 5, 0 µL Blowout

6. Shuck Tips at 1

7. Pause

8. Load Tips at 1

9. Aspirate 25 µL [stage height 0] from 5, 0 µL Air Gap

10. Aspirate 190 µL from Sample Plate in 5, 25 µL Air Gap

11. Aspirate 10 µL [stage height 0] from 5, 0 µL Air Gap

12. Dispense 250 µL to Transfer Plate at 6, 0 µL Blowout

13. Shuck Tips at 1

14. Pause

15. Load Tips at 1

16. Aspirate 150 µL from Reconstitution Solution in 4, 25 µL Air Gap

17. Dispense 175 µL to Transfer Plate at 6, 0 µL Blowout

18. Shuck Tips at 1

19. Quit at 2

The Quadra 96 program is run with one tip exchange step after the precipitant solution (acetonitrile) is delivered into the sample plate (since the tips touch the sample). Two off-line steps exist; one for the initial seal, vortex mix and centrifuging of the extraction plate and another for the dry-down and reconstitution steps.

The reconstitution procedure aspirates a fixed volume from a reagent reservoir containing mobile phase compatible solvent and delivers into the plate containing dried extracts. If desired, this reconstitution step can be added to the end of the Quadra 96 program. However, it is suggested to maintain the reconstitution as a separate program so that the instrument is not occupied and unavailable for the length of time that it takes to dry down the supernatant, typically 15–30 min. When the Quadra 96 is used daily by multiple analysts, there is often a waiting list or queue to perform other sample preparation assays. Closing the program at the end of the supernatant transfer step simply allows more efficient utilization of the instrument. The analyst who needs to reconstitute dried extracts then walks up to the instrument at the next available time opening, performs the procedure, and leaves the Quadra 96 available for the next procedure or analyst. Some tips and tricks in using the Quadra 96 for performing LLE in microplates are discussed next.

7.1.3 Tips and Tricks

7.1.3.1 Using the Polypropylene Tips

Many times, the aspiration of an organic solvent using the polypropylene tips in their dry state will result in solvent leaking from the tips due to weak surface tension. When pipetting organic solvents, several sequential steps are recommended:

1. Mix 3 times using the volume to be delivered and dispense back into the reagent reservoir (at the same stage height)
2. Aspirate the desired volume
3. Aspirate a 10 µL air gap at the tip ends (use a stage height setting of 0) and then move the deck to the desired position for the dispense step
4. After dispensing, remove solvent vapors from above the 96-piston head by using a time dispense of 0 µL 5-times; make sure that the air compressor is turned on.

The tips hold a maximum volume of 450 µL. Therefore, when a volume greater than 425 µL is to be dispensed (a 25 µL air gap in addition to the solvent volume is suggested), use multiple aspirate and dispense steps within a loop procedure. For example, when a 600 µL volume needs to be dispensed, specify a volume of 300 µL (plus a 25 µL air gap equals 325 µL) to be dispensed two times.

Tips are usually exchanged after any step that touches drug or analyte in solution to avoid contaminating a reagent reservoir in subsequent steps. The tips can be removed, or "shucked," into the "tip jig" but sometimes it may be convenient to shuck them into a used microplate. For example, after transfer of the supernatant layer to a clean microplate, the tips may be shucked into the used precipitation plate containing the protein pellet left behind. Then, the plastic plate and tips can be removed in one step and disposed into a proper biohazardous waste receptacle. Shucking the tips into a different position than the tip jig provides a visual clue that new tips are to be loaded since that position remains empty.

Note that using the Quadra 96 to add internal standard to the sample plate delivers this solution to all 96 wells. In those instances where one or more double blanks (zero analyte level and no internal standard) are designated in the plate, the pipet tips corresponding to those exact well locations can simply be removed from the tip rack prior to pipetting. This selective tip removal approach can also be used for the organic solvent transfer and reconstitution steps to keep selected wells free of solvent. System checks, blank mobile phase or other solution can be manually delivered into dry wells of the final analysis plate following reconstitution. Several tips may be removed manually from the tip rack but removal of too many will cause the assembly to become unbalanced and an error may occur on loading.

7.1.3.2 Determining Well Depth for Supernatant Removal

The depth that the tips are submerged into the precipitated sample wells for the supernatant aspirate step is determined by trial and error using blank matrix. Different stage height settings are evaluated and a specified volume of supernatant is aspirated. The goal is to remove as much supernatant as possible yet leave the pellet at the well bottom undisturbed. Note that liquid level sensing is not applicable for the Quadra 96, so the tips move to a specific position within the well and remain there. If the volume defined for aspirating is greater than that available, air will be aspirated at the tip bottom which is not detrimental.

7.1.3.3 Using Move Steps to Change Deck Position

The use of move steps, or simply mix steps with no volume defined, is beneficial to move the deck position without performing any function. Access to the precipitation plate after solvent delivery is better achieved when that deck position is not directly underneath the pipetting head. Movement of the deck thus allows better access to a specific position. A move step is also beneficial when loading tips; moving the deck to Position 3 and then executing a pause step permits easy access to Position 1 to load/unload tips.

7.1.4 Applications

A representative list of semi-automated applications performed using a 96-tip workstation for PPT is provided in Chapter 5, Table 5.2; both collection plate and filtration plate examples are included within the listed references. More detail about analytes, sample volumes and precipitant volumes reported in published applications can be found in Chapter 6, Table 6.5. Note that while the liquid handling procedures performed for protein precipitation at first appear to be easily accomplished manually, the Quadra 96 can perform these pipetting steps with greater consistency and speed. One very valuable advantage provided by a workstation is the ability to aspirate the supernatant at a fixed depth from the extraction plate; once the exact depth is chosen, the instrument reliably transfers the supernatant without disruption of the protein pellet. It is very difficult to accomplish this task manually in a reliable manner.

7.1.5 Automating Method Development

Method development strategies for protein precipitation have been described in Chapter 6, Section 6.5. Some of these tasks can be automated using the Quadra 96 but keep in mind that selective solvent aspiration and delivery to specific wells or regions of wells is not possible. When all 96 tips enter a reagent reservoir, all 96 tips will deliver that same reagent. Also note that variable volume delivery cannot be performed with the Quadra 96.

In order to designate specific grids of tips to aspirate specific solvents (*e.g.*, an evaluation of three precipitant solutions for their extraction efficiency placed in rows of 12 wells each), the reagent reservoir needs to be customized. Instead of an open one compartment polypropylene reservoir, a multiple component reservoir can be used (shown in Chapter 10, Figure 10.6). These custom reservoirs are available from Tomtec in any configuration desired, *e.g.*, by row, by column, every two rows or every four columns. When one of these custom reservoirs is not available, a 2 mL microplate provides the ultimate custom reservoir; simply place different solutions into corresponding wells to achieve the pattern desired and use it as the reagent reservoir in the interim. Method development in this manner can be directed toward the evaluation of different precipitant solutions added to the sample or varying ratios of sample to precipitant solution.

7.2 Automation of Protein Precipitation in Filter Plates Using a 96-Tip Workstation

7.2.1 Strategies

7.2.1.1 Introduction

The use of a filtration plate offers the advantage of eliminating the manual off-line steps of mixing and centrifuging, as used with a collection microplate. A protocol for using a filtration microplate involves the addition of plasma (already spiked with internal standard) and precipitant solution. These two liquids are (ideally) delivered simultaneously inside the wells of the filtration plate. The precipitation occurs immediately and protein is trapped on the surface of the filter in each well. Vacuum is used to process liquid through the filter plate. A manifold for the Quadra 96 is supplied by Tomtec for this purpose. More information about the use of filtration plates for this application is found in Chapter 6, Section 6.4.2.

7.2.1.2 Deck Layout and Labware

A typical deck layout of the Quadra 96 for protein precipitation in a filter plate is shown in Figure 7.2. The layout is similar to that for collection plates, except that a vacuum manifold with a filtration microplate is placed in Position 5. Position 1 contains the tips, Position 2 contains the precipitant solution, and Position 6 contains the sample plate which was prepared off-line prior to this procedure. Note that Position 3 is unused and Position 4 is reserved for a subsequent reconstitution or dilution procedure.

TIPS in 1 *Tip Jig*	Precipitant Solution in 2 *Tall Nest*	(empty)
Sample Plate in 6 *Short Nest*	Filter Plate mated with Collection Plate in 5 *Vacuum Manifold*	Reconstitution Solution or Aqueous Diluent in 4 *Tall Nest*

Figure 7.2. Typical deck layout of the Quadra 96 liquid handling workstation as configured for performing protein precipitation using a filtration microplate.

7.2.1.3 Performing the Filtration Procedure

The Quadra 96 can perform a multiple component aspiration when used with a filtration plate. The liquid handling workstation first aspirates an aliquot of acetonitrile from Position 2, followed by an air gap, and then aspirates plasma from the sample plate in Position 6. The plasma and solvent are forcibly dispensed into wells of the filtration plate, thus avoiding a manual intervention to vortex the plate. The dispensing speed can be increased above the default value to accomplish vigorous mixing within the well. Plasma hits the well first, followed by the precipitant solution, and precipitation occurs instantly. Vacuum is applied to collect the filtrate. Proteins remain on top of the filter and are discarded with the plate. The underlying collection microplate containing filtrate is removed and prepared for injection.

At this point in the procedure when the filtrate is contained within the microplate, there are commonly three options prior to analysis: (1) seal the plate and direct inject, (2) dilute with aqueous, mix, seal and inject, or (3) evaporate to dryness and reconstitute, seal and inject. The Quadra 96 can be used for the dilution procedure as well as the reconstitution procedure following the evaporation step.

This automated approach can also be used with centrifugation as the liquid processing method of choice. However, a manual step is introduced into the procedure. Using this method, a deep well collection plate sits in the base of the vacuum manifold in Position 5 and the filtration microplate is placed on top; the lid of the vacuum manifold is not used and the base provides a rigid support. Liquids are delivered into the wells of the filter plate as in the vacuum based application described above; instead of activating vacuum, the plate combination is manually removed and centrifuged to collect the filtrate for subsequent analysis.

7.2.2 Applications

Applications and methodology using filtration microplates are described thoroughly in Chapter 6 and the reader is referred to Section 6.4.4 for details. A representative list of semi-automated applications performed using a 96-tip workstation with filtration microplates is provided in Chapter 5—Table 5.2 lists published references for automation of the protein precipitation method; both collection plate and filtration plate examples are included within the listed references. More detail about analytes, sample volumes and precipitant volumes used with filtration microplates is found in Chapter 6, Table 6.9.

7.3 Automation of Protein Precipitation in Microplates Using a 4-/8-Probe Workstation

7.3.1 Strategies

7.3.1.1 Introduction

The strategies for automating protein precipitation in the collection microplate format are thoroughly discussed in Section 7.1. Although this previous section focused on use of a 96-tip liquid handling workstation, the strategies presented are also pertinent for 4-/8-probe workstations. This section will focus on the unique procedural aspects of performing automated protein precipitation using a 4-/8-probe workstation and is more concise than earlier sections. Typical examples of 4-/8-probe liquid handling workstations include the MultiPROBE II (Packard Instruments, Meriden, CT USA, now a part of PerkinElmer Life Sciences) and Biomek 2000 (Beckman Coulter, Fullerton, CA USA). The discussion of automation strategies will focus on these instruments, although the general approach is similar for other 4-/8-probe workstations. Features, specifications and capabilities for these instruments as a group are described in Chapter 5.

7.3.1.2 Deck Layout and Labware

A typical MultiPROBE deck configuration for protein precipitation in microplates consists of the following labware components:

- Test tube or vial rack to contain the source samples
- Three solvent troughs to contain the internal standard solution, precipitant solution (*e.g.*, acetonitrile) and mobile phase compatible solvent (for reconstitution) or aqueous diluent
- Sample microplate that becomes the precipitation plate
- Disposable tip rack with tip chute for disposal of used tips
- Analysis microplate to contain the final reconstituted samples for injection

Disposable tips are used for all sample transfers and fixed probes are used for solvent transfers. Typically, disposable tips having a 200 µL volume capacity are used; for very small sample volumes, a 20 µL tip is available and for larger sample volumes a 1 mL tip can be used. Note that the use of 1 mL tips with the MultiPROBE requires the 4-panel extended deck platform, not the 2-panel regular deck.

7.3.1.3 Preparation of the Sample Plate

Since 4- and 8-probe liquid handling workstations accommodate tubes, the source samples can be placed directly onto the deck of the instrument in the appropriate rack. The probes are able to change their center-to-center spacing from tube width to microplate width (9 mm). Note that a recommended pretreatment step before using automation for pipetting is to first centrifuge the sample tubes to pellet any fibrin clots and particle debris at the bottom. The freeze/thaw cycle for plasma is known to introduce these clots and proper efforts such as centrifugation must be performed so that pipetting errors are not introduced. It is also possible to work with source plasma samples that are contained in a microtube rack rather than individual test tubes; in some studies the specification is made in the protocol to collect plasma in the microplate format and these individually prenumbered microtubes are a common choice.

An aliquot is aspirated, using unique disposable tips, from each set of sample tubes (4 or 8 at a time) and dispensed into the desired positions within the microplate wells. Depending on the final volume in each well, either a deep well plate or a shallow well plate is commonly used. The well locations for the aspirating and dispensing steps are specified in the well maps for the source (test tube rack) and destination (microplate) labware. Once all the samples are transferred into the microplate, the internal standard (IS) is delivered into those same wells. The IS can be dispensed using fixed tips in multiple dispense mode but user preference is to use unique disposable tips for each transfer set of 4 or 8 samples. More accurate pipetting is achieved for small volumes of 50 µL or less using disposable tips. A recommended option after IS delivery is to perform a post-dispense mix to ensure adequate sample integration before the precipitant solution is added to each well.

Note that often times the internal standard is placed within the precipitant solution when an organic solvent such as methanol or acetonitrile is used as the precipitant. However, this approach is not recommended. **Internal standards should be added directly to the plasma sample before the addition of acetonitrile, so that the binding of the analyte can be mimicked by the internal standard.** If internal standard is pre-mixed with the solution of acetonitrile, inaccurate quantitation can result.

7.3.1.4 Addition and Transfer of Precipitant Solution

Once the standards, QC samples and unknowns are delivered into the sample plate, the precipitant solution is delivered into each well. Commonly this step is performed using multiple dispensing, in which a set of four (or eight) probes aspirates enough liquid from the solvent reservoir to dispense a given volume repeatedly to multiple wells without refilling. Note that the height of the probes within the wells upon dispensing can be adjusted, so that there is enough clearance above the plasma upon solvent delivery. If the probes are too close to the well bottom, the potential of analyte contamination exists from splashing. Also, the speed at which the solvent is delivered into the wells can be adjusted. A more forceful delivery may sometimes be desired. Since off-line mixing and centrifugation are to be performed for this application, adjustment of the dispensing speed is usually not necessary. However, it can be a useful parameter to modify when using filtration microplates.

Following the addition of precipitant solution, the plate is manually removed, sealed, vortex mixed and centrifuged. The seal is removed and the plate is placed back onto the deck of the instrument for transfer of supernatant into the plate that will be used for analysis. The microplate wells now contain pelleted protein at their bottom. The next objective is to aspirate a fixed volume of supernatant from above the pellet without disturbing this mass. Note that the pellet will be of varying hardness or compactness, depending on which precipitant was used and how much force was used for the centrifugation. Within this step, the depth that the probes descend into the wells is precisely defined by adjusting the dispense height. Instead of using the labware default value within the software, the height can be defined as a "% of well height" or as "mm above well bottom." The desired height is determined by trial and error using a blank set of samples of exactly the same volumes and types.

An important precaution during the transfer step is to decide whether or not liquid level sensing (LLS) should be utilized. By default, LLS is turned on and in use. Here, the tips descend until they sense liquid, then stop and aspirate; the probes descend within the well accordingly as the top level of the liquid is reduced. LLS is usually desirable in this situation, otherwise the probes can be specified to descend to a certain depth within the well, stop and aspirate a defined volume. Note that if the probes are defined to stop at a position right above the pellet, the upward flow from aspirating liquid may disrupt the pellet, depending on its compactness.

Once these considerations for aspiration from the sample well are decided upon and optimized for the sample type and volumes used, disposable tips transfer a volume of supernatant to a clean, unused destination microplate in a simple plate-to-plate transfer. The microplate chosen for the destination plate will be the one used for injection. The microplate containing the transferred supernatant can be further processed in several ways, as described in Section 7.1.1.6.

7.3.2 Applications

A thorough and illustrative example of the automation of protein precipitation in the collection plate format has been published by Watt *et al.* [4], using fluoxetine, imipramine and L-775,606 as test analytes. Automated liquid handling operations were performed by a Biomek 2000 Workstation. This equipment uses various pipetting tools for liquid transfers, detailed as follows: a 1000 µL tool (P1000L) with liquid level sensing (LLS), a 200 µL tool (P200L) with LLS, an 8-channel 200 µL tool (MP200), a 20 µL tool (P20), and an 8-channel 20 µL tool (MP20). Disposable tip capacities used were 1000 µL, 250 µL and 20 µL.

In the reported procedure [4], the plasma samples to be analyzed (50 µL) are transferred from a source rack of 96 microcentrifuge tubes to Positions B1 through G9 in the analysis plate using a 200 µL single dispense tool with disposable tips. Two sets of 12 calibration samples are contained in rows A and H of the plate. Acetonitrile (100 µL) is transferred from a solvent reservoir to all wells using an 8-channel tool and the program is paused for manual intervention. The plate is removed, heat sealed, and vortex mixed for 1 min. The plate is centrifuged (2000g x 15 min), the seal is removed, and the plate is placed back onto the deck of the Biomek. A fresh collection plate (to receive supernatant) is also placed on the deck. Using the 8 channel tool, 50 µL of 25mM ammonium formate is added to each well of this collection plate. Into this same plate is added 100 µL of supernatant from each well of the precipitation plate. The plate containing diluted supernatant is removed, heat sealed, vortex mixed, and placed on an autosampler for LC-MS/MS analysis.

The analysis plate was injected from top left to bottom right such that a calibration curve was analyzed at the beginning and end of the analytical run, and samples were analyzed in six groups of nine each followed by a set of three QC standards. A follow-up report from the same laboratory presented a generic and flexible methodology for automation of PPT using the Biomek 2000, such that variations in the nature and number of samples to be analyzed can be easily accommodated [5]. This report also details how the liquid handling workstation is used to prepare dilutions of analyte stock solutions and spiked plasma to generate analytical standards

The issue of manual versus automated sample preparation was addressed by Watt *et al.* [4] in order to demonstrate correlation and acceptability for the automated procedure. The three test analytes were processed by both the manual and automated routes, including preparing the calibration and quality control samples. All assays were linear from 1–2000 ng/mL and showed similar intercepts, slopes and high correlation coefficients. The accuracy measurements did not differ by more than 20% from the nominal concentration, deemed acceptable for a drug discovery environment. No differences in precision were observed between the manual and automated methods. This methodology also reported many gains in throughput. A full plate of samples was prepared for LC-MS/MS analysis in 2 h with minimal manual intervention, compared with a full working day if all steps were performed manually using multichannel pipettors. The gain in efficiency of sample preparation and analysis was reported to be a 4-fold increase to 400 samples per day per LC-MS/MS instrument.

7.4 Automation of Protein Precipitation in Filter Plates Using a 4-/8-Probe Workstation

7.4.1 Strategies

The strategies for automating protein precipitation in the filtration plate format have been thoroughly discussed in Section 7.2. Although this previous section focused on use of a 96-tip liquid handling workstation, the strategies presented are also pertinent for 4-/8-probe workstations. This section will focus on the unique procedural aspects of performing automated protein precipitation using a 4-/8-probe workstation and is more concise than earlier sections.

The deck layout and labware items are similar to those described in Section 7.3.1.2, except that a vacuum manifold is placed onto the deck and configured with a filter plate on top and a collection plate inside to isolate filtrate. Most manifolds are designed with interchangeable collars and/or adapters to accommodate collection of filtrates in microplates having different heights (both shallow well and deep well plates). An advantage of a 4-/8-probe workstation is that it can work directly with sample tubes and deliver volumes directly into the filtration microplate. Software control of the vacuum parameters allows a fully automated procedure without manual intervention.

7.4.2 Applications

Applications and methodology using filtration microplates are described thoroughly in Chapter 6 and the reader is referred to Section 6.4.4 for details. A representative list of semi-automated applications performed using a 4-/8-probe workstation with filtration microplates is provided in Chapter 5—Table 5.2 lists published references for protein precipitation that specifically used these liquid handling workstations; both collection plate and filtration plate examples are included within the listed references.

The liquid handling workstation first aspirates acetonitrile, followed by an air gap, and then aspirates plasma in the same disposable tip, as reported by Biddlecombe and Pleasance [6]. This mixture is then forcibly dispensed into wells of the filtration plate, thus avoiding a manual intervention to vortex the plate. In this step, the dispensing speed can be increased above the default value to accomplish vigorous mixing within the well. Plasma hits the well first, followed by the precipitant solution, and precipitation occurs instantly. Vacuum is applied to collect the filtrate. The MultiPROBE software accommodates full vacuum control via switching valves. Proteins remain on top of the filter and are discarded with the plate. The underlying microplate containing filtrate is removed and prepared for injection.

A condition of the liquid handler used for this filtration approach is that it be able to work with a vacuum manifold on its deck and, optionally, be able to sequentially aspirate solvents separated by an air gap. In the reported optimization of this procedure using a MultiPROBE instrument with three types of filter plates [6], it was found that the sequential aspiration of first acetonitrile, followed by an air gap and then plasma, was the preferred order of reagent addition. The amount of protein carried over into the filtrate was quantitated and formed the basis for this preference. Other schemes examined were the aspiration of plasma, an air gap, then acetonitrile, as well as a separate transfer of plasma to the well, followed by acetonitrile. If it is chosen to transfer plasma first, followed by a separate transfer of acetonitrile, it is advised to first wet the filter with water as this technique has been shown to enhance the reproducibility of the results [7].

This automated approach can also be used with centrifugation as the liquid processing method of choice. However, a manual step is then introduced into the procedure. Using this method, the filtration microplate (mated with a deep well collection plate) is placed on the deck of the liquid handling workstation. Liquids are delivered into the wells of the filter plate as in the vacuum based application described above; instead of activating vacuum, the plate combination is manually removed and centrifuged to collect the filtrate for subsequent analysis. In this scenario, the combination of filter plate on top of a collection plate needs to be defined as a unique labware item and its default dispensing height needs to be properly determined.

The throughput using filtration microplates in this manner has been reported as 20 min to process a complete plate of 96 samples [6, 7], plus another 20–30 min for the dry-down step [6]. Two plates can then comfortably be processed within 1 h and each evaporated in parallel. Using filtration microplates manually has been reported to require from 1–1.5 h for the full procedure [8].

7.4.3 Additional Automation Capabilities

Assembly and disassembly of the vacuum manifold in an automated system is an important task required for a walk away solution. The common approach to full automation utilizes a gripper arm within the instrument for assembly and disassembly of the vacuum manifold. The following liquid handling workstations present typical examples which incorporate a gripper arm: Biomek 2000, Speedy (Zinsser Analytics, Frankfurt, Germany), MICROLAB® STAR and MPH (Hamilton Company, Reno, NV USA), as well as the TALON option for the MultiPROBE Series. Note that use of the TALON requires the extended deck version of the MultiPROBE.

The Hamilton Company offers a unique approach for microplate filtration with the MICROLAB STAR pipetting workstation in combination with the MICROLAB Automated Vacuum System (AVS), described in Chapter 5, Section 5.4.4.2. Also, a MICROLAB SWAP plate handling robot can automatically add a filtration plate to the manifold from a storage location off to the side of the instrument. This plate handler performs a similar function to the Twister™ II (Zymark Corporation, Hopkinton, MA) which is the most widely used universal microplate handler.

7.5 Full Robotics Integration with Multiple Task Modules

Strategies for the automation of protein precipitation procedures using liquid handling workstations have been described. Full automation of this application is achieved by adopting multiple task modules so that the collection plate containing supernatant (or filtrate) can be forwarded by a robotic or gripper arm to perform the next unattended function, such as centrifugation. Once centrifugation is completed, the plate is shuttled back to a liquid handling workstation for aspiration of the supernatant and transfer into a clean microplate. After this transfer step, the plate is forwarded to another station for evaporation. The plate can be removed from the evaporation station and placed back onto a liquid handling workstation. It is reconstituted with mobile phase compatible solvent and taken to another station for sealing.

This fully automated approach requires the ability to interface with multiple components and represents the ultimate in a custom configured application. An example of this approach is reported by Depee [9] that utilized the following modules with a Zymate™ XP robotics system (Zymark Corporation) on an 8 ft robot track: Genesis RSP 8051 liquid handler (Tecan US, Research Triangle Park, NC USA), Zymark RapidPlate™, Zymark microplate centrifugation module, Zymark reagent addition station with three reagent addition modules, a microplate evaporator unit and an orbital plate shaker. An additional report by Biddlecombe and Pleasance details a similar fully automated setup for protein precipitation using both collection and filtration microplates [10].

Acknowledgments

The author is appreciative to the bioanalytical support staff at Millennium Pharmaceuticals for their helpful discussions of protein precipitation techniques and automation strategies which helped shape the organization and focus of this chapter.

References

[1] Y. Hsieh, M.S. Bryant, J.-M. Brisson, K. Ng and W.A Korfmacher, J. Chromatogr. B 767 (2002) 353-362.

[2] W.Z. Shou, H.-Z. Bu, T. Addison, X. Jiang and W. Naidong, J. Pharm. Biomed. Anal. 29 (2002) 83-94.

[3] N. Barbarin, D.B. Mawhinney, R. Black and J. Henion, J. Chromatogr. B *in press.*

[4] A.P. Watt, D. Morrison, K.L. Locker and D.C. Evans, Anal. Chem. 72 (2000) 979-984.

[5] D. O'Connor, D.E. Clarke, D. Morrison and A.P. Watt, Rapid Commun. Mass Spectrom. 16 (2002) 1065-1071.

[6] R.A. Biddlecombe and S. Pleasance, J. Chromatogr. B 734 (1999) 257-265.

[7] M.C. Rouan, C. Buffet, F. Marfil, H Humbert and G.Maurer, J. Pharm. Biomed. Anal. 25 (2001) 995-1000.

[8] R.E. Walter, J.A. Cramer and F.L.S. Tse, J. Pharm. Biomed. Anal. 25 (2001) 331-337.

[9] S. Depee, Proceedings International Symposium on Laboratory Automation and Robotics, Boston, MA USA (2001).

[10] R.A. Biddlecombe and S. Pleasance, Proceedings International Symposium on Laboratory Automation and Robotics, Boston, MA USA (1998).

Chapter 8

Liquid-Liquid Extraction: High Throughput Techniques

Abstract

This chapter introduces liquid-liquid extraction (LLE), an efficient sample preparation technique that has commonly been performed using individual test tubes. High throughput LLE is performed when a collection microplate is used in place of multiple test tubes. An alternative approach to LLE is the use of solid-supported diatomaceous earth particles contained in tubes or cartridges. High throughput solid-supported LLE (SS-LLE) is performed when this material is packed in flow-through microplates. The sample preparation techniques for performing high throughput LLE using both approaches are presented in this chapter. Details are provided about the accessory products needed to perform each procedure, and typical examples of bioanalytical applications are introduced. Strategies for method development and automation of LLE procedures are discussed in following chapters.

8.1 Understanding the Technique

8.1.1 Fundamental Principles

Liquid-liquid extraction (LLE), also called solvent extraction, is a technique used to separate analytes from interferences in the sample matrix by partitioning the analytes between two immiscible liquids. A given volume of aqueous sample solution (*e.g.*, plasma) containing analytes is mixed with an internal standard in solution. A volume of buffer at a known pH (or a strongly acidic or basic solution that will adjust the pH) is then added to maintain the analytes in their unionized (uncharged) state. The resulting solution is then vigorously mixed with several ratio volumes of a water immiscible organic solvent or mixtures of two or more solvents [*e.g.*, hexane, diethyl ether, methyl tert-butyl ether (MTBE) or ethyl acetate]. Analytes distribute between the two liquid phases (aqueous and organic) and partition preferentially into the organic phase when the analytes (1) are unionized and (2) demonstrate solubility in that

277

organic solvent. Agitation increases the available surface area for this interaction to occur and aids in the mass transfer process.

When the LLE procedure is optimized, the hydrophilic compounds in the sample matrix will prefer to remain in the aqueous phase and the more hydrophobic compounds (ideally analytes to the exclusion of interfering substances) will migrate into the organic phase. The organic phase is isolated and, since it is typically not compatible with solvents used in liquid chromatography (LC), is concentrated by evaporation. Reconstitution in a mobile phase compatible solvent is then performed prior to analysis. The overall procedure for LLE is outlined in Figure 8.1.

Note that it is possible to perform a back extraction or re-extraction of the organic solvent once analytes have partitioned into it. This procedure is sometimes performed when coeluting interferences cross over into the organic phase with the analyte. A basic analyte will be back extracted from the organic to the aqueous phase when the aqueous environment is acidic (analyte becomes

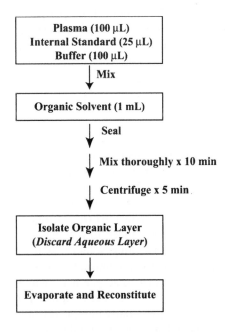

Figure 8.1. Schematic diagram of the liquid-liquid extraction procedure in which analyte partitions from an aqueous phase into an organic phase.

ionized and partitions into the aqueous environment); neutral species remain in the organic phase. Likewise, an acidic analyte will be back extracted from the organic to the aqueous phase when the aqueous environment is alkaline; neutral species remain in the organic phase. This back extraction allows the analytes to reenter the aqueous phase. Prior to analysis, the usual organic solvent evaporation and reconstitution steps are unnecessary when the aqueous phase chosen is compatible with mobile phase for direct injection (Figure 8.2).

A detailed theory of sample partitioning between two immiscible liquids is explained by the Nernst Distribution Law which states that a chemical species will distribute between two immiscible solvents such that the ratio of the concentrations remains constant, as expressed by the equation:

$$K_D = C_0/C_{aq} \tag{1}$$

In equation (1), K_D represents the distribution constant, C_0 is the concentration of analyte in the organic phase, and C_{aq} is the concentration of analyte in the aqueous phase. A more detailed discussion of this theory is available [1–3].

An important concept in liquid-liquid extraction is that for quantitative recovery to occur, $C_0 \gg C_{aq}$ and thus K_D should be a large value that is $\gg 1$. The extraction efficiency or the recovery is related to both the partition coefficient (K_D) and to the volume of organic solvent used for the extraction. **High recoveries will be obtained when large volumes of organic solvent relative to aqueous are used.** Typically, a sample to solvent ratio of from 1:7 to 1:10 (v/v) is used in LLE. This ratio is often reduced for microplate applications due to reservoir volume limitations, as discussed later in this chapter. When the ratio of sample to solvent is lower than about 1:7 and the partition coefficient K_D is also small, less than complete recoveries may result.

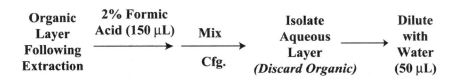

Figure 8.2. A neutral analyte is back extracted from an organic into an aqueous phase when pH is manipulated to cause ionization of the analyte. The aqueous layer is isolated, diluted (optional) and then analyzed.

8.1.2 Advantages

Liquid-liquid extraction is a very popular technique for sample preparation due to its many advantages. A major benefit is that LLE is of wide general applicability for many drug compounds. With proper selection of organic solvent and adjustment of sample pH, very clean extracts can be obtained with good selectivity for the target analytes. Inorganic salts are insoluble in the solvents commonly used for LLE and remain behind in the aqueous phase along with proteins and water soluble endogenous components. These interferences are excluded from the chromatographic system and a cleaner sample is prepared for analysis. The removal of these unwanted matrix materials provides potential benefits of extending LC column lifetime and minimizing the downtime of the mass spectrometer caused by interface fouling. The time required for method development is relatively short; methods can usually be developed within two days, an advantage when time is at a premium as it is in pharmaceutical research laboratories. The method development process for LLE is discussed in Chapter 9. Also, considered among many other sample preparation techniques, LLE methods can be run at relatively low cost.

8.1.3 Disadvantages

Several disadvantages exist with the liquid-liquid extraction technique:

- It is a very labor intensive procedure because of multiple transfer steps and the need to frequently cap and uncap tubes
- It requires large volumes of organic solvents which can be expensive to purchase and presents added costs for disposal as hazardous waste
- Exposure of these solvents to personnel can present health hazards
- The procedure has been difficult to fully automate using traditional liquid handling instruments
- Evaporative losses may sometimes occur upon dry-down with volatile or oxygen labile reactive analytes
- Emulsion formation is a potential problem

Emulsion formation is a recurring problem that happens without warning. It is commonly caused when samples contain enough lipids or surfactants to disrupt the two phase separation and prevent a distinct boundary from forming between the aqueous and organic phases; instead, a colloidal suspension results. When an emulsion forms, attempts must be made to break the emulsion or the extraction results will be poor. The sample is not recoverable for further analysis. Some common techniques for breaking emulsions [1] include adding salt to the aqueous phase to decrease the solubility of organic liquids, filtering through phase separation paper [*e.g.*, Whatman 1PS® (Whatman Inc., Clifton, NJ USA)], extending centrifugation times, and adding a small amount of an organic solvent having a higher density than the one used for the extraction.

8.1.4 Accessory Products

The accessory products needed to perform liquid-liquid extraction in the test tube format include the following:

- Test tubes, capped and placed into a rack
- Pipettors
- Multiple tube vortex unit
- Centrifuge
- Evaporation device

Test tubes used for LLE are commonly made of polypropylene or borosilicate glass. Screw caps or snap caps are used to tightly seal the tubes. The size of the tube depends on the total liquid volume with allowance for headspace above the liquid level for efficient vortex mixing; tube sizes usually range from 4–15 mL. Any pipettor that aspirates and dispenses the required volumes can be used for sample and solvent delivery into tubes. Repeater pipettors are especially useful for repeated delivery into multiple tubes.

Vortex mixing for 10–15 min is accomplished using a multiple tube vortex unit with a rack of capped test tubes. Sometimes an orbital or horizontal shaker unit may be utilized. A sufficient time for centrifugation of tubes for the extraction procedure is typically 10 min. Centrifugation speeds generally range from 2000 rpm to 3000 rpm. The isolated organic layers are evaporated using any of several models of evaporation units that accommodate tubes, such as N-EVAP® (Organomation Associates, Berlin, MA USA) or TurboVap™ (Zymark Corporation, Hopkinton, MA USA).

8.1.5 Applications

Liquid-liquid extraction remains an effective and widely utilized sample preparation technique because it provides efficient sample cleanup and enrichment. Reports of liquid-liquid extraction used as a sample preparation technique for bioanalysis are ubiquitous in the literature. A general review of LLE applications can be found within a discussion of sample enrichment techniques as reported by Kataoka and Lord [4]. Summaries of LLE applications can be found within literature reviews; these summaries commonly survey published applications for specific drug classes, such as benzodiazepines [5] and metalloproteinase inhibitors [6], and disciplines such as systematic toxicological analysis [7]. These reviews are also focused on a particular analytical detection technique, such as the use of liquid-liquid extraction in capillary electrophoresis in publications by Pedersen-Bjergaard [2] and Lloyd [8].

Some selected applications utilizing liquid-liquid extraction for bioanalysis are summarized in Table 8.1. These reports all used traditional test tubes as part of the method and were performed manually, as is typical when using test tubes. Take particular note of the volumes ratios—organic solvent to sample; this ratio ranges from 2.5:1 to 12.7:1 for the applications listed. The average ratio for high throughput applications using microplates typically stays toward the lower end of this range because the well reservoir volume is much smaller.

The traditional LLE technique is labor intensive when performed using screw capped test tubes because several pipetting and aspirating steps are involved. In addition, tubes must be labeled, capped, vortex mixed, centrifuged and uncapped during this process. Transfer steps add to the tedium of performing this procedure; each transfer presents the possibility of introducing sample loss. An additional transfer step is necessary after the evaporation step as the sample is reconstituted, vortex mixed, and then transferred into an autosampler vial for injection and analysis. Use of several of the high throughput tools, including automation, has been shown to reduce the labor intensive aspect of liquid-liquid extraction. Also, by reducing the number of transfer steps, better analyte retention can be achieved. Procedures for the use of liquid-liquid extraction in the 96-well microplate format are introduced next.

8.2 High Throughput Techniques Using Collection Microplates

8.2.1 Introduction

A high throughput approach to performing LLE is to use the microplate format instead of test tubes for most of the procedural steps, recognizing that sample sources are usually contained in single tubes or vials and must be reformatted into the 96-well configuration. The volume capacity of traditional microplates (2 mL or less) is much smaller than test tubes (*e.g.*, 4–15 mL). Sample and solvent volumes must be reduced accordingly to fit within this 2 mL volume and leave sufficient air space above the liquid for mixing. Single and multiple

Table 8.1
Selected bioanalytical applications using liquid-liquid extraction in the traditional tube format without automation

Analytes	Sample Matrix and Volumes	pH Modifier and Volumes	Extraction Solvent and Volumes	Ratio (Org:Aq)	Analytical Method	Reference
Proprietary and 2 metabolites	Dog plasma 0.5mL IS 50µL	0.1M HCl 0.5mL	MTBE 3mL	2.9:1	LC-MS ESI SIM	[9]
Thrombin inhibitor & 2 metabolites	Human plasma 0.5mL IS 50µL	2N NaOH 0.25mL	EtOAc/MTBE (1:3) 4mL; back extracted into 2% HCOOH	5.0:1	LC-MS/MS	[10]
Proprietary	Rat plasma 0.1mL IS 25µL	0.1N Na$_2$CO$_3$ 0.2mL	EtOAc 4mL	12.3:1	LC-MS/MS	[11]
Proprietary	Rat brain homogenate 0.5mL, IS 25µL	0.1N Na$_2$CO$_3$ 0.2mL	EtOAc 4mL	5.5:1	LC-MS/MS	[11]
Topiramate	Human plasma 0.1mL IS 10µL	(none)	Chloroform 1.4mL	12.7:1	LC-MS/MS	[12]
Rofecoxib	Human plasma 1.0mL IS 100µL	Carbonate buffer pH 9.8, 1mL	MTBE 8mL	3.8:1	LC-MS/MS	[13]
Clozapine and 2 metabolites	Human plasma 0.5mL IS 100µL	Na$_2$CO$_3$ (ss) pH ~10, 0.5mL	EtOAc/CH$_2$Cl$_2$/Pentane (5:3:2) 7mL	6.4:1	LC-MS/MS	[14]

EtOAc, Ethyl acetate; ss, saturated solution

probe liquid handlers can be utilized for sample and solvent aspirate and dispense steps. These units are especially convenient to reformat samples from source tubes into a microplate. Once samples are in the microplate format a 96-tip benchtop pipetting unit can be used to realize greater throughput. Automation strategies for performing liquid-liquid extraction are presented in Chapter 10.

An important safety consideration with all LLE applications is to vent the volatile (and often toxic) organic solvent vapors away from the work area. Performing the procedure inside a fume hood or locating trunks from the ceiling near the work area are two common approaches to contain and remove these vapors.

8.2.2 Procedure

8.2.2.1 Sample Reformatting

In order to perform liquid-liquid extraction using collection microplates, samples must be reformatted from individual tubes into a microplate with its 9.0 mm well-to-well spacing. While this transfer can be done manually with single pipettors or with an articulating (variable tip spacing) multichannel pipettor, it is tedious to perform and may lead to errors. These errors are often a result of placing aliquots into the incorrect wells or into the same wells as other samples. However, with practice and concentration, an articulating 8-channel pipettor (as well as a single channel pipettor) can be used with success to place samples. Pipetting aids are available to assist in this manual delivery. One such accessory is a 96-well plate row indexer and column indexer, described in Chapter 4, Section 4.7. A liquid handling workstation can also effectively be utilized for this sample reformatting step.

8.2.2.2 Sample Pretreatment and Sample Plate Preparation

Aliquots of sample, spiking solution and quality control samples are transferred into microplate wells, followed by internal standard. An aqueous buffer (or strongly acidic or basic solution) is added to adjust the sample pH; the well contents are mixed for a few seconds. Once these samples are in the microplate format (deep well plate or microtubes), any of the following automated instruments can be used for the subsequent aspirating and dispensing of solvents: multiple probe benchtop pipetting systems, 1-, 4- or 8-probe liquid handling workstations, or 96-tip workstations. The throughput of each of these automated systems varies but each is a viable option. An overview of liquid handling systems can be found in Chapter 5. A discussion of the applications used to perform automated liquid-liquid extraction procedures is presented in Chapter 10.

8.2.2.3 Organic Solvent Delivery, Mixing and Centrifugation

A volume of organic solvent is delivered into the wells to be extracted. The microplate is then securely sealed using a tightly fitting silicone cap mat or a heat sealable aluminum film. Sealing of microplates is discussed further in Section 8.2.3.3. Vigorous vortex mixing or horizontal shaking of the sealed microplate is commonly performed for 10–15 min. Following mixing, the plate is centrifuged to efficiently separate the organic and aqueous layers. The microplate seal is removed in preparation for the transfer step.

8.2.2.4 Transfer of Organic Layer, Evaporation and Reconstitution

The organic layer is isolated and transferred into a clean collection plate. The solvent in this plate is evaporated to dryness and reconstituted in a mobile phase compatible solvent. An aliquot is injected into a chromatographic system for analysis. A back extraction of the isolated organic solvent is sometimes performed. In this case, instead of evaporating the organic solvent, a small volume of aqueous formic acid solution, *e.g.*, is added and the mixture is vigorously shaken for 10–15 min to repeat the extraction. The formic acid solution is then isolated and made compatible with mobile phase for direct injection by the addition of water or an organic/water mixture. An aliquot is then analyzed.

8.2.2.5 Efficiency and Productivity Considerations

In order for the extraction procedure to remain efficient when the volume ratio of organic to aqueous is low, the partition coefficient K_D (equation 1) must be relatively large. Two ways to increase this ratio when using microplates are to use a microplate having a larger well volume (*e.g.*, 4 mL instead of 2 mL) for the extraction and to use a smaller sample volume. However, the sensitivity needs of the analytical assay usually govern the selection of sample volume and it quite often may not be possible to use a lower volume. Note that once the sample aliquots are delivered into the microplate wells, the 96-well format is retained throughout the extraction and analysis steps since autosamplers accommodating the 96-well format are used.

8.2.3 Accessory Products

8.2.3.1 Introduction

The proper choices of microplate design and accessory products (especially sealing systems) can be very important to achieve success in performing liquid-liquid extraction in the microplate format. Microplates are reviewed in Chapter 3 and the assortment of accessory products needed to perform a 96-well liquid-liquid extraction procedure is described in Chapter 4. Table 8.2 contains an outline of these various accessory products for reference. While 96-well LLE can be performed manually using multichannel pipettors, maximum laboratory productivity is achieved through successful implementation of automated liquid handling workstations. Chapter 5 reviews the commercial automation systems available and techniques specifically for liquid-liquid extraction are reviewed in Chapter 10. Important notes on microplate selection and sealing systems are provided next.

8.2.3.2 Microplates

The most commonly used microplate for LLE is the deep well collection plate having square wells, since the well volume (2 mL) is greater than that of most round well plates (1 mL). Note that the shared wall technology introduced by Nalge Nunc International (Rochester, NY USA) allows a 2 mL volume in a round well plate. Given the choice of a square well or round well 2 mL plate, when vortex mixing is required the round well geometry is preferable for its

Table 8.2
Outline of accessory products that assist in performing 96-well liquid-liquid extraction

Accessory Products	Reference
Collection plates	Chapter 3, Section 3.4
Multichannel pipetting systems	Chapter 4, Section 4.1
Small footprint liquid dispensing units	Chapter 5, Section 5.3.2
Automated liquid handling workstations	Chapter 5, Section 5.4
Sealing systems for microplates	Chapter 4, Section 4.3
Mixing liquids within collection devices	Chapter 4, Section 4.4.2
Centrifugation of microplates	Chapter 4, Section 4.4.4
Solvent evaporation systems for dry-down of eluates	Chapter 4, Section 4.5
Autosamplers accepting the microplate format	Chapter 4, Section 4.6

efficiency in mixing. However, when the plate is placed sideways in a horizontal shaker for mixing, the well geometry is not as important.

A second convenient collection plate format is the microtube rack, which is a set of 96 polypropylene tubes (usually 1–2 mL) fitted into a base plate. Microtubes retain the same 9.0 mm well-to-well spacing and outer dimensions as found in standard collection plates. These tubes can be obtained as individual tubes or as strips of eight or twelve. Microtubes are also available having preprinted alphanumeric labeling on each tube bottom. Larger volume tubes are available having square wells with rounded bottom.

When a 2 mL well volume is not sufficient for a method, some taller 96-well modular microplates having volumes of 4 mL can be used. Also, 48-well microplates with 7.5 and 10 mL volume capacities exist, which may be beneficial when performing the extraction step. However, these larger volume plates may be too tall for some styles of microplate evaporation units; the manufacturer should be consulted concerning compatibility issues. Plates having 48-wells will need to be reformatted into the 96-well configuration for continued use in the sample preparation procedure, as autosamplers and other accessory tools are commonly designed for the 96-well format. Note that an isolated organic volume greater than 2 mL can be dried down in a traditional 96-well plate in two sequential portions, when necessary.

8.2.3.3 Sealing Systems

A very important choice for performing LLE successfully in microplates is the selection of a tightly fitting cap mat or heat sealing system. The sealed plate will undergo vigorous mixing for as long as 15 min; if the cap mat does not seal tightly, leakage may occur and cross-contamination between sample wells may result. Two sealing choices are preferred for LLE— a heat sealable film and a silicone cap mat.

Heat sealing of microplates is performed by placing a cover sheet of aluminum film onto the surface of the microplate and applying heat evenly for several seconds using a rectangular iron or a semi-automated benchtop sealer unit. The underside of the aluminum film is coated with polypropylene; upon heating a strong bond is formed between it and the polypropylene plate. **Important Note: A microplate having a raised rim around each well is important for heat sealing and preventing cross contamination between wells.** Details can be found in Chapter 4, Section 4.4.3.

After the LLE procedure is completed, the coated aluminum film needs to be punctured in order to allow pipettor tips or automated workstation tips to enter and remove the liquid. An efficient tool to perform this operation is a 96-well piercing plate (Orochem Technologies, Westmont, IL USA). This plate contains 96 solid tips on the underside of a plate lid; a single forceful press on the top of a sealed microplate rapidly pierces each well.

A preferred cap mat choice is one made of silicone that forms a tight fit in the wells of the microplate. The cap mats made from polypropylene or polyethylene are designed for storage purposes to protect the well contents from dust and the environment but do not seal as tightly as the silicone variety. Potential problems from solvent interaction with the underside of the cap mats are prevented by using PTFE coated silicone cap mats. Caps for microtubes generally seal well and it is more convenient to use these caps in strips of eight or twelve.

Note that removal of the cap mat (or seal) can introduce another source of contamination as drops of liquid on its underside can splash if removal is rapid and forceful. It is recommended to first centrifuge the plate after extraction in order to bring all the liquid down the sides of the wells and to the bottom.

Once a microplate and sealing system are chosen, it is advisable to test how well this combination of plate and seal retains the liquid inside the wells, when subjected to the mixing conditions chosen. Cross-contamination can be evaluated using the technique described in Chapter 4, Section 4.4.3. If the test reveals no contamination into adjacent wells, then that particular combination of plate, sealing mat and solvent, when subjected to the same mixing conditions, is likely a good match.

8.2.4 Tips and Tricks

8.2.4.1 Plate Selection

Two microplates are commonly used in the liquid-liquid extraction procedure. One is the extraction plate and the other is the transfer plate which ultimately becomes the sample injection plate. If a back extraction is performed then a third plate is required. The plate format chosen can be important for the success of the extraction. Rather than have multiple deep well plate varieties available in the laboratory, it may be wise to standardize on just one plate format to eliminate the possibility for selecting the incorrect plate. The V-bottom is attractive because it allows adequate mixing. However, more importantly, the conical bottom permits good concentration of analyte on dry-down and also allows deeper penetration of the autoinjector needle. Small reconstitution volumes can then be effectively utilized. Vendors offer a wide selection of glass vials that are contained in the microplate format. Frequently, glass is preferable to plastic for extractions with certain analyte chemistries. The usefulness of glass lined wells or glass vial inserts as a collection plate should not be overlooked.

8.2.4.2 Pipetting Organic Solvents

While automated liquid handling workstations are favored for removing the analyst time commitment to the extraction, there may be preferences or situations when manual pipetting of organic solvents is desirable. With practice, organic solvents can be effectively transferred using 8-channel pipettors. It is most useful to aspirate a small air gap after aspirating the organic solvent to prevent dripping from the tips due to low surface tension and solvent volatility. Priming the pipettor by aspirating and dispensing several volumes also aids in elimination of the tip drip. These suggestions are appropriate for both manual and automated pipetting tasks. It is very important to contain and vent the solvent vapors when a solvent reservoir is placed into the working laboratory environment.

8.2.4.3 Clamping the Extraction Plate Prior to Mixing

Once the extraction microplate is sealed and ready to undergo vigorous mixing, a practical aid to ensure tight sealing is to add an aluminum tile to the top of the plate and tightly clamp it down with a U-clamp or equivalent. A similar approach is achieved by using a reaction block holder as a securing chamber for a sealed microplate, as described in Chapter 4, Section 4.7 (Figure 4.15). This clamped microplate combination is then inserted into a multi-tube vortex unit or equivalent.

8.2.4.4 Repeating the Extraction

The intensity of mixing within the extraction well is an important variable that contributes to the success of the technique. Inadequate mixing can be compensated somewhat by a longer duration of mixing. The influence of mixing intensity and duration can be tempered by repeating the extraction a second time in a routine manner. For example, 100 µL sample is extracted with 500 µL MTBE; 400 µL of the extracted solvent is transferred to a second microplate. The remaining aqueous sample is re-extracted with another 500 µL aliquot of MTBE. Again, 400 µL extracted solvent is transferred and added to the first isolated organic solvent; the total 800 µL volume is then evaporated to dryness as the method specifies. Reports have confirmed the improved efficiency of this two step extraction approach [15, 16].

8.2.4.5 Conversion from Tubes to Plates

Conversion of a tube method to a microplate method may lead to lower than expected recovery or performance. In this case, method troubleshooting is required to determine the cause, usually a result of an inadequate ratio of organic solvent to sample. However, issues also arise that are caused by factors such as the mechanics of mixing, transfer loss or analyte volatility on dry-down.

When a method does not convert from the original tube method to one using a microplate, the actual causes for failure are not always obvious. A suggestion in troubleshooting a method conversion is to perform the microplate method in tubes using the exact same volumes as used for the microplate method. When the tube method in small volumes works, but the microplate method does not, then well geometry or mixing mechanics may be at fault, rather than the chemistry of the extraction. When the small volume tube method also fails, perhaps the fundamentals should be examined such as the volume ratio of organic solvent to sample.

8.2.5 Applications

Many laboratories have demonstrated the capabilities for performing high throughput LLE in the microplate format using manual and semi-automated techniques. Selected applications are listed in Table 8.3 along with pertinent information about the sample preparation schemes. The volume ratios of organic to aqueous in the microplate or microtube wells for these applications range from 0.9:1 to 3.6:1. Recall that when LLE is performed in tubes, ratios range from 2.5:1 to 12.7:1 (Table 8.1). A smaller ratio is used in microplates because of the smaller well volume capacity.

Rapid methods of analysis have been investigated to provide fast sample turnaround for bioanalytical programs, especially those that support the drug discovery process in pharmaceutical research. Semi-automated liquid-liquid extraction methods provide efficient and reliable sample preparation toward this goal. Several pertinent reports are described next.

Table 8.3.
Selected bioanalytical applications using liquid-liquid extraction in the microplate format with automation

Analytes and Extraction Format	Sample Matrix and Volumes	pH Modifier and Volumes	Extraction Solvent and Volumes	Ratio (Org:Aq)	Automation	Analytical Method	Reference
Benzodiazepines; deep well plate	Human urine 0.5mL	$(NH_4)_2CO_3$ 0.1M, 50µL	Chloroform 0.5mL	0.9:1	Quadra® 96	LC-MS/MS	[17]
Proprietary; microtubes	Human plasma 0.25mL, IS 25µL	0.5N HCl 100µL	MTBE 0.5mL	1.3:1	MultiPROBE® II	LC-MS/MS	[18]
Proprietary; microtubes	Human plasma 0.1mL, IS 50µL	0.1M HCl 100µL	MTBE 0.6mL	2.4:1	MultiPROBE II	LC-MS/MS	[19]
Methylphenidate; deep well plate	Human plasma 0.35mL, IS 50µL	1M $NaHCO_3$ pH 10, 50µL	Cyclohexane 1mL	2.0:1	Quadra 96	LC-MS/MS	[20]
Fluoxetine enantiomers; deep well plate	Human plasma 0.2mL, IS 5µL	NH_4OH 20µL	Ethyl acetate 0.8mL	3.6:1	Quadra 96	LC-MS/MS	[21]
Diphenhydramine, desipramine, chlorpheniramine and trimipramine; microtubes	Rat plasma 0.1mL, IS 25µL	0.1M KOH/K_2CO_3, pH 12, 100µL	Chloroform, MTBE, EtOAc or MTBE/ethanol (95:5); 0.8mL each	3.6:1	Quadra 96	LC-MS/MS API TOF MS	[22] [23]
Cyclosporin A and Everolimus	Human blood 0.3mL, IS 50µL	NH_4OH pH 9.5, 50µL	MTBE 0.75mL (twice)	1.9:1	Quadra 96	LC-MS/MS	[15]

Analytes and Extraction Format	Sample Matrix and Volumes	pH Modifier and Volumes	Extraction Solvent and Volumes	Ratio (Org:Aq)	Automation	Analytical Method	Reference
Reserpine; deep well plate	Mouse plasma 0.1mL, IS 50µL	NH$_4$OAc 1M pH 9.5, adj with NH$_4$OH 50µL	MTBE 0.5mL	2.5:1	96-Channel Personal Pipettor	LC-MS/MS	[24]
Six benzodiazepines; deep well plate	Human urine 0.4mL IS 100µL	0.1M Na$_2$CO$_3$ 50µL	Chloroform 0.5mL	0.9:1	Quadra 96	LC-MS/MS	[25]
Methotrexate and metabolite; deep well plate	Human plasma 0.2mL, spiking solution 20µL	(none)	Chloroform 0.5mL; after PPT	0.7:1	Quadra 96	SRM LC/MS	[26]
Dextromethorphan; microtubes	Human plasma 0.2mL, IS 20µL	1M Na$_2$CO$_3$ pH 10.5 20µL	Ethyl ether 0.6mL	2.5:1	MICROLAB® AT Plus 2 and Quadra 96	LC-MS/MS and pcSFC-MS/MS	[27] [28]
Tamoxifen, raloxifene, nafodixine and idoxifene; deep well plate	Human plasma 0.1mL, IS 50µL ACN 25µL	(none)	Hexane/isoamyl alcohol (96:4, v/v) 0.4mL	2.3:1	Quadra 96	SRM LC/MS	[29] [30] [31]

PPT, Protein Precipitation; EtOAc, Ethyl Acetate; MTBE, Methyl tert-Butyl Ether

Steinborner and Henion [26] reported a semi-automated LLE procedure (following protein precipitation using acetonitrile) for the quantitative analysis of the anticancer drug methotrexate (MTX) and its major metabolite from human plasma. The calibration curves for MTX and its metabolite were linear over the range 0.5 to 250 ng/mL and 0.75 to 100 ng/mL, respectively. Sample preparation throughput was reported as 4 sample plates (384 samples) processed in 90 min by one person. The analytical throughput was reported as 768 samples analyzed within 22 h (maximum 820 samples per 24 h), using a selected reaction monitoring (SRM) LC/MS method with a 1.2 min analysis time per sample.

Zweigenbaum and Henion [25] also reported the high throughput determination of six benzodiazepines in human urine using a selected reaction monitoring LC/MS method with semi-automated LLE sample preparation. When four autosamplers were interfaced to one chromatographic column and one tandem mass spectrometer, 1152 samples (twelve 96-well plates) were analyzed in less than 12 h. A custom designed electronic switching box synchronized the autosamplers with the mass spectrometer so that injections were made as soon as the mass spectrometer was ready to collect data. Multiple component separations were performed in less than 30 s; separations in less than 15 s have been reported as well [17].

Another SRM LC-MS application reported the determination of Selected Estrogen Receptor Modulators (SERM) in human plasma where more than 2000 samples were analyzed in 24 h [29]; LC separation of five analytes was performed in less than 30 s. The time for one person to prepare these 2000 plasma extractions using LLE in a semi-automated mode was reported as 7 h. A related application for rapid analysis for idoxifene (an SERM) and its metabolite, using semi-automated liquid-liquid extraction, cited an average analysis time of 23 s/sample. It took about 37 min to analyze one 96-well plate and a throughput of over 3700 samples per day was achieved [30].

The extraction of cyclosporine A from whole blood was performed using LLE in microplates. Sample throughput was documented as four 96-well plates prepared in less than 5.5 h [15]. With a run time of 3.5 min per sample, four plates (384 samples) could be prepared and the analysis completed within 28 h.

The throughput advantages of using a volatile, low viscosity mobile phase (supercritical CO_2) in packed column supercritical fluid chromatography mass spectrometry (pcSFC-MS/MS) were reported for the bioanalysis of dextromethorphan from plasma [28]. The low viscosity of supercritical CO_2

allows the use of higher flow rates than in typical aqueous mobile phases for LC-MS/MS, and the volatile nature of the phase permits the entire effluent to be directed into the mass spectrometer interface. The analysis time for a 96-well plate was reported as 10 min 12 s, using a multiplexed autosampler, a short column (2 x 10 mm), and a flow rate of 7.5 mL/min. Semi-automated LLE was used for the sample preparation step.

8.3 High Throughput Solid-Supported Techniques Using Flow-Through Microplates

8.3.1 Introduction

A variation of liquid-liquid extraction uses flow-through tubes or 96 well microplates filled with inert diatomaceous earth particles. This technique is referred to as solid-supported LLE. The diatomaceous earth particle is a white or cream colored, friable porous rock composed of the fossil remains of diatoms (small water plants with silica cell walls). These fossils build up on the ocean floors to form diatomite, and in some areas have become dry land or diatomaceous earth. This material is chemically inert, having a rough texture with high surface area and thus great adsorption potential [32]. One common brand name for diatomaceous earth is Hydromatrix® and it is known by other names such as Celite, Extube and Kieselguhr. Diatomaceous earth is available from numerous vendors (*e.g.*, Varian Inc., Harbor City, CA USA; Orochem Technologies; IST Ltd., Hengoed, United Kingdom, now an Argonaut Technologies Company).

When used in LLE applications, the high surface area of the diatomaceous earth particle facilitates efficient, emulsion-free interactions between the aqueous sample and the organic solvent. Essentially, the diatomaceous earth with its treated aqueous phase behaves as the aqueous phase of a traditional liquid-liquid extraction, yet it has the characteristics of a solid support. In addition to this application for LLE, analytical laboratories use diatomite for flash chromatography, for water removal from reaction mixtures, for purification of parallel synthesis reactions in combinatorial chemistry [33], and for an inert support material used in accelerated solvent extraction and supercritical fluid extraction [34]. Diatomaceous earth is also used for several scientific and industrial purposes. Some of these applications include its use as a filtering agent, an abrasive, a building material, and as insulation material used against heat, cold and sounds [32].

8.3.2 Procedure

The procedure for the use of solid-supported particles to perform liquid-liquid extraction (SS-LLE) follows. A mixture of sample (*e.g.*, plasma), internal standard, and buffer solution to adjust pH is prepared. This aqueous mixture is then added to the dry particle bed (no conditioning or pretreatment of the bed is necessary). The mixture is allowed to partition for about 3–5 min on the particle surface under gravity flow. The analyte in aqueous solution is now spread out among the particles and occupies a high surface area. A hydrophobic filter on the bottom of each well prevents the aqueous phase from breaking through into the collection vessel placed underneath. Organic solvent is then added to the wells. The analyte, under appropriate conditions of pH, will partition from the adsorbed aqueous phase into the organic solvent as this solvent slowly flows through the particle bed under gravity. The organic eluate is collected in wells or tubes underneath the tips of the plate (Figure 8.3). A second addition of organic solvent is performed and the combined eluates are evaporated to dryness. The residue is reconstituted in a mobile phase compatible solution and an aliquot is injected into the chromatographic system for analysis.

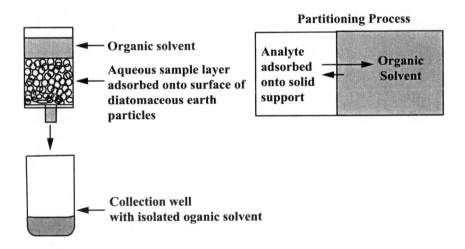

Figure 8.3. Schematic diagram of the solid-supported liquid-liquid extraction technique using diatomaceous earth particles in a flow-through column or well. Adapted from [35] and reprinted with permission. Copyright 2001 American Chemical Society.

Solutions generally pass through the particle bed using gravity flow. This slow flow rate maximizes the time for interaction and partitioning of analyte from the aqueous into the organic phase, so little or no vacuum is necessary. However, depending on the SS-LLE product configuration, low vacuum (about 3–5 in Hg; 100–170 mbar) may sometimes be necessary. When the diatomaceous earth is packed in wells *without* a top frit, gravity flow is usually adequate. However, when a top frit is in place (such as a polyethylene or PTFE frit or mesh screen), resistance to gravity flow may develop at the hydrophobic frit surface. In this case, a small burst of vacuum (~5 in Hg or 170 mbar) or positive pressure may be necessary during the load step to bring the sample through the frit and into the dry particle bed.

8.3.3 Advantages

One advantage of using solid-supported LLE is that solutions are sequentially added to the tube or wells, eliminating the need to vortex mix. The procedure is now more automatable since each step is simply a liquid transfer or addition. The many capping and mixing steps as used for traditional LLE are unnecessary and less hands-on time is required from the analyst. There is no need to condition or wash the sorbent bed as with solid-phase extraction. Also, the potential formation of emulsions is eliminated using this technique. A special note with the use of these particle beds is that the aqueous phase is not retrievable.

8.3.4 Disadvantages

The diatomaceous earth particles used in SS-LLE procedures demonstrate a capacity limit for adsorption of aqueous sample matrix. It is possible to overload the particle with too much sample volume; the end result of overloading the column is poor recovery. Columns for SS-LLE are identified by manufacturers as having a sample capacity volume of, *e.g.*, 0.3 mL, 1 mL, 3 mL, 5 mL or 10 mL. Large volumes of organic solvent are sometimes required to exchange and elute adsorbed analytes from the particle bed. If insufficient organic volume is used, poor recovery may result.

Method transfer from an established LLE procedure to one utilizing SS-LLE is not always straightforward. An investment in time is often required in order to optimize the method in terms of sample volume loaded onto the particle bed, the mass of particle in each well or column, the volume of organic solvent required and/or the mechanics of collecting and working with the isolated organic solvent. There have also been isolated reports of mass spectrometer fouling with the use of particle sources that are presumably not sufficiently cleaned by the vendor at the column filling stage. Filled tubes and microplate products are sealed by the vendor when shipped to avoid adsorption of potential contaminants from the atmosphere before use.

8.3.5 Product Configurations

8.3.5.1 Columns or Cartridges

The common product configurations for diatomaceous earth include polypropylene syringe style tubes (known as columns or cartridges) in various sizes, as well as flow-through microplates. Bulk material is also available for those analysts who pack their own extraction tubes or plates. A reported application [35] used a hydrophobic glass fiber filter plate (GF/C, Whatman) as the base plate and packed diatomaceous earth particles into the plate wells (~260 mg/well).

Solid-supported LLE columns (*e.g.*, ChemElut™, Extube™ ChemElut, Varian; ISOLUTE® HM-N, IST Ltd., now Argonaut Technologies) are often used with a rack that allows gravity flow of liquids through the particle bed. The racks display enough clearance underneath for a set of eluate collection tubes. The individual SS-LLE columns are classified and ordered in terms of sample capacity, *e.g.*, 0.3 mL, 1 mL, 3 mL, 5 mL, 10 mL and larger). Elution volumes for these tube sizes are recommended as 3, 8, 12, 16 and 24 mL, respectively. Diatomaceous earth in tubes is also available prefilled with a buffer; Varian's product is named ToxElut™. Two buffer choices are supplied; pH 4.5 is used for the extraction of acidic analytes and pH 9.0 is used for the extraction of basic analytes.

8.3.5.2 Microplates

The 96-well plate configuration containing diatomaceous earth (Figure 8.4) is available in both the single molded plate and modular plate formats. Details about these plate varieties are found in Chapter 3, Section 3.6. Typical examples of these diatomaceous earth plates are listed in Table 8.4. While the information in this table is meant to be instructive, it should not be considered exhaustive (some vendors did not respond to requests for information about their products). Individual suppliers should always be contacted to obtain more complete and current information about their SS-LLE products. Note that many vendors will also custom fill plates to customer specifications.

Table 8.4
Typical examples of diatomaceous earth filled microplates for solid-supported liquid-
liquid extraction

Plate Format	Vendor	Brand Name	Sorbent Bed Mass†	Top Frit	Well Geometry
One piece molded plate	Varian	Combilute™	200 mg	No*	Square, 2 mL
	Orochem Technologies	Hydromatrix	260 mg	Yes	Square, 2 mL
		Hydromatrix	260 mg	No*	Square, 2 mL
	Macherey-Nagel	Chromabond® Multi96 XTR	150 mg	Yes	Round, 1.2 mL
Modular Plate	Varian	VersaPlate™ Hydromatrix	260 mg	Yes	Round, 1.8 mL
	IST Ltd. (now Argonaut)	ISOLUTE® HM-N Array®	200 mg	Yes	Square, 2 mL
	Orochem Technologies	Hydromatrix	260 mg	Yes	Square, 2 mL

†Contact vendor for a custom sorbent bed mass
*Shipped with a sealing cap mat

The general recommendation from manufacturers of these products that contain 200–260 mg mass per well is that no more than 200 µL total sample volume (sample, internal standard and aqueous buffer) should be loaded. Otherwise, the capacity of the particle bed can be exceeded. Regarding the elution volume, two successive aliquots of 0.5 mL each are generally advised by vendors. When sample volumes >200 µL must be loaded, a plate containing a larger mass of particles can be used. Plates are available as 500 mg in a 96-well plate or 1000 mg in a 48-well plate (*e.g.*, Horizon Instrument, King of Prussia, PA and Chemical Separation Corporation, Phoenixville, PA USA).

Careful consideration must be made regarding the procedure for the collection and evaporation of the organic eluate. When a 260 mg bed mass of particle is used in a plate, two successive aliquots of 0.5 mL are generally recommended

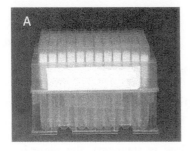
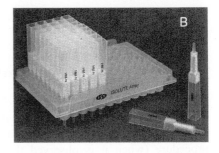

Figure 8.4. Typical examples of a flow-through microplate containing diatomaceous earth particles packed within wells. Plate varieties include (A) single molded plate and (B) modular plate. Photos reprinted with permission from (A) Horizon Instrument and (B) IST Ltd., now an Argonaut Technologies Company.

for elution. Note that the situation may arise, in certain instances, when 4 aliquots of 0.5 mL are necessary for complete elution of all analytes from the particle bed due to gradual release. The largest volume collection plate that can fit inside a vacuum manifold, and which fits in evaporators and autosamplers, is the 2 mL deep well plate. If this plate is filled with 2 mL organic liquid, there is not enough headspace to place it in an evaporator unit where the needles are lowered to deliver nitrogen. It may be practical to elute the first 1 mL aliquot, dry down that volume, and then repeat the elution with the second aliquot, again followed by a dry-down step from the same wells. In the case when many successive aliquots are required from a sorbent bed, *e.g.*, 10 aliquots of 0.5 mL solvent, a 48-well SS-LLE plate can be used [36]. In this case, the plate sits on top of a 48-well collection plate (5 mL well volume) and gravity flow is used for liquid processing.

8.3.6 Accessory Products

8.3.6.1 Introduction

A list of the accessory products needed to perform solid-supported LLE in the microplate format is identical to those listed in Table 8.2, with the addition of two items. A vacuum manifold is used to process liquid through the plate (in the event that gravity flow is not sufficient) and an evaporation unit is needed to dry down the organic eluate collected from the columns. Details about these two items are found in Chapter 4, Sections 4.2.1 and 4.5, respectively.

8.3.6.2 Vacuum Manifold with Collection Plate

A solid-supported LLE plate can be conveniently placed on a vacuum manifold and a deep well 2 mL microplate is inserted to collect the organic eluate. Gravity is used for flow processing, but if vacuum is needed during the procedure (when it is observed that flow is stagnant), the plate is already configured. While an SS-LLE plate could be placed directly on top of a collection plate on a lab bench for manual pipetting functions, it is preferred to place it on a vacuum manifold to more easily allow automated pipetting functions. Manifolds serve as a secure containment rack for the microplate, permit eluate collection underneath, and are already defined in the workstation software.

8.3.6.3 Liquid Transfers

Before loading the sample mixture onto the extraction plate and adding organic solvent, samples first must be mixed with internal standard (IS) and a pH adjustment solution (*e.g.*, a defined buffer or a strong acidic or basic solution) to make the analyte neutral. Depending on the equipment available in the laboratory, liquid transfers can be accomplished in several ways. Single channel and repeater pipettors can be used to manually aliquot sample, IS and buffer into test tubes or into a microplate (deep well collection plate or microtube rack). The tubes or sample plate undergo a brief vortex mix step. A single channel pipettor or an articulating multichannel pipettor can be used to manually transfer samples from tubes into the SS-LLE plate; if samples are in a microplate, manual transfer is best accomplished with a multichannel pipettor.

The organic solvent can be transferred to the extraction plate using a repeater or a multichannel pipettor. An automated approach for liquid delivery is the use of a single or multiple probe liquid handling workstation to aliquot sample, IS and buffer into microplate wells, mix the contents and then transfer samples directly into the SS-LLE plate configured on a vacuum manifold. Organic solvent can be delivered from a solvent trough directly into the SS-LLE plate. Also, a 96-tip workstation can transfer in one step the contents of all the sample wells into the SS-LLE plate (configured on a vacuum manifold). In a second step, organic solvent is simultaneously delivered to all the wells of the SS-LLE plate. An overview of these liquid handling systems can be found in Chapter 5 and automation strategies for LLE are discussed further in Chapter 10.

8.3.7 Tips and Tricks

8.3.7.1 Choosing the SS-LLE Plate Format

Many varieties of diatomaceous earth plates are available, distinguished by sorbent mass loading, the presence or absence of a top frit and plate format of square or round wells. Particles may be specially treated or their performance is enhanced by some manufacturers and not others. With the presence of these variables, it has been noticed by users of this technique that some plate brands/formats perform better than others for the same method. When an SS-LLE plate is evaluated and the results are less than satisfactory, the chemistry of the extraction process is the first item to troubleshoot. Sorbent bed mass can be increased to aid in the performance, but keep in mind that larger elution volumes are required. However, do not overlook the influence of plate geometry and manufacturer product specifications. If a square well plate is evaluated and initially fails, try a round well plate from another manufacturer, or vice versa.

8.3.7.2 Securing the Collection Plate

The SS-LLE plate can be mated directly with a collection plate and placed on the lab bench. Liquids flow via gravity through the sorbent bed and collect in the plate wells underneath. However, this plate combination is not secure and may be accidentally bumped in the normal course of multiple users working at a laboratory bench. A suggestion for a more secure footing for the collection plate is to put it inside a vacuum manifold. Then, if necessary, gravity flow can be augmented by the use of a short burst of vacuum at ~3–5 in Hg (100–170 mbar) or pulsing of the vacuum.

8.3.7.3 Exhaustive Elution

The use of SS-LLE plates, with certain sample preparation methods, may require organic solvent volumes larger than 1.5 mL for elution. If only 0.75 mL is used, and incomplete recovery results, it may be that analyte is still retained on the particle bed. When evaluating an SS-LLE method for the first time, it is advised to perform at least three successive elutions to ensure that analyte has been completely removed from the sorbent bed. Since a deep well collection plate only holds a maximum volume of 2 mL, and some air space at the top of the well is needed for a dry-down step, fill the collection plate only to about 1.5 mL with a total elution volume. Label 3 plates for elution as #1, #2 and #3 and successively elute from the sorbent bed into these three plates, one after the other. Dry down the solvent from each plate, reconstitute and inject an aliquot into the chromatographic system. The amount or percentage of analyte recovered in elution fractions #1, #2 and #3 can be determined and the efficiency of elution calculated.

8.3.8 Applications

Most reported bioanalytical applications using solid-supported liquid-liquid extraction products have used the cartridge or tube format. Some of these applications include the determination of: mexiletine [37], amiodarone [38] and other antiarrhythmic drugs [39], organophosphate pesticide residues [40], proxyphylline [41], 16β-hydroxystanozolol [42], dextromethorphan [43], ondansetron, buproprion, doxorubicin and diclofenac [44] and simvastatin [45] from biological fluids. Microplate applications for SS-LLE include three reports [33, 35, 36]. Selected bioanalytical applications using solid-supported liquid-liquid extraction are listed in Table 8.5. The overall number of applications using SS-LLE is expected to increase as more vendors offer this format and the clear advantages that automation provides for this technique become better known.

Three sample preparation methods for the determination of a ß3-agonist (a potential anti-obesity agent) in human plasma were developed and compared using traditional liquid-liquid extraction, individual 1 mL ChemElut™ cartridges, and a 48-well diatomaceous earth plate [36]. In all three procedures, 0.5 mL human plasma was extracted; IS, spiking solution and 1M sodium carbonate were premixed with plasma before application. MTBE was used as the immiscible extraction solvent.

In the original LLE procedure, 6 mL MTBE was used and the volume ratio of MTBE to aqueous was 8:1 (v/v). The identical load volume was used for ChemElut cartridges and elution was performed with 2 x 4 mL MTBE (8 mL total). For the plates, the method was identical with the exception that the load volume was 0.6 mL total (instead of 0.75 mL); 0.5 mL plasma plus IS and basic solution. Using ChemElut cartridges, the assay concentration range was 0.5–100 ng/mL. Using diatomaceous earth plates, the range was extended to 0.5–200 ng/mL. For all three sample preparation methods, the statistics for precision and accuracy were comparable. Intraday precision ranged from 2.1–9.8%, 2.2–11% and 2.4–7.7% for LLE, tubes and plates, respectively. Vast differences in throughput were reported.

Table 8.5
Selected bioanalytical applications using solid-supported liquid-liquid extraction

Analytes and SS-LLE Format	Sample Matrix and Volumes	pH Modifier and Volume	Extraction Solvent and Volume	Reference
Protease inhibitors; 96-well plate	Rat plasma 0.1mL, IS 25μL	Formate buffer pH 3.5, 200μL	Methyl ethyl ketone 0.8mL	[35]
Topoisomerase I inhibitor; 96-well plate	Human plasma 0.25mL, IS 25μL	(None)	Methyl tert-butyl ether (MTBE)/-isopropyl alcohol (91:9, v/v), 2mL x 2 aliquots	[48]
ß3-Agonists; 48-well plate and 1mL tubes	Human plasma 0.5mL, IS 50μL	0.5M Sodium carbonate 50μL	MTBE 0.5mL x 10 aliquots	[36]
ß3-Agonists; 1mL tubes	Human plasma 0.5mL, IS 100μL	0.5M Sodium carbonate 100μL	MTBE 4mL x 2	[36]
Simvastatin lactone and acid; 1mL tubes	Human plasma 0.5mL, IS 50μL	0.1M Ammonium formate pH 4.5 300μL	MTBE 4mL x 3	[45]
SR 49059; 1mL tubes	Human plasma 1mL	(None)	Dichloro-methane/-hexane (1:1) 4mL x 2	[46]
1,25-Dihydroxy-vitamin D_3; 5mL tubes	Human plasma (0.5–2mL)	0.01M Na_2HPO_4 1mL	Dichloro-methane/-acetone (9:1) 30mL total	[47]

The total time required for traditional LLE was reported as three times longer than for the solid-supported plate method. The additional time required for tube labeling, single channel pipetting, capping of tubes and autosampler vials, and vortex mixing all contribute to a time consuming, labor intensive procedure. Throughput was reported as 9, 15 and 30 samples processed per hour for LLE, SS-LLE in tubes and SS-LLE in plates, respectively. Note that these three sample prep procedures (LLE, SS-LLE in tubes, SS-LLE in 48-well plates) were all performed without benefit of a liquid handling workstation; only multichannel pipetting units were used. Multiple probe and 96-tip liquid handlers can automate this entire extraction procedure using diatomaceous earth filled plates. Details are presented in Chapter 10.

Acknowledgments

The author is appreciative to Shane Needham and Ken Ruterbories for helpful discussions on microplate liquid-liquid extraction techniques that contributed to the content of this chapter. The author also thanks David Ehresman for his careful review of the manuscript and valuable suggestions. Line art illustrations were kindly provided by Woody Dells.

References

[1] R.E. Majors, LC-GC 14 (1996) 936-943.
[2] S. Pederson-Bjergaard, K.E. Rasmussen and T.G. Halvorsen, J. Chromatogr. A 902 (2000) 91-105.
[3] D.J. Peters, J.M. Hayes and G.M. Hieftje, A Brief Introduction to Modern Chemical Analysis, W.B. Saunders Co., Philadelphia (1976).
[4] H. Kataoka and H. Lord, In: J. Pawliszyn, Ed., Sampling and Sample Preparation for Field and Laboratory: Fundamentals and New Directions in Sample Preparation, Elsevier, Amsterdam (2002).
[5] O.H. Drummer, J. Chromatogr. B 713 (1998) 201-225.
[6] S.X. Peng,J. Chromatogr. B,764 (2001) 59-80
[7] O.H. Drummer, J. Chromatogr. B 733 (1999) 27-45.
[8] D.K. Lloyd, J. Chromatogr. A 735 (1996) 29-42.
[9] M. Jemal, R.B. Almond and D.S. Teitz, Rapid Commun. Mass Spectrom. 11 (1997) 1083-1088.
[10] K.A. Riffel, H. Song, X. Gu, K. Yan and M.W. Lo, J. Pharm. Biomed. Anal. 23 (2000) 607-616.
[11] C.S. Tamvakopoulos, L.F. Colwell, K. Barakat, J. Fenyk-Melody, P.R. Griffin, R. Nargund, B. Palucki, I. Sebbat, X. Shen and R.A. Stearns, Rapid Commun. Mass Spectrom. 14 (2000) 1729-1735.
[12] S. Chen and P.M. Carvey, Rapid Commun. Mass Spectrom. 15 (2001) 159-163.

[13] C.M. Chavez-Eng, M.L. Constanzer and B.K. Matuszewski, J. Chromatogr. B 748 (2000) 31-39.

[14] M. Aravagiri and S.R. Marder, J. Pharm. Biomed. Anal. 26 (2001) 301-311.

[15] N. Brignol, L.M. McMahon, S. Luo and F.L.S. Tse, Rapid Commun. Mass Spectrom. 15 (2001) 898-907.

[16] A. Staab, S. Scheithauer, H. Fieger-Buschges, E. Mutschler and H. Blume, J. Chromatogr. B 751 (2001) 221-228.

[17] K. Heinig and J. Henion, J. Chromatogr. B 732 (1999) 445-458.

[18] D.S. Teitz, S. Khan, M.L. Powell and M. Jemal, J. Biochem. Biophys Meth. 45 (2000) 193-204.

[19] M. Jemal, D. Teitz, Z. Ouyang and S. Khan, J. Chromatogr. B 732 (1999) 501-508.

[20] L. Ramos, R Bakhtiar and F.L.S. Tse, Rapid Commun. Mass Spectrom. 14 (2000) 740-745.

[21] Z. Shen, S. Wang and R. Bakhtiar, Rapid Commun. Mass Spectrom. 16 (2002) 332-338.

[22] N. Zhang, K.L. Hoffman, W. Li and D.T Rossi, J. Pharm. Biomed. Anal. 22 (2000) 131-138.

[23] N. Zhang, S.T. Fountain, H. Bi and D.T. Rossi, Anal. Chem. 72 (2000) 800-806.

[24] J. Ke, M. Yancey, S. Zhang, S. Lowes and J. Henion, J. Chromatogr. B 742 (2000) 369-380.

[25] J. Zweigenbaum, K. Heinig, S. Steinborner, T. Wachs and J. Henion, Anal. Chem. 71 (1999) 2294-2300.

[26] S. Steinborner and J. Henion, Anal. Chem. 71 (1999) 2340-2345.

[27] R.D. Bolden, S.H. Hoke II, T.H. Eichhold, D.L. McCauley-Myers and K.R. Wehmeyer, J. Chromatogr. B 772 (2002) 1-10.

[28] S.H. Hoke II, J.A. Tomlinson II, R.D. Bolden, K.L. Morand, J.D. Pinkston and K.R. Wehmeyer, Anal. Chem. 73 (2001) 3083-3088.

[29] J. Zweigenbaum and J. Henion, Anal. Chem. 72 (2000) 2446-2454.

[30] J.M. Onorato, J.D. Henion, P.M. Lefebvre and J.P. Kiplinger, Anal. Chem. 73 (2001) 119-125.

[31] H. Zhang and J. Henion, J. Chromatogr. B 757 (2001) 151-159.

[32] The Carnegie Library of Pittsburgh, Minerals and Other Materials, In: The Handy Science Answer Book, Visible Ink Press, Detroit (1994).

[33] S.X. Peng, C. Henson, M.J. Strojnowski, A. Golebiowski and S.R. Klopfenstein, Anal. Chem. 72 (2000) 261-266.

[34] M.L. Hopper and J.W. King, J. Assoc. Off. Anal. Chem. 74 (1991) 661-666.

[35] S.X. Peng, T.M. Branch and S.L. King, Anal. Chem. 73 (2001) 708-714.

[36] A.Q. Wang, A.L. Fisher, J. Hsieh, A.M. Cairns, J.D. Rogers and D.G. Musson, J. Pharm. Biomed. Anal. 26 (2001) 357-365.

[37] F. Susanto, S. Humfeld and H. Reinauer, Chromatographia 21 (1986) 41-43.

[38] T.A. Plomp, M. Engels, E.O. Robles de Medina and R.A. Maes, J. Chromatogr. 273 (1983) 379-392.

[39] J.F. Wesley and F.D. Lasky, Clin. Biochem. 15 (1982) 284-290.

[40] M.L. Hopper, J. Assoc. Off. Anal. Chem. 71 (1988) 731-734.

[41] M. Ruud-Christensen, J. Chromatogr. 491 (1989) 355-366.

[42] M. Van de Wiele, K. De Wasch, J. Vercammen, D. Courtheyn, H. De Brabander and S. Impens, J. Chromatogr. A 904 (2000) 203-209.

[43] E. Bendriss, N. Markoglou and I.W. Wainer, J. Chromatogr. B 754 (2001) 209-215.

[44] J.R. Allianti, C.M. Grosse and G.A. Smith, Proceedings International Symposium on Laboratory Automation and Robotics, Boston, MA USA (1998).

[45] J.J. Zhao, I.H. Xie, A.Y. Yang , B.A. Roadcap and J.D. Rogers, J. Mass Spectrom. 35 (2000) 1133-1143.

[46] R. Burton, M. Mummert, J. Newton, R. Brouard and D. Wu, J. Pharm. Biomed. Anal. 15 (1997) 1913-1922.

[47] N. Kobayashi, T. Imazu, J. Kitahori, H. Mano and K. Shimada, Anal. Biochem. 244 (1997) 374-383.

[48] A.Q. Wang, W. Zeng, D.G. Musson, J.D. Rogers and A.L. Fisher, Rapid Commun. Mass Spectrom. 16 (2002) 975-981.

Chapter 9

Liquid-Liquid Extraction: Strategies for Method Development and Optimization

Abstract

The fundamentals of liquid-liquid extraction (LLE) were presented in Chapter 8 along with general protocols for using the technique in collection microplates and flow-through solid-supported plates. An important step toward achieving success with LLE is learning how to develop new methods and improve existing ones by applying the fundamentals of extraction chemistry. Two of the most important variables influencing the recovery of an analyte from a biological sample matrix are the sample pH and the choice of organic solvent. These fundamental concepts are discussed and examples of their use are given. Strategies are introduced for developing methods rapidly in microplates and for improving existing methods using optimization techniques. The ultimate goal in developing and using 96-well LLE methods is to achieve high throughput with time and labor savings. Efficient use of automation is the key to achieving this goal. Automation strategies for performing the method development approaches presented here are discussed in Chapter 10.

9.1 Extraction Chemistry: Importance of Sample pH

9.1.1 Understanding the Influence of Ionization on the Partitioning Process

The choice of successful extraction conditions for liquid-liquid extraction (LLE) results in preferential partitioning of analytes from the aqueous (sample) phase into the organic phase. Two very important considerations in selecting appropriate extraction conditions are (1) sample pH and (2) characteristics of the organic solvent. Analytes that are unionized (neutral) will preferentially extract into an organic solvent when it is soluble in that particular solvent. Ideally, the solvent should exhibit sufficient selectivity to exclude potentially interfering substances from the sample matrix. Analytes that are neutral at the sample pH usually do not require strong buffering to remain neutral in preparation for LLE.

307

When an analyte is ionized, it will **not** extract into a nonpolar organic solvent and remains in the aqueous phase. In some cases, this information can be used to preferentially exclude unwanted interfering substances by making them ionized, ideally leaving the analyte neutral so only it partitions into the organic phase.

Analytes that are acidic or basic require pH adjustment in order to become neutral for the extraction to occur. Acidic analytes, *e.g.*, carboxylic acids, require the addition of acid or an acidic buffer to the aqueous sample to promote their equilibrium to the unionized state. The pH for an acid should be adjusted to a value that is two units less than its pKa. Likewise, basic analytes, *e.g.*, amines, require the addition of base or a basic buffer to the aqueous sample to promote their equilibrium to the unionized state. The pH of a base should be adjusted to a value that is two units greater than the pKa of the analyte.

The successful use of LLE relies on the proper utilization of pH during the extraction process. The unionized (neutral) form will be extracted, and the ionized form is left behind in the aqueous phase. Reprinted from Chapter 2, Section 2.3.1 is a summary of the important pH adjustment guidelines to promote the ionization of acids and bases to their fully ionized or unionized species.

**For > 99% conversion the pH should be adjusted 2 pH units
above or below the pKa value of the analyte**

Acids	pH < pKa = R–COOH	unionized
	pH > pKa = R–COO⁻	ionized
Bases	pH < pKa = R–NH3+	ionized
	pH > pKa = R–NH₂	unionized

An example of the importance of pH adjustment is shown in Table 9.1. A series of eleven proprietary discovery compounds containing various amine functional groups was extracted using methyl tert-butyl ether (MTBE), with and without the addition of a basic 0.1M Na_2CO_3 solution (pH ~10.5). Raising the pH of the sample formed more unionized species and assisted in the partitioning of analytes into the organic phase. However, when the pH was neutral a certain percentage of the amines were positively charged and did not fully extract into the nonpolar organic phase. The extent of ionization can be determined by the Henderson-Hasselbach equation (Chapter 2, Section 2.3.1).

Table 9.1
Extraction recoveries for a series of 11 discovery compounds comparing the effect
of pH adjustment

	Extraction Recovery (%)										
	A	B	C	D	E	F	G	H	I	J	K
Water	64	70	71	65	76	71	68	65	64	55	76
0.1M Na$_2$CO$_3$	86	89	90	100	96	94	79	94	84	99	89

Extractions contained 100 μL plasma, 50 μL IS and 100 μL 0.1M Na$_2$CO$_3$ or water;
500 μL methyl tert-butyl ether was the extracting solvent.

9.1.2 Further Manipulation of Sample pH via Back Extraction

Note that it is possible to perform a back extraction or re-extraction of the
organic solvent once analytes have partitioned into it. This procedure is
sometimes performed when coeluting interferences cross over into the organic
phase with the analyte. A basic analyte will be back extracted from the organic
to the aqueous phase when the aqueous environment is acidic (analyte becomes
ionized and partitions into the aqueous phase); neutral species remain in the
organic phase. Likewise, an acidic analyte will be back extracted from the
organic to the aqueous phase when the aqueous environment is alkaline; neutral
species remain in the organic phase. This back extraction allows the analytes to
reenter the aqueous phase. Prior to analysis, the usual organic solvent
evaporation and reconstitution steps are unnecessary when the aqueous phase
chosen is compatible with mobile phase for direct injection.

Two selected examples from the literature that employed the back extraction
technique are now described. The basic analyte dextromethorphan (**1**) was
extracted from plasma (1.5 mL) using 6 mL heptane/ethyl acetate (1:1, v/v).
Back extraction was performed into 100 μL of an aqueous acidic buffer (25mM
phosphate buffer, pH 2.6) [1]. Analysis was by capillary zone electrophoresis.

1

A proprietary discovery compound (Figure 9.1), a basic analyte, was extracted from human plasma (0.5 mL) using 1 mL ethyl acetate. Sample pretreatment involved the addition of 50 μL IS and 250 μL 2M NaOH. Samples were vortex mixed for 5 min and MTBE (3 mL) was added to this mixture; sample rotation for 15 min at 60 rpm was followed by centrifugation at 2060 x *g*. The isolated organic layer was back extracted using 150 μL of a 2% formic acid solution. Following vortex mixing and centrifuging, the aqueous layer was frozen in acetone containing dry ice; the organic layer was discarded. The aqueous layer quickly thawed, samples were transferred into autosampler vials, and aliquots were analyzed by LC-MS/MS [2].

Note that the back extraction process can also be carried further in one or more additional extraction steps for even greater analyte purity. Once the analyte has been back extracted into the aqueous phase, this solution can be pH adjusted in the opposite direction and again extracted with organic solvent. For greater selectivity, a different organic solvent can be chosen for the second extraction step. An example of this application is reported for the analysis of reserpine from plasma with LC separation and fluorescence detection [3]. LLE of reserpine from plasma (3 mL) was performed at basic pH by adjustment with 1.0 mL carbonate buffer (0.6M, pH 9.5) using 1.5% isoamyl alcohol in n-heptane (10 mL). The organic layer was isolated and then back extracted using 0.1M HCl (1.2 mL). The aqueous layer was made basic using 0.5 mL of the carbonate buffer and extracted with MTBE (0.5 mL). The dried and reconstituted extracts were oxidized to form a fluorophor for fluorescence detection. The lower limit of quantitation (LLOQ) achieved was 0.3 ng/mL using 3 mL plasma. This multiple step procedure was also adopted for analytical determination by LC-MS/MS where the LLOQ was lowered to 0.05 ng/mL using 2 mL plasma [4].

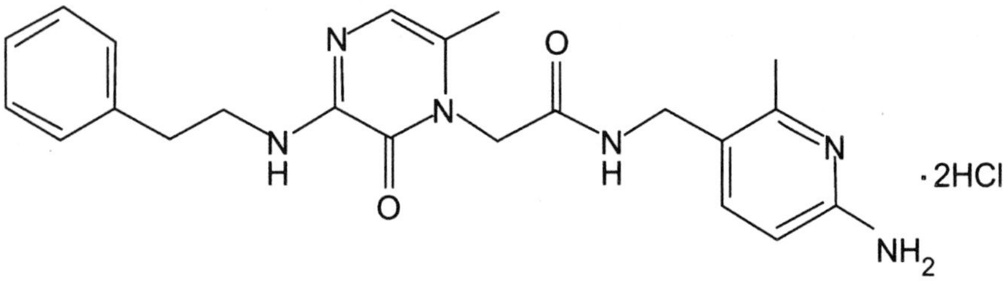

Figure 9.1. Chemical structure for a potent orally active thrombin inhibitor that was extracted under basic conditions. Reprinted with permission from [2]. Copyright 2000 Elsevier Science.

A variation of the traditional back extraction procedure is to evaporate the extracted organic solvent, reconstitute in a different organic solvent, back extract that second solvent and analyze the isolated aqueous portion. This approach is described in the determination of the ergot alkaloid cabergoline from plasma using LC-MS/MS [5]. Plasma (2 mL) and a pH 10 borate buffer (2 mL) were mixed and extracted with ethyl acetate (8 mL). The organic layer was removed following centrifugation and the extraction repeated. The combined ethyl acetate portions were evaporated to dryness. The residue was dissolved in dichloromethane (0.35 mL), 1 mL of 1% acetic acid was added, and the solution was mixed and centrifuged. The aqueous portion was isolated and analyzed. The limit of quantitation was ~60 pg/mL using a large sample volume introduction technique (analyte focusing) into the mass spectrometer.

9.1.3 Assisting the Extraction of Hydrophilic Analytes

Very hydrophilic analytes may not yield acceptable recoveries when they present a low partition coefficient under commonly used extraction conditions. In this instance, the performance of the LLE procedure may be improved by using a very high ratio of organic solvent to sample volume. Alternately, the addition of an ion pairing agent (*e.g.*, tetrabutyl ammonium bromide) assists the extraction process by masking the charge on an analyte, making it neutral overall while also introducing some nonpolar character. Another method is to add a saturated salt solution (*e.g.*, sodium sulfate) to the aqueous phase to decrease the concentration and solubility of analytes in the sample, promoting partitioning into the organic phase. An example of this salting out approach is reported in the extraction of cimetidine from human plasma (0.25 mL) using ethyl acetate (1 mL). Sample pretreatment involved the addition of 2.5M NaOH (20 µL) and 100 µL of a saturated solution of potassium carbonate [6].

9.2 Organic Solvent Selection: Influence of Polarity, Volatility, Selectivity and Density

9.2.1 Solvent Polarity

Organic solvents used for liquid-liquid extraction vary greatly in terms of their polarity and some solvents demonstrate a small miscibility with water. These physical characteristics are known values. However, the solubility of organic solvents for specific analytes must be predicted and the solvent evaluated for analyte recovery in an effort to find an optimal choice. The important parameters that must be considered when selecting an appropriate solvent for an extraction method include polarity, volatility, selectivity and density.

A Polarity Index (Table 9.2) is a useful guide for selecting organic solvents for use in liquid-liquid extraction procedures. This index [7] lists solvents in order of polarity from nonpolar (*e.g.*, pentane; Polarity Index = 0) to polar (*e.g.*, water; Polarity Index = 10). The Polarity Index provides a relative ranking that can be used to predict the solubility of analytes based on their own polar or nonpolar functional group chemistry. For example, analytes that have more polar overall character as a result of hydrogen bonding may be extracted best with a more polar solvent such as methyl ethyl ketone, while analytes with predominantly nonpolar character (having many aromatic groups and hydrocarbon chains) may extract better using toluene. Note that some solvents are not appropriate for LLE since they have appreciable solubility and miscibility with water and do not form a two phase liquid boundary. Examples

Table 9.2
Polarity index of organic solvents, listed in ascending order

Solvent	Index	Solvent	Index
Pentane	0.0	Isobutyl alcohol	4.0
Trichlorofluoroethane	0.0	2-Methoxyethyl acetate	4.0
Cyclopentane	0.1	Methyl isoamyl ketone	4.0
Heptane	0.1	n-Propanol	4.0
Iso-hexanes	0.1	Tetrahydrofuran (THF)	4.0
Petroleum ether	0.1	Chloroform	4.1
Trimethylpentane	0.1	Methyl isobutyl ketone	4.2
Cyclohexane	0.2	Ethyl acetate	4.4
Hexadecane	0.5	Methyl n-propyl ketone	4.5
n-Butyl chloride	1.0	Methyl ethyl ketone (MEK)	4.7
Trichloroethylene	1.0	Dioxane	4.8
Carbon tetrachloride	1.6	2-Ethoxyethanol	5.0
Toluene	2.4	beta-Phenethylamine	5.0
Methyl t-butyl ether (MTBE)	2.5	Acetone	5.1
ortho-Xylene	2.5	Methanol	5.1
Benzene	2.7	Pyridine	5.3
Chlorobenzene	2.7	Diethyl carbonate	5.5
ortho-Dichlorobenzene	2.7	2-Methoxyethanol	5.5
Ethyl ether	2.8	Acetonitrile	5.8
Methylene chloride	3.1	Propylene carbonate	6.1
Ethylene dichloride	3.5	Dimethyl formamide (DMF)	6.4
Butanol-1	3.9	Dimethyl acetamide	6.5
Propanol-2	3.9	N-Methylpyrrolidone (NMP)	6.7
Butanol-2	4.0	Dimethyl sulfoxide (DMSO)	7.2
n-Butyl acetate	4.0	Water	10.2

of **unsuitable solvents** include some low molecular weight alcohols, some ketones, acetonitrile, methanol, dioxane and N-methylpyrrolidone.

9.2.2 Solvent Volatility

Other considerations that may aid in the selection of an organic solvent for a liquid-liquid extraction procedure include volatility (which influences evaporation time) and solvent acquisition and disposal costs (chlorinated and fluorinated solvents are especially expensive). An important practical concern is that the volatile organic solvents used will contaminate the air in the laboratory, causing a potential inhalation health hazard to employees. **Liquid-liquid extraction procedures should always be performed within a fume hood or, if using semi-automated liquid handling equipment, adequate ventilation with fume removal is required.** Another safety concern is the entry of toxic solvents through the skin. For example, adequate safety precautions must be taken with halogenated hydrocarbons and hexane because the nonpolar character of these solvents allows them to easily pass through the skin when contact is made. Note that many of the halogenated hydrocarbon solvents and benzene are known to be toxic and/or carcinogenic. Also, be aware that some solvents (*e.g.*, diethyl ether) can form peroxides under certain storage conditions and are potential explosion hazards.

9.2.3 Solvent Selectivity

9.2.3.1 Relative Characteristics

When solubility toward an analyte is shown to be equal for two solvents, the more selective one, *i.e.,* the one isolating analytes to the exclusion of interferences, is preferred. Selectivity characteristics result from a solvent's ability to function as a proton acceptor or a proton donor, or to interact via a dipole mechanism. For example, if extraction recovery were demonstrated to be equal between ethyl acetate and MTBE, the latter solvent may be preferred because ethyl acetate is more polar (with some water miscibility) and often extracts more endogenous compounds from the sample matrix. These interferences are frequently carried over into the analysis and result in a higher background as well as ionization suppression with certain mass spectrometry techniques. Conversely, note that ethyl acetate may be preferred in the case when a range of structurally diverse polar analytes is extracted from a sample matrix. The polarity of ethyl acetate may allow it to yield higher recoveries for a greater number of compounds than the less polar MTBE, so selectivity in terms of analyte attraction becomes the predominant factor. Solvent selection requires tradeoffs between matrix and analyte selectivity.

The relative ability of organic solvents to carry over matrix components can generally be estimated by comparing the solubility of water in an organic solvent [7]. For example, the solubility of water in benzene, methylene chloride, MTBE, ethyl acetate and methyl ethyl ketone at 20° C is 0.06, 0.24, 1.5, 3.3 and 10.0 % (w/w), respectively. Water is more soluble in ethyl acetate (3.3%, w/w) than in MTBE (1.5%, w/w) and so a higher matrix background is carried over and seen in the analysis. Note that the amount of coextracted matrix components can sometimes be increased in LLE procedures simply by using a large volume of polar organic solvent relative to sample volume.

9.2.3.2 Solvent Mixtures

A mixture of two (or sometimes three) organic solvents is often used for liquid-liquid extraction as it has been shown to enhance the range of solubilities and selectivities displayed for multiple analytes. For example, in drug discovery applications the bioanalytical chemist may have a series of compounds for extraction and analysis, and each compound may have certain key functional groups that affect its ability for hydrogen bonding or its solubility. One organic solvent may not be optimal for the range of chemistries displayed by the compound series. Table 9.3 presents data demonstrating the improved extraction ability of a mixture of ethyl acetate with MTBE compared with either solvent alone, with the group of nine analytes considered as a whole. Ethyl acetate introduced a polar character to the solvent mixture that improved the solubility of many of the proprietary discovery compounds in this series.

Table 9.3
Extraction recoveries for a series of 9 discovery compounds comparing two organic solvents alone and in a 1:1 mixture

	Extraction Recovery (%)								
	A	B	C	D	E	F	G	H	I
Methyl tert-butyl ether (MTBE)	7	28	7	12	35	11	81	20	31
Ethyl acetate (EtOAc)	43	78	43	48	81	53	83	74	80
MTBE/EtOAc (1:1, v/v)	66	76	66	55	80	52	75	66	79

Extractions contained 100 µL plasma, 50 µL IS and 100 µL 0.1M Na_2CO_3 with 1 mL organic solvent

9.2.4 Solvent Density

The density of organic solvents is another consideration in selection. Those solvents that are lighter than water remain at the top of a well or tube after the LLE procedure is finished. It is convenient to quickly freeze the bottom aqueous layer by submerging the container in an acetone/dry ice bath to a height of about one-third that of the total volume. The upper organic layer is decanted and transferred to a clean tube for subsequent evaporation and reconstitution.

Note that solvents that are denser than water (Table 9.4) will remain at the bottom of a well or tube after the LLE procedure is finished. It is more difficult to isolate the organic layer when the aqueous layer remains on top. Aspirating the aqueous layer to waste is performed, leaving the organic layer behind in the well or tube. However, there may be some contamination introduced from drops of the aqueous remaining at the interface of the two layers. Strictly in terms of convenience, analysts generally prefer that the organic layer be lighter than water and remain on top of the well or tube after LLE. This preference is a factor that can guide and sometimes limit the solvent choices for a liquid-liquid extraction method.

9.3 Rapid Method Development Strategies

9.3.1 Solvent Screening Experiment

When a defined strategy for method development is followed, a selective bioanalytical method can be quickly developed and optimized. The scheme presented here evaluates multiple solvents and solvent combinations for extraction efficiency at one pH. The determination of an optimal pH for sample

Table 9.4
Examples of solvents that are heavier than water, listed in order of increasing density

Solvent	Density (g/mL)
Chlorobenzene	1.106
Ethylene Dichloride	1.253
o-Dichlorobenzene	1.306
Methylene Chloride	1.326
Trichloroethylene	1.476
Chloroform	1.480
Carbon Tetrachloride	1.594

adjustment is usually straightforward since the goal is to ensure that analytes are neutral—adjust amines to basic pH, acids to acidic pH and neutral species to neutral pH. The following discussion presents a strategy from the author that has proved to be a rapid and successful approach to defining a selective bioanalytical LLE method.

The microplate format presents an 8-row by 12-column grid of 96 wells. Consider this arrangement of wells an empty canvas or template for designing a method development experiment that can simultaneously evaluate multiple variables in replicate sets. In one test the goal is to obtain as much information as possible about what organic solvent or solvent combination extracts analytes best and how that attraction is modified by changes in sample pH. Appendix I contains different layouts of a microplate grid of 96 wells that the reader can photocopy and use to help design a method development experiment. The following grids are provided: samples 1–96 numbered by row and by column, and an empty plate of 96 wells.

Figure 9.2 illustrates a plate layout that evaluates 12 solvent extractions at one sample pH. Eight wells are available per solvent and these can be arranged according to the analyst's preferences; *e.g.*, 4 replicates, 2 blanks to spike post-extraction for recovery determination, 1 double blank and 1 empty well for calculation of the matrix effect. Alternately, 6 solvents can be chosen and evaluated at each of two pH values (*e.g.*, pH 10 and 12 for a basic compound). Essentially, any combination of solvents and pH values can be assessed as long as there are enough wells to accommodate the number of replicates desired for spiked samples, matrix blanks and post-extracted blanks for determination of recovery and matrix effect.

The specifics of a method development experiment are described. A series of 8 proprietary discovery analytes, each very hydrophilic and displaying basic character, was evaluated for extraction efficiency among 8 organic solvents and solvent combinations, selected from the list in Table 9.5. Rat plasma samples (100 µL) were mixed with IS (50 µL) and a basic solution of 0.1M sodium carbonate (100 µL) prior to the addition of organic solvent (500 µL).

Previous LLE experience with some of these analytes by another chemist in the laboratory resulted in poor recovery from MTBE. Repeated in this screening experiment, poor MTBE performance was confirmed; 7 analytes showed <8% recovery, while 1 analyte showed 23% recovery (Table 9.6). A more polar solvent, ethyl acetate, improved recoveries for every analyte (now 9–36% recovery), but not high enough to the values desired (>40%). The combination

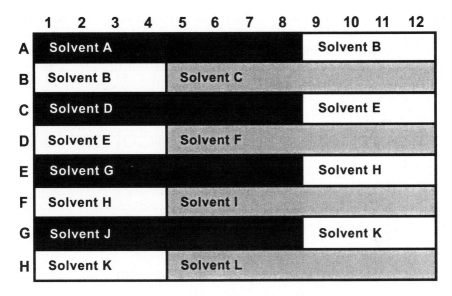

Figure 9.2. Arrangement of replicate sample sets within wells of a microplate to evaluate the effect of 12 organic solvents on analyte recovery at one sample pH. Any combination of solvents and pH values can be assessed as long as there are enough wells to accommodate the number of replicates and blanks needed for accurate determination of analyte recovery and matrix effect.

Table 9.5
Solvent suggestions for development of a liquid-liquid extraction method

Solvents lighter than water	Solvents denser than water
MTBE	Dichloromethane
EtOAc	Dichloromethane/MTBE (1:1)
MTBE/EtOAc (1:1)	Dichloromethane/toluene (1:1)
Toluene	Dichloromethane/toluene/EtOAc (3:4:3)
Hexane	
Acetone/EtOAc/MTBE (1:2:2)	
Dichloromethane/MTBE/EtOAc (1:2:2)	
Dichloromethane/toluene/MTBE (3:4:3)	

All ratios are by volume
MTBE, methyl tert-butyl ether; EtOAc, ethyl acetate

Table 9.6
Percent extraction recoveries for a series of 8 discovery compounds comparing the influence of 8 organic solvents

	A	B	C	D	E	F	G	H
MTBE	4	4	8	1	1	23	6	1
Ethyl acetate (EtOAc)	16	18	27	11	13	36	28	9
EtOAc/MTBE (1:1)	5	6	8	1	3	21	9	2
Acetone/EtOAc/MTBE (1:2:2)	34	27	24	22	39	23	24	40
CH_2Cl_2/MTBE/EtOAc (1:2:2)	7	10	15	2	2	26	21	4
CH_2Cl_2/toluene/MTBE (3:4:3)	10	6	6	3	19	13	6	2
Toluene	ND	ND	ND	ND	ND	ND	ND	ND
Hexane	ND	ND	ND	ND	ND	ND	ND	ND

ND, none detected
Ratios are by volume
Extractions contained 100 µL plasma, 50 µL IS and 100 µL 0.1M Na_2CO_3 with 500 µL organic solvent

of MTBE and ethyl acetate was less effective than ethyl acetate alone, revealing that MTBE is detrimental to the beneficial influence of the more polar ethyl acetate. Very nonpolar and aromatic solvents (hexane and toluene) performed poorly and area counts were not detected.

The most promising from this screening evaluation is a mixture of the polar solvents acetone and ethyl acetate with modification by MTBE; using this combination, recoveries from 22–40% were achieved for all 8 analytes. Since it is shown that MTBE is not a great contributor to recovery, it may be exchanged with another solvent (*e.g.*, ethanol or diethyl ether) in a follow-up experiment using additional solvent choices. The effect of pH could also be examined more closely (*e.g.*, pH 10 and 12) with two or three solvent choices, if desired.

9.3.2 Tips and Tricks

When different solvents are evaluated in parallel within the microplate format, it is important to only use solvents that are all lighter than water or all denser than water; never mix the two in the same microplate. Problematic organic isolations occur if solvent densities are mixed in a plate, *e.g.*, isolation of the top layer yields organic in some wells but aqueous in the wells that used a denser solvent than water. Using automation for liquid transfers, tips cannot selectively aspirate from the top of the well for some samples and from the bottom for others.

The all or none liquid transfer applies with a workstation or a multichannel pipettor, as one common set of actions is performed for all sample wells. However, it is possible to perform this selective approach to aspirating and dispensing using multiple probe (4 or 8) liquid handlers when the software allows such customization. The automation procedure as described for mixing solvents that are lighter and denser than water in the same plate should only be done if it is to be performed often. As described in Chapter 5, the time should be invested to automate a complex procedure only if that procedure is to be used routinely. Exact placement of tubes and solvents in certain positions would be required.

Another practical note is that the analyst cannot look into each well to view the two phase interface layer, as can be done with individual tubes. If emulsions form in selected wells, it may be very difficult to notice within a 96-well extraction plate; it would become more evident during the transfer step when the organic layer is isolated from the mixture.

9.4 Method Optimization and Troubleshooting

9.4.1 Introduction

Many parameters must be evaluated and optimized during the method development process. Each parameter individually can be important to the success of an extraction. Proper development of a method, as well as optimization and troubleshooting, requires an understanding of the influence that each of these parameters plays in the overall process.

Each of the following parameters will now be discussed:

a) Choice of organic solvent and solvent combinations
b) Sample pH adjustment
c) Volume of organic solvent
d) Mixing mode and time
e) Cleanliness of extracts after dry-down and reconstitution
f) Analyte volatility on dry-down

9.4.2 Choice of Organic Solvent or Solvent Combinations

The choice of organic solvents to investigate in an experiment is a very important one. As previously discussed, the chemical characteristics and polarities of the analytes guide in making this decision. The modification of pH is very important; acidic analytes are pretreated with acid (*e.g.*, 0.5N HCl or 2% formic acid) while basic analytes are adjusted with base (*e.g.*, 0.5M NaHCO$_3$, 2–5% NH$_4$OH or 0.5M NaOH). At least four different organic solvents and/or combinations should be evaluated. Once data are obtained using these four solvents, the relative extraction recoveries influence final selection. However, other factors influence this decision as well, such as the matrix effect, safety concerns with solvent volatility or toxicity, and a reasonable evaporation time.

An example of the importance of solvent selection to matrix effect is shown by the work of Wang [8]. In this report, MTBE did not show a matrix effect but a substantial one was demonstrated by substitution of MTBE with ethyl acetate, ethyl acetate/pentane (9:1 or 8:2, v/v) or ethyl acetate/hexane (8:2, v/v). Interpretation of these data suggests that the polar component ethyl acetate carried over some matrix components in the extraction by nature of its water miscibility, introducing a matrix effect.

Another method development experiment evaluated recovery of analytes from four organic solvents: MTBE, MTBE with 5% ethanol, ethyl acetate and chloroform [9]. Four basic analytes were spiked into rat plasma: trimipramine (**2**), chlorpheniramine (**3**), desipramine (**4**) and diphenhydramine (**5**). The extraction protocol pretreated rat plasma (100 µL) and IS (25 µL) with 0.1M KOH/K$_2$CO$_3$ pH 12 (100 µL) to ensure that the analytes were in the neutral state.

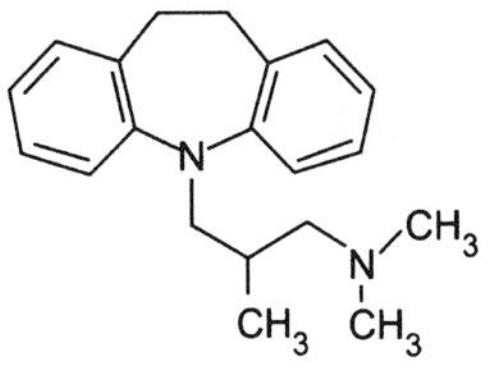

2

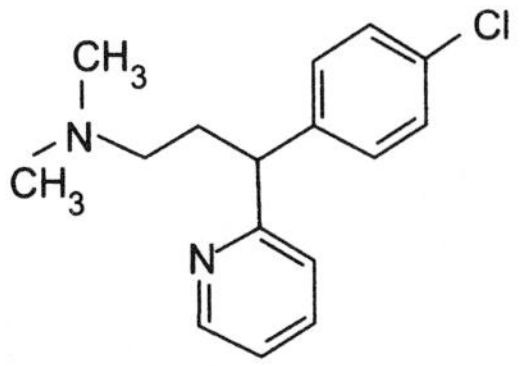

3

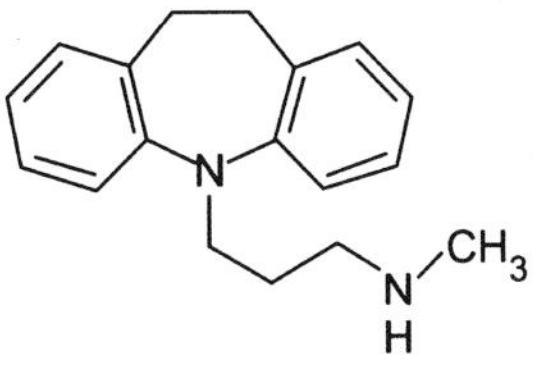

4

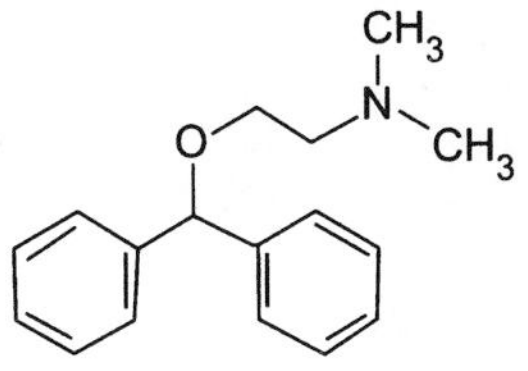

5

A summary of recoveries of these four analytes using the four extraction solvents is shown in Figure 9.3. The solvent yielding acceptable recoveries (45–60%) for all four compounds was ethyl acetate. Another acceptable solvent in terms of recovery was MTBE/Ethanol (95:5, v/v). Note the recovery of trimipramine from MTBE compared with MTBE/Ethanol (95:5, v/v); the addition of only 5% ethanol improved recovery an additional 10%. While the use of chloroform yielded the highest recovery for one analyte (chlorpheniramine), its recovery for the other three analytes was poor. Additionally, given the choice of two solvents that perform equally well, convenience chooses the immiscible solvent that is lighter than water.

9.4.3 Sample pH Adjustment

When the ionization characteristics of an analyte can be predicted with reasonable accuracy, *e.g.*, a nonpolar basic amine, then it is clear that the sample pH should be adjusted to the basic side of the scale. However, note that pH values of 9.0 and 11.0 can yield two very different extraction results. Since the pKa values for most discovery compounds and many development compounds are unknown or unavailable to the analyst, it is recommended to always evaluate at least two pH modifications of the sample solution. For example, a basic analyte can be evaluated at both pH 9.0 and 11.0; a neutral analyte can be evaluated at pH 6.0 and 7.5; an acidic analyte can be evaluated at pH 2.5 and 4.0. It has sometimes been observed that use of different pHs can

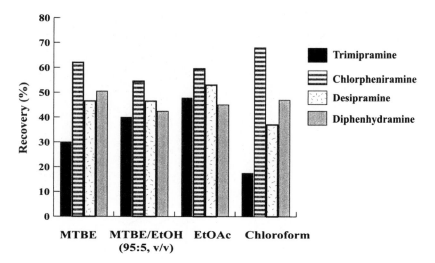

Figure 9.3. Summary of extraction recoveries of four analytes evaluated with four organic solvents. Averages are shown (n=3). Reprinted with permission from [9]. Copyright 2000 Elsevier Science.

yield extracts, after dry-down, that are slightly dirtier looking than those from other pHs. This phenomenon is likely a result of more or less mass of coextractable contaminants from the sample matrix carried over with the analyte in the extraction process. The pH value for the extraction can sometimes influence this occurrence, as does the solvent itself by its degree of water solubility.

9.4.4 Volume of Organic Solvent

The volume of organic solvent used should be relatively large, such as a 7:1 ratio (organic:aqueous) or greater. When using individual test tubes, typically a 15 mL screw capped tube is employed and there is more than adequate headspace when using, *e.g.*, a 1 mL aqueous volume to 7 mL organic volume. Note that the 1 mL aqueous volume in this discussion includes the sum of volumes for the sample matrix, internal standard working solution and aqueous solution used to adjust pH. Recall that when the ratio of organic to aqueous is low and when the partition coefficient KD is also small, less than complete recoveries may result. **Using a larger volume of organic solvent can compensate for a small partition coefficient.** A 7:1 ratio or greater is desired when using test tubes for LLE.

When method optimization or troubleshooting is required, the solvent volume is one of the most common variables that influences recovery. As the volume of the tube is reduced (*e.g.*, to a 4 mL tube instead of 15 mL, or to 2 mL as in a microplate), and the sample size stays the same (*e.g.*, 0.5 mL), the ratio of organic to aqueous decreases; solvent volume is then of greater importance. The incomplete transfer of the full volume of organic solvent can also influence recovery and it may be a parameter to examine during method troubleshooting.

A successful approach to aspirating maximum organic solvent volume from a tube is to freeze the aqueous (bottom) layer in a mixture of dry ice/acetone and remove the upper organic layer via decanting or aspirating. If an organic solvent more dense than water is used, freezing is not applicable; solvent removal from the bottom of the tube can be accomplished by inserting a pipette or tip straight to the bottom. A small volume of air is contained within the tip as it passes to the bottom to discourage aqueous solution from entering. The organic layer is then aspirated from the bottom of the tube.

9.4.5 Mixing Mode and Time

The partitioning of analyte from the aqueous into the organic phase is influenced strongly by the completeness of mixing. The mixing process should be vigorous so that the surface area for this interaction is maximized, and it should also be of a sufficient duration. Mixing is generally performed for 10–15 min using a multi-tube vortex mixer or an orbital or horizontal shaker unit. This procedural step is an important one and proper mixing for a sufficient duration is an important component of the overall LLE sample preparation method.

9.4.6 Cleanliness of Extracts after Dry-Down and Reconstitution

The visual and analytical cleanliness of extracts for analysis can sometimes be the determining factor in selecting between two promising LLE schemes. For example, matrix interferences can sometimes be more prevalent following a basic extraction or from using a more polar organic solvent having greater water solubility than another solvent. A visual examination of the test tube bottoms following the dry-down procedure can sometimes yield clues to identify cleanliness issues; if particles and debris are seen at the tube bottom before reconstitution solvent is added, those materials may dissolve and remain, potentially contributing to matrix interferences. Note that occasionally a residue after dry-down can be caused by low purity or contaminated solutions used for either the pH adjustment or the organic extraction; check labels for purity data and examine expiration dates carefully. At other times, a careful evaluation of the chromatography from extracted matrix blanks may be the determining factor in deciding the specifics of a sample preparation method.

Insoluble particles remaining in the reconstituted extract can be efficiently removed by passing the dissolved extract through a 0.45 μm filtration plate before the injection process. A filtration plate is mated on top of a deep well collection plate and the extract is delivered into the wells of the filtration plate. A 96-tip liquid handling workstation can perform this transfer in one step. Centrifugation of the extract through the filter plate/collection plate combination is preferred to vacuum, in this case, as the volumes are generally small and holdup volume in the filter plate is usually much greater with vacuum. The higher force attainable with centrifugation ensures that most of the volume entering the filter is passed though and into the receiving plate. Filtration plates are discussed in Chapter 3, Section 3.7.1.

A change in the analysis conditions can sometimes detect an extract cleanliness issue that was not noticeable with another technique. In the extraction of the cyclooxygenase inhibitor rofecoxib from human plasma, the sample preparation procedure used was one developed earlier [10] by an associate in the laboratory. A volume of 1 mL plasma (without pretreatment) was extracted with 8 mL of a mixture of hexane/methylene chloride (1:1, v/v). The isolated organic phase was evaporated to dryness and reconstituted in 1 mL acetonitrile followed by 1 mL water; an aliquot of 150 μL was injected into the LC-MS/MS system using positive ionization mode. However, when the ionization was later modified to the negative mode, the extracts were found to be dirtier than desired. A change in extraction solvent was made to MTBE and the sample was also made basic before extraction. Cleaner extracts were obtained using this revised procedure. A volume of 1 mL plasma, containing internal standard working solution (100 μL), was added to 1 mL of a pH 9.8 carbonate buffer, followed by 8 mL MTBE. The solution was mixed using a rotator for 15 min. The isolated organic layer was evaporated and reconstituted in 50 μL acetonitrile. Following vortex mixing, a 50 μL aliquot of water was added, the solution was briefly mixed again and subjected to sonication. Aliquots of 25 μL were injected into the LC-MS/MS system for analysis [11].

9.4.7 Analyte Volatility on Evaporation

One additional consideration influencing recovery of an analyte as part of an extraction method is the volatility of that analyte during an evaporation process. Ideally, the solvent should be volatile enough to evaporate to dryness within a reasonable time (20–35 min) under conditions of heat (30–40° C) and nitrogen gas. The analyte should be able to withstand these same conditions without degradation or loss. However, poor analyte recovery is sometimes caused by its volatility on dry-down which can be influenced by the residual solvent. The influence of the evaporation step on loss of trace analyte from masking by solvent contaminants is discussed by Crowley *et al.* [12].

A procedure to examine analyte volatility in a given solvent follows. Into four wells of a microplate, add a known volume of analyte that has been previously prepared in organic solvent (same solvent as used for the extraction) so that a known mass is delivered. Add blank solvent to the tubes to bring the total volume to that evaporated as part of the LLE procedure (*e.g.*, dilute to 1 mL). Into four other plate wells add blank solvent (no analyte), *e.g.*, 1 mL. All eight wells then undergo evaporation under the usual conditions of nitrogen and heat. Reconstitute the four spiked analyte wells with a mobile phase compatible solvent. The four wells with solvent blanks receive a previously prepared reconstitution solution spiked with analyte so that the volume delivered for reconstitution contains the same mass of analyte as originally spiked into the wells that underwent evaporation. The peak areas (or heights) from the four extracted standards are compared with those from the non-extracted standards. If the analyte is volatile on dry-down in this particular solvent, the data will confirm analyte loss. It is sometimes found that analyte volatility can be reduced by use of a different solvent, and perhaps using a second best solvent as determined in a method development experiment may yield better results. If not, perhaps the LLE procedure should be modified to back extract into a small volume of aqueous solution; this approach adds additional extraction, mix, centrifuge and transfer steps but it does avoid the dry-down step entirely.

The temperature at which evaporation occurs can also influence analyte volatility. It is important to operate at a controlled temperature for the routine procedure, whether it is 30°, 35° or 40° C. However, it sometimes happens that the temperature will rise higher than the set point on the evaporator. The temperature can actually exceed the safety zone in which the analyte has shown temperature stability. Excessive temperature ramping may be the fault of a bad fuse, an occluded or inoperative sensor, or simply the result of multiple users adjusting the instrument by changing the set point each time. Additionally, when multiple brands of evaporation units are in the laboratory, one may operate slightly differently than the other and minor operational adjustments are not made for this fact. For example, one brand may heat the plate from the bottom as well as from the top via heated nitrogen gas, but the other evaporation unit does not heat from the bottom. Evaporation systems are discussed in Chapter 4, Section 4.5.

Acknowledgments

The author is appreciative to David Ehresman for his critical review of the manuscript, helpful discussions and contributions to this chapter. Line art illustrations were kindly provided by Woody Dells.

References

[1] H.T. Kristensen, J. Pharm. Biomed. Anal. 18 (1998) 827-838.
[2] K.A. Riffel, H. Song, X. Gu, K. Yan and M.W. Lo, J. Pharm. Biomed. Anal. 23 (2000) 607-616.
[3] R.F. Suckow, T.B. Cooper and G.M. Asnis, J. Liq. Chrom. 6 (1983) 1111-1122.
[4] M.A. Anderson, T. Wachs and J.D. Henion, J. Mass Spectrom. 32 (1997) 152-158.
[5] B.A. Kimball, T.J. DeLiberto and J.J. Johnston, Anal. Chem. 73 (2001) 4972-4976.
[6] M.T. Kelly, D. McGuirk and F.J. Bloomfield, J. Chromatogr. B 668 (1995) 117-123.
[7] P.A. Krieger, Ed., High Purity Solvent Guide, Burdick & Jackson Laboratories, Inc. (1994).
[8] A.Q. Wang, A.L. Fisher, J. Hsieh, A.M. Cairns, J.D. Rogers and D.G. Musson, J. Pharm. Biomed. Anal. 26 (2001) 357-365.
[9] N. Zhang, K.L. Hoffman, W. Li and D.T Rossi, J. Pharm. Biomed. Anal. 22 (2000) 131-138.
[10] E.J. Woolf, I. Fu and B.K. Matuszewski, J. Chromatogr. B 730 (1999) 221-227.
[11] C.M. Chavez-Eng, M.L. Constanzer and B.K. Matuszewski, J. Chromatogr. B 748 (2000) 31-39.
[12] X.W. Crowley, V. Murugaiah, A. Naim and R.W. Giese, J. Chromatogr. A 699 (1995) 395-402.

Chapter 10

Liquid-Liquid Extraction: Automation Strategies

Abstract

The traditional liquid-liquid extraction (LLE) technique for sample preparation utilizing individual test tubes is very labor intensive and its automation has been a challenge met with limited success. The ability to more completely automate this procedure has been advanced by the microplate format and the use of liquid handling workstations. Liquid-liquid extraction procedures can now be performed in a semi-automated mode, freeing analyst time and improving overall productivity. This chapter introduces three techniques for the automation of LLE in microplates using a variety of workstation solutions. One approach automatically detects the boundary between the organic and aqueous phases and selectively removes these solvents. The 4-/8-probe liquid handling workstations operate using fixed and variable volumes, defined aspirate heights and/or liquid level sensing to remove one liquid from the other in tubes and microplates, and also reformat samples from tubes into a microplate. A 96-tip liquid handling unit offers fixed volume aspirating and dispensing at a defined aspirate height for liquid removal. The automation of solid-supported LLE in microplates is also described in this chapter.

10.1 Automation of LLE Using Phase Boundary Sensing

An example of a single probe workstation performing automated LLE using proprietary phase boundary sensing technology is the Allex™ (Mettler-Toledo Autochem, Vernon Hills, IL USA). The Allex™ (Automated Liquid-Liquid EXtraction, shown in Figure 10.1) is a fully integrated benchtop system that automates the entire process of extraction and workup. The following functions are performed:

- Addition of aqueous and organic solutions
- Sample mixing
- Phase separation
- Phase distribution

327

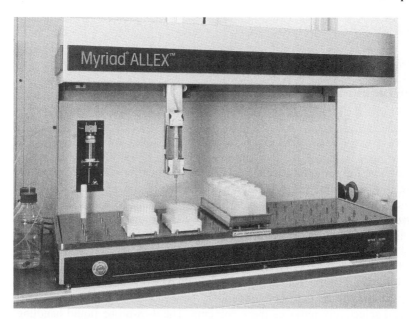

Figure 10.1. The Allex workstation automates the liquid-liquid extraction procedure in the test tube or 96-well format. Photo reprinted with permission from Mettler-Toledo Autochem.

Vials are placed into racks on the deck of the Allex™. The system accommodates 7, 14 and 28 mL sample vials, centrifuge tubes (24 and 25 mm), scintillation vials, as well as Gilson and Zymark tube racks. Custom racks can be configured for other tube sizes and formats. A microplate rack holder is also available for 2 mL square well microplates. The dimensions of the system (width, 1350 mm; depth, 650 mm; height, 1400 mm) allow the entire unit to fit inside a fume hood. A system enclosure can be built around the unit when positioned on a lab bench.

The procedure for performing LLE with this instrument follows. The probe adds solvent to a sample tube. Vigorous mixing occurs in the vessel via aspirating and dispensing. The actual phase separation occurs in the settling chamber (Figure 10.2). Once the two phases are settled, the unique sensing technology detects the boundary between the two layers. The phases are then isolated and distributed to the destination vials, similar to the manual use of a separatory funnel. Wash steps following each sample extraction ensure that the probe, chambers and nozzle are adequately cleaned before the next sample. Extractions are performed in either serial mode or batch mode. Two 1-liter

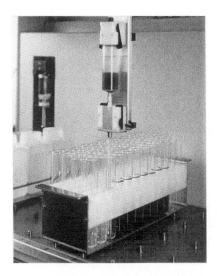

Figure 10.2. Phase separation occurs in the settling chamber of the Allex workstation. Photo reprinted with permission from Mettler-Toledo Autochem.

solvent flasks are configured in-line for commonly used dilutions or washes. The unit is software controlled and the extraction processes are built by defining the deck layout and combining single action steps into a customized user program.

Although the system was originally designed and marketed for medicinal chemistry applications in solution phase synthesis, it is applicable for bioanalytical LLE in tube or microplate format. Three probe sizes are available—micro, standard and macro. Bioanalytical applications would commonly use either the micro or the standard size probe for maximum sample volumes (aqueous plus organic) of 2.5 mL or 10 mL, respectively.

The power of the Allex™ instrument is that two phases can be detected even when the user cannot visually see the interface. Error handling tools are built into the system so that if foaming or an emulsion occurs, the sample can be flagged and skipped, or a preset height definition can be activated in which the probe will aspirate at a user supplied depth. This unit also displays multiple and back extraction capability. Its throughput is reported by the manufacturer to be up to 60 separations per hour.

10.2 Automation of LLE Using a 96-Tip Workstation

10.2.1 Strategies

10.2.1.1 Introduction

A typical example of a 96-tip liquid handling workstation is the Quadra® 96 (Tomtec Inc., Hamden, CT USA). The discussion of automation strategies here will focus on this instrument, although the general approach is similar for other 96-tip workstations and pipetting units. More information about 96-tip workstations can be found in Chapter 5.

A summary of the tasks performed for a liquid-liquid extraction procedure in the microplate format are listed in Table 10.1. Note that a fair number of

Table 10.1
Summary of the tasks performed for a liquid-liquid extraction procedure in the microplate format

Task	Manual	Liquid Handling Workstation
Prepare sample plate		
• Aliquot samples into microplate	X*	X*
• Deliver aliquot of internal standard		X
• Add pH adjustment solution		X
• Mix solutions		X
Deliver organic solvent		X
Seal	X	
Vortex mix	X	
Centrifuge	X	
Unseal or pierce foil	X	
Transfer organic layer		X
Evaporate organic extract	X	
Reconstitute		X
Mix	X	
Seal for injection	X	

*Can be performed either manually or by a 4-/8-probe liquid handling workstation

manual tasks in this procedure cannot be automated; however, the productivity of the overall procedure is great enough to justify these manual steps. Since manual steps are necessary, the overall procedure is considered semi-automated.

10.2.1.2 Identify Tasks and Roles for Automation

In designing an automation strategy, the user must decide at what part of the method sequence the liquid handler first becomes involved. Preparation of the sample plate for extraction (sample plus internal standard addition and brief mixing) can be performed off-line either manually or using a multiple probe liquid handling workstation. At this point, the prepared microplate can be placed onto the 96-tip workstation deck for rapid pipetting. The following discussion is based on the plasma samples being reformatted into the microplate format off-line and then the 96-tip liquid handling workstation performs all subsequent pipetting tasks.

Four main liquid transfer functions with respect to liquid handling automation are performed during a liquid-liquid extraction procedure in the microplate format. These general functions are:

1. Liquid transfers to finish preparing the sample plate (IS, buffer; mix)
2. Addition of organic solvent to create the extraction plate
3. After mixing and centrifugation, transfer of the organic layer from the extraction plate into a clean collection microplate
4. Reconstitution of the extracts following evaporation

Once the extraction plate containing samples is placed onto the deck of the instrument, the specific functions that the 96-tip liquid handling workstation performs in this sample preparation procedure are listed below.

- Add internal standard to sample plate
- Add pH adjustment buffer or solution to sample plate
- Mix contents of wells (after dispensing buffer)
- Deliver organic solvent into sample wells
- Transfer organic solvent (following the mix step and centrifugation) into a clean microplate
- Reconstitute in mobile phase compatible solvent (following evaporation)

10.2.1.3 Plan the Deck Layout

The deck of the Quadra 96 has six positions that will accommodate the hardware (tips), extraction plate, transfer plate and all reagent reservoirs required to perform this procedure. This deck moves to the left, right, forward and backward relative to a fixed position under a 96-tip aspirating/dispensing head. When one of the six positions is directly under the tips, the stage is raised to a programmed height to meet the tips and the aspirate or dispense step is performed. Labware components sit on either a short or a tall nest, or placeholder, in each position. The actual nest type used influences the height that the stage is raised. A typical arrangement of these components on the Quadra 96 deck, as configured for LLE, is shown in Figure 10.3.

10.2.1.4 Preparation of the Sample Plate

In order to use a 96-tip workstation, samples must be reformatted from individual source tubes into a microplate. This sample plate can be prepared off-line either manually or using a multiple probe liquid handler. Once the samples are arranged in this 8-row by 12-column array, all subsequent pipetting tasks can be performed by the liquid handling workstation. The ideal scenario is to have a 4-/8-probe liquid handling workstation prepare the full sample block (*e.g.*, plasma, internal standard and buffer), although not all laboratories are able to afford both a 96-tip and a 4-/8-probe liquid handling workstation.

10.2.1.5 Addition and Transfer of Organic Solvent

Once the sample plate has been prepared and placed on the deck of the Quadra 96, organic solvent is added into all the wells from a reagent reservoir.

TIPS in 1 *Tip Jig*	Organic Solvent in 2 *Tall Nest*	Internal Standard in 3 *Tall Nest*
Transfer Plate in 6 *Short Nest*	Extraction Plate in 5 *Short Nest*	Buffer in 4 *Tall Nest*

Figure 10.3. Typical deck layout of the Quadra 96 liquid handling workstation as configured for performing liquid-liquid extraction in the microplate format.

Note that the tip capacity is only 450 μL (actually, only 425 μL when using a 25 μL air gap). Volumes larger than this amount must be delivered using repeated pipetting steps. When the total volume has been dispensed, the plate is removed from the deck, sealed, mixed and centrifuged to cleanly separate the aqueous and organic layers.

The extraction plate is placed back into its position on the deck of the Quadra 96. The next step aspirates a fixed volume of organic solvent from the wells of the extraction plate and dispenses that volume into a clean microplate. Note that the Quadra 96 does not utilize liquid level sensing technology. In this case, the tips descend to a fixed depth and aspirate a preset volume. The volume aspirated can be larger than the actual organic volume, in which case air will be aspirated at the end of the cycle because the tips stay at a fixed depth.

When the organic solvent is denser than water and is at the bottom of the tube, the procedure for pipetting using the workstation follows. The tips are programmed to go to the bottom of the well, remain at that fixed point, and begin to aspirate liquid. Organic solvent will be removed while the upper aqueous layer descends; the aspirating is stopped at a volume that does not greatly disrupt the interface layer between the two phases, leaving some organic behind. The probes ascend and dispense the organic solvent into another plate for the next step of the procedure. See Tips and Tricks in Section 10.2.3.5 for more information on this technique.

10.2.1.6 Evaporation and Reconstitution

The microplate containing the transferred organic solvent is evaporated to dryness. After this evaporation step, the plate is placed back onto the deck of the workstation for the reconstitution procedure. A mobile phase compatible solution is placed into a solvent reservoir. The desired volume is aspirated from the reservoir and delivered into the destination wells of the microplate. The plate is removed from the deck; it is sealed, manually vortex mixed and then placed in an autosampler for injection into a chromatographic system and analysis.

10.2.1.7 Throughput Considerations

Very fast sample turnaround is needed for bioanalytical programs that support the drug discovery process in pharmaceutical research. Semi-automated liquid-liquid extraction methods are a cost effective and reliable means to reach throughput goals. Although an analyst must remain at or near the instrument when methods are run, 96-tip workstations still yield short sample preparation times. The following examples of throughput were previously mentioned in Chapter 8, Section 8.2.5, but they are repeated here for completeness within this chapter and because they are such dramatic illustrations of the potential of LLE automation.

Steinborner and Henion [1] reported a semi-automated LLE procedure (following a quick protein precipitation using acetonitrile) for the quantitative analysis of the anticancer drug methotrexate and its major metabolite from human plasma. Sample preparation throughput was reported as 4 sample plates (384 samples) processed in 90 min by one person. The analytical throughput was reported as 768 samples analyzed within 22 h (maximum 820 samples per 24 h), using a selected reaction monitoring (SRM) LC/MS method with a 1.2 min analysis time per sample.

Zweigenbaum and Henion [2] reported the high throughput determination of six benzodiazepines in human urine using a selected reaction monitoring LC/MS method with sample preparation using semi-automated LLE. When four autosamplers were connected to one chromatographic column and one tandem mass spectrometer, 1152 samples (twelve 96-well plates) were analyzed in less than 12 h.

Another reported SRM LC-MS application was the determination of Selected Estrogen Receptor Modulators (SERM) in human plasma where more than 2000 samples were analyzed in 24 h [3]; LC separation of five analytes was performed in less than 30 s. The time for one person to prepare these 2000 plasma extractions using liquid-liquid extraction in a semi-automated mode was reported as 7 h. A related application for rapid analysis for idoxifene (an SERM) and its metabolite, using semi-automated LLE for sample preparation, cited an average analysis time of 23 s/sample. It took about 37 min to analyze one 96-well plate and a throughput of over 3700 samples per day was achieved [4].

The extraction of cyclosporine A from whole blood was performed using LLE in microplates and sample throughput was documented as four 96-well plates prepared in less than 5.5 h [5]. With a run time of 3.5 min per sample, the four plates (384 samples) could be prepared and the analysis completed within 28 h.

The throughput advantages of using a volatile, low viscosity mobile phase (supercritical CO_2) in packed column supercritical fluid chromatography mass spectrometry (pcSFC-MS/MS) were reported for the bioanalysis of dextromethorphan from plasma [6]. The low viscosity of supercritical CO_2 allows the use of higher flow rates than in typical aqueous mobile phases for LC-MS/MS, and its volatility allows the entire effluent to be directed into the MS interface. The analysis time for a 96-well plate was reported as 10 min 12 s, using a multiplexed autosampler, a short 2 x 10 mm column, and a flow rate of 7.5 mL/min.

10.2.2 Method Example

A typical example of a liquid-liquid extraction method that can be automated on the Quadra 96 in the microplate format is one developed for the liquid-liquid extraction of a proprietary compound and its internal standard (Figure 10.4) from human plasma [7]. Note that the original published method was not automated on the Quadra 96 in this report but is used here for illustration of how such a procedure would be performed. The complete LLE method is listed in Table 10.2.

This LLE method must be translated into a series of sequential actions for the Quadra 96 to execute. A typical Quadra 96 program for performing this method is detailed in Table 10.3. Each item, or numbered line in the program, involves an action to be performed at one of the six positions that is moved under the 96-tip aspirating/dispensing head. When pipetting, each line specifies a volume of liquid, an air gap volume, and the stage height setting as a number. Note that the stage height setting for each pipetting step depends on the specific nest type used and the preference of the user. Stage height settings are not included in this table as individual variations apply.

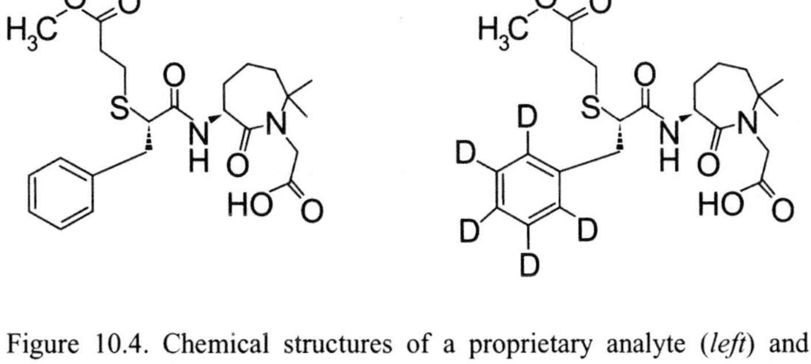

Figure 10.4. Chemical structures of a proprietary analyte (*left*) and its $^2H_5$-labeled internal standard (*right*) that were extracted under acidic conditions using MTBE. Reprinted with permission from [7]. Copyright 2000 Elsevier Science.

Table 10.2
Typical example of a microplate liquid-liquid extraction procedure

Prepare sample plate

- Aliquot 250 µL plasma from vials and transfer into 2 mL round well microplate
- Add IS 25 µL and briefly mix
- Add 0.5N HCl 100 µL and briefly mix

Dispense MTBE 500 µL

Seal microplate with heat sealable film

Vortex mix 10 min

Centrifuge 3000 rpm x 10 min

Optional step*

Transfer organic layer to 2 mL microplate

Evaporate to dryness

Reconstitute in 75 µL [1mM formic acid solution in acetonitrile/water (1:2, v/v)]

Seal plate with pierceable cap mat

Inject aliquot using 96-well autosampler

*Optional step: Immerse microplate in acetone/dry ice bath to freeze bottom aqueous layer when an organic solvent lighter than water is used

The Quadra 96 program is run with several tip exchange steps and a couple pause steps. These pause steps allow time for the off-line functions of (a) sealing the prepared extraction plate, mixing, centrifuging, and unsealing, and (b) evaporation of organic solvent after the extraction. Following the evaporation step, a reconstitution program is run.

The reconstitution procedure aspirates a fixed volume from a reagent reservoir containing mobile phase compatible solvent and delivers into the plate containing dried extracts. If desired, this reconstitution step can be added to the end of the Quadra 96 program. However, it is suggested to maintain the reconstitution as a separate program so that the instrument is not occupied and unavailable for the length of time that it takes to dry down the eluate, typically 15–45 min. When the Quadra 96 is used daily by multiple analysts, there is often a waiting list or queue to perform other sample preparation assays. Closing the program at the end of the organic transfer step simply allows more

Table 10.3
Typical Quadra 96 program for performing liquid-liquid extraction in the microplate format. See Figure 10.3 to identify the labware positions defined for this program. Stage height settings are not included as individual variations apply.

1. Load Tips at 1

2. Aspirate 25 µL from 3, 25 µL Air Gap

3. Dispense 50 µL to 5, 0 µL Blowout

4. Shuck Tips at 1

5. Pause

6. Load Tips at 1

7. Aspirate 100 µL from 4, 25 µL Air Gap

8. Dispense 125 µL to 5, 0 µL Blowout

9. Mix 250 µL 3 times at 5

10. Shuck Tips at 1

11. Pause

12. Load Tips at 1

13. Mix 250 µL 3 times at 2

14. Loop 2 times

15. Aspirate 250 µL from 2, 25 µL Air Gap

16. Aspirate 10 µL [stage height 0] from 2, 0 µL Air Gap

17. Dispense 285 µL to 5, 0 µL Blowout

18. End Loop

19. Dispense 5 times at 2 [Time Dispense of 0 µL]

20. Pause

21. Mix 250 µL 3 times at 2

22. Loop 2 times

23. Aspirate 250 µL from 5, 25 µL Air Gap

24. Aspirate 10 µL from 5, 0 µL Air Gap

25. Dispense 285 µL to 6, 0 µL Blowout

26. End Loop

27. Dispense 5 times at 6 [Time Dispense of 0 µL]

28. Shuck Tips at 1

29. Quit at 1

efficient utilization of the instrument. The analyst who needs to reconstitute dried extracts then walks up to the instrument at the next available time opening, performs the procedure, and leaves the Quadra 96 available for the next procedure or analyst. Some tips and tricks in using the Quadra 96 for performing LLE in microplates are discussed next.

10.2.3 Tips and Tricks

10.2.3.1 Containing the Solvent Vapors

The most important factor to consider when using the Quadra 96 for LLE is that placing an organic solvent into an open reagent reservoir holding about 300 mL volume will allow volatile solvent vapors to rapidly escape and flood the surrounding atmosphere. In the interest of worker and laboratory safety, the Quadra 96 must be totally contained within a fume hood or surrounded with an efficient enclosure to contain and vent the solvent vapors.

10.2.3.2 Using the Polypropylene Tips

Many times, aspirating an organic solvent using the polypropylene tips in their dry state will result in leaking solvent due to weak surface tension. When pipetting organic solvents, several sequential steps are recommended:

1. Mix 3 times using the volume to be delivered and dispense back into the reagent reservoir
2. Aspirate the desired volume
3. Aspirate a 10 µL air gap at the tip ends (at a stage height setting of 0) and then move the deck to the desired position for the dispense step
4. After dispensing, remove solvent vapors from above the 96 piston head by using a time dispense of 0 µL five times

The tips hold a maximum volume of 450 µL. When a volume greater than 425 µL is to be dispensed (a 25 µL air gap in addition to the solvent volume is suggested), use multiple aspirate and dispense steps within a loop procedure. For example, when a volume of 1.5 mL is to be dispensed, specify a volume of 375 µL (plus a 25 µL air gap equals 400 µL) to be dispensed four times.

Tips are usually exchanged after any step that touches drug or analyte in solution to avoid contaminating a reagent reservoir in subsequent steps. The tips can be removed, or "shucked," into the "tip jig" but sometimes it may be convenient to shuck them into a used microplate. For example, after transfer of the organic layer to a clean microplate, the tips may be shucked into the used extraction plate containing the aqueous matrix left behind. Then, the plastic plate and tips can be removed in one step and disposed into a proper biohazardous waste receptacle. Shucking the tips into a different position than the tip jig provides a visual clue that new tips are to be loaded since that position remains empty.

Note that using the Quadra 96 to add internal standard to the sample plate delivers this solution to all 96 wells. In those instances where one or more double blanks (zero analyte level and no internal standard) are designated in the plate, the pipet tips corresponding to those exact well locations can simply be removed from the tip rack prior to pipetting. This selective tip removal approach can also be used for the organic solvent transfer and reconstitution steps to keep selected wells free of solvent. System checks, blank mobile phase or other solution can be manually delivered into dry wells of the final analysis plate following reconstitution. Several tips may be removed manually from the tip rack but removal of too many will cause the assembly to become unbalanced and an error may occur on loading.

10.2.3.3 Centrifuging the Microplate

The extraction plate should be centrifuged after the mixing step to bring all solvent down the sides of the wells to the bottom. Otherwise, droplets remaining near the surface or on the underside of the seal may potentially introduce well-to-well contamination as the seal is removed.

10.2.3.4 Piercing the Plate Seal for Solvent Removal

Following the mixing and centrifugation steps, the Quadra 96 tips need to enter the wells to aspirate organic solvent and dispense into a clean microplate. When a silicone/PTFE cap mat is used, it can be carefully removed following centrifugation. However, when an aluminum film (coated with polypropylene) is used, it is not straightforward to remove the foil from the plate as it is securely heat sealed to the polypropylene plate surface. Puncturing each well is necessary to break the seal and allow an entry point for the Quadra 96 tips. A convenient means to break the seal is to use a piercing plate; a single forceful press on the top of a sealed microplate rapidly pierces each well.

10.2.3.5 Determining Well Depth for Organic Solvent Removal

The depth that the tips are submerged into the extracted sample wells for the organic solvent aspirate step is determined by trial and error using blank matrix. Different stage height settings are evaluated and a specified volume of organic solvent is aspirated. The goal is to remove as much organic solvent as possible yet leave the aqueous layer behind and undisturbed. Note that liquid level sensing is not applicable for the Quadra 96, so the tips move to a specific position within the well and remain there. If the volume specified for aspirating is greater than that available, air will be aspirated at the tip bottom which is not detrimental.

When the organic solvent used is heavier than water and remains at the well bottom, the tips are submerged to the well bottom to aspirate solvent gently away beneath the aqueous layer. Successful use of this approach requires first aspirating about 50 µL air in the tips, lowering the tips into the wells to reach the bottom, and then dispensing about 25 µL air to expel any aqueous contamination as the tips pass through.

10.2.3.6 Using Move Steps to Change Deck Position

The use of move steps, or simply mix steps with no volume specified, is beneficial to move the deck position without performing any function. Access to the extraction plate after solvent delivery is better achieved when that deck position is not directly underneath the pipetting head. Movement of the deck will allow better access to a specific position, *e.g.*, more clearance is available when the head is moved to Position 3 so that access can be provided to Position 5. A move step is also beneficial when loading tips; move the deck to Position 3 and then execute a Pause step so that easy access to Position 1 is permitted to load/unload tips.

10.2.4 Applications

A representative list of semi-automated applications performed using a 96-tip workstation for LLE is provided in Table 10.4. While the Quadra 96 can perform all pipetting steps once the sample plate is prepared, some researchers prefer to manually add all liquids, including organic solvent, into the microplate using a multichannel or repeater pipettor [8, 9]. After mixing and centrifugation, the Quadra 96 is first used to transfer the organic from the extracted microplate into a clean microplate for the dry-down step. This approach is underutilizing the capabilities of the available automation.

Table 10.4

Representative applications of automated liquid-liquid extraction in microplates using a 96-channel liquid handling workstation with analysis by LC-MS/MS

Analytes and Extraction Block	Sample Matrix and Volumes	Extraction Solvent and Volumes	Ref.
Methotrexate and metabolite; deep well plate	Human plasma 0.2mL	Chloroform 0.5mL	[1]
Six benzodiazepines; deep well plate	Human urine 0.4mL	Chloroform 0.5mL	[2]
Tamoxifen, raloxifene, nafodixine, idoxifene; deep well plate	Human plasma 0.1mL	Hexane/isoamyl alcohol (96:4, v/v) 0.4mL	[3] [4]
Cyclosporine A and everolimus; deep well plate	Human blood 0.3mL	MTBE 0.75mL (twice)	[5]
Dextromethorphan; microtubes	Human plasma 0.2mL	Ethyl ether 0.6mL	[6]
Methylphenidate; deep well plate	Human plasma 0.35mL	Cyclohexane 1mL	[8]
Fluoxetine enantiomers; deep well plate	Human plasma 0.2mL	Ethyl acetate 0.8mL	[9]
Diphenhydramine, desipramine, chlorpheniramine, trimipramine; microtubes	Rat plasma 0.1mL	MTBE 0.8mL	[10] [13]
Reserpine; deep well plate	Mouse plasma 0.1mL	MTBE 0.5mL	[11][a]
Benzodiazepines; deep well plate	Human urine 0.5mL	Chloroform 0.5mL	[12]
Idoxifene; deep well plate	Human plasma 0.1mL	Hexane 0.6mL	[14]

[a]Personal Pipettor-96 (Apricot Designs, Monrovia, CA USA)
All other applications used Quadra® 96 (Tomtec Inc., Hamden, CT USA)

The full capability of using a 96-tip liquid handling workstation for LLE has been nicely demonstrated by Zhang, Rossi and coworkers [10]. An example of the placement of labware components within the six position deck is shown in Figure 10.5. Note that any of the components can be in any position; within the software the user identifies the identity of each component and its location. In this application, aliquots of plasma samples are placed into microtubes and positioned on the deck of the Quadra 96 (Position 6). A clean set of 96 pipet tips is put into the tip jig (Position 1), along with a clean deep well plate in Position 5 (to receive the organic solvent after extraction). Three solvent troughs on the deck contain internal standard (Position 3), pH adjustment solution (Position 4) and organic solvent (Position 5).

The procedure reported by Zhang, Rossi and coworkers consist of the following actions:

a) Aliquots of internal standard (25 µL) are transferred into the sample plate (microtube rack)
b) Aliquots (100 µL) of pH adjustment solution are transferred into the sample plate
c) Two aliquots of organic solvent (400 µL each) are transferred into the sample plate (tip capacity is 450 µL; two sequential transfers are needed to deliver an 800 µL volume)
d) Sample plate is removed from the deck, capped, shaken for 10 min, centrifuged at 4000 rpm for 5 min, uncapped, and returned to the same position on the deck
e) The organic layer is transferred (2 x 350 µL) from the microtubes into a clean deep well plate for dry-down

TIPS in 1 *Tip Jig*	Organic Solvent in 2 *Tall Nest*	Internal Standard in 3 *Tall Nest*
Sample Plate in 6 *Short Nest*	Receiving Plate in 5 *Short Nest*	pH Adjustment Solution in 4 *Tall Nest*

Figure 10.5. Deck layout for a semi-automated 96-well liquid-liquid extraction procedure using the Quadra 96 liquid handling workstation. Reprinted with permission from [10]. Copyright 2000 Elsevier Science.

The transfer step is performed by raising the stage height to an exact position, carefully determined in advance, that aspirates most of the organic solvent from the top layer yet leaves the lower aqueous layer undisturbed. It is accepted that some organic is left behind but a tradeoff is made between analyte recovery and keeping the isolated solvent free of matrix particles near the phase interface [10]. Since the method is automated, individual sample well assessments of tip depth cannot be made and, as the authors reported, it is good practice to err on the side of caution when dealing with the phase separation interface.

Following the dry-down step, the deep well plate containing dried extracts is placed onto the deck and a reagent reservoir containing mobile phase compatible solution is placed in another position. Clean tips are loaded and a transfer step is performed from the reagent reservoir into the 96-well plate. The plate is then sealed, briefly vortex mixed, and is ready for injection into the chromatographic system. The time required for processing one 96-well plate using this semi-automated approach was reported as approximately 1.5 h, which is about one-third the time required for manual operation [10].

10.2.5 Automating Method Development Using a 96-Tip Workstation

Method development strategies for liquid-liquid extraction have been described in Chapter 9. Some of these tasks can be automated using the Quadra 96 but keep in mind that selective solvent aspirating and dispensing is not possible. When all 96 tips enter a reagent reservoir, all 96 tips will deliver that same reagent. Also note that variable volume delivery cannot be performed with the Quadra 96.

In order to designate specific grids of tips to aspirate specific solvents (*e.g.*, an evaluation of four organic solvents for extraction efficiency), the reagent reservoir needs to be customized. Instead of an open one compartment polypropylene reservoir, a multiple component reservoir can be used as shown in Figure 10.6. These custom reservoirs are available from Tomtec in any configuration desired; *e.g.*, by row, by column, every two rows or every four columns. When one of these custom reservoirs is not available, a 2 mL square well microplate provides the ultimate custom reservoir; simply place different solutions into corresponding wells to achieve the pattern desired and use it as the reagent reservoir in the interim.

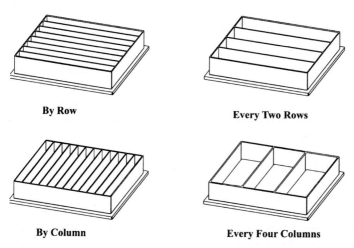

By Row   **Every Two Rows**

By Column   **Every Four Columns**

Figure 10.6. Custom reagent reservoirs for the Quadra 96 allow the placement of specific solutions or solvents into separate compartments.

10.3 Automation of Solid-Supported LLE Using a 96-Tip Workstation

10.3.1 Strategies

10.3.1.1 Introduction

The automation strategies for solid-supported liquid-liquid extraction (SS-LLE) are very similar to those previously discussed in this chapter for traditional liquid-liquid extraction in microplates (Section 10.2). The primary difference between the two sample preparation procedures, relating to automation, is that with SS-LLE the organic solvent is delivered into a flow-through extraction plate system (containing particles onto which the analytes are already adsorbed) and no vigorous mixing and off-line centrifugation steps are necessary. The procedure is more fully automatable for SS-LLE than for LLE.

When a flow-through microplate is used, a vacuum manifold is practical in those cases when gravity flow is not sufficient to pass the organic solvent through the sorbent bed within the microplate wells. A quick, short burst of vacuum (about 15 s at 3–5 in Hg; 100–170 mbar) usually brings the solvent through the bed and into the collection plate underneath. In cases of severe flow restriction, a higher pressure for a longer time may be necessary.

10.3.1.2 Identify Tasks and Roles for Automation

The tasks and roles for automation for SS-LLE are similar to those discussed in Section 10.2.1.2 for traditional LLE. Specifically for SS-LLE in the microplate format, the liquid transfer steps required are summarized below:

1. Liquid transfers to finish preparing the sample plate (IS, buffer; mix)
2. Liquid transfer of the sample mixture into the wells of the SS-LLE plate
3. Liquid transfer of organic solvent from a reagent reservoir into the SS-LLE plate
4. Reconstitution of the extracts following evaporation

Once the extraction plate containing samples is placed onto the deck of the instrument, the specific functions that the 96-tip liquid handling workstation will perform in this sample preparation procedure are listed below:

- Add internal standard to sample plate
- Add pH adjustment buffer or solution to sample plate
- Mix contents of wells in the sample plate
- Aspirate contents of sample wells and dispense into the corresponding wells of a flow-through microplate containing diatomaceous earth particles
- Deliver organic solvent into the SS-LLE plate
- Reconstitution following the evaporation step

10.3.1.3 Plan the Deck Layout

The deck layout has been introduced in Section 10.2.1.3 for traditional LLE. A typical arrangement of the components used on the deck of the Quadra 96 when configured for SS-LLE is shown in Figure 10.7. The SS-LLE plate in Position 5 takes the place of the extraction microplate for traditional LLE.

10.3.1.4 Preparation of the Sample Plate

In order to use a 96-tip workstation, samples must be reformatted from individual source tubes into a microplate. This sample plate can be prepared off-line either manually or using a multiple probe liquid handler. Once the samples are arranged in this 8-row by 12-column array, all subsequent pipetting tasks can be performed by the liquid handling workstation, such as addition of

TIPS in 1 *Tip Jig*	Organic Solvent in 2 *Tall Nest*	Internal Standard in 3 *Tall Nest*
Sample Plate in 6 *Short Nest*	SS-LLE Plate in 5 *Vacuum Manifold*	Buffer in 4 *Tall Nest*

Figure 10.7. Typical deck layout of the Quadra 96 liquid handling workstation as configured for performing solid-supported liquid-liquid extraction in a flow-through microplate.

internal standard to the sample plate, addition of pH adjustment buffer or solution and subsequent mixing of the well contents. The ideal scenario is to have a 4-/8-probe liquid handling workstation prepare the full sample block (*e.g.*, plasma, internal standard and buffer), although not all laboratories are able to afford both a 96-tip and a 4-/8-probe liquid handling workstation.

10.3.1.5 Sample Loading

Note that in traditional LLE the organic solvent is added directly to the wells containing sample mixture; in SS-LLE, the entire volume of sample mixture is aspirated from the collection wells and dispensed into wells of the SS-LLE plate. The SS-LLE plate sits on top of a vacuum manifold with a deep well collection plate inside. A sufficient time (5–10 min) is allowed for the samples to adsorb onto the surface of the particles via gravity. Note that a manifold is not required but is advised for two reasons—to provide some needed support to the SS-LLE plate and collection plate combination and, when necessary, a quick burst of vacuum can be applied without manipulating the components.

10.3.1.6 Addition of Organic Solvent

Organic solvent is delivered into the wells of the SS-LLE plate and also allowed to gravity flow through the sorbent bed. If necessary, a short burst of vacuum is applied to completely pass the entire volume of solvent through the particle bed. Since the available volume above the sorbent bed is not great, a second (and sometimes third) aliquot of organic solvent may be added; the total volume should not exceed 2 mL (the capacity of the collection plate underneath).

10.3.1.7 Evaporation and Reconstitution

Once the total volume of organic extraction solvent has been isolated in the collection microplate, the solvent is evaporated to dryness. After this evaporation or dry-down step, the plate is placed back onto the deck of the workstation for reconstitution. A mobile phase compatible solution is placed in a reagent reservoir. The desired volume is aspirated from the trough and delivered into the destination wells in the microplate. The plate is removed from the deck; it is sealed, manually vortex mixed and then placed into an autosampler for injection into a chromatographic system.

10.3.1.8 Throughput Considerations

Although an analyst must remain at or near the instrument when methods are run, 96-tip workstations still yield rapid sample preparation times. Generally, the pipetting steps are very rapid and can be accomplished within 10 min. Additional time is added to the method by the wait time for gravity flow of sample and organic solvent through the beds of the microplate wells (10–20 min) and the time for the evaporation process (15–30 min). Only about 1 h is required to complete this method once the workstation is utilized. Prior to performing the procedure, additional time is required to reformat samples from source tubes or vials into a microplate. The throughput of SS-LLE methods using a 96-tip workstation is slightly faster than that of LLE using collection microplates (as discussed in Section 10.2.1.7) since fewer steps are required that involve manual intervention.

10.3.2 Method Example

A typical example of a solid-supported liquid-liquid extraction method that can be automated on the Quadra 96 in the microplate format is one developed for the extraction of a proprietary ß3-adrenergic receptor agonist (A) and its internal standard (B), shown in Figure 10.8, from human plasma [15]. Note that the original published method compared three sample preparation techniques: LLE in tubes, SS-LLE in tubes (ChemElut™) and SS-LLE in 48-well plates; all methods were performed manually. Described here is a typical example of how to automate such a method using the Quadra 96. Listed in Table 10.5 is a proposed SS-LLE method using 96-well plates (adapted from the reported 48-well plate method).

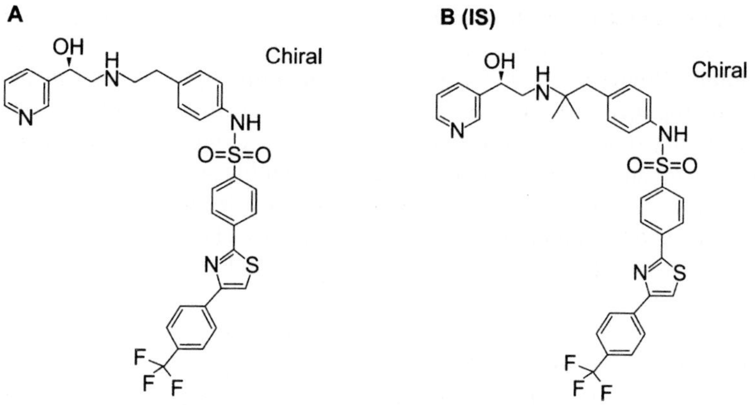

A

B (IS)

Figure 10.8. Chemical structures for a chiral proprietary ß3-adrenergic receptor agonist (A) and its chiral internal standard (B). Reprinted with permission from [15]. Copyright 2001 Elsevier Science.

Table 10.5
Typical example of a 96-well microplate solid-supported liquid-liquid extraction (SS-LLE) procedure

Prepare sample plate
• Aliquot 200 μL plasma from vials into deep well plate
• Add IS 20 μL
• Add 40 μL 0.5M Na_2CO_3
Load sample mixture into SS-LLE plate
Wait 5–10 min; gravity flow
Deliver MTBE 0.75 mL
Wait 5–10 min; gravity flow
Deliver MTBE 0.75 mL
Wait 5–10 min; gravity flow
Evaporate to dryness
Reconstitute in 125 μL acetonitrile/water (50/50, v/v)
Seal plate with pierceable cap mat
Inject aliquot using 96-well autosampler
Adapted from [15]

This SS-LLE method must be translated into a series of sequential actions for the Quadra 96 to execute. A typical Quadra 96 program for performing this method is detailed in Table 10.6. Each item, or numbered line in the program, involves an action to be performed at one of the six positions that is moved under the 96-tip aspirating/dispensing head. When pipetting, each line specifies a volume of liquid, an air gap volume, and the stage height setting as a number. Note that the stage height setting for each pipetting step depends on the specific nest type used and the preference of the user. Stage height settings are not included in this table as individual variations apply.

The Quadra 96 program is run with little manual intervention; one stopping point allows time for the off-line function of evaporating the organic solvent following its collection in the microplate. After the evaporation step, a reconstitution program is run and then the samples are ready for injection.

The reconstitution procedure aspirates a fixed volume from a reagent reservoir containing mobile phase compatible solvent and delivers into the plate containing dried extracts. If desired, this reconstitution step can be added to the end of the Quadra 96 program. However, it is suggested to maintain the reconstitution as a separate program so that the instrument is not occupied and unavailable for the length of time that it takes to dry down the eluate, typically 15–45 min. When the Quadra 96 is used daily by multiple analysts, there is often a waiting list or queue to perform other sample preparation assays. Closing the program at the end of the organic transfer step allows more efficient utilization of the instrument. The analyst who needs to reconstitute dried extracts then walks up to the instrument at the next available time slot, performs the procedure, and leaves the Quadra 96 available for the next procedure or analyst. Some hardware issues and tips and tricks in using the Quadra 96 for performing LLE in microplates are discussed next.

10.3.3 Hardware Issues

10.3.3.1 Configuring the Vacuum Manifold

The Tomtec vacuum manifold (#196-503) is actually composed of three pieces—a top, a bottom and a middle. The thin middle portion has the word TOMTEC printed on its face. Note that the middle portion is necessary only when using the short skirt extraction plates; it raises the top of the plate to compensate for the extra depth that the tips descend into the manifold. The tall skirt plates do not need this manifold insert. Short skirt and tall skirt plates are described in more detail in Chapter 3, Section 3.6.

Table 10.6
Typical Quadra 96™ program for performing solid-supported liquid-liquid extraction (SS-LLE) in the microplate format. The labware positions defined for this program are provided in Figure 10.7. Stage height settings are not included as individual variations apply.

 1. Load Tips at 1
 2. Aspirate 20 µL from 3, 25 µL Air Gap
 3. Dispense 45 µL to 6, 0 µL Blowout
 4. Shuck Tips at 1
 5. Pause
 6. Load Tips at 1
 7. Aspirate 40 µL from 4, 25 µL Air Gap
 8. Dispense 65 µL to 6, 0 µL Blowout
 9. Mix 150 µL 3 times at 6
10. Aspirate 265 µL from 6, 25 µL Air Gap
11. Aspirate 10 µL from 6 (stage height 0), 0 µL Air Gap
12. Dispense 300 µL to 5, 0 µL Blowout
13. Shuck Tips at 6
14. Pause
15. Load Tips at 1
16. Mix 400 µL 3 times at 2
17. Loop 2 times
18. Aspirate 375 µL from 2, 25 µL Air Gap
19. Aspirate 10 µL from 2, 0 µL Air Gap
20. Dispense 410 µL to 5, 0 µL Blowout
21. End Loop
22. Pause
23. Loop 2 times
24. Aspirate 375 µL from 2, 25 µL Air Gap
25. Aspirate 10 µL from 2, 0 µL Air Gap
26. Dispense 410 µL to 5, 0 µL Blowout
27. End Loop
28. Dispense 5 times at 2 [Time Dispense of 0 µL]
29. Shuck Tips at 1
30. Quit at 1

It is important that the tips from the extraction plate are in close contact with the wells of the collection plate inside the manifold, underneath the plate tips. The use of shims is sometimes necessary to raise the height of the collection plate inside the vacuum manifold to reach below the tips of the extraction plate. Also ensure that the tips do not descend too far into the well of the collection plate and reduce the available volume of the well; this may occur with a microplate having a small volume *e.g.*, a 0.35 mL deep well plate or any shallow well plate. The many varieties of collection microplates are described in more detail in Chapter 3, Section 3.3).

10.3.3.2 Working with SS-LLE Plates that are too tall

Some varieties of diatomaceous earth plates, when placed on top of the Tomtec vacuum manifold, are too tall and the Quadra 96 tips do not clear the top of the plate. For example, the 2 mL square well Array® Plate (IST Ltd., Hengoed United Kingdom; now an Argonaut Technologies Company) and the VersaPlate™ (Varian Inc., Harbor City, CA USA) are too tall for the Quadra 96 when using the Tomtec two piece manifold. In order to make these tall plate formats compatible with the z-axis clearance of the Quadra 96 two general approaches are commonly used.

One approach is to use a lower profile collection plate inside the Tomtec manifold, such as a 750 µL plate (Whatman Inc., Clifton, NJ USA) or a 1 mL plate having the shared wall technology from Nalge Nunc International (Rochester, NY USA). However, SS-LLE requires larger elution volumes than these small volume plates can hold so many sequential plates will be necessary, making this approach somewhat impractical. A 2 mL plate is preferred when using SS-LLE. Note that this lower profile collection plate works with the VersaPlate (tips extend much lower than the base plate) but not with the Array Plate (tips are shorter below the base plate).

The second approach toward compatibility (allowing the use of a 2 mL deep well collection plate inside the manifold) is to use a manifold that sits lower on the deck. Varian supplies one for use with its VersaPlate on the Quadra 96; it is designed to directly replace the Tomtec unit. The bottom of this manifold fits into the recess in the Quadra 96 deck. By setting the manifold lower, the workstation tips can now pass over the VersaPlate without encountering resistance.

10.3.4 Tips and Tricks

10.3.4.1 Quick Review

Several tips and tricks were discussed in an earlier section but are also pertinent for SS-LLE. The reader is encouraged to review the following sections: "Containing the Solvent Vapors" in Section 10.2.3.1, "Using the Polypropylene

Tips" in Section 10.2.3.2, and "Using Move Steps to Change Deck Position" in Section 10.2.3.6.

10.3.4.2 Exhaustive Elution

The use of SS-LLE plates, with certain sample preparation methods, may require organic solvent volumes for elution that are larger than 2 mL. If only 1.5 mL is used for elution, and incomplete recovery results, it may be that analyte is still retained on the particle bed. When evaluating an SS-LLE method for the first time or two, it is advisable to perform three successive elutions to ensure that analyte has been completely removed from the sorbent bed. Since a deep well collection plate holds a maximum volume of 2 mL, and some air space at the top of the well is needed for a dry-down step, fill the collection plate only to about 1.5 mL with each elution volume. Label 3 plates for elution as #1, #2 and #3 and successively elute from the sorbent bed into these three plates. Dry down the solvent from each, reconstitute and inject an aliquot into the chromatographic system. The amount or percentage of analyte recovered in elution fractions #1, #2 and #3 can be determined.

10.3.5 Applications

A fully automated application using 96-well plates filled with diatomaceous earth is reported by Peng [16]. These plates were actually prepared from bulk material (Varian) by placing about 260 mg/well into glass fiber filter plates (GF/C; Whatman); this filter at the well bottom is hydrophobic, ensuring retention of the aqueous phase until vacuum or pressure is applied. A 96-channel liquid handling workstation (Quadra 96) contributed to the high throughput of this method.

The SS-LLE plate was placed on top of a vacuum manifold and a collection plate inserted. A sample plate was prepared by the addition of 25 μL IS to wells containing 100 μL plasma. The Quadra 96 loaded 200μL of an aqueous buffer into the SS-LLE plate, followed by transfer of a 100 μL aliquot from the sample plate. Elution was performed with methyl ethyl ketone (two elutions, 400 μL each); this solvent is reported to work well for the extraction of polar compounds displaying high levels of plasma protein binding. Gentle vacuum (<1 in Hg or <34 mbar) was used. The time for the extraction was reported as 10 min. The contents of the collection plate were evaporated to dryness and the Quadra 96 was used for reconstitution of the dried extract in a mobile phase compatible solvent prior to injection into the chromatographic system.

A related application using the identical variety of 96-well SS-LLE plates was reported by the same authors for an initial purification of crude combinatorial libraries from butyl acetate [17]. In this case, excess amines were removed from the butyl acetate by protonation and retention on the solid support using a hydrochloric acid solution as the aqueous phase. Modification of the method results in removal of acidic and neutral water soluble components by changing the aqueous extraction solvent and adjusting its pH.

10.4 Automation of LLE Using a 4-/8-Probe Workstation

10.4.1 Introduction

Liquid-liquid extraction using individual tubes is one of the most labor intensive sample preparation schemes. Although the specific task of capping and uncapping tubes can be automated using a benchtop instrument that grips four tubes at a time (*e.g.*, Horizon Instrument Inc., King of Prussia, PA USA), procedures to reduce the total hands-on analyst time are desirable. Typical 1-, 4- and 8-probe workstations (See Chapter 5 for more information) can automate the pipetting steps of a liquid-liquid extraction procedure that uses test tubes and transform it into a semi-automated procedure. These (single and) multiple probe units can accommodate different tube to tube diameters for both the aspirate and dispense steps and are quite versatile.

A typical example of a 4-/8-probe liquid handling workstation is the MultiPROBE II (Packard Instruments, Meriden, CT USA, now a part of PerkinElmer Life Sciences). The discussion of automation strategies will focus on this instrument, although the general approach is similar for other 4-/8-probe workstations. This discussion also is focused at the automation of LLE in tubes—not microplates—to provide the reader with information on how to automate this common procedure. The 4-/8-probe liquid handlers can, at any step in the procedure, reformat samples from tubes into plates and perform tasks with the identical objectives discussed here.

10.4.2 Identify Tasks and Roles for Automation

The steps required to perform a typical liquid-liquid extraction procedure are listed in Table 10.7. The liquid handling workstation is able to perform only the pipetting tasks, leaving a large number of manual duties to the analyst. These operations include capping and uncapping of tubes, vortex mixing, centrifugation and evaporation. These tasks are performed many times and contribute to the hands-on analyst involvement with this sample preparation

Table 10.7
Comparison of manual vs. automated capabilities in performing LLE functions

Task	Manual	Liquid Handling Workstation
Uncap sample source tube	X	
Pipet an aliquot of sample into clean tube	X	X
Cap sample source tube	X	
Pipet aliquot of internal standard	X	X
Pipet pH adjustment solution	X	X
Briefly vortex mix tube	X	
Pipet organic solvent	X	X
Cap tube	X	
Vigorously mix for 10–15 min	X	
Centrifuge for 10 min	X	
Uncap tubes	X	
Pipet organic layer into a clean tube	X	X
Evaporate tube organic layer to dryness	X	
Reconstitute	X	X
Vortex mix	X	

procedure. The role for automation in this case is to perform the pipetting functions involving solvent delivery and tube to tube transfer.

10.4.3 Deck Layout

A typical MultiPROBE deck configuration for liquid-liquid extraction in tubes consists of the following labware components:

- Test tube rack for the extraction step
- Test tube rack to collect the organic solvent after extraction (usually a designated area within one rack is sufficient)
- Four solvent troughs to contain the internal standard solution, aqueous buffer, organic solvent and reconstitution solution
- Disposable tip rack with tip chute for disposal of used tips
- Microplate to contain the final reconstituted samples for injection

10.4.4 Preparation of the Sample Tubes

The analyst must decide at what part of the method sequence the liquid handler first becomes involved. Commonly, users prefer to manually aspirate a sample aliquot from the original thawed source tube (following its centrifugation to pellet clots) into a clean test tube for the sample preparation procedure. A rack of these prepared tubes is then placed onto the deck of the liquid handling workstation. The workstation then dispenses a fixed volume of the internal standard into each sample tube. Ideally, the sample and IS should be mixed at this point so that the IS interacts with the sample matrix in the same manner as the analyte; a post-dispense mix (aspirate and dispense multiple times within the tube) can be performed using disposable tips. A volume of aqueous buffer is then added from a solvent trough to adjust the sample pH. Brief mixing is performed in another post-dispense step using disposable tips. The sample tube rack has now been prepared using minimal hands-on analyst time.

Even though the workstation will perform all pipetting tasks, many analysts prefer to pipet the sample, internal standard and pH adjustment solution manually into each tube because they are used to doing it that way. Regardless, the rack of prepared sample tubes is placed onto the deck of the workstation for organic solvent to be dispensed. When samples are pretreated manually, automation in this case begins at the organic solvent delivery step.

At times, analysts may prefer to manually deliver the organic solvent into the test tubes because they can do it faster than the instrument. Note that this latter approach may be valid, but two of the important roles that automation provides are elimination of the human component (save analyst time so that other tasks can be performed) and introduction of pipetting consistency and accuracy into a procedure. Speed is not always the most important factor when automating a sample preparation method. Note that it is also possible for the analyst to begin the automated tasks at the point of liquid transfer of organic after the extraction, but clearly an opportunity is missed to reduce analyst hands-on time during all earlier steps in the procedure.

10.4.5 Elimination of Capping/Uncapping Tubes Using Septum Piercing

The task of uncapping and capping 96 sample tubes is a time consuming procedure that has been reported to take 30 min [7]. Also, this procedure is potentially hazardous as it exposes the analyst to the biological samples via accidental skin exposure upon dripping or via inhalation from droplet aerosol formation (a concern with samples from virus infected patients). Another consideration is that physical stress on the human hands may potentially lead to an occupational hand injury when these repetitive tasks are performed daily over an extended time. The use of screw capped sample tubes having pierceable caps is one solution for this problem.

A report of pierceable caps on polypropylene sample tubes (56 x 16 mm) has described how the uncapping of these tubes during transfer has been eliminated [7]. The tube caps were modified by cutting a portion of the top of each cap and retrofitting it with a nonleaking resealable polymer septum. These sample tubes with modified caps were placed in a custom rack and secured on the deck of a MultiPROBE liquid handling workstation. The sample rack had a removable 96-hole cover that could be screwed down over the capped plasma sample tubes to hold them in place. This cover was required to prevent the tips of the workstation from lifting the tubes during retraction.

Some procedural notes are important to mention from this septum piercing application [7]. The probes of the liquid handling workstation encounter resistance when piercing septa for the first time, resulting in z-motor errors. Since by default the MultiPROBE software pierces during an aspirating step, when an error is encountered it stops further attempts; however, several attempts will be made during a dispense step. To circumvent this problem, a separate cap piercing step was created in the software while "dispensing" air into the tubes so that multiple attempts were made to pierce the septum during sample transfer from the capped tubes. Also, injection of air into each tube was beneficial to partially compensate for the vacuum that could result from pipetting a sample volume from a closed tube. Another point is that the standard rinse/wash cycle for the probes involves rinsing through the full inside of the probes but only around the outer bottom area. In order to rinse the length of the probes on their full outside surface, an additional program step was created to dip the probes into a wash solution (25% methanol in water). This software step is accomplished by aspirating from the bottom of the solvent trough and dispensing to waste.

10.4.6 Addition and Transfer of Organic Solvent

Once the samples have been prepared, capped, vigorously mixed and centrifuged, they are put back onto the deck of the workstation for transfer of the organic solvent into a rack of clean tubes. This transfer step can be accomplished by using liquid level sensing technology. When the organic solvent represents the top layer in a tube, the instrument pipetting probes descend and, upon sensing liquid, aspirate a programmed volume. As the task proceeds, the probes descend further into the tubes, stopping when the full volume has been aspirated. It is wise to first practice on tubes with blank samples (without analyte) and solvents to determine exactly how much organic volume can be aspirated and dispensed into a separate tube without greatly disturbing the interface layer between the two phases and accidentally aspirating some aqueous volume along with the organic. Once this value is determined, it can be tested on 20–30 tubes for confirmation of reproducibility before routine use.

Note that the liquid level sensing technology cannot sense methyl tert-butyl ether. In this case, programming the workstation involves having the probes descend into the tubes to a fixed depth, stay at that point, and then aspirate a known volume. The volume aspirated can be larger than the organic volume, in which case air will be aspirated at the end of the cycle because the probes stay at a fixed depth.

When the organic solvent is denser than water and is at the <u>bottom</u> of the tube, the procedure for pipetting using a workstation follows. The probes are programmed to go to the bottom of the tube, remain at that fixed point, and then begin to aspirate liquid. Organic solvent will be removed and the upper aqueous layer descends; aspirating is stopped at a volume that does not greatly disrupt the interface layer between the two phases, leaving some organic behind. The probes ascend and dispense the organic solvent into clean tubes for the next step of the procedure. See Tips and Tricks in Section 10.2.3.5 for more information on this technique.

10.4.7 Evaporation and Reconstitution

After the evaporation step, tubes are placed back onto the deck of the workstation for the reconstitution procedure. A mobile phase compatible solution is placed into a solvent trough. The desired volume is aspirated from the trough and delivered into the destination tubes. The tubes are removed from the deck, manually vortex mixed and then placed back onto the deck in their test tube rack. A transfer step is performed from the tubes into autosampler vials contained in a rack or into a microplate.

10.4.8 Throughput Considerations

The throughput of this procedure can be improved in several ways, depending on the hardware used in the laboratory. When an autosampler is available that accommodates 96-well plates, the reconstituted volumes can be transferred directly from each tube into a microplate well in the desired sample order, rather than placed into individual autosampler vials. The microplate is simply sealed with a pierceable cap mat or heat sealable film and placed in the autosampler tray for injection into the chromatographic system. If the volume of organic solvent used for the extraction is small enough (about 1.5 mL or less) it can be dispensed directly into a microplate prior to the evaporation step, rather than afterward. In this scenario, the microplate format is retained from the evaporation step to reconstitution and through injection.

This section of the chapter has shown how to introduce some automated tasks into a liquid-liquid extraction procedure in the tube format by using a liquid handling workstation. An analysis of the time requirements was made for LLE performed manually and via a semi-automated procedure using a 4-probe liquid handling workstation [18]. The total time required for the manual procedure was about three times longer than for the semi-automated procedure (4 h, 50 min manual; 1 h 43 min semi-automated). An evaluation from a different perspective, that of analyst hands-on time, revealed numbers of 4 h 10 min for the manual procedure, but <10 min for the semi-automated procedure, as shown in Table 10.8. However, note that this reported manual procedure used tubes exclusively throughout, whereas the semi-automated procedure being compared used smaller sample and solvent volumes in tubes and then the organic extracts were transferred from tubes into the microplate format for dry-down. In this case, the microplate format was retained from the point of reconstitution.

Table 10.8
Time analysis of liquid-liquid extraction comparing manual preparation with semi-automated preparation using a 4-probe liquid handling workstation

	Manual LLE	Semi-Automated LLE
Standard curve preparation	25 min	8 min
Labeling of tubes	15 min; two sets of tubes, one set of autosampler vials	5 min for one set of tubes
Sample transfers and extraction	180 min	65 min
Drying	40 min	20 min
Reconstitution and transfer	30 min	5 min
Total time	4 h 50 min	1 h 43 min
Analyst time	4 h 10 min	<10 min

Reprinted with permission from [18]. Copyright 1999 Elsevier Science.

Clearly, automation offers time savings and frees the analyst from many of the repetitive tasks of tube manipulation. Automation using the tube format at the beginning of a procedure is feasible as reported [18, 19], but note that it is also quite convenient to reformat samples from tubes into a microplate at either the extraction step or the organic isolation step. At each juncture, time savings and conveniences from using the microplate format result, *e.g.*, crimping or sealing 96 individual autosampler vials can be replaced with a single sealing procedure by securing a pierceable cap mat onto a collection microplate.

Acknowledgments

The assistance of the following vendor representatives who contributed information and/or reviewed text is appreciated: Robert Speziale (Tomtec) and Susan Williams (Mettler-Toledo Autochem). Line art illustrations were kindly provided by Woody Dells.

References

[1] S. Steinborner and J. Henion, Anal. Chem. 71 (1999) 2340-2345.
[2] J. Zweigenbaum, K. Heinig, S. Steinborner, T. Wachs and J. Henion, Anal. Chem. 71 (1999) 2294-2300.
[3] J. Zweigenbaum and J. Henion, Anal. Chem. 72 (2000) 2446-2454.
[4] J.M. Onorato, J.D. Henion, P.M. Lefebvre and J.P. Kiplinger, Anal. Chem. 73 (2001) 119-125.
[5] N. Brignol, L.M. McMahon, S. Luo and F.L.S. Tse, Rapid Commun. Mass Spectrom. 15 (2001) 898-907.
[6] S.H. Hoke II, J.A. Tomlinson II, R.D. Bolden, K.L. Morand, J.D. Pinkston and K.R. Wehmeyer, Anal. Chem. 73 (2001) 3083-3088.
[7] D.S. Teitz, S. Khan, M.L. Powell and M. Jemal, J. Biochem. Biophys Meth. 45 (2000) 193-204.
[8] L. Ramos, R Bakhtiar and F.L.S. Tse, Rapid Commun. Mass Spectrom. 14 (2000) 740-745.
[9] Z. Shen, S. Wang and R. Bakhtiar, Rapid Commun. Mass Spectrom. 16 (2002) 332-338.
[10] N. Zhang, K.L. Hoffman, W. Li and D.T Rossi, J. Pharm. Biomed. Anal. 22 (2000) 131-138.
[11] J. Ke, M. Yancey, S. Zhang, S. Lowes and J. Henion, J. Chromatogr. B 742 (2000) 369-380.
[12] K. Heinig and J. Henion, J. Chromatogr. B 732 (1999) 445-458.

[13] N. Zhang, S.T. Fountain, H. Bi and D.T. Rossi, Anal. Chem. 72 (2000) 800-806.

[14] H. Zhang and J. Henion, J. Chromatogr. B 757 (2001) 151-159.

[15] A.Q. Wang, A.L. Fisher, J. Hsieh, A.M. Cairns, J.D. Rogers and D.G. Musson, J. Pharm. Biomed. Anal. 26 (2001) 357-365.

[16] S.X. Peng, T.M. Branch and S.L. King, Anal. Chem. 73 (2001) 708-714.

[17] S.X. Peng, C. Henson, M.J. Strojnowski, A. Golebiowski and S.R. Klopfenstein, Anal. Chem. 72 (2000) 261-266.

[18] M. Jemal, D. Teitz, Z. Ouyang and S. Khan, J. Chromatogr. B 732 (1999) 501-508.

[19] W. Naidong, W. Shou, Y.-L. Chen and X. Jiang, J. Chromatogr. B 754 (2001) 387-399.

Chapter 11

Solid-Phase Extraction: High Throughput Techniques

Abstract

Solid-phase extraction (SPE) is a versatile and selective method of sample preparation in which analytes are bound onto a solid support, interferences are washed off and analytes are selectively eluted for further workup and analysis. The fundamental principles of this technique are presented in this chapter. The many different choices for sorbent chemistry that make SPE a very powerful method for sample preparation also create many different ways in which sorbents can be used. In order to help the reader understand SPE methodology, many extraction protocols are presented as examples, arranged by sorbent chemistry. The basics of high throughput SPE are introduced by reviewing packed particle and disk formats, plate configurations, and the mechanics of using 96-well plates to perform sample preparation methods. Published applications for solid-phase extraction in microplates are also reviewed. Subsequent companion chapters present method development and optimization principles for SPE (Chapter 12) and high throughput automation strategies using the microplate configuration (Chapter 13).

11.1 Understanding the Technique

11.1.1 Fundamental Principles

Solid-phase extraction (SPE) is a specific type of sample preparation in which an analyte, contained in a liquid phase, comes in contact with a solid phase (sorbent particles contained within a packed bed or a disk) and is selectively adsorbed onto the surface of that solid phase by chemical attraction. All other materials not adsorbed remain in the liquid phase and pass through the sorbent particle bed to waste. A wash solution is passed through the sorbent bed to ideally remove adsorbed endogenous contaminants from the sample matrix, yet retain the analyte of interest on the solid phase. A selective elution step is then performed in which the analytes partition away from the solid support and into another solvent in which there is a greater affinity than for the sorbent bed. The

361

overall procedure for solid-phase extraction is illustrated in Figure 11.1.

The key points to successful use of solid-phase extraction for sample preparation are:
1. **Proper choice of sorbent phase chemistry to attract the analyte of interest**
2. **Efficient utilization of pH to manipulate the analyte into the desired ionic or neutral form**
3. **Solubility of analyte in the solutions used for the extraction method**

Adjustments in pH are used to vary retention and selectivity of the solid-phase extraction process. The pH of the solution is very important during the **adsorption** step to promote analyte retention. Likewise, pH adjustments are often also made during the **wash** step for selective removal of interferences. During the **elution** step, pH can be used to either ionize the analyte or make it neutral; likewise, in some cases the solid sorbent particle to which the analyte is bound can be made ionized or neutral to facilitate analyte release. Note that an SPE procedure that is not using pH control during at least one of the steps is

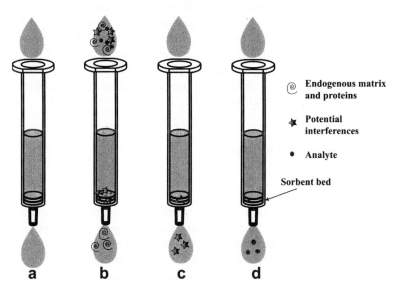

Figure 11.1. The basic steps for solid-phase extraction involve (a) conditioning the sorbent bed, (b) loading analytes, (c) washing away interferences and (d) selective elution for further workup and analysis. The SPE product format shown is for a disk in a cartridge or column although the procedure is generally similar regardless of format.

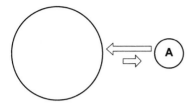

Figure 11.2. An analyte (A) is adsorbed onto the surface of a sorbent particle by a chemical attraction.

not fully exploiting the utility of this powerful technique. The reader is encouraged to review the influence of sample pH on ionization as discussed in Chapter 2, Section 2.3.1 to reinforce the concepts of pH manipulation and the basics of extraction chemistry.

Three fundamental concepts are required in understanding and using SPE:

1. A strong chemical attraction must exist (*e.g.*, nonpolar, ion exchange or polar) between the analyte and the sorbent chosen so that the analyte preferentially adsorbs to the solid sorbent as it passes through the bed (Figure 11.2).

2. Adsorption requires that sufficient residence time is permitted for the interaction of analyte and sorbent to occur (Figure 11.3). The sorbent particles have a certain distance between them, and the analyte in solution flows through the sorbent bed at a certain linear velocity. Traditional geometries of sorbents packed in columns have large diffusion distances, thus the flow rate of solution through the bed is performed at slow rates. The advent of particles packed more tightly into a disk format has reduced this need for slow flow rates during adsorption.

3. The elution solvent chosen (*e.g.*, an organic solvent such as methanol or acetonitrile, with or without pH modification) must be strong enough to disrupt the attraction between analyte and the solid phase, causing desorption, or elution from the particle (Figure 11.4).

The volume of elution solvent necessary is typically larger than is desirable as a result of the excess particle mass used in the SPE procedure. It is usually necessary to perform a solvent exchange before injection into a chromatographic system for analysis and detection.

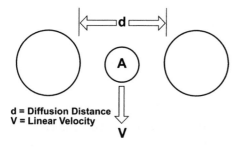

Figure 11.3. An analyte (A) passes through a sorbent particle bed at a certain linear velocity (V). The particles are separated within the bed at a certain distance (d).

11.1.2 Advantages

Numerous advantages to the solid-phase extraction technique for sample preparation include: very selective extracts (reduce the potential for ionization suppression from matrix materials); wide variety of sample matrices accepted; analytes can be concentrated; high recoveries with good reproducibilities; improved throughput via parallel processing; low solvent volumes; suitable for full automation; and no emulsion formation as seen with liquid-liquid extraction (LLE). In addition, so many formats and chemistry choices are available that nearly all requirements for sample preparation can be met. The technology has been improved in recent years with the introduction of more selective but also more generic solid sorbent chemistries, improved disk based SPE devices, smaller bed mass sorbent loading, improved plate formats for using smaller volumes, and faster and more efficient automation hardware.

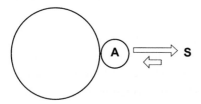

Figure 11.4. An analyte (A) is desorbed, or eluted, from a solid particle when its attraction is disrupted by a particular solvent (S) that is passed through the particle bed. Note that a change in pH and/or ionic strength can also facilitate analyte release.

11.1.3 Disadvantages

Many of the disadvantages commonly mentioned for SPE are directed toward the perceived difficulty to master its usage. The great selectivity and the many choices for manipulating pH and solvent conditions make it difficult to grasp the chemistry of the technique, although this chapter and the next are intended to educate the reader to be successful. As a result of this perceived complexity, generic sorbents and methods have been developed to make it easier to perform SPE. The procedure itself has many sequential steps, which result in increased time requirements. However, efficient utilization of a lower sorbent mass product is the key to improved throughput in this regard.

Occasionally issues arise related to slow or difficult flow of sample through the particle bed; precipitated fibrins, thrombins or other particles in the biological sample are often at fault. In order to avoid this potential problem, samples are usually centrifuged or filtered prior to extraction and this step is especially important with the use of automated methods. Sometimes a larger particle size packing (*e.g.*, 100 µm instead of 40 µm) is used to improve flow characteristics. The batch to batch variation of bonded silica sorbents had infrequently been a cause of concern for users, although in recent years manufacturers have improved the quality control of their products.

Also related to the time issue is the historical perception that the SPE method development process takes too much time. Again, the focus of the next chapter presents strategies for rapidly arriving at a selective method in only a few experiments. The cost of performing a solid-phase extraction procedure can be greater than that of other techniques, such as protein precipitation (PPT) and liquid-liquid extraction. In return for its greater selectivity, which is of paramount importance in the bioanalysis of clinical samples which strive to reach the lowest quantifiable concentrations (pg/mL), the cost per sample may be greater. However, any estimate of cost should not include only the consumable items needed to perform the extraction but also must factor in the time savings that the approach offers.

Some users perceive the off-line approach to SPE as a disadvantage because a batch of samples must be prepared independently of the analysis and then taken to the instrument, rather than performed in a connected (on-line) step. On-line SPE techniques are presented in Chapter 14, Section 14.3.6. Other users may cite as a disadvantage that utilization of SPE in a high throughput manner requires an additional investment in automation hardware and training.

11.2 Sorbent Chemistries and Attraction Mechanisms

11.2.1 Nature of the Sorbent Particle

11.2.1.1 Silica Sorbents

Two major types of sorbent particles are used for SPE—silica and polymer. Raw, underivatized silica is an amorphous, porous solid that contains polysiloxane (Si–O–Si) and silanol groups (Si–OH). The presence of these silanol groups allows polar adsorption sites and makes the surface weakly acidic. Silica is covalently bonded to a functional group through a chemical process. These groups are commonly alkyl chains of varying length, such as octadecyl (C18), octyl (C8), butyl (C4), ethyl (C2) and methyl (C1). The base silica used to produce a bonded sorbent has an impact on the performance of the finished product. An illustration of an alkyl chain bonded to silica is shown in Figure 11.5.

The chemical bonding reaction with silica can result in a monomeric species or a polymeric species. The polymeric bonding process is most often used to achieve a greater degree of alkyl loading with increased stability toward hydrolysis. However, this process is less reproducible batch to batch and can

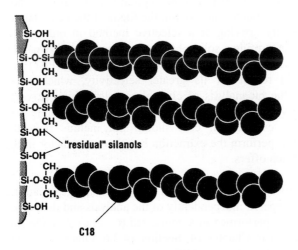

Figure 11.5. Illustration of the surface of silica bonded with an octadecyl (C18) alkyl chain. Note the presence of residual silanol groups on the silica surface. Illustration reprinted with permission from LC Resources.

result in a high degree of residual terminal silanol groups. One cause of this variability is that reagents in the synthesis tend to react with trace amounts of water and then polymerize in solution before bonding to the surface hydroxyls in ways that may or may not conform to the original surface [1]. The monomeric process provides better control of the reaction (better reproducibility) and higher silica activity.

After the bonding process, the endcapping of residual unreacted silanols is performed through a reaction with a derivatized silane, shown in Figure 11.6. However, an important point to remember is that endcapping never completely covers all the silanols. This remaining percentage has been found to vary among manufacturers of bonded silica particle chemistries and can play an important role in the selectivity for many basic analytes.

Some common specifications for bonded silica particles are listed in Table 11.1. Surface area is a function of the average pore diameter—the smaller the pore diameter, the higher the surface area. A wide range of surface areas and pore diameters can be prepared. The %carbon loading is influenced by the functional group bonded to silica (*e.g.*, C18 bonded silica contains a higher %carbon than C2 bonded silica).

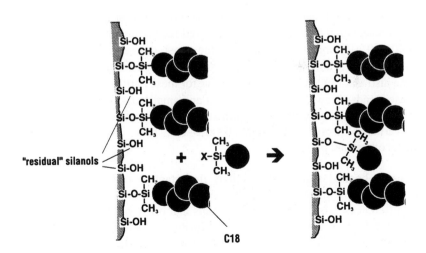

Figure 11.6. An endcapping reaction to mask the residual silanols on the silica surface is performed after a functional group has been bonded to the silica particle. Illustration reprinted with permission from LC Resources.

Table 11.1
Typical specifications for bonded silica particles used for solid-phase extraction

Particle size	8–12 µm, 12–20 µm or 40–60 µm
Particle shape	Irregular
Surface area	350–450 m²/g
Pore size	60–80 Å
Bonded phase preparation	Trifunctional
Percent carbon	C18: 18%, C8: 12%, C2: 6%

The particle size used for solid-phase extraction has been shown to affect flow characteristics and performance. The traditional 40–60 µm particle size range offers a good flow rate for biological fluids through a particle bed at a reasonable price. In the case of biological matrices that are extremely viscous (*e.g.*, horse urine, blood, tissues and breast milk) and difficult to pass through these beds, a larger 100 µm particle size in a packed bed is useful to provide better flow characteristics. An example of this larger particle sized product is XtrackT® (United Chemical Technologies, Bristol, PA USA).

Some cases warrant use of a smaller particle size. These smaller particles (10–20 µm) are packed tightly in a bed and have generally been shown to offer improved extraction kinetics and performance (but can offer more resistance to flow). A presentation by Roach *et al.* compared recoveries for a set of analytes using 40, 20, 10 and 5 µm particles; the most dramatic improvement in performance was seen when the particle size decreased from 20 µm to 10 µm [2]. Note also that raw particle sources differ for these particle sizes; 20 µm and larger use standard raw silica, while 10 µm and smaller use HPLC grade silica which has higher purity, greater symmetry and less residual metals content. These variables present additional factors that may influence extraction efficiency other than particle size.

When using bonded silica sorbents it is important to realize that the same chemistries among manufacturers do not perform in an identical manner. Subtle differences exist in their specifications that often yield slightly different performance for the same analyte. The following characteristics are known to vary: particle size distribution; pore size distribution; surface area; mono-, di- or trifunctional bonding process; %carbon loading; and the type and degree of end capping (which in turn influences the amount of residual silanols).

Exchanging one variety of C18 bonded silica for another may either improve or reduce the analyte recovery. As an example, the reader is referred to a report by Martin *et al.* [3] which describes the effect of carbon loading on the extraction properties of an acidic (anisic acid) and basic (propranolol) analyte using a range of C18 bonded silica products. In this report, an intermediate %carbon loading of 15% was preferred for these two analytes. End capping effects were only seen for the basic analyte whose affinity was mediated by silanol interactions. A similar investigation from the same laboratory was reported using radiolabeled β-blockers (practolol, propranolol, ICI 118,551 and epanolol) as the analytes [4]; again, a moderate %carbon was shown to be optimal and residual silanols on the silica surface were shown to play an important role. Further support for the role of ionic interactions of β-blockers with residual silanols was reported in a subsequent study [5]. A base-deactivated octyl bonded silica was shown to interact with radiolabeled propranolol primarily by hydrophobic interactions; secondary silanol effects were suppressed [6]. Comparative studies of bonded silica varieties are frequently described in the literature for specific analyte extractions, such as the report by Papadoyannis [7]. An evaluation model using analyte probes is useful to quickly characterize the silanol content and influence among sorbent brands [8].

11.2.1.2 Polymer Sorbents

An alternative to the traditional bonded silica sorbent is a polymer sorbent. The first polymer sorbent for SPE was a cross-linked poly(styrene divinylbenzene) as shown in Figure 11.7. The advantages of a polymer sorbent include the following:

- 100% Organic composition (chemically synthesized)
- Stable across the entire pH range from 0–14
- Predictable attraction (no silanols)
- More consistent manufacturing process batch to batch
- Greater capacity per gram than bonded silica

The sorbent poly(styrene divinylbenzene) is commonly abbreviated as SDB, SDVB or PSDB and is available from sorbent manufacturers under many different brand names. This polymer imparts a different selectivity than C18 bonded silica by nature of the aromatic rings and vinyl groups in its chemical structure. Generally, it is able to adsorb a wider range of analytes; some more polar analytes are able to be captured compared with traditional reversed phase bonded silica sorbents. A common misconception is that C18 groups are bonded onto polymers; while C18 bonded to a polymer is reported [9], most

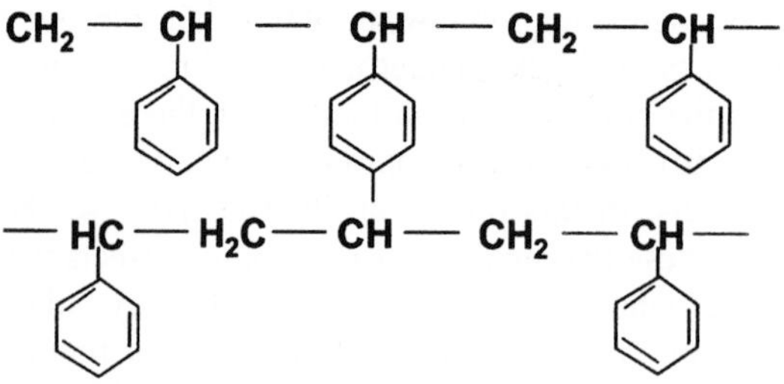

CH₂ — CH   —   CH — CH₂ — CH —

— HC — H₂C— CH — CH₂ — CH —

Figure 11.7. An original polymer sorbent useful for solid-phase extraction is poly-(styrene divinylbenzene).

polymer sorbents impart their selectivity through a different reversed phase attraction using an affinity for their aromatic rings and attached functional groups.

A modified and improved version of the SDB polymer sorbent was introduced in 1996 as Oasis® HLB (Waters Corporation, Milford MA USA). This sorbent is a synthetic polymer of divinylbenzene and N-vinylpyrrolidone. The letters HLB refer to "Hydrophilic Lipophilic Balance," describing hydrophilicity in terms of its wetting properties and lipophilicity for analyte retention (Figure 11.8). The typical particle size range for this polymer is 25–35 µm. The Oasis HLB sorbent has shown great utility for the solid-phase extraction of a wide variety of analytes having diverse chemical structures, as reported [10] for six structurally diverse substrates and their metabolites.

While this sorbent can perform selective extractions when a method has been optimized for particular analytes, its great appeal is its ability to extract analytes with success using a generic methodology (condition with methanol then water, load sample, wash using 5% methanol in water, and elute with methanol). This common approach is especially appealing for drug discovery bioanalysis where time constraints are great. Summaries of the characterization of a hydrophilic lipophilic balanced SPE sorbent have been published by Bouvier *et al.* [11, 12] and useful discussion of this same polymer sorbent is contained within a review of sample preparation methods by Gilar, Bouvier and Compton [13]. Many published reports have shown the utility of such a generic sorbent for solid-phase extraction methods in both individual columns [10, 14–33] and 96-well plates [34–57].

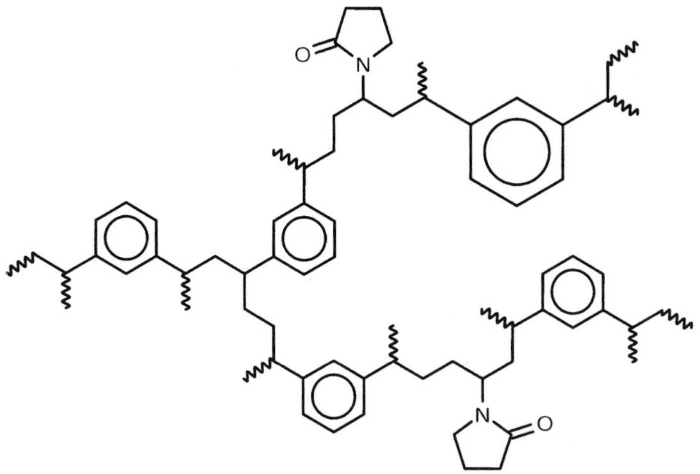

Figure 11.8. Chemical structure of the Oasis® HLB sorbent, a synthetic polymer of divinylbenzene and N-vinylpyrrolidone. Illustration reprinted with permission from Waters Corporation.

The success of the Oasis product, in terms of both performance and marketing, sparked competitors to match and improve upon the polymer approach to SPE to meet the same or unmet needs. The exact chemical nature of these other polymer species is sometimes guarded as proprietary. Some examples of polymer choices from other manufacturers include FOCUS™, NEXUS™, LMS and PPL (Varian Inc., Harbor City, CA USA), Strata™ X (Phenomenex, Torrance, CA USA), ISOLUTE® 101 (SDB) and ENV+ (hydroxylated SDB; IST Ltd., Hengoed, United Kingdom, now an Argonaut Technologies Company), Universal Resin (3M Corporation, St. Paul, MN USA) and POLYCROM™ (CERA Inc., Baldwin Park, CA USA).

A functionalized styrene divinylbenzene polymer called FOCUS™ (Varian) is noteworthy for its ability to provide three mechanisms of polar interaction—proton donor, proton acceptor and dipole-dipole—in addition to a hydrophobic attraction on the base polymer. These polar interactions can occur at sites on the sorbent under neutral conditions. According to the manufacturer, a strict control of pH is therefore unnecessary. This FOCUS polymeric sorbent has demonstrated strong retention for a variety of polar molecules, as shown in Figure 11.9. Some model drugs described in promotional materials from the manufacturer as especially interactive with FOCUS sorbent are atenolol, ranitidine and pseudoephedrine.

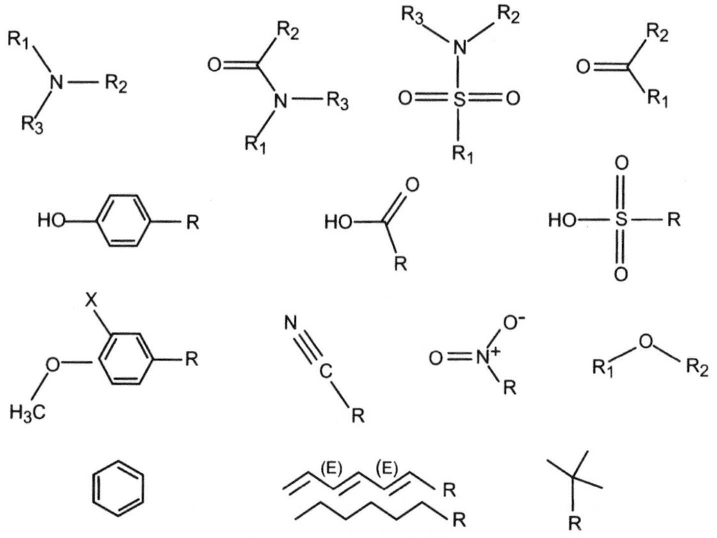

Figure 11.9. The FOCUS™ polymer sorbent (Varian) provides polar drug interactions, as well as traditional hydrophobic interactions, at neutral pH conditions. Examples of interacting analyte functional groups are shown above. Illustration reprinted with permission from Varian Inc.

A typical extraction procedure using FOCUS follows. After the load step, a solution of 10–20% acetonitrile in water (v/v) is tolerated as a wash solvent; methanol is also effective at these and higher concentrations. The retention capability for polar compounds allows the use of these stronger solvents in the wash step. A more efficient wash step in this manner provides cleaner extracts by eliminating matrix constituents from biological samples and reduces the ionization suppression often seen with LC-MS and LC-MS/MS analyses. A generic elution solvent recommended by the manufacturer is methanol/acetonitrile/water/acid (60:30:10:0.1, by volume). The attractive features of the FOCUS chemistry include the ability to eliminate pH control and yet achieve enhanced retention for polar analyte species.

A recent trend has been to derivatize a polymer sorbent by adding an additional functional group (*e.g.*, to allow for cation exchange or anion exchange) and altering its selectivity. A sulfonated version of the Oasis polymer is named Oasis MCX and it is used for performing cation exchange; derivatization of the Oasis polymer with quaternary amine functionality yields Oasis MAX for anion exchange (Waters Corporation). These two derivatized sorbents act as mixed mode ion exchangers, enabling the retention of a range of acidic, neutral and basic drugs. Modification of the extraction protocol enables Oasis MCX to selectively retain basic drugs, and Oasis MAX to retain acidic drugs. Note that the performance of Oasis HLB, MCX and MAX are not compromised if the particle bed runs dry after conditioning. Mixed mode attraction mechanisms are discussed further in Section 11.2.2.4.

A modest improvement to existing polymer sorbent chemistries is the ability to eliminate the conditioning step and simply load the sample onto a dry particle bed. This approach simplifies the overall procedure. An example of such a sorbent is Oasis MCX (a strong cation exchange version of Oasis HLB). A related and useful feature is the ability to dry the cartridge bed after sample loading without an adverse effect on recovery or other performance parameters.

A review of polymer based sorbents for solid-phase extraction has been prepared by Huck and Bonn [58]. Several reports of polymer derivatives for SPE are available. High capacity carboxylic acid functionalized resins for the extraction of a range of organic compounds from water are reported by Ambrose *et al.* [59]. The utility of a lightly sulfonated SDB has been reported for isolation of a sulfoxide metabolite of 2'-deoxy-3'-thiacytidine in human urine [60]; synthetic details of this resin's sulfonic acid capacities are reported [61, 62]. A review of chemically modified polymeric SDB sorbents is provided by León-González and Pérez-Arribas [63].

11.2.1.3 Additional Sorbent Varieties

11.2.1.3.1 Dual-Zone

The capacity of traditional bonded silica and polymer sorbents can be adversely affected by the amount of protein present in the sample matrix. As sample is loaded onto the sorbent bed, proteins are known to occupy binding sites and make those sites unavailable for analyte; the presence of protein can also alter sorbent selectivity. A recurring problem with proteins is that some amount may potentially remain adsorbed following the wash step and be eluted with analyte under favorable conditions. Dual-Zone™ SPE sorbents (Diazem Corporation, Midland, MI USA) offer a solution to overcome these difficulties with proteins in the sample matrix.

Dual-Zone sorbents are unique in that the outer and inner particle surfaces are bonded with different materials. The outer particle surface is bonded with a hydrophilic, nonadsorptive ligand that repels proteins while the interior surface of the pores is bonded with an adsorptive partitioning phase such as C18, C8, *etc.* Proteins and other large biomolecules cannot enter the hydrophobic inner pores and are efficiently washed from the particle bed during the load and wash steps. This patented bonding process also allows better control in the reduction of residual silanols. Dual-Zone chemistries include reversed phase, ion exchange and mixed mode. In general, particles which allow permeation and partitioning of small molecules on bonded phases that are protected from contamination by macromolecules are referred to as restricted access media [64] and these are discussed in more detail in Chapter 14, Section 14.2.3.

11.2.1.3.2 Phenylboronic Acid

Specialty phases have been developed that are bonded to silica and/or polymer sorbents resulting in unique selectivities. Also, alternative chemical sorbents are used for solid-phase extraction methods. While not as widely used or generally applicable for a range of analytes, these additional sorbents have been found useful in certain bioanalytical applications. An example of a specialty bonded phase is phenylboronic acid.

A phenylboronic acid (PBA) phase is very specific for isolating coplanar vicinal hydroxyl molecules such as found in catecholamines (*e.g.*, epinephrine, norepinephrine and dopamine). Under slightly alkaline load conditions (pH 8.7), these catecholamines are extracted onto the PBA; a selective retention occurs due to the formation of a cyclic ester between the boronic acid bound to the stationary phase and the 1,2-diol functional group on the catecholamines. Following a wash step to remove residual components, the pH is made acidic, causing the cyclic boronic acid ester to hydrolyze and the catecholamines to be eluted. [65, 66]. A book chapter by Wilson and Martin provides a review of this specific extraction affinity [67]. In addition to PBA, alumina as a solid phase sorbent has been shown useful for catecholamine extraction [68, 69].

11.2.1.3.3 Graphitized Carbon

Graphitized carbon presents different chemical properties than C18 bonded silica and has been investigated as an alternative reversed phase sorbent for SPE. A variety of basic β-blocker drugs was efficiently extracted from dog plasma and recovered using an elution solvent of chloroform/methanol (80:20, v/v) [70]. The same study extracted metabolites of ibuprofen from human urine. A related communication describes the use of graphitized carbon as a liquid chromatographic sorbent for the analysis of polar drug metabolites in urine [71]. Carbon presents no silanols and offers a strongly hydrophobic phase for adsorption of analytes. However, a difficulty can be encountered for desorption of strongly retained analytes. Graphitized carbons for SPE have been reviewed by Matisová and Škrabáková [72] and Hennion [73].

11.2.1.3.4 Silicalite

A molecular sieve known as silicalite was used for the solid-phase extraction of organic analytes from aqueous samples. This molecular sieve is unique in that its intricate system of channels is able to retain organic compounds via hydrophobic mechanisms. An additional advantage is that many low molecular weight hydrophilic compounds are also retained [74].

11.2.1.3.5 Diatomaceous Earth

Diatomaceous earth is used in a manner similar to SPE but it is also closely related to liquid-liquid extraction. The untreated biological sample in aqueous liquid is adsorbed onto the surface of the polar diatomaceous earth, which acts as a support for the aqueous phase. A large surface area is provided for partitioning into an elution solvent (100% organic) that is then passed through the sorbent bed using gravity flow. The elution is a continuous process [75]. This technique is also referred to as solid-supported liquid-liquid extraction and is discussed in more detail in Chapter 8, Section 8.3.

11.2.1.3.6 Molecularly Imprinted Polymers

Molecularly imprinted polymer (MIP) materials for solid-phase extraction provide a potential new source of highly selective chemistries. The molecular imprinting technique utilizes a highly selective binding with antibody-like affinities for a target analyte on a stationary phase. Further discussion of MIP is outside the scope of this text and the reader is referred to the book edited by Sellergren [76], a review [77] and various applications of the technology as used to extract analytes from biological matrices [78–84].

11.2.2 Attraction Mechanisms

11.2.2.1 Introduction

Many different choices of sorbent chemistries are available in solid-phase extraction to accomplish the same objective. The number of these chemistries can be daunting. These choices are better understood by grouping them according to their mechanism of chemical attraction. Four primary modes of analyte attraction to a solid sorbent particle are:

1. Reversed phase (also called nonpolar)
2. Ion exchange
3. Mixed mode (also called mixed phase)
4. Normal phase (also called polar)

The two most prevalent attraction mechanisms utilized in bioanalytical applications are reversed phase and mixed mode interactions (a combination of reversed phase and ion exchange). An observation by the author is that, in practice, ion exchange is used more often as a component of the mixed mode interaction rather than as ion exchange alone. One reason that pure ion exchange is not used much is that biological samples naturally contain a high concentration of ions (from salts) that do not allow the desired attraction to occur on an ion exchange sorbent. Samples must be diluted several-fold in order to work with ion exchange.

11.2.2.2 Reversed Phase

The attraction of analytes onto a reversed phase sorbent is primarily attributed to weak van der Waals forces (also called induced dipole interaction). This type of electrostatic attraction occurs between the nonpolar portions of molecules (*e.g.*, hydrocarbon and aromatic systems) and is brought about by a mutual distortion of electron clouds making up the covalent bonds. The hydrophobic bonding energy is very weak (0.5–1.0 kcal/mole for each atom involved) and is temperature dependent (important at low temperatures but not at high). The attraction occurs only over a short distance and steric factors strongly influence van der Waals attraction [85].

An important note when using reversed phase bonded silica sorbents is that silanols can offer a secondary mechanism of attraction, which can increase or decrease retention efficiency, depending on the protocol and pH conditions used for the extraction method. Generally, above pH 4 the silanol sites will be ionized and available to interact with one or more positively charged nitrogen molecules on the analytes of interest.

Reversed phase sorbents are found in the polymer series and the bonded silica series. Some typical examples of reversed phase polymer sorbents include SDB, FOCUS, NEXUS, PPL, ENV+, Strata X, Universal Resin and Oasis HLB. Some typical examples of reversed phase bonded silica sorbents include C18, C8, C2 (Figure 11.10), phenyl, cyclohexyl and cyanopropyl. Different

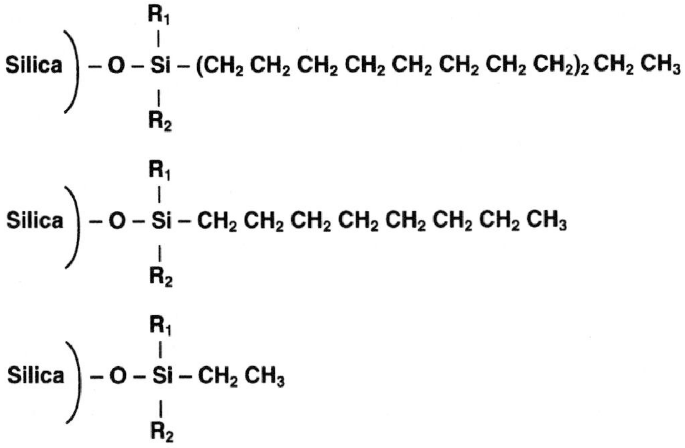

$$\text{Silica} \Big) - O - \underset{\underset{R_2}{|}}{\overset{\overset{R_1}{|}}{Si}} - (CH_2\ CH_2\ CH_2\ CH_2\ CH_2\ CH_2\ CH_2\ CH_2)_2\ CH_2\ CH_3$$

$$\text{Silica} \Big) - O - \underset{\underset{R_2}{|}}{\overset{\overset{R_1}{|}}{Si}} - CH_2\ CH_2\ CH_2\ CH_2\ CH_2\ CH_2\ CH_2\ CH_3$$

$$\text{Silica} \Big) - O - \underset{\underset{R_2}{|}}{\overset{\overset{R_1}{|}}{Si}} - CH_2\ CH_3$$

Figure 11.10. Illustration of silica bonded with alkyl chains of varying lengths. Shown from top to bottom are C18 (octadecyl), C8 (octyl) and C2 (ethyl) bonded silica.

degrees of hydrophobicity are exhibited by these bonded silica sorbents. Octadecyl silica is very nonpolar, and can be made even more lipophilic by increasing its carbon loading above 18% to about 21–22%. Shorter alkyl chain moieties such as C2 (ethyl), C1 (methyl) and cyanopropyl are at the opposite end of the polarity range. Since these chemistries are much less nonpolar than C18, they are often referred to as polar sorbents. Note that these polar sorbents are still used with aqueous sample matrices in a reversed phase mode.

11.2.2.3 Ion Exchange

Ionic attraction is common for inorganic molecules and salts of organic molecules and involves the attraction of a negative atom for a positive atom. The ionic bond formed has a strong attractive force of 5 kcal/mole or more and is least affected by temperature and distance [85]. This ionic bond is more difficult to break compared with van der Waals forces. The manipulation of solvent pH and ionic strength are necessary to promote conditions favoring retention and elution of analyte to the sorbent.

Another important ionic chemical bond formed is the ion-dipole bond, seen when an organic salt is dissolved in water. This bond is formed by the association of either a cation or an anion with a dipole such as is found in water. A cation bonds to a region of high electron density (the oxygen atom in water) and an anion bonds to an electron deficient region (the hydrogen atom in water). The attraction between ion and dipole is strong and is insensitive to temperature and distance [85]. Examples of this type of attraction include the formation of an ionic salt when a basic compound (*e.g.*, amine) is added to an aqueous acidic medium (pH below 7.0) or when an acidic compound (*e.g.*, carboxylic acid, phenol, unsubstituted sulfonamide or imide) is added to an aqueous basic medium (pH above 7.0).

A variety of cation and anion exchange sorbents is available, having both strong and weak affinities. The strong ion exchange sorbents are always ionized to promote elution; the charge on the analyte is eliminated through pH manipulation. The functional groups bonded to weak ion exchange sorbents can be made ionized or unionized through the manipulation of pH and analytes can be eluted without modification of their own existing charge.

Ion exchange sorbents are found in the polymer series and the bonded silica based series. Some typical examples of ion exchange sorbents are listed in Table 11.2. Shown in Figure 11.11 are chemical structures for a strong cation and a strong anion exchanger (Oasis MCX and MAX, respectively).

11.2.2.4 Mixed Mode

A mixed mode interaction relies on dual mechanisms of attraction—ion exchange (cation or anion exchange) and reversed phase or hydrophobic affinities. Silanol interactions also play a role with mixed mode bonded silica sorbents. Note that some manufacturers use a proprietary process to bond unique functional groups that minimize the silanol influence. The chemical

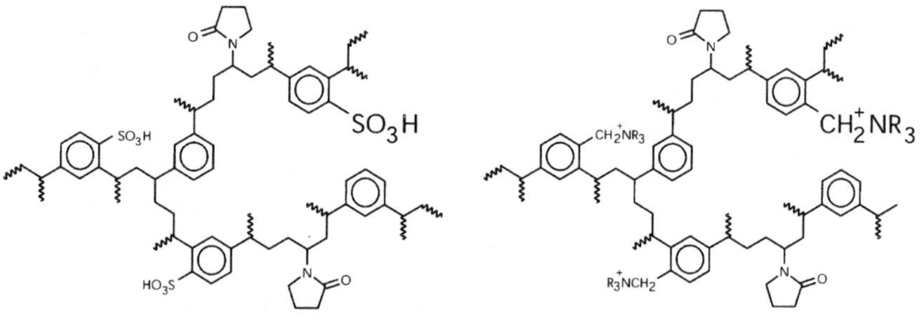

Figure 11.11. Chemical structures for a strong cation exchanger (*Left*, Oasis® MCX) and a strong anion exchanger (*Right*, Oasis® MAX) used for ion exchange SPE. Illustrations reprinted with permission from Waters Corporation.

Table 11.2.
Typical examples of anion and cation exchange chemistries used for solid-phase extraction.

Mechanism	Strong Ion Exchanger*	Weak Ion Exchanger*
Anion Exchange	Quaternary amine	Aminopropyl pKa=9.8
		Diethylaminopropyl pKa=10.7
		Primary/secondary amine pKa=10.1, 10.9
Cation Exchange	Benzene sulfonic acid	Carboxylic acid pKa=4.8
	Propyl sulfonic acid	

*Weak ion exchangers are defined as those whose ionization can be controlled via manipulation of pH; strong ion exchangers are almost always ionized

nature of the hydrophobic ligand component of mixed mode chemistries is variable but is described by manufacturers as a "chain length consistent with an octyl chain" [86]. By using ion exchange and reversed phase approaches simultaneously, more selective extractions can be developed since the analytes are held by strong ionic bonds. The careful manipulation of pH can retain analytes yet wash off potentially interfering matrix components and endogenous compounds.

Since the majority of drugs contain one or more amine groups, a mixed mode cation exchange is of great utility in bioanalysis; analytes in the charged state are loaded onto the sorbent, remain adsorbed while interferences are washed away, and are eluted from the sorbent by a mixture of an organic solvent with a small percentage of a basic aqueous solution. The basic pH of the eluting solution neutralizes the charge on the analyte and the organic solvent disrupts hydrophobic interactions. The greater selectivity of mixed mode cation sorbent chemistry was shown by Rose *et al.* [87] who documented fewer endogenous extracts than typical alkyl chain bonded silica. Acetonitrile was used as a wash solvent in this application, which aided in the cleanup process.

The chemical structure for typical silica based mixed mode cation sorbents is shown in Figure 11.12. One of the earliest copolymerized mixed mode cation bonded silica sorbents introduced was Clean Screen® DAU (United Chemical

Technologies) which has shown great versatility for the extraction of drugs of abuse [86] and individual analytes [88] from biological fluids as well as tissues [89]. Some other brand names include Mixed Phase Cation (MPC; 3M Corporation), MP-1 (nonpolar/SCX) and MP-3 (slightly polar/SCX; Ansys Technologies, Lake Forest, CA, now a part of Varian Inc.), ISOLUTE HCX (C8/SCX), HCX-3 (C18/SCX) and HCX-5 (C4/SCX; IST Ltd., now an Argonaut Technologies Company), Certify™ (Varian) and narc™-2 (Mallinckrodt Baker, Phillipsburg, NJ USA).

While the original mixed mode sorbents introduced were based on silica, polymer versions have been introduced and are also available, *e.g.*, Oasis MCX and others (POLYCROM™ series, CERA Inc.) The polymer based mixed mode chemistries usually offer superior wetting properties and so some sample matrices such as urine can be directly loaded onto the polymer sorbent bed without a prior conditioning step.

Mixed mode anion sorbents are complementary to the mixed mode cation sorbents and are used for the dual extraction of analytes having both nonpolar and acidic functional groups. Some examples of silica based mixed mode anion sorbents include Clean Screen® THC (United Chemical Technologies), ISOLUTE HAX (C8/SAX), Certify™ II (Varian) and narc™-1 (Mallinckrodt Baker).

It is generally recognized that the two particle chemistries perform better when they are copolymerized onto the same sorbent particle, rather than simply

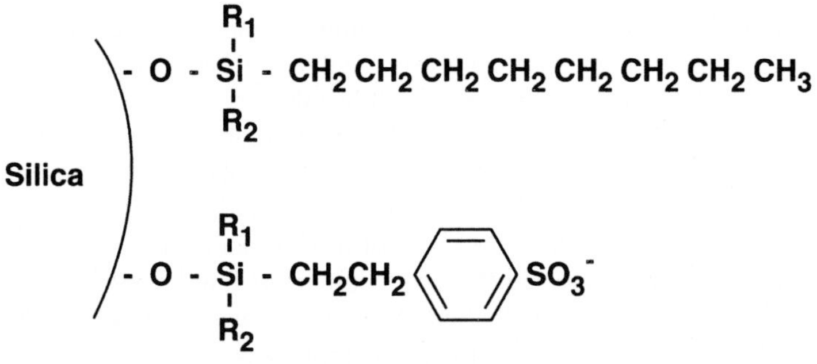

Figure 11.12. Illustration of a typical mixed mode sorbent in which two functional groups are copolymerized onto silica. Shown above is a mixed mode cation sorbent comprised of an alkyl chain and a strong cation exchange group.

blending (mixing) two separate derivatized particles. When a blended particle mix is used, desorption from the reversed phase particle must be followed by adsorption on a neighboring cation exchange particle for retention; physical proximity then plays an extremely important role [90].

11.2.2.5 Normal Phase

Chemical attraction of analytes onto a normal phase sorbent arises from a dipole-dipole chemical bond (*e.g.*, hydrogen bonding). A dipole results from unequal sharing of a pair of electrons making up a covalent bond when the two atoms differ significantly in electronegativity. A partial ionic character develops in this portion of the molecule, leading to a permanent dipole, with the compound described as a polar compound. Dipole-dipole attraction between two molecules arises from the negative end of one dipole being electrostatically attracted to the positive end of a second dipole. The hydrogen bond can occur when at least one dipole contains an electropositive hydrogen (*e.g.*, a hydrogen covalently bonded to an electronegative atom such as oxygen, sulfur, nitrogen or selenium), which in turn is attracted to a region of high electron density. Atoms with high electron density are those with unshared pairs of electrons such as amine nitrogens, ether or alcohol oxygens, and thioether or thiol sulfurs.

Note that while hydrogen bonding is an example of dipole-dipole bonding, not all dipole-dipole bonding is hydrogen bonding. The energy of hydrogen bonding is 1.0–10.0 kcal/mole for each interaction [85]. Some typical examples of normal phase sorbents include silica, cyanopropyl, diol and aminopropyl.

11.2.3 Further Reading

The first two sections of this chapter have introduced the fundamental principles of solid-phase extraction and provided an overview of sorbent chemistries and their attraction mechanisms. The subject of solid-phase extraction can be studied in much greater detail then presented here. As a start, the reader is referred to some concise general summaries of this technique as presented by Zief and Kiser [91], McDowall *et al.* [92–94], Krishnan and Ibraham [95], Hennion [96] and Fritz and Macka [97]. More advanced theoretical discussion of SPE is found in the book chapter by Henry [98]. Trends, formats, chemistries and techniques are reviewed by Majors [99]. A detailed study of retention mechanisms for basic compounds on silica under "pseudo reversed phase" conditions is reported by Cox and Stout [100]. A review by Cox summarizes the deleterious effects of silanol groups on reversed phase separations, details analytical methods for their study, and suggests methods to minimize their effects [101]. Interferences in SPE columns or cartridges have been investigated in an early report (1988) by Junk *et al.* [102].

Experimental factors which establish the breakthrough volume in SPE have been interpreted using a solvation parameter model [103, 104]. A related report presents a method for the determination of capacity factor on SPE cartridges [105]. Useful literature reports and book chapters review the techniques of ion-pair solid-phase extraction [106], SPE procedures in peptide analysis [107] and systematic toxicological analysis [75, 108], SPE for combinatorial libraries [109], SPE of lipid classes in biological fluids and tissues [110], solid-phase analytical derivatization [111] and immunoaffinity solid-phase extraction [112, 113].

Analytical techniques are available to probe the surface of silica and provide theoretical explanations of sorbent characteristics. The use of proton NMR as a method for studying the solid-phase extraction of biofluids is introduced [114]. Reports of the NMR analysis of phenyl bonded silica [115] and C18 bonded silica [116] relate structure to extraction performance. The kinetic and retention properties of a cyanopropyl bonded phase are described [117]. The basic variables and conditions in the synthesis of a bonded silica packing are presented by Unger *et al.* [118]. Forced flow planar chromatographic methods allow the study of kinetic and thermodynamic properties of SPE devices and these methods were applied toward a C18 bonded particle-loaded membrane and glass fiber disk [104, 119–122]. Characterizations of silica based octyl bonded phases of different bonding densities have been reported [123, 124].

Some books have been published on the subject of solid-phase extraction that the reader may find of interest. The book edited by Simpson is a recent compilation that discusses: SPE theory, nature of particle chemistry and bonding, method development, a range of specialty applications such as phenylboronic acid and immunosorbents, the automation of SPE cartridges and the integration of SPE with analytical techniques [125]. The original *Handbook of Solid-Phase Extraction* remains a useful reference guide for novices as it introduces the basics of SPE, the chemistries and applications [126]. Books by Thurman and Mills [127] and Fritz [128] discuss solid-phase extraction but are strongly focused on environmental applications. Forensic and clinical applications of SPE, with an emphasis on mixed mode interactions, are reviewed in a book by Telepchak *et al.* [129].

11.3 Solid-Phase Extraction Protocols

11.3.1 Introduction

The nature of sorbent particle chemistries and the several types of attraction mechanisms have been introduced in previous sections. It is appropriate now to review the methods for using these various sorbent chemistries. General extraction protocols are presented in the following sections as a useful summary. Note that multiple methods exist for the same sorbent chemistry when dual mechanisms of attraction are present; manipulation of pH is used to influence the desired interaction.

The general procedure for performing solid-phase extraction consists of several sequential steps, as shown in Figure 11.13. **Generally, the four major steps are: condition, load, wash and elute.** Sample pretreatment steps are commonly used to adjust sample pH and/or disrupt protein binding. Post-elution steps may include evaporation and reconstitution, or dilution in mobile phase compatible solvent followed by direct injection.

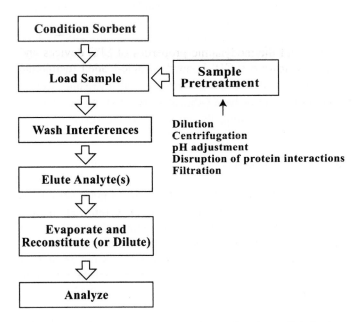

Figure 11.13. The general scheme for performing solid-phase extraction consists of several sequential steps. Sample pretreatment is important to ensure analyte retention.

11.3.2 Simple Reversed Phase (Polymer)

A simple reversed phase method involves a nonpolar attraction between analyte and sorbent. Polymer sorbents provide this reversed phase affinity in a straightforward, predictable manner and are the focus of this section. Note that bonded silica sorbents perform similarly, but always have the potential for secondary interactions via residual silanol groups. Section 11.3.3 discusses reversed phase extraction using bonded silica sorbents.

In a simple reversed phase extraction, the analyte contains one or more hydrophobic functional groups and sample pH is manipulated to ensure that the analyte remains neutral when loaded onto the polymer sorbent. If the analyte does not contain any ionizable groups, pH adjustment will not have any effect (although pH may be used to promote matrix components to their ionized form so that they are not retained and/or are washed away).

A general scheme for performing a simple reversed phase method is shown in Figure 11.14. Following the condition and sample load steps, a two step wash is recommended—a 100% aqueous wash to displace residual proteins from the sorbent bed, followed by a small percentage of organic in water to remove impurities but still retain analyte. The elution is accomplished with an organic solvent (*e.g.*, acetonitrile or methanol) to disrupt the nonpolar attraction of analyte to sorbent. A strongly hydrophobic retention of analyte to sorbent may benefit from an elution solvent such as dichloromethane, ethyl acetate or methyl tert-butyl ether. Elution can also be performed with organic solvent containing a small amount of acid or base (*e.g.*, 0.5–2%, v/v); the pH change may ionize the analyte, thus eliminating its attraction to the reversed phase sorbent.

11.3.3 Reversed Phase (Bonded Silica)

All bonded silica sorbents display some degree of residual silanol activity as discussed in Section 11.2.1. This activity from silanols is referred to as a secondary interaction. Bonded silica sorbents are commonly used in the reversed phase mode and their silanol effects can be minimized. However, when analyte recovery using a reversed phase bonded silica is not as high as desired, the silanols can be used in a more direct manner to improve retention, as discussed next in Section 11.3.4.

The general scheme for reversed phase attraction using bonded silica sorbents is identical to that for polymer sorbents, as shown in Figure 11.14. The analyte

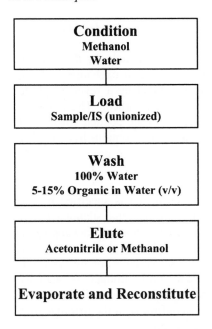

Figure 11.14. General scheme for a simple reversed phase extraction method using a polymer sorbent or bonded silica. Analyte is loaded as a neutral species; two sequential wash solutions are processed; analyte is displaced from the sorbent bed by an organic solvent. A small amount of acid or base added to the elution solvent may be helpful.

contains one or more hydrophobic functional groups and sample pH is manipulated to ensure that the analyte remains neutral when loaded onto the polymer sorbent. A two step wash is recommended; first a 100% aqueous wash followed by a small percentage of organic in water to remove impurities yet retain analyte. Elution is accomplished with an organic solvent to disrupt the nonpolar attraction of analyte to sorbent.

When analyte recovery using an elution solvent of acetonitrile or methanol is lower than desired, it may help to add a small amount (*e.g.*, 0.5–2%, v/v) of acid or base. If desired, concentrations up to 5% are used as well. The change in pH may ionize the analyte, eliminating its nonpolar attraction to the reversed phase sorbent. In the case of silanols, addition of acid in the elution solvent may protonate them, removing their negative charge and essentially neutralizing their influence.

11.3.4 Reversed Phase with Primary Silanol Interactions (Bonded Silica)

A reversed phase bonded silica method that relies on silanols for the primary mechanism of attraction essentially performs as a cation exchange sorbent, although there is not as much ionic capacity exhibited by a reversed phase bonded silica as compared with a pure ion exchange sorbent. The nonpolar affinity between functional groups on the analyte and the bonded silica alkyl chain behaves as a secondary interaction. An analyte that contains hydrophobic functional groups as well as one or more ionizable nitrogen groups is a good candidate for this dual mechanism of retention.

Typically, a C18 (C8, C4 or C2) sorbent is chosen that is specified by the manufacturer as unendcapped which means that it has maximal silanol activity. Most products are endcapped by definition if no mention of endcapping is made, *e.g.*, the name C18 implies endcapped and C18-OH describes that the sorbent is unendcapped (Varian). However, this assumption is not always valid. It is known that one or two manufacturers' C18 product is actually unendcapped by definition, and to get an endcapped version a different part number must be used. It is suggested to always confirm with the manufacturer whether or not the sorbent selected is actually endcapped or not.

The general scheme for using primary silanol interactions on bonded silica sorbents is illustrated in Figure 11.15. The sample pH is manipulated to ensure that the analyte is positively charged when loaded onto the bonded silica sorbent so that it will be attracted to the negatively charged silanols. Two wash steps are utilized—first 100% aqueous to efficiently displace residual proteins and matrix materials, followed by a small percentage of organic in water (*e.g.*, 5–50% acetonitrile, v/v). When the analyte is held by cation exchange interactions, a higher percentage of organic (up to 100%) can be tolerated in the wash step compared with a typical reversed phase attraction (*e.g.*, 5–15% organic in water, v/v). A nonpolar solvent like acetonitrile often performs better for the wash than the more polar solvent methanol because the silanol interaction is not disrupted as much; a polar solvent tends to displace the analyte, as discussed next.

The elution step can be manipulated in several ways when silanol attraction is predominant. A polar elution solvent such as 100% methanol in some cases may fully displace the analyte from its attraction to silanols. However, it may be more effective to use: (a) methanol containing a small percentage of acid (*e.g.*, 0.5–2% HCOOH) which floods the environment with protons, displacing analyte, or (b) a combination of methanol/water (e.g., 40–90% methanol, v/v).

When methanol/water is used for elution in this manner, the percentage should be titrated in such a way that maximal recovery of analyte is obtained yet the nonpolar matrix interferences remain on the sorbent bed. Several combinations should be examined, and the recovery and cleanliness of the extracts obtained are assessed.

Recall that a polar sorbent such as cyanopropyl (unendcapped) can also be used in a reversed phase extraction scheme with aqueous sample matrices. The short alkyl group provides adequate steric access to the silanol sites and the cyanopropyl group provides a weak alkyl affinity. An application isolating basic drugs from plasma with a cyanopropyl bonded phase is reported by Musch and Massart [130].

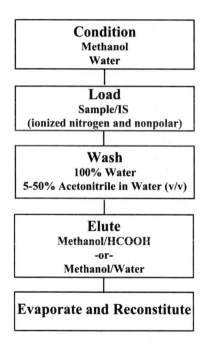

Figure 11.15. General scheme for a reversed phase (bonded silica) SPE method with primary silanol interactions. Analyte is loaded as a positively charged species and held by ionic and nonpolar attraction; two sequential wash solutions are used. Elution is accomplished with polar solvents such as methanol with 0.5–2% acid or a combination of methanol in water (40% methanol up to 90%, v/v.).

11.3.5 Mixed Mode

11.3.5.1 Introduction

A mixed mode sorbent contains two bonded functional groups for dual mechanisms of analyte attraction to sorbent. Usually these sorbents are comprised of reversed phase functionality with an ion exchange component.

Mixed mode chemistry allows three extraction scenarios:

1. A *selective extraction* can be designed which is very specific for a charged analyte or group of structurally similar analytes

2. A *general fractionation* can be performed in which neutral/acidic (or neutral/basic) components are eluted in one fraction and then the basic (or acidic) analytes, respectively, are eluted in a second fraction by modifying solution pH.

3. A *permanent retention* of unwanted charged materials can be designed. The sorbent is used as a simple reversed phase sorbent and the unwanted cationic (or anionic) interferences are adsorbed by manipulation of pH. Elution of analyte occurs in pure organic solvent (leaving charged components on the sorbent bed).

The successful use of mixed mode interactions relies on the proper utilization of pH during the extraction process. Reprinted from Chapter 2, Section 2.3.1 is a summary of the important pH adjustment guidelines to promote the ionization of acids and bases to their fully ionized or unionized species.

For > 99% conversion the pH should be adjusted 2 pH units above or below the pKa value of the analyte

Acids	pH < pKa = R–COOH	unionized
	pH > pKa = R–COO$^-$	ionized
Bases	pH < pKa = R–NH3+	ionized
	pH > pKa = R–NH$_2$	unionized

11.3.5.2 Mixed Mode Cation Exchange of a Basic Analyte

The general scheme for performing a selective mixed mode extraction of a basic analyte is shown in Figure 11.16. Note that the overall procedure uses more solutions than that of a simple reversed phase method (Figure 11.14). For example, an additional conditioning solution is used for mixed mode extractions to prepare the sorbent bed at the same pH as the sample.

The following sequential wash steps are used:
1. Water (100%) to remove unwanted proteins and salts
2. Acetic acid 1M (or 0.5% formic acid) to maintain analyte ionization and act as a proton lock by regenerating the cation exchange functionality which may have lost some of its capacity when the sample was loaded [90]
3. Methanol (or acetonitrile) to remove interferences

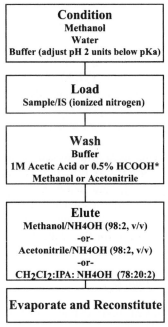

*0.1N HCI can be used for low pKa amines

Figure 11.16. General scheme for a mixed mode cation SPE method. Analyte is loaded as a positively charged species. An acid wash regenerates the ion exchange capacity and pure organic solvent follows as another wash solvent. Analyte is eluted in organic solvent with base (to neutralize the cationic retention mechanism); three choices for elution are shown, depending on analyte characteristics. IPA=Isopropanol. The pH of the buffer used should be at least two units **below** the pKa of the analyte.

Note that 0.5% formic acid can be used instead of acetic acid and acetonitrile can be used in place of methanol, as reported by Rose *et al.* [87]. Also note that acidic methanol can be used in place of methanol to not only regenerate cation exchange capacity but also keep an amine in its cationic form [90].

Typically, for a reversed phase extraction, methanol shows partial or full elution of the analyte of interest. However, when the analyte is retained by the stronger ion exchange attraction, methanol is not able to elute the analyte. Using methanol as a wash solvent provides a strong solvent wash to remove potentially interfering substances from the sorbent bed. Elution from the always charged sorbent bed (when using strong cation exchange) is accomplished by making the pH basic to neutralize the positive charge on the analyte (2 pH units above the pKa of the amine).

Ammonium hydroxide is a typical basic additive that is used with organic solvents (acetonitrile or methanol) for the elution step. Usually a 2% addition of this base (by volume) is adequate but note that percentages up to 5% are sometimes used. An alternative to NH_4OH is triethylamine (TEA). Aromatic and tertiary amines often yield better recovery using TEA rather than NH_4OH, with Oasis based sorbents (Waters Corporation). This observation can be exploited for a selective extraction of a secondary amine (*e.g.*, amphetamine) by using TEA as a component in the final wash solvent (*e.g.*, methanol/TEA, 98:2, v/v). Other examples of basic elution solvents used with mixed mode cation sorbents include dichloromethane/isopropanol/NH_4OH (78:20:2, v/v/v) [88], chloroform/isopropanol/NH_4OH (78:20:2) [131], and ethyl acetate/-NH_4OH (98:2) [132]. An important note regarding the elution solvent is that it should be made fresh each week; the basic additive is volatile and repeated exposure to the atmosphere may reduce its concentration in the reagent over time, resulting in lower performance.

11.3.5.3 Mixed Mode Anion Exchange of an Acidic Analyte

The general scheme for performing a selective mixed mode extraction of an acidic analyte is shown in Figure 11.17. The procedure for using a mixed mode anion exchange sorbent is similar to that of mixed mode cation exchange, with the exception that sample and solvent pH values are adjusted to the opposite end of the pH scale. When using an anion exchange sorbent, samples are acidic so loading at a neutral pH of about 7 is usually enough to ionize these functional groups (*e.g.*, carboxylic acids) to promote adsorption. The wash solvents are also kept at neutral pH to retain the analyte negative charge. Elution is performed using an acidic solution of methanol or acetonitrile

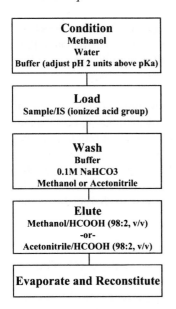

Figure 11.17. General scheme for a mixed mode anion SPE method. Analyte is loaded as a negatively charged species. A basic wash regenerates the ion exchange capacity and pure organic solvent follows as another wash solvent. Analyte is eluted in organic solvent with acid (to neutralize the anionic retention mechanism); two choices for elution are shown. The pH of the buffer used should be at least two units **above** the pKa of the analyte to ensure complete ionization.

(*e.g.*, containing 0.5–2% formic acid) to protonate the acid, neutralizing its charge to break the ionic attraction. The organic solvent also helps to disrupt the nonpolar attraction.

11.3.6 Traditional Ion Exchange

The successful use of traditional silica based ion exchange sorbent chemistries is dependent on pH manipulation during the solid-phase extraction procedure. This careful control of pH is achieved using buffers, typically in concentrations from 0.1–0.25 M (100–250 mM). It is important to understand that the strong ion exchange chemistries are always ionized, while the degree of ionization of the weak ion exchange chemistries is influenced by pH. For example, the weak cation exchanger CBA (carboxylic acid) has a pKa of 4.8. At a pH ≥ 6.8, the sorbent phase is fully ionized and attraction to a positively charged amine analyte occurs. Elution can be accomplished at an acidic pH by neutralizing the charge on the sorbent (pH ≤ 2.8) or at a basic pH by neutralizing the charge on the analyte. In the case of a strong ion exchanger, the sorbent charge cannot be neutralized and elution is entirely dependent on neutralizing the charge on the analyte.

The general scheme for traditional ion exchange extractions is very similar to that discussed in Section 11.3.5 for mixed mode interactions, except that a strong hydrophobic attraction mechanism is eliminated. Elution can be performed in a 100% aqueous buffered solution. Note that, in practice, the addition of some organic to the aqueous buffered solution may be desirable. This eluate solution is at the proper pH to either neutralize the charge on the analyte and/or neutralize the charge on the sorbent bed to disrupt the ionic attraction.

In addition to pH, another important component of traditional ion exchange chemistry is the use of counter ions at a high ionic strength (≥0.5 M) to displace the analyte. The choice of a counter ion with a great affinity allows more complete and rapid displacement of the analyte; a range of affinities is exhibited by various counter ions and tables of relative selectivities are available in reference texts of ion exchange chemistry. For example, a 1M citrate buffer shows great selectivity as an anion and is often used as the eluting solution for a strong anion exchange sorbent. Strong counter ions for cation exchangers include Ag(+), Pb(2+) and Hg(2+) salts. A general summary of the goals for maintaining sample and solution pH during ion exchange SPE procedures is presented in Table 11.3.

Traditional ion exchange methods are not performed often in bioanalysis for two main reasons:

1.   The ionic strength of biological fluids is very high and competes for retention on the ion exchange sorbent bed
2.   An elution solvent consisting of a totally aqueous salt solution (with metal ions) is not desirable for injection into LC-MS/MS systems.

Further discussion of the ion exchange technique is outside the scope of this text. A book chapter by Thurman and Mills is recommended reading for additional information on the use of ion exchange chemistries [133]; although its discussion of ion exchange emphasizes environmental applications, the applications chosen are straightforward and exemplary for the novice user.

Table 11.3
Summary of the goals for maintaining sample and solution pH during traditional strong and weak ion exchange SPE procedures.

Sorbent	Load Conditions	Elute Conditions	Reference
Strong cation exchange	2 pH units below pKa of analyte	2 pH units above pKa of analyte and/or high ionic strength	[134]
Weak cation exchange (carboxylic acid pKa=4.8)	pH ≥ 6.8 and must also be 2 pH units below pKa of analyte	pH ≤ 2.8 neutralizes sorbent charge or adjust pH to 2 pH units above pKa of analyte	[135]
Strong anion exchange	2 pH units above pKa of analyte	2 pH units below pKa of analyte and/or high ionic strength	[136]
Weak anion exchange (diethyl aminopropyl pKa=10.7)	pH ≤ 8.7 and must also be 2 pH units above pKa of analyte	pH ≥ 12.7 neutralizes sorbent charge or adjust pH to 2 pH units below pKa of analyte	[137]

11.3.7 Normal Phase

A normal phase method involves a polar attraction between analyte and sorbent. Instead of aqueous, the analyte is contained within a nonpolar matrix to promote a polar adsorption to the sorbent particle. Raw, underivatized silica is often used for normal phase methods, as well as other polar sorbents such as cyano, diol and aminopropyl. Residual silanols are present on the silica surface and often aid in the polar attraction of analyte to sorbent.

A general scheme for performing a normal phase SPE method is shown in Figure 11.18. Since the analyte is contained within an organic solvent, it is neutral to ensure solubility. Promoting ionization usually results in the analyte precipitating out of solution since an ionic species is insoluble in an organic solvent. The wash step often utilizes the same organic solvent as sample loading. Elution is accomplished with a very polar solvent such as methanol or ether/hexane. Sometimes, solvents are used in series to fractionate species (*e.g.,* 40% methanol in water, 60% methanol in water, and 100% methanol).

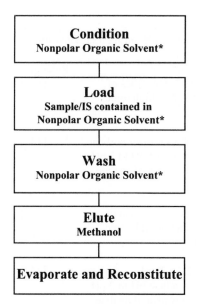

*e.g., Hexane, Hexane/Ethyl Acetate (4:1, v/v) or Dichloromethane

Figure 11.18. General scheme for a normal phase (bonded silica) SPE method. Analyte is loaded in an organic solvent matrix and eluted using a more polar solvent.

An application for the use of normal phase sorbents for SPE includes the isolation and extraction of preservatives from cosmetic products [138]. Sometimes, the use of a normal phase sorbent can be added to an existing method to provide added selectivity. For example, a nonpolar organic solvent such as methylene chloride, collected following a C18 or polymer reversed phase extraction, can be applied to a normal phase column. Subsequent wash with a similar nonpolar solvent and elution with methanol yields a purified extract. Normal phase SPE can follow a liquid-liquid extraction procedure as well. A book chapter by Thurman and Mills is recommended reading for additional information on the use of normal phase chemistries [139].

11.4 Sorbent Presentation and Product Formats

11.4.1 Introduction

The sorbent particles used for solid-phase extraction are packaged or presented for use in primarily two different geometries—particles loosely packed in a column or well and particles packed into a thin disk or membrane. Each of these varieties can be found in one of two microplate product formats, grouped for discussion as a single piece molded plate and a modular plate. A modular plate is characterized as having individual wells or strips of wells (removable) assembled into a base plate. The following sections describe for the reader the many varieties in which sorbents are presented within a well and introduce the available product formats. The accessory products required for use of these SPE microplates are briefly reviewed. Also, since packed particle beds and disks have slightly different recommendations for optimal usage, these differences are explained.

11.4.2 Packed Particle Beds

The traditional format for SPE has been single disposable columns (commonly 1, 3 and 6 mL syringe barrels), also called cartridges, filled with solid sorbent particles (from 25–500 mg) held between two polyethylene frits. Particles (typically 40–60 µm in size) are loosely packed into these polypropylene columns. The kinetic aspects of the process favor slow flow rates for maximal interaction of analyte with the sorbent bed. However, in this packing geometry voids or spaces can form between the dry particles and less intimate analyte contact is achieved. Also, when the flow rate is fast, the efficiency of the extraction process is reduced and this occurrence is known as channeling.

Individual columns or cartridges held a stronghold as the preferred SPE format until the late 1990s. At this time, several advances occurred in parallel that brought about a change from columns to microplates as the preferred sample and collection format. Faster analytical techniques such as liquid chromatography interfaced with tandem mass spectrometry (LC-MS/MS) became affordable and widespread, allowing researchers to analyze samples more quickly than ever before. Liquid handling workstations became more prevalent and affordable, and replaced most manual pipetting tasks. Within the pharmaceutical industry, drug development cycles were shortened, favoring a more rapid identification of lead candidates with a more rapid development pace. When the sample preparation step became rate limiting in this process, the need for improved efficiency and throughput became evident.

The pharmaceutical industry responded to the challenge of higher throughput sample preparation by utilizing solid-phase extraction in a 96-well format. This microplate format presents many efficiencies of operation such as ease of labeling, sealing and manipulation and, when combined with multiple tip workstations, dramatically faster pipetting throughput via parallel processing of

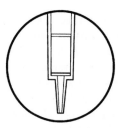

Figure 11.19. Particles are packed into columns within 96-well SPE plates and held between two frits; different bed mass loadings are available. Column geometries vary slightly among the various plate configurations.

samples. The microplate format clearly offered the increased proficiency and productivity in sample preparation processes and analyses that were sought. More information on the history and development of this format has been reviewed by the author [140] and is discussed in Chapter 3, Section 3.5.

The geometry of sorbent particles packed into wells of a microplate is similar to that of the syringe barrel; particles are packed between two polyethylene frits (Figure 11.19). The lower reservoir volume of microplates (from 1–2 mL) allows a maximum packing mass of about 300 mg. More commonly, the range of particle bed masses available in 96-well SPE plates is from about 25–100 mg; larger packing masses are available as custom products. Note that sorbent mass loadings smaller than 25 mg are also available from some manufacturers; Waters Corporation, using proprietary plate designs, offers an exemplary selection in 2, 5 and 10 mg particle bed mass loadings per well (see Section 11.4.4 for details).

11.4.3 Disks

11.4.3.1 Introduction

Large particle masses (≥100 mg) for SPE require the use of large volumes (≥1 mL) for the various condition, wash and elute steps of an extraction method. Large elution volumes represent a great dilution of sample concentration, and evaporation and reconstitution procedures must be utilized. Time requirements for evaporation are increased as well when using large volumes. SPE in the disk format was introduced as a means to overcome these limitations. The physical geometry of particle packaging within disks is improved, compared with packed particle beds, and the limitations in flow kinetics are removed. Volume usage is reduced for disks, and of special interest is the ability to elute analytes in volumes small enough for direct injection when the solution is compatible with mobile phase, eliminating the time consuming evaporation and reconstitution step. A summary of disk technologies and applications for bioanalytical SPE has been published in a book chapter by Lensmeyer [141]; the benefits of using SPE disks are highlighted in this work. Three main varieties of disks grouped for discussion here are: (1) laminar and sintered disks, (2) glass fiber disks and (3) PTFE particle-loaded membranes (disks).

11.4.3.2 Laminar and Sintered Disks

A laminar disk consists of micro particles (particles smaller than the traditional 40–60 µm used in SPE packed particle beds) sandwiched above and below by filters and screens. A guard filter (also called prefilter) is placed above the thin bed of micro particles to resist plugging and a support filter is placed immediately below. Instead of using polyethylene frits, glass fiber screens support the configuration on top and bottom (Figure 11.20). Essentially, this arrangement represents an improved method of packing particles and, since a smaller particle mass is used, advantages are realized in reduced reagent volumes and superior flow kinetics. This patented design (CEREX™ by CERA Inc.) is called the Speedisk® by Mallinckrodt Baker. It is also found in the individual column format from CERA (distributed by SPEware, San Pedro, CA USA) and United Chemical Technologies (Reduced Solvent Volume columns).

Speedisk columns from Mallinckrodt Baker contain either silica based (10 µm) or polymer based (15 µm) micro particles in a range of bonded chemistries. These columns are placed into a base plate that allows them to be used with traditional microplate processors and liquid handling workstations.

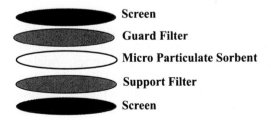

Screen

Guard Filter

Micro Particulate Sorbent

Support Filter

Screen

Figure 11.20. Illustration of the laminar disk configuration containing particles sandwiched between filters and screens.

A sintered disk specially manufactured with silica chemistries is available from Porvair Sciences (Shepperton, United Kingdom) under the name Generic Silica Vyon®. Generic Silica Vyon represents a proprietary construction of ultrahigh molecular weight polypropylene with various silica chemistries (*e.g.*, C18, C8, mixed mode cation and mixed mode anion). The resulting disk has an open, porous structure and displays good flow characteristics with biological matrices. The 20 μm particles are enmeshed within the polypropylene matrix so the risk of channeling is eliminated. These disks are mounted into the Porvair Microlute™ 96-well plates for high throughput SPE. Note that Vyon represents a family of high molecular weight frit products within Porvair Sciences; another configuration of Vyon is specifically optimized for protein precipitation in a filtration application (see Chapter 6, Section 6.4).

11.4.3.3 Glass Fiber Disks

Glass fiber materials have been used for many years as an efficient filtration medium and are available in a variety of configurations. Glass fiber disks with embedded bonded silica particles for SPE were first introduced by Ansys Technologies. This system is called SPEC•® (Solid-Phase Extraction Concentrator) and is available in both individual columns and microplates with a wide range of bonded silica chemistries [142–144]. Sorbent mass loadings in this rigid disk format are about 15–30 mg. The lower bed mass allows reduced reagent and elution volumes. This disk format is also reported to eliminate channeling observed with traditional SPE packed particle beds. Many published reports have shown the applicability of SPEC disk products for bioanalysis [136, 145–154] and provided side by side comparisons with traditional packed bed SPE products [155, 156]. Forced flow planar chromatography was used to determine the kinetic and retention properties of the C18 bonded and glass fiber disks [119]. The use of glass fiber disks as a trapping technique prior to supercritical fluid chromatography is reported [157].

Ansys Technologies was the innovator in glass fiber disk products for SPE. Today these products are also available from Orochem Technologies (Westmont, IL USA) in a wide range of sorbent chemistries. Applied Separations (Allentown, PA USA) and Whatman (Clifton, NJ USA) offer glass fiber disks in C18 bonded silica only. A notable product addition from Orochem Technologies is a glass fiber disk containing a reversed phase polymer sorbent, the first to market in the glass fiber disk format.

11.4.3.4 PTFE Particle-Loaded Membranes (Disks)

A patented process by the 3M Corporation uses fibrillated polytetra-fluoroethylene (PTFE) to create a dense particle-loaded membrane that consists of 90% particles (by weight). The particles are tightly held together within the inert PTFE matrix and retain their full sorbent activity. This Empore® technology results in a denser, more uniform extraction medium than can be achieved in a traditional packed bed SPE column. As seen with other disks, the thin Empore disk also allows reduced reagent volumes, smaller elution volumes, elimination of channeling and the ability to elute in a volume small enough for direct injection into a chromatographic system.

This fabrication process has been manipulated to create membranes of different densities. Empore Disk Products for bioanalytical SPE are available in standard density (SD) and high density (HD) versions. The SD membrane (~0.75 mm thickness) is optimized for use with biological fluids, contains standard particle size sorbents (approx. 40–60 μm), and processes a wide range of sample types (Figure 11.21). The 96-well SPE plates from 3M and their single syringe barrel columns contain this SD membrane. The HD membrane (~0.50 mm thickness)

Figure 11.21. Micrographs of PTFE particle-loaded membranes (~0.50 mm thickness) containing irregular shaped bonded silica particles. *Left*: Standard Density (SD), 40–60 μm; *Right*: High Density (HD), 10–20 μm. Photos reprinted with permission from 3M Corporation.

contains smaller particle size sorbents (approx. 10–20 μm); packing small particles tightly results in a high back pressure so the HD version is best used with clean sample matrices. This HD membrane is available in single syringe barrel columns for bioanalytical applications. Note that the standard density products are recommended as the first choice for most applications.

Many published reports have shown the applicability of Empore disk products for bioanalysis in the single column [14, 17, 60, 158–177] and 96-well format [35, 45, 87, 153, 178–197]. Some side by side comparisons with traditional packed bed SPE products have been reported [155, 156, 198] and a general overview of particle-loaded membranes for sample concentration and/or cleanup in bioanalysis has been prepared [199]. Forced flow planar chromatography was used to determine the kinetic and retention properties of Empore C18 membranes [104, 120–122, 200]. An interesting note is that this particle-loaded membrane technology is quite versatile and has been used in related analytical applications such as drug extraction from large volumes of urine [201, 202], membrane preconcentration capillary electrophoresis mass spectrometry (mPC-CE-MS) [203–211], thin layer chromatography [212–217] and environmental analyses [218–227]. An early and interesting report by the inventor of the Empore membrane technology, L.A. Errede, discusses the physical and chemical nature of these reactive microporous composite membranes in applications as liquid absorbent membranes, Merrifield resin membranes and catalytic membranes [228].

A special note is made of another PTFE membrane (disk) that also enmeshes sorbent particles (60% polymeric resin and 40% PTFE, w/w); it is enclosed in a large diameter (25 and 47 mm) medical grade polypropylene housing (Alltech Associates Inc., Deerfield, IL USA). Since this disk is intended for sample cleanup prior to ion chromatography of various types of water samples, its discussion is outside the scope of this text. The reader is referred to reports that describe its configuration and provide details of an application to remove matrix interferences from water prior to analysis by capillary electrophoresis [229, 230]. It is interesting to note that all three configurations of particle-embedded disks (laminar, glass fiber and PTFE) have also been shown useful for environmental analyses of water samples in the larger diameter format, extending the applicability of these disk products into additional markets to meet needs for sample preparation.

11.4.4 Plate Configurations

11.4.4.1 Introduction

The variety in 96-well SPE plate configurations is discussed in Chapter 3, Section 3.6. Only a very brief summary is provided here for convenience; however, the reader is encouraged to review Chapter 3 for important details. The configurations of 96-well SPE plates can generally be classified by the manner in which they are molded—a single block or a multiple piece block composed of a base plate having removable wells. Note that each of these plates can also be distinguished as having square wells (~2 mL volume) or round wells (~1 mL).

Within the category of one piece molded plates, the two varieties seen are distinguished by the height of the sides of the microplate, described as short and tall skirt plates. Likewise, there is variation in the height of tubes or wells used in the modular plates which, in some cases, presents an incompatibility with the use of liquid handling workstations. Compatibility with specific brands of automation should be confirmed before use of any modular plate product.

11.4.4.2 Single Piece Molded Plates

A list of typical examples of 96-well solid-phase extraction plates in the single piece molded plate format is provided in Table 11.4. Representative images matching each plate design are indicated within the table. While the information is meant to be informative, it should not be considered comprehensive (some companies did not respond to requests for information), and individual manufacturers should always be contacted to obtain more complete and current information about their products. For example, missing from Table 11.4 is information about the single piece molded plate offered by Mallinckrodt Baker, as details of this new product entry were not available before publication of this book. Note that many vendors offer custom products; these can be comprised of specialty sorbents, custom mass loadings or a unique arrangement of several particle chemistries within a plate.

11.4.4.3 Modular Plates with Removable Wells

A list of typical examples of 96-well solid-phase extraction plates in the modular (removable well) format is provided in Table 11.5. Representative images matching each plate design are indicated within the table. Modular

Table 11.4
Typical examples of solid-phase extraction plates in the one piece molded format

Vendor	Product Name	Packed Particle Bed or Disk	Total No. of Sorbents*	Are Any Sorbents Polymer?	Sorbent Bed Mass, Range*†	Photo
Agilent Technologies	Accubond II™ SPE 96-Well Plates	Particle Bed	4	Yes (1)	25–100 mg	A
Alltech Associates Inc.	96-Well Plates	Particle Bed	30	Yes (7)	25–200 mg	A
Ansys (now Varian)	SPEC®•96-Well	Disk, Glass Fiber	15	No	15 mg	B
Applied Separations	Spe-ed™ 96	Particle Bed	>30	Yes (6)	10–200 mg	C
		Disk, Glass Fiber	1	No	15 mg	B
Diazem Corporation	Diazem®	Particle Bed	14	No	10–300 mg	A
IST Ltd. (now Argonaut Technologies)	ISOLUTE®-96	Particle Bed	36	Yes (2)	10–100 mg	A
Macherey-Nagel Inc.	Chromabond® Multi 96-SPE	Particle Bed	40	Yes (7)	25–100 mg	B
Orochem Technologies	Orpheus™	Particle Bed	>20	Yes (1)	15–200 mg	B, C
	SuPErScreen	Disk, Glass Fiber	12	Yes (1)	14 mg	B, C
Phenomenex	Strata™ 96-Well Plates	Particle Bed	15	Yes (2)	10–100 mg	A
Porvair Sciences	Microlute™	Particle Bed	>56	Yes (1)	10–100 mg	A
		Disk, Silica Vyon®	4	No	20 mg	A

Supelco Inc.	Discovery® SPE-96	Particle Bed	12	Yes (1)	25–100 mg	C
3M Corporation	Empore™ Extraction Disk Plates	Disk, PTFE	6	Yes (2)	14 mg	D
United Chemical Technologies Inc.	96-Well Plates	Particle Bed	>35	No	50–300 mg	C
Varian Inc.	BondElut Matrix™	Particle Bed	14	Yes (3)	10–100 mg	A
Waters Corporation	Oasis® Plate	Particle Bed	3	Yes (3)	5–60 mg	E
	Oasis® µElution Plate	Particle Bed	3	Yes (3)	2 mg	F
Whatman	SPE Bioplate	Particle Bed	10	No	50–500 mg	n/a
	SPE Bioplate	Disk, Glass Fiber	1	No	9 mg	n/a

* Contact vendor for custom sorbent varieties or a custom sorbent bed mass outside this range
† Sorbent bed mass for polymers may be slightly different than listed here; n/a= photo not available (2mL/well and 0.8mL/well)

Table 11.5
Typical examples of solid-phase extraction plates in the modular format having removable wells

Vendor	Product Name	Packed Particle Bed or Disk	Total Number of Sorbents*	Are Any Sorbents Polymer?	Sorbent Bed Mass, Range*†	Well Geometry and Volume (Photo)
IST Ltd. (now Argonaut Technologies)	ISOLUTE Array®	Particle Bed	36	Yes (2)	10 – 100 mg	Square 1 mL and 2 mL (G)
Mallinckrodt Baker	Speedisk® 96 Columns	Disk, Laminar	>20	Yes (6)	20 mg	Round 1.25 mL (H)
Orochem Technologies	Oro-Sorb III	Particle Bed	>20	Yes (1)	25 – 100 mg	Square 2 mL (G)
	Oro-Sorb III	Disk, Glass Fiber	12	Yes (1)	14 mg	Square 2 mL (G)
Varian Inc.	VersaPlate™	Particle Bed	20	Yes (3)	10 – 100 mg	Round 1.8 mL (I)

* Contact vendor for custom sorbent varieties or a custom sorbent bed mass outside this range
† Sorbent bed mass for polymers may be slightly different than listed here

H

G

plates are especially useful during the method development process as different sorbents can be located in a custom well configuration and multiple variables (sorbent chemistry, bed mass, *etc.*) can be examined simultaneously [231]. More information on method development and the use of modular plates for this purpose is found in Chapter 12.

11.4.5 Accessory Products

The assortment of accessory products needed to perform a 96-well solid-phase extraction procedure is described in Chapter 4. The use of accessory products for 96-well SPE has also been briefly summarized in another publication by the author [232]. Table 11.6 contains an outline of these various accessory products. While 96-well SPE can be performed manually using multichannel pipettors, maximum laboratory productivity is achieved through successful implementation of automated liquid handling workstations. Chapter 5 reviews the commercial automation systems available and strategies for their use.

Table 11.6
Outline of accessory products that aid in performing 96-well solid-phase extraction

Accessory Products	Reference
Multichannel pipetting systems	Chapter 4, Section 4.1
Small footprint liquid dispensing units	Chapter 5, Section 5.3.2
Automated liquid handling workstations	Chapter 5, Section 5.4
Options for processing liquids through 96-well plates	
Vacuum	Chapter 4, Section 4.2.1
Centrifugation	Chapter 4, Section 4.2.2
Positive displacement	Chapter 4, Section 4.2.3
Collection devices for receiving eluates	Chapter 3, Section 3.4
Sealing systems for collection devices	Chapter 4, Section 4.3
Mixing liquids within collection devices	Chapter 4, Section 4.4.2
Centrifugation of collection devices	Chapter 4, Section 4.4.4
Solvent evaporation systems for dry-down of eluates	Chapter 4, Section 4.5
Autosamplers accepting the microplate format	Chapter 4, Section 4.6

11.4.6 Liquid Processing: Packed Particle Beds vs. PTFE Disks

The fundamental principles in using solid-phase extraction, as discussed in Section 11.1, are similar regardless of which product is used. An analyte is chemically attracted to a solid support and adsorbs to that support as it passes through the column, interferences are washed off the support leaving analyte, and an elution solvent is passed through disrupting the attraction of analyte and sorbent. However, there is an important difference with regard to liquid processing between packed particle beds and a certain type of disk (PTFE particle-loaded membrane). If the distinction is not followed, poor performance as well as frustration may result when using these disks.

The primary caution in using packed particle beds is that low flow rates should be employed for essentially all of the liquid processing steps, *e.g.*, use a starting vacuum setting of ≤5 in Hg (≤170 mbar) and adjust upward only if flow restriction results. Recall that voids or spaces form between the particles as they are packed within a column. The kinetic aspects of the extraction process favor slow flow rates for maximal interaction of analyte with the sorbent bed as the sample flows through the extraction device. Failure to use slow flow rates may lead to channeling, reducing the efficiency of the extraction process and resulting in low recovery. This same occurrence is not usually observed with glass fiber disks, laminar disks and sintered particle disks due to the tighter particle packing even though they are porous disks. One exception where the use of a higher flow rate is warranted is following passage of the last wash solvent; residual liquid should be eliminated before elution and so high vacuum (or high positive pressure) is applied.

PTFE particle-loaded membranes represent the main exception to what has been taught regarding the use of low flow rates for solid-phase extraction. Since the particles are tightly packed within a network of PTFE fibrils, there is more resistance to flow. Higher force or vacuum must be applied to pass liquids through this particle bed. The kinetics of the extraction process are usually not affected by these higher flow rates. While packed particle beds and most other disks are routinely used at a vacuum setting of 5–10 in Hg (170–340 mbar), Empore disks may require higher vacuum (*e.g.*, 10–20 in Hg or 340–680 mbar) in order to pass liquids through the thin, densely packed particle bed.

The greatest resistance to flow with PTFE disks is seen during the sample load step; cleaner (less viscous, more dilute) sample matrices may pass through at 10 in Hg (340 mbar), while dirtier matrices may require 20 in Hg (680 mbar) or more to ensure complete flow. These flow characteristics can also be volume dependent, important for volumes 0.25 mL or larger, but unimportant at smaller sample volumes such as 100 μL. Also note the distinction between standard density and high density PTFE disks (Section 11.4.3.4). High density disks, as found in the individual disk cartridge format, require the highest vacuum or pressure since they contain smaller (~12 μm) particles but the standard density disks (~40 um) display improved flow characteristics. The Empore 96-well SPE plates contain only the standard density version.

11.5 High Throughput Applications for 96-Well SPE

11.5.1 Early Reports

11.5.1.1 Introduction

Beginning in about 1995, several advances occurred in parallel in pharmaceutical bioanalysis that brought about a change from individual SPE columns to microplates for higher throughput sample preparation. Faster analytical techniques such as LC-MS/MS became more widespread and affordable, allowing researchers to analyze samples more quickly than ever before. Liquid handling workstations became more prevalent and feature rich, and replaced most manual pipetting tasks. Within the pharmaceutical industry, drug development cycles were shortened, favoring a more rapid identification of lead candidates with a faster pace of development. When the sample preparation step became rate limiting in this overall process, a more efficient and faster way to work with the greater number of samples was sought and the 96-well SPE plate was developed. Details on the historical development of SPE in the 96-well format can be found in Chapter 3, Section 3.5.1 and a review of the early models and prototypes of plates is found in Section 3.5.2.

11.5.1.2 First Published Application—Manual

The first published application used an early model of a patented 96-well SPE plate (before licensing and commercial introduction by Porvair Sciences) and was reported by Kaye *et al.* in 1996 [233]. An LC-MS/MS method for the analysis of darifenacin in human plasma (1 mL), with deuterated darifenacin as internal standard (IS), was developed and validated over the concentration range 25–2000 pg/mL. Accuracy and precision were considered satisfactory at 0.6–4.6% and 3.6–18.8%, respectively. Recovery from C18 bonded silica was about 50%. A comparison of time requirements was reported as 1.5 h to manually prepare 96 samples in microplate versus 3 h to manually prepare 50–60 samples in the traditional single column format.

11.5.1.3 First Published Applications—Automated

Another experience using this novel plate design was reported by Allanson *et al.* [234] who combined the format with a MultiPROBE® liquid handling workstation (Packard Instruments, Meriden, CT USA, now a part of PerkinElmer Life Sciences). The MultiPROBE performed all liquid handling steps for the extraction of a novel 5-hydroxytryptamine receptor agonist and a metabolite from human plasma (0.5 mL); the program was paused to allow removal of the waste tray from inside the vacuum manifold and replacing it with a deep well collection plate. The concentration range evaluated for both analytes was 0.1–15 ng/mL and recovery was 80–90%. Since higher throughput was one of the goals of using this format, time savings were again documented; 1 h to prepare 96 samples in semi-automated mode compared with 2–3 h to prepare 60 samples manually.

This same research group continued to refine and further automate the 96-well extraction process by interfacing the system with a Zymark robot (Zymark Corporation, Hopkinton, MA USA). The potential to analyze a complete clinical study (~800 h) in under 48 h using LC-MS/MS techniques was presented at the 1996 ISLAR conference [235] and summarized in a book chapter [236]. In order to automate the entire procedure, a customized SPE station was developed to condition the 96-well SPE plates (blocks), wash and elute into the appropriate collection plate. Conditioned blocks were transferred to the MultiPROBE for sample loading and then back to the SPE station for the wash and elution steps. Eluate in the collection plate was transferred to a cooled storage carousel in an autoinjector which could initiate 8 consecutive runs. Throughput of the developed system was reported as 12–16 blocks prepared in one working day, enough to supply two mass spectrometers. A refined perspective on this fully automated approach, including a discussion of enhancements made to the system, was presented by this same research group two years later at another ISLAR conference [237].

11.5.1.4 Reduced Sorbent Mass Plates

The potential for even greater throughput using 96-well SPE plates was explored by Plumb *et al.* [178] who examined the use of reduced sorbent bed mass (10–35 mg) and particle-loaded membranes (Empore disks in a prototype plate mold prior to commercial introduction) for the analysis of pharmaceutical compounds in biological fluids. A reduction of the elution volume to 100 μL or less with a high analyte recovery allows the elimination of the evaporation and reconstitution steps. The majority of eluant is injected into a narrow bore LC column interfaced to a tandem mass spectrometer utilizing electrospray ionization. This report documented improvements by the use of small sorbent masses and disks in terms of reduced solvent volumes and sample preparation time; capacity of these small sorbent masses was found adequate and performance results were very satisfactory.

11.5.1.5 High Throughput Using a 96-Tip Workstation

Subsequent early reports on the use of 96-well solid-phase extraction described experiences in using and automating Empore disk plates. Janiszewski *et al.* [185] reported the first microplate SPE application on the Tomtec Quadra® 96 (Tomtec Inc., Hamden CT USA), an established liquid handling workstation used in high throughput screening programs in drug discovery laboratories. Elution in small volumes using the disk plate allowed direct injection of eluate, saving time. Use of the 96-tip liquid pipetting unit further extended the time savings; note that biological samples were reformatted off-line from tubes into a collection plate for processing on the Quadra 96. A block of 96 samples was reported to be extracted in 10 min, which was about 30 times faster than manual single cartridge SPE methods.

Another innovation reported from this same laboratory was simultaneous conditioning and sample delivery to the disk plate [185]. Sequential aspirating was performed using the Quadra 96 tips in this order: 300 μL serum, 25 μL air gap, and 50 μL methanol. The Quadra 96 dispensed this total volume into the disk plate (vacuum on) so that methanol was delivered first, followed by the sample. The air gap in the pipet tip separated the two liquids. A subsequent aspirate and dispense procedure was performed using the remainder of the serum sample (100 μL) since the maximum volume of the tips is only 450 μL. Analyte (ziprasidone) precision of detection using this single conditioning protocol was reported to be better than traditional two step conditioning and loading steps. An intermediate aqueous conditioning step between the organic solvent and the sample, as is commonly performed in SPE, was found to be unnecessary using this particular approach.

11.5.1.6 Optimization of Disk Plate Usage with Workstation

Simpson *et al.* [186] discussed their results of the further development, characterization and optimization of the Empore disk plate format using the

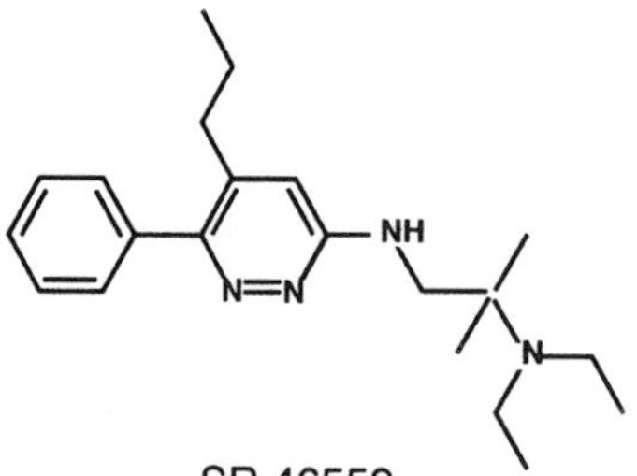

SR 46559 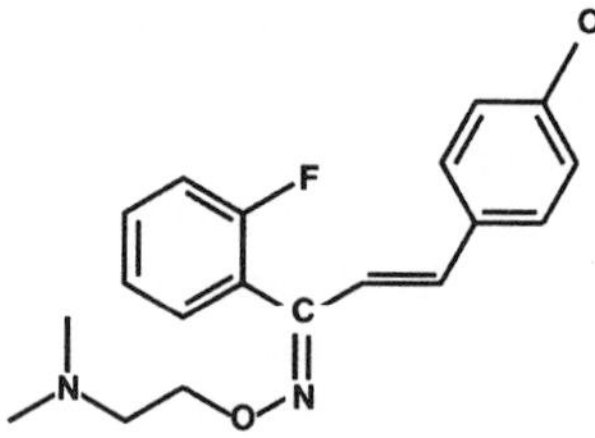   SR 46349

Figure 11.22. Chemical structures of clinical drug candidate compounds SR46559 and SR46349. Reprinted with permission from [186]. Copyright 1998 John Wiley & Sons, Inc.

Packard MultiPROBE 4-probe liquid handling workstation. Several developmental compounds from clinical trials (two are shown in Figure 11.22) were extracted from human plasma (0.5–0.8 mL) using C8 bonded silica disk plates. Three important topics were addressed in this report: (1) efficient utilization and control of vacuum by the MultiPROBE instrument; (2) optimization of issues relating to use of the Empore disk for SPE in the plate format; and (3) the exploration of robotic issues in bioanalysis of plasma and other biofluids.

A control valve was configured to permit the generation of vacuum pulses during the different liquid processing steps of the SPE method. In the initial conditioning step, only a brief 2 s vacuum was used to pass the 100 µL volume of methanol through the disk. The sample loading step used maximum vacuum held for 5 min to remove a large 1 mL plasma volume. A slowly rising profile was used during the elution with organic solvents. Repeated pulses could be applied using this technique. Vacuum pulses are very useful during the sample load step when the most viscous fluid is processed through the disk.

It was found that a 100 µL volume of methanol added into each extraction well was sufficient to spread out over the diameter of the disk and soak below the top surface (within about 15 s), thoroughly wetting the particle-loaded disk. A vacuum step was unnecessary; 200 µL water was added directly onto the disk to displace the methanol, followed by brief vacuum. For a fixed 1 mL volume, dilution of plasma (up to 50%) with buffer (for the load step) did not affect mass recovery of analyte and did not alter flow rate. Following the load step, an aqueous wash was preferred first to displace proteins, followed by an organic/aqueous wash to more completely remove interfering substances; omitting the initial aqueous wash could lead to protein precipitation on the surface of the disk, reducing the available surface area and adversely affecting flow characteristics.

One of three elution solvents was used, depending on the analyte:
  (a) 200 µL acetonitrile containing 0.1% trifluoroacetic acid (TFA),
  (b) 75 µL acetonitrile containing 0.1% TFA, or
  (c) 200 µL aqueous mobile phase/acetonitrile (1:1 v/v, both containing 2mM ammonium acetate and 0.2% formic acid)

Elution fractions were diluted with 200 µL buffer, 165 µL aqueous mobile phase (2 mM ammonium acetate and 0.2% formic acid), and 300 µL aqueous mobile phase for (a), (b) and (c), respectively. Injection volumes were 50 µL or 100 µL. The time consuming steps of evaporation and reconstitution were eliminated using this approach.

Binding of analyte to the protein in plasma is a possible cause for altered adsorption to the sorbent particle used for solid-phase extraction and may lead to lower recovery. This binding may also be sensitive to freeze/thaw cycles during normal sample handling procedures. In order to promote release of drug bound to plasma protein, a 5% (v/v) acetonitrile (final concentration) was found sufficient; this concentration was low enough to avoid a large degree of precipitation. Samples were then loaded onto the SPE disk without adverse effects. Plugging of the SPE well can be a potential concern, especially with densely loaded membranes, as reported in [45]. Plasma samples treated with heparin can form visible amounts of fibrin upon repeated freeze/thaw cycles. This laboratory recommended that heparinized plasma samples be vortex mixed prior to sample preparation to break fibrin into smaller fragments, and then centrifuged before use. This procedure was found to reduce the frequency of plugged wells during the extraction procedure.

The use of fixed tips by the MultiPROBE was shown to present a carryover issue for a particular analyte. In addition to washing the fixed tips with system liquid (water) following each sample transfer, an acetonitrile wash step (1 mL, twice after each transfer) was sufficient to remove the carryover issue. Note that the tips should be washed with water before and after acetonitrile to avoid potential precipitation of plasma protein.

Overall, the time required for sample preparation was about 1 h for 96 samples using the MultiPROBE, compared with 4–5 h for traditional methods that are fully attended. In addition to time and labor savings, a comparison in consumable costs between 96 disk cartridges and a disk plate showed a savings when using the plates; another application also reported cost savings [238]. One of the major advantages of using the 96-well disk SPE plate was cited as the small elution volume (75–200 µL) which eliminated the need for an evaporation step. Further discussion of this subject is found in Chapter 12, Section 12.3.2.3.3.

11.5.2 Applications using Polymer Sorbents

The role of polymer sorbents in solid-phase extraction was introduced in Section 11.2.1.2. An advantage of the polymer sorbents is their ability to be used in a generic method that is expected to retain most analytes, regardless of polarity: condition with methanol followed by water, load sample in unionized form, wash with 5% methanol in water and elute in 100% methanol. The Oasis HLB sorbent is discussed here as a representative example of versatile polymer chemistry since there is a wealth of published information available about its use. By no means should it be considered the only choice for a polymer sorbent. Development of new and improved sorbents with additional unique affinities (*e.g.*, FOCUS™ by Varian) continues to be an active area of research and market growth.

An application demonstrating a generic methodology for an Oasis HLB polymer is reported by Cheng *et al.* [34] who used 96-well plates for the extraction of verapamil and norverapamil from porcine plasma; ultraviolet (UV) detection at 230 nm was employed. While this generic method isolated the analytes of interest, it also extracted other plasma components that interfered with peak quantitation of the polar metabolite norverapamil, which was eluted first by liquid chromatography (LC). While LC conditions could be optimized, the simpler approach is to make the SPE method more selective for the analytes of interest.

Experiments were performed to optimize the composition of the wash solvent (to remove interferences yet retain analytes) and the elution solvent. Different combinations of wash and elution solutions were examined (wash, 2% ammonium hydroxide with from 0–90% methanol; elution, 2% acetic acid with from 0–65% methanol). It was found that the wash solvent could tolerate up to 70% methanol containing 2% base before analytes started to elute from the sorbent bed. Regarding elution, greater than 60% methanol containing 2% acid was required.

Comparison of chromatograms from the generic with the more selective method showed an improved selectivity from the latter, resulting in a reduction in the mass of interfering analytes. While polymer sorbents are marketed as useful for generic analyses, they can also be used in a more selective manner to extract the analytes of interest from biological matrices. The choice of which approach to use depends on the objectives of the assay and the time involved for method development. Further discussion of SPE method development issues and protocols is found in Chapter 12.

The wide applicability of the Oasis HLB polymer sorbent in the 96-well SPE format is demonstrated by the number of publications reporting its use [33–57]. The 30 mg sorbent bed mass is the most frequently used format and its applicability includes serum and plasma, urine and *in vitro* microsomal incubation media (as discussed in Section 11.5.4). The use of lower bed mass polymers has been shown to offer adequate performance in several applications. The reduction in bed mass allows reduced solvent volumes and the dry-down and reconstitution steps can be eliminated. Cheng *et al.* demonstrated use of a 10 mg Oasis HLB plate for the extraction of methadone and a metabolite in human urine [36]. An elution volume of 200 μL of 80% methanol containing 2% acetic acid was used; dilution with 600 μL water was made and an aliquot was analyzed by LC with UV detection at 210 nm. This study also discusses wash and elution solvent optimization procedures. A 10 mg Oasis HLB plate was used to extract *in vitro* microsomal incubation solution but elution volumes were large at 800μL [46]. The extraction of betamethasone from 50 μL rat plasma also used excess elution volume at 500 μL [54]. Another application reported high throughput biopolymer desalting using a 5 mg Oasis HLB plate prior to mass spectrometric analysis; elution was performed using 100 μL of 70% acetonitrile in water [47].

11.5.3 Processing Options and Throughput Considerations

Solid-phase extraction in microplates can be performed manually using a multichannel pipettor and a vacuum manifold. Selected applications that used manual processing are summarized in Table 11.7. Additional gains in throughput are achieved using liquid handling workstations such as the MultiPROBE or Quadra 96. When both instruments are acquired, the Multi-PROBE is used to reformat samples from tubes or vials into plates and this reconfigured sample plate is processed on the Quadra 96. The many alternatives to the MultiPROBE and liquid handling workstations are discussed in detail in Chapter 5. Selected applications for 96-well SPE that used automation for sample processing are summarized in Table 11.8.

Table 11.7
Selected bioanalytical applications using solid-phase extraction in the microplate format without automation

Analytes	Sample Matrix and Volumes	SPE Sorbent and Plate Type	Elution Solvent and Volume	LOQ	Analytical Method	Reference
Verapamil and norverapamil	Porcine plasma 1mL IS, H_3PO_4	Oasis® HLB 30mg	Methanol or 65% methanol with 2% acetic acid 1mL	340ng/mL	LC-UV	[34]
Oxcarbazepine	Plasma 50µL IS 50µL	C18 Empore™ disk	Methanol 150µL	0.2µmol/L	LC-UV	[180]
Rufinamide	Plasma 50µL diluent 50µL	C18 Empore disk	Acetonitrile 80µL	50ng/mL	LC-UV	[35]
Alachlor and metabolites	Rat plasma 1mL 5M HCl 10µL	Oasis HLB 30mg	Acetonitrile with 2% acetic acid 1mL	2.3ng "per injection"	LC-MS	[53]
Cyclooxygenase II inhibitor	Human plasma 1mL ACN 50µL, IS 50µL 17% ACN mL	Oasis HLB 30mg	Acetonitrile 500µL	5ng/mL	LC-UV	[45]
Exemestane	Human plasma 50µL IS 50µL water 500µL	C2 50mg	Acetonitrile with 0.1% trifluoroacetic acid 300µL	0.05ng/mL	LC-MS/MS	[239]
Alpha-1a receptor antagonist	Human plasma 1mL IS 2x50µL Buffer 1mL	MPC Empore Disk	CH_2Cl_2:IPA: NH_4OH 78:20:2 300µL	0.2ng/mL	LC-MS/MS	[197]

Table 11.8
Selected LC-MS/MS bioanalytical applications using solid-phase extraction in the microplate format with automation

Analytes	Sample Matrix and Volumes	SPE Sorbent and Plate Type	Elution Solvent and Volume	LOQ	Automation	Ref.
Neurokinin-1 receptor antagonist and metabolite	Human plasma or urine 0.5mL, IS 0.5mL	Oasis® HLB 30mg	0.04% HCOOH in methanol 300μL add 300μL buffer	0.1ng/mL each	Genesis RSP200 with RoMA	[44]
Indinavir and L-756423	Human plasma 0.5mL IS 25μL, 50% ACN 25μL, Buffer pH 2, 0.5mL	MPC Empore™ disk	ACN/water/-NH₄OH 58/40/2, 225μL	1 and 5ng/mL	MultiPROBE and Quadra96	[87]
Morphine and its 3- and 6-glucuronides	Human Plasma 250μL IS 25μL	C18 50mg Versaplate™	Methanol/water 1:1, 0.8mL	0.5, 10 and 1.0ng/mL	MultiPROBE and Quadra96	[240]
Proprietary analyte	Human Plasma 100μL IS 50μL, Buffer pH 3.5 250μL	C18 Empore disk	ACN 200μL	50pg/mL	MultiPROBE	[179]
Chlorambucil and metabolite	Human serum or plasma 200μL, IS 150μL	C18 15mg Microlute™	Mobile Phase 100μL	4ng/mL	MultiPROBE	[238]
Everolimus and cyclosporine A	Supernatant after PPT of blood, 800μL	C18 AR SPEC• PLUS™ disk	ACN 100μL	0.375 and 6.95ng/mL	Quadra96	[147]

Analytes	Sample Matrix and Volumes	SPE Sorbent and Plate Type	Elution Solvent and Volume	LOQ	Automation	Ref.
Fenfluramine, temazepam, tamoxifen	Plasma 200µL IS 200µL	Oasis HLB 30mg	Methanol 1mL	10ng/mL	Quadra96	[37]
Fluticasone propionate	Plasma 500µL IS 500µL	C18 50mg Microlute	Methanol 200µL	20pg/mL	Zymark custom robotics	[241]
Five estrogen sulfates	Human urine 150µL IS 150µL	C18 Empore disk	Methanol 200µL	0.2ng/mL per 100µL	Quadra96	[182]
Fentanyl	Human plasma 250µL IS 25µL	Certify™ 25mg Versaplate	CHCl3/IPA/-NH$_4$OH 78:20:2, 750µL	0.05ng/mL per 250µL	MultiPROBE and Quadra96	[131]
Rosuvastatin	Human Plasma 500µL Acetate buffer 500µL IS 50µL, 1M Acetic acid 750µL	Oasis HLB 30mg	Methanol with 0.5% acetic acid 1mL	0.1ng/mL	Genesis RSP100	[52]
Three proprietary analytes	Plasma 250µL, IS 25µL 0.1N HCl 250µL	Oasis HLB 30mg	70% Acetonitrile in water 0.7mL	2ng/mL	Quadra96	[43]
(R)- and (S)-Ketoprofen	Human plasma 1mL HCOOH 50µL IS 25µL	Oasis HLB 30mg	Methanol/water/-HCOOH 90:10:0.1	0.05ng/mL	Biomek 2000 and Multimek	[49]

Analyte	Sample	Sorbent	Elution	Concentration	Instrument	Ref
Salbutamol (racemic)	Human plasma or urine 400µL, Buffer pH 10, 400µL, IS 50µL	C2 50mg Microlute	Methanol 200µL	5pg/mL based on 0.5mL	Zymark custom robotics	[50]
Betamethasone	Rat plasma 50µL IS 50µL, water 250µL	Oasis HLB 10mg	ACN:0.1% TFA (9:1), 500µL	2ng/mL	MultiPROBE	[54]
	Dog plasma 50µL IS 50µL, Buffer 50µL	Oasis HLB 10mg	Methanol with 1% HCOOH 600µL	25ng/mL	Quadra96	[55]
Melphalan	Human serum or plasma 200µL IS in water 150µL	C18 15mg Microlute	(0.1% HCOOH in ACN):(0.1% HCOOH) [6:4] 200µL	2ng/mL	MultiPROBE	[242]
Oxazepam	Human plasma 200µL IS 300µL	Oasis HLB 30mg	Methanol 900µL	10nM	Zymark track robot system	[48]

Generally, in terms of throughput for semi-automated SPE procedures using 96-well plates, from 2–2.5 plates per day (192–240 samples) can be prepared and analyzed per analyst [192]. Extraction times using the MultiPROBE are reported generally as from 50–60 min, and for the Quadra 96 from 10–15 min. The practical limitations to improving throughput one step further using Quadra 96 automation were evaluated by Rule and Henion [184]. One analyst was able to prepare and analyze 384 samples within a 24 h period. The interday and intraday accuracy and precision obtained over the course of these samples was within 8% coefficient of variation when analyzed by atmospheric pressure chemical ionization mass spectrometry using positive ion detection.

The combination of automated sample preparation with high speed chromatography using a monolithic column provides tremendous throughput and is useful in drug discovery applications. Wu *et al.* [37] developed a monolithic chromatography method for three analytes with a run time of only 1 min. Sample preparation was performed using a Quadra 96 instrument. Three components were measured in 600 plasma extracts during an overnight run.

In an extension of this work by the same laboratory, Deng *et al.* [48] described a component system capable of extracting and analyzing 1152 plasma samples (12 plates) within 10 h. A Zymark track robot was interfaced with a Tecan Genesis (Tecan US, Research Triangle Park, NC USA) liquid handling workstation for the simultaneous and fully automated solid-phase extraction of four 96-well plates. The extracts were injected onto four parallel monolithic columns for separation via a four-injector autosampler. The effluent from each of the four columns was directed into an LC-MS/MS system equipped with an indexed four-probe electrospray ionization source (MUX®, Micromass, Beverly, MA USA). An overall throughput was reported as 30 s per sample using this novel four-channel parallel format.

11.5.4 SPE of in vitro Incubation Media

High throughput screening methods for cytochrome P450 isoform activity are used to evaluate the *in vitro* potential of drug candidates to inhibit drug metabolizing enzymes. An isoform activity assay reported by Yin *et al.* [51] used human liver microsomes with generic conditions for incubation, reaction termination, metabolite extraction using a 96-well plate system and LC-MS/MS quantitation. The assay used pooled human liver microsomes and seven probe substrates for the major human hepatic CYP isoforms 1A2, 2A6, 2C9, 2C19, 2D6, 2E1 and 3A4. Reactions were performed in 96-well collection plates and all liquid handling steps were completed by a Quadra 96 liquid handling workstation. Reactions were initiated by the addition of NADPH after a 5 min preincubation at 37°C. Following a 10 min incubation time, a 450 µL aliquot of the reaction mixture was loaded onto a conditioned Oasis HLB plate so that the NADPH cofactor was removed from the mixture and the reactions were stopped. A wash step was performed with 1 mL water and metabolites were eluted from the sorbent bed using 700 µL methanol. An aliquot from each of the seven eluents was pooled into one well. Evaporation and reconstitution in mobile phase compatible solvent were performed. Internal standard was added prior to evaporation. This combination of reaction termination and metabolite extraction within an SPE sorbent bed is novel and superior to other sample preparation methods (*e.g.*, protein precipitation or liquid-liquid extraction) [51].

The automated use of Oasis HLB plates (10 mg) was reported for another *in vitro* experiment involving the extraction and analysis of 27 highly diversified pharmaceutical compounds from a human microsomal incubation, using a Zymark robotic system with a Tecan Genesis workstation [46]. A reduction in matrix effect was noted by the use of this SPE procedure. Note that a novel approach to the solid-phase extraction of *in vitro* microsomal incubations is to extract all 96 wells simultaneously from the incubation microplate using an SPE Card, and elute each well in sequence directly into a mass spectrometer, as discussed further in Chapter 15, Section 15.1.

11.5.5 Comparisons with Other Sample Preparation Techniques

Sample throughput of manual solid-phase extraction methods using individual cartridges was compared with 96-well SPE using automation. Souppart *et al.* [180] documented a threefold increase in sample throughput and a twofold decrease in required plasma volume. The time required to analyze a human pharmacokinetic study with about 1000 clinical samples prepared using a fully automated 96-well SPE procedure [8-probe Genesis with RoMA (Robotic Manipulator Arm)] was compared with automated single SPE cartridge processing (Zymark RapidTrace™). An evaluation of the cumulative hours for sample preparation showed a reduction in time from 7 man-days to 2 man-days using the 96-well approach.

Three extraction procedures were compared for the quantitative determination of a carboxylic acid containing analyte in human plasma by LC-MS/MS: (1) manual LLE of acidified plasma with methyl tert-butyl ether (MTBE); (2) automated LLE in 96-well collection microplates; and (3) automated 96-well SPE [179]. While a lower limit of quantitation (LLOQ) of 50 pg/mL was achieved using all three approaches, the total time requirements varied greatly.

The processing of 96 samples by manual LLE was three times longer than that required for 96-well LLE or SPE (4 h, 50 min vs. 1 h, 43 min). Actual hands-on analyst time requirements were 4 h, 10 min for manual LLE but <10 min for 96-well LLE or SPE. Total time requirements for SPE could be reduced by 50% if the eluate dry-down and reconstitution steps were eliminated.

Solid-phase extraction in the 96-well format was compared with protein precipitation, also in the same format [39]. It was found that the SPE procedure removed nonvolatile salts and matrix interferences, allowing faster LC runs compared to PPT. Also, since SPE was automated using a Quadra 96, significant time savings were realized. The PPT procedure required an additional vortex and centrifugation step to separate plasma proteins from analyte even with the use of the Quadra 96 for automation.

The quantitative analysis of several candidate compounds from plasma was compared using on-line turbulent flow chromatography (TFC), column switching after PPT, a micro column switching assay, and semi-automated 96-well disk SPE [193]. The limit of quantitation for these analytes using TFC was found to be five-fold higher compared with the other methods examined. Disk SPE was fast, semi-automated and minimized matrix effects while providing reliable LC-MS/MS methods. Column switching after PPT was fast, accurate and sensitive; microbore LC allowed fast analysis with low sample consumption. Zimmer *et al.* compared TFC with automated 96-well SPE and LLE [40]. One 96-well plate with 96 plasma samples is analyzed within 5.25 h—mass spec time. Since no off-line sample preparation step is performed for TFC, the 5.25 h is from start to finish. SPE adds an additional 1 h off-line time to prepare the samples; 6.25 h total. LLE and analysis of 96 samples takes about 16 h total, requiring hands-on analyst time for much of the working day. All three techniques displayed accuracy and precision within guidelines.

Wu *et al.* compared the performance of dual-column TFC with SPE for ten structurally diverse drug compounds and found that the dynamic range, accuracy, and precision were very similar between the two methods [41]. A report from Niggebrugge concluded that TFC used with an initial 96-well filtration of plasma provided a simple and cost effective assay that is as dependable as semi-automated 96-well SPE [243].

This chapter has focused on off-line solid-phase extraction in the high throughput microplate format, performed independent of the analysis. SPE can also be efficiently performed on-line using individual disposable cartridges. On-line SPE in this manner is discussed in Chapter 14, Section 14.3.6.

Acknowledgments

The author is appreciative to Gary Lensmeyer for his critical review of the manuscript, helpful discussions and contributions to this chapter. Line art illustrations were kindly provided by LC Resources and Varian, as indicated; Willy Lee and Woody Dells provided the remainder of the illustrations. The assistance of the following vendor representatives who reviewed portions of the text is appreciated: Linda Alexander (Macherey-Nagel), Paul Bouise (Mallinckrodt Baker), Mike Brown and Tony Castleman (Porvair Sciences), Claire Desbrow (IST Ltd., now an Argonaut Technologies Company), Michael Early (Waters Corporation), Ron Majors (Agilent Technologies), Rich Matner (3M Corporation), Asha Oroskar (Orochem Technologies), Roger Roberts (Ansys Technologies, now a part of Varian), Rob Stubbs (Varian), Bethany Telepchak (United Chemical Technologies), James Teuscher (Phenomenex), An Trinh (Supelco), Sarah Turkington (Alltech Associates) and Yale West (Applied Separations).

References

[1] R.D. Golding, A.J. Barry and M.F. Burke, J. Chromatogr. 384 (1987) 105-116.

[2] J.P. Roach, D.M. Marchisin, K.D.W. Roth, P.P. Davis, S. Parchman, R. Roberts and R.N. Hayes, Proceedings 49th American Society for Mass Spectrometry Conference, Chicago, IL USA (2001).

[3] P. Martin, E.D. Morgan and I.D. Wilson, Anal. Chem. 69 (1997) 2972-2975.

[4] P. Martin, J. Taberner, A. Fairbrother and I.D. Wilson, J. Pharm. Biomed. Anal. 11 (1993) 671-677.

[5] R.J. Ruane and I.D. Wilson, J. Pharm. Biomed. Anal. 5 (1987) 723-727.

[6] P. Martin, E.D. Morgan and I.D. Wilson, Anal. Proc. 32 (1995) 179-181.

[7] I. Papadoyannis, J. Liq. Chrom. & Rel. Technol. 16 (1993) 3827-3845.

[8] G.L. Lensmeyer and A. Oroskar, Proceedings 50th American Society for Mass Spectrometry Conference, Orlando, FL USA (2002).

[9] R.M. Patel, J.R. Benson, D. Hometchko and G. Marshall, Amer. Lab. 22 (1990) 92-99.

[10] R.J. Scott, J. Palmer, I.A. Lewis and S. Pleasance, Rapid Commun. Mass Spectrom. 13 (1999) 2305-2319.

[11] E.S.P. Bouvier, P.C. Iraneta, U.D. Neue, P.D. McDonald, D.J. Phillips, M. Capparella and Y.-F. Cheng, LC-GC 16 (1998) S53-S57.

[12] E.S.P. Bouvier, D.M. Martin, P.C. Iraneta, M. Capparella, Y.-F. Cheng and D.J. Phillips, LC-GC 15 (1997) 152-158.

[13] M. Gilar, E.S.P. Bouvier and B.J. Compton, J. Chromatogr. A 909 (2001) 111-135.

[14] R. Bonfiglio, R.C. King, T.V. Olah and K. Merkle, Rapid Commun. Mass Spectrom. 13 (1999) 1175-1185.

[15] C.R. Mallet, J.R. Mazzeo and U. Neue, Rapid Commun. Mass Spectrom. 15 (2001) 1075-1083.

[16] J. Ding and U.D. Neue, Rapid Commun. Mass Spectrom. 13 (1999) 2151-2159.

[17] R. King, R. Bonfiglio, C. Fernandez-Metzler, C. Miller-Stein and T. Olah, J. Amer. Soc. Mass. Spectrom. 11 (2000) 942-950.

[18] R. Oertel, K. Richter, J. Fauler and W. Kirch, J. Chromatogr. A 948 (2002) 187-192.

[19] S.K. Teo, R.S. Chandula, J.L. Harden, D.I. Stirling and S.D. Thomas, J. Chromatogr. B 767 (2002) 145-151.

[20] T. Iwasa, T. Takano, K. Hara and T. Kamei, J. Chromatogr. B 734 (1999) 325-330.

[21] Q. Song and L. Putcha, J. Chromatogr. B 763 (2001) 9-20.

[22] K. Vishwanathan, M.G. Bartlett and J.T. Stewart, Rapid Commun. Mass Spectrom. 14 (2000) 168-172.

[23] J.S. Millership, L.G. Hare, M. Farry, P.S. Collier, J.C. McElnay, M.D. Shields and D.J. Carson, J. Pharm. Biomed. Anal. 25 (2001) 871-879.

[24] X. Tong, I.E. Ita, J. Wang and J.V. Pivnichny, J. Pharm. Biomed. Anal. 20 (1999) 773-784.

[25] M. Sarasa, N. Riba, L. Zamora and X. Carne, J. Chromatogr. B 746 (2000) 183-189.

[26] E.J. Woolf and B. Matuszewski, J. Chromatogr. A 828 (1998) 229-238.

[27] W. Naidong, H. Bu, Y.-L. Chen, W.Z. Shou, X. Jiang and T.D.J. Halls, J. Pharm. Biomed. Anal. 28 (2002) 1115-1126.

[28] M. Kollroser and C. Schober, J. Pharm. Biomed. Anal. 28 (2002) 1173-1182.

[29] P. Metz, S.J. Kohlhepp and D.N. Gilbert, J. Chromatogr. B 773 (2002) 159-166

[30] Y.-F. Cheng, D.J. Phillips, U. Neue and L. Bean, J. Liq. Chrom. & Rel. Technol. 20 (1997) 2461-2473.

[31] Y.-F. Cheng, D.J. Phillips and U. Neue, Chromatographia 44 (1997) 187-190.

[32] N. Hanioka, Y. Saito, A. Soyama, M. Ando, S. Ozawa and J.-I. Sawada, J. Chromatogr. B 774 (2002) 105-113.

[33] J. Yawney, S. Treacy, K.W. Hindmarsh and F.J. Burczynski, J. Anal. Toxicol. 26 (2002) 325-332.

[34] Y.-F. Cheng, U.D. Neue and L. Bean, J. Chromatogr. A 828 (1998) 273-281.

[35] M.C. Rouan, C. Buffet, L. Masson, F. Marfil, H. Humbert and G. Maurer, J. Chromatogr. B 754 (2001) 45-55.

[36] Y.-F. Cheng, U.D. Neue and L.L. Woods, J. Chromatogr. B 729 (1999) 19-31.

[37] J.-T. Wu, H. Zeng, Y. Deng and S.E. Unger, Rapid Commun. Mass Spectrom. 15 (2001) 1113-1119.

[38] S.H. Hoke II, J.D. Pinkston, R.E. Bailey, S.L. Tanguay and T.H. Eichhold, Anal. Chem. 72 (2000) 4235-4241.

[39] S.X. Peng, S.L. King, D.M. Bornes, D.J. Foltz, T.R. Baker and M.G. Natchus, Anal. Chem. 72 (2000) 1913-1917.

[40] D. Zimmer, V. Pickard, W. Czembor and C. Müller, J. Chromatogr. A 854 (1999) 23-35.

[41] J.-T. Wu, H. Zeng, M. Qian, B.L. Brogdon and S.E. Unger, Anal. Chem. 72 (2000) 61-67.

[42] J. Ayrton, G.J. Dear, W.J. Leavens, D.N. Mallet and R.S. Plumb, J. Chromatogr. B 709 (1998) 243-254.

[43] R.C. King, C. Miller-Stein, D.J. Magiera and J. Brann, Rapid Commun. Mass Spectrom. 16 (2002) 43-52.

[44] D. Schütze, B. Boss and J. Schmid, J. Chromatogr. B 748 (2000) 55-64.

[45] R.S. Mazenko, A. Skarbek, E.J. Woolf, R.C. Simpson and B. Matuszewski, J. Liq. Chrom. & Rel. Technol. 24 (2001) 2601-2614.

[46] J.J. Zheng, E.D. Lynch and S.E. Unger, J. Pharm. Biomed. Anal. 28 (2002) 279-285.

[47] M. Gilar, A. Belenky and B.H. Wang, J. Chromatogr. A 921 (2001) 3-13.

[48] Y. Deng, J.-T. Wu, T.L. Lloyd, C.L. Chi, T.V. Olah and S.E. Unger, Rapid Commun. Mass Spectrom. 16 (2002) 1116-1123.

[49] T.H. Eichhold, R.E. Bailey, S.L. Tanguay and S.H. Hoke II, J. Mass Spectrom. 35 (2000) 504-511.

[50] K.B. Joyce, A.E. Jones, R.J. Scott, R.A. Biddlecombe and S. Pleasance, Rapid Commun. Mass Spectrom. 12 (1998) 1899-1910.

[51] H. Yin, J. Racha, S.-Y. Li, N. Olejnik, H. Satoh and D. Moore, Xenobiotica 30 (2000) 141-154.

[52] C.K. Hull, A.D. Penman, C.K. Smith and P.D. Martin, J. Chromatogr. B 772 (2002) 219-228.

[53] L.-Y. Zang, J. DeHaven, A. Yocum and G. Qiao, J. Chromatogr. B 767 (2002) 93-101.

[54] C.S. Tamvakopoulos, J.M. Neugebauer, M. Donnelly and P.R. Griffin, J. Chromatogr. B 776 (2002) 161-168.

[55] P.H. Zoutendam, J.F. Canty, M.J. Martin and M.K. Dirr, J. Pharm. Biomed. Anal. 30 (2002) 1-11.

[56] M. Gilar and E.S.P. Bouvier, J. Chromatogr. A 890 (2000) 167-177.

[57] D. Fraier, E. Frigerio, G. Brianceschi, M. Casati, A. Benecchi and C. James, J. Pharm. Biomed. Anal. 22 (2000) 505-514.

[58] C.W. Huck and G.K. Bonn, J. Chromatogr. A 885 (2000) 51-72.

[59] D.L. Ambrose, J.S. Fritz, M.R. Buchmeiser, N. Atzl and G.K. Bonn, J. Chromatogr. A 786 (1997) 259-268.

[60] R.S. Plumb, R.D.M. Gray, A.J. Harker and S. Taylor, J. Chromatogr. B 687 (1996) 457-461.

[61] T.K. Chambers and J.S. Fritz, J. Chromatogr. A 797 (1998) 139-147.

[62] P.J. Dumont and J.S. Fritz, J. Chromatogr. A 691 (1995) 123-131.

[63] M.E. León-González and L.V. Pérez-Arribas, J. Chromatogr. A 902 (2000) 3-16.

[64] T.C. Pinkerton, J. Chromatogr. 544 (1991) 13-23.

[65] J.R. Mazzeo and I.S. Krull, BioChromatogr. 4 (1989) 124-130.

[66] P. Martin, B. Leadbetter and I.D. Wilson, J. Pharm. Biomed. Anal. 11 (1993) 307-312.

[67] I.D. Wilson and P. Martin, In: N.J.K. Simpson, Ed., Solid-Phase Extraction: Principles, Techniques and Applications, Marcel Dekker, New York (2000) 331-347.

[68] P.F. Maycock, Clin. Chem. 33 (1987) 286-287.

[69] A.H.B. Wu and G. Gornet, Clin. Chem. 31 (1985) 298-302.

[70] D.W. Roberts, R.J. Ruane and I.D. Wilson, J. Pharm. Biomed. Anal. 7 (1989) 1077-1086.

[71] J. Ayrton, M.B. Evans, A.J. Harris and R.S. Plumb, J. Chromatogr. B 667 (1995) 173-178.

[72] E. Matisová and S. Škrabáková, J. Chromatogr. A 707 (1995) 145-179.

[73] M.-C. Hennion, J. Chromatogr. A 885 (2000) 73-95.

[74] D.L. Mayer and J.S. Fritz, J. Chromatogr. A 771 (1997) 45-53.

[75] J.P. Franke and R.A. de Zeeuw, J. Chromatogr. B 713 (1998) 51-59.

[76] B. Sellergren, Ed., Molecularly Imprinted Polymers: Man-made Mimics of Antibodies and their Applications in Analytical Chemistry, Elsevier, Amsterdam (2001); Techniques and Instrumentation in Analytical Chemistry, Volume 23.

[77] K. Ensing, C. Berggren and R.E. Majors, LC-GC 19 (2001) 942-954.

[78] W.M. Mullett and E.P.C. Lai, Anal. Chem. 70 (1998) 3636-3641.

[79] C. Crescenzi, S. Bayoudh, P.A.G. Cormack, T. Klein and K. Ensing, Analyst 125 (2000) 1515-1517.

[80] C. Berggren, S. Bayoudh, D. Sherrington and K. Ensing, J. Chromatogr. A 889 (2000) 105-110.

[81] L.I. Andersson, J. Chromatogr. B 739 (2000) 163-173.

[82] W.M. Mullett and E.P.C. Lai, J. Pharm. Biomed. Anal. 21 (1999) 835-843.

[83] R.F. Venn and R.J. Goody, Chromatographia 50 (1999) 407-414.

[84] P. Martin, I.D. Wilson, D.E. Morgan, G.R. Jones and K. Jones, Anal. Commun. 34 (1997) 45-47.

[85] T.L. Lemke, Review of Organic Functional Groups: Introduction to Organic Medicinal Chemistry, 3rd. Ed., Lea & Febiger, Philadelphia (1992).

[86] E. Heebner, M. Telepchak and D. Walworth, In: D. Stevenson and I.D. Wilson, Eds., Sample Preparation for Biomedical and Environmental Analysis, Plenum Press, New York (1994) 155-161.

[87] M.J. Rose, S.A. Merschman, R. Eisenhandler, E.J. Woolf, K.C. Yeh, L. Lin, W. Fang, J. Hsieh, M.P. Braun, G.J. Gatto and B.K. Matuszewski, J. Pharm. Biomed. Anal. 24 (2000) 291-305.

[88] D. Rogowsky, M. Marr, G. Long and C. Moore, J. Chromatogr. B 655 (1994) 138-141.

[89] J. Scheurer and C.M. Moore, J. Anal. Toxicol. 16 (1992) 264-269.

[90] M.S. Mills, E.M. Thurman and M.J. Pedersen, J. Chromatogr. 629 (1993) 11-21.

[91] M. Zief and R. Kiser, Amer. Lab. 22 (1990) 70-83.

[92] R.D. McDowall, J.C. Pearce and G.S. Murkitt, J. Pharm. Biomed. Anal. 4 (1986) 3-21.

[93] R.D. McDowall, J. Chromatogr. 492 (1989) 3-58.

[94] R.D. McDowall, E. Doyle, G.S. Murkitt and V.S. Picot, J. Pharm. Biomed. Anal. 7 (1989) 1087-1096.

[95] T.R. Krishnan and I. Ibraham, J. Pharm. Biomed. Anal. 12 (1994) 287-294.

[96] M.-C. Hennion, J. Chromatogr. A 856 (1999) 3-54.

[97] J.S. Fritz and M. Macka, J. Chromatogr. A 902 (2000) 137-166.

[98] M. Henry, In: N.J.K. Simpson, Ed., Solid-Phase Extraction: Principles, Techniques and Applications, Marcel Dekker, New York (2000) 125-182.

[99] R.E. Majors, LC-GC 16 (1998) S8-S15.

[100] G.B. Cox and R.W. Stout, J. Chromatogr. 384 (1987) 315-336.

[101] G.B. Cox, J. Chromatogr. A 656 (1993) 353-367.

[102] G.A. Junk, M. J. Avery and J.J. Richard, Anal. Chem. 60 (1988) 1347-1350.

[103] M.L. Larrivee and C.F. Poole, Anal. Chem. 66 (1994) 139-146.

[104] C.F. Poole, S.K. Poole, D.S. Siebert and K.G. Miller, In: E. Reid, H.M. Hill and I.D. Wilson, Eds., Biofluid Assay for Peptide-related and Other

Drugs, The Royal Society of Chemistry, Cambridge (1996) 194-208; Methodological Surveys in Bioanalysis of Drugs Vol. 24, E. Reid, Ed.

[105] A. Gelencser, G. Kiss, Z. Krivacsy, Z. Varga-Puchony and J. Hlavay, J. Chromatogr. A 693 (1995) 217-225.

[106] M.C. Carson, J. Chromatogr. A 885 (2000) 343-350.

[107] T. Herraiz and V. Casal, J. Chromatogr. A 708 (1995) 209-221.

[108] R.A. de Zeeuw and J.P. Franke, In: N.J.K. Simpson, Ed., Solid-Phase Extraction: Principles, Techniques and Applications, Marcel Dekker, New York (2000) 243-271.

[109] U.J. Nilsson, J. Chromatogr. A 885 (2000) 305-319.

[110] V. Ruiz-Gutierrez and M.C. Perez-Camino, J. Chromatogr. A 885 (2000) 321-341.

[111] J.M. Rosenfeld, J. Chromatogr. A 843 (1999) 19-27.

[112] D. Stevenson, J. Chromatogr. B 745 (2000) 39-48.

[113] D. Stevenson, B.A.A. Rashid and S. Jamaleddin, In: N.J.K. Simpson, Ed., Solid-Phase Extraction: Principles, Techniques and Applications, Marcel Dekker, New York (2000) 349-360.

[114] I.D. Wilson and J.K. Nicholson, In: D. Stevenson and I.D. Wilson, Eds., Sample Preparation for Biomedical and Environmental Analysis, Plenum Press, New York (1994) 37-52.

[115] R. Brindle, K. Albert, E.D. Morgan, P. Martin and I.D. Wilson, J. Pharm. Biomed. Anal. 13 (1995) 1305-1312.

[116] K. Albert, R. Brindle, P. Martin and I.D. Wilson, J. Chromatogr. A 665 (1994) 253-258.

[117] D.S. Seibert and C.F. Poole, J. High Resol. Chromatogr. 18 (1995) 226-230.

[118] K.K. Unger, N. Becker and P. Roumeliotis, J. Chromatogr. 125 (1976) 115-127.

[119] M.L. Mayer, C.F. Poole and M.P. Henry, J. Chromatogr. A 695 (1995) 267-277.

[120] M.L. Mayer, S.K.Poole and C.F. Poole, J. Chromatogr. A 697 (1995) 89-99.

[121] W.P.N. Fernando, M.L. Larrivee and C.F. Poole, Anal. Chem. 65 (1993) 588-595.

[122] L. Botz, S. Nyiredy, E. Wehrli and O. Sticher, J. Liq. Chrom. 13 (1990) 2809-2828.

[123] Y. Bereznitski and M. Jaroniec, J. Chromatogr. A 828 (1998) 51-58.

[124] Y. Bereznitski, M. Jaroniec and M.E. Gangoda, J. Chromatogr. A 828 (1998) 59-73.

[125] N.J.K. Simpson, Ed., Solid-Phase Extraction: Principles, Techniques and Applications, Marcel Dekker, New York (2000).

[126] N. Simpson and K.C. Van Horne, Eds., Handbook of Sorbent Extraction Technology, Varian Associates, Los Angeles (1993).

[127] E.M. Thurman and M.S. Mills, Solid-Phase Extraction Principles and Practice, John Wiley & Sons, New York (1998); Chemical Analysis Vol. 147, J.D. Winefordner, Ed.

[128] J.S. Fritz, Analytical Solid-Phase Extraction, Wiley-VCH, New York (1999).

[129] M.J. Telepchak, R. Chaney and T.F. August, Forensic and Clinical Applications of Solid Phase Extraction, Humana Press, Totowa, New Jersey (2003) in press.

[130] G. Musch and D.L. Massart, J. Chromatogr. 432 (1988) 209-222.

[131] W.Z. Shou, X. Jiang, B.D. Beato and W. Naidong, Rapid Commun. Mass Spectrom. 15 (2001) 466-476.

[132] S.A. Westwood and M.C. Dumasia, In: D. Stevenson and I.D. Wilson, Eds., Sample Preparation for Biomedical and Environmental Analysis, Plenum Press, New York (1994) 163-166.

[133] E.M. Thurman and M.S. Mills, Solid-Phase Extraction Principles and Practice, John Wiley & Sons, New York (1998) 123-159; Chemical Analysis Vol. 147, J.D. Winefordner, Ed.

[134] B.K. Logan, D.T. Stafford, I.R. Tebbett and C.M. Moore, J. Anal. Toxicol. 14 (1990) 154-159.

[135] J.S. Janiszewski, H.G. Fouda and R.O. Cole, J. Chromatogr. B 668 (1995) 133-139.

[136] R.L. Sheppard and J. Henion, Electrophoresis 18 (1997) 287-291.

[137] J.A. Tørnes and B.A. Johnsen, J. Chromatogr. 467 (1989) 129-138.

[138] E. Mikami, T. Goto, T. Ohno, H. Matsumoto and M. Nishida, J. Pharm. Biomed. Anal. 28 (2002) 261-267.

[139] E.M. Thurman and M.S. Mills, Solid-Phase Extraction Principles and Practice, John Wiley & Sons, New York (1998) 105-121; Chemical Analysis Vol. 147, J.D. Winefordner, Ed.

[140] D.A. Wells, LC-GC 17 (1999) 600-610.

[141] G. Lensmeyer, In: S. Wong and I. Sunshine, Eds., Handbook of Analytical Therapeutic Drug Monitoring and Toxicology, CRC Press, Boca Raton (1997) 137-148.

[142] G.M. Hearne and D.O. Hall, Amer. Lab. 25 (1993) 28H-28M.

[143] D.D. Blevins and M.P. Henry, Amer. Lab. 27 (1995) 32-35.

[144] D.D. Blevins and D.O. Hall, LC-GC 16 (1998) S16-S21,

[145] S. Giandinoto, Presented at American Association for Clinical Chemistry Annual Meeting, Chicago, IL USA (1992).

[146] A.H.B. Wu, N. Liu, J. Yoon, K.G. Johnson and S.S. Wong, J. Anal. Toxicol. 17 (1993) 215-217.

[147] L.M. McMahon, S. Luo, M. Hayes and F.L.S. Tse, Rapid Commun. Mass Spectrom. 14 (2000) 1965-1971.

[148] M.M. Kushnir, J. Crossett, P.I. Brown and F.M. Urry, J. Anal. Toxicol. 23 (1999) 1-6.

[149] J. Liu and J.T. Stewart, J. Chromatogr. B 692 (1997) 141-147.

[150] M.A. Anderson, T. Wachs and J.D. Henion, J. Mass Spectrom. 32 (1997) 152-158.

[151] T. Breindahl, J. Chromatogr. B 746 (2000) 249-254.

[152] B.A. McCue, M.M. Cason, M.A. Curtis, R.D. Faulkner and D.C. Dahlin, J. Pharm. Biomed. Anal. 28 (2002) 199-208.

[153] P.T. Vallano, R.S. Mazenko, E.J. Woolf and B.K. Matuszewski, J. Chromatogr. B (2002) in press.

[154] R.A. de Zeeuw, J. Wijsbeek and J.P. Franke, J. Anal. Toxicol. 24 (2000) 97-101.

[155] L. Ye, J.T. Stewart and H. Zhang, J. Pharm. Biomed. Anal. 13 (1995) 1185-1188.

[156] L. Ye and J.T. Stewart, Anal. Lett. 29 (1996) 395-407.

[157] J.L. Ezzell, Anal. Meth. Instrument. 2 (1995) 48-51.

[158] G.L. Lensmeyer, D.A. Wiebe and T. Doran, Ther. Drug Monit. 13 (1991) 244-250.

[159] G.L. Lensmeyer, D.A. Wiebe and B.A. Darcey, J. Chrom. Sci. 29 (1991) 444-449.

[160] G.L. Lensmeyer, C. Onsager, I.H. Carlson and D.A. Wiebe, J. Chromatogr. A 691 (1995) 239-246.

[161] M. Nakajima, S. Yamato and K. Shimada, Biomed. Chromatogr. 12 (1998) 211-216.

[162] J. Singh and L. Johnson, J. Anal. Toxicol. 21 (1997) 384-387.

[163] L.V. Rao, J.R. Petersen, M.G. Bissell, A.O. Okorodudu and A.A. Mohammad, J. Chromatogr. B 730 (1999) 123-128.

[164] M.J. Rose, E.J. Woolf and B.K. Matuszewski, J. Chromatogr. B 738 (2000) 377-385.

[165] N. Ohashi and M. Yoshikawa, J. Chromatogr. B 746 (2000) 17-24.

[166] G.L. Lensmeyer and M.A. Poquette, Ther. Drug Monit. 23 (2001) 239-249.

[167] M.D. Green, Y. Bergqvist, D.L. Mount, S. Corbett and M.J. D'Souza, J. Chromatogr. B 727 (1999) 159-165.

[168] M.J. Rose, S.A. Merschman, E.J. Woolf and B.K. Matuszewski, J. Chromatogr. B 732 (1999) 425-435.

[169] G.L. Lensmeyer, T. Kempf, B.E. Gidal and D.A. Wiebe, Ther. Drug Monit. 17 (1995) 251-258.

[170] M. Mizugaki, T. Hishinuma and N. Suzuki, J. Chromatogr. B 729 (1999) 279-285.

[171] M.D. Green, D.L. Mount, G.D. Todd and A.C. Capomacchia, J. Chromatogr. A 695 (1995) 237-242.

[172] A. Pauwels, D.A. Wells, A. Covaci and P.J.C. Schepens, J. Chromatogr. B 723 (1999) 117-125.

[173] Y. Wu, L.Y.-T. Li, J.D. Henion and G.J. Krol, J. Mass Spectrom. 31 (1996) 987-993.

[174] G.L. Lensmeyer, D.A. Wiebe abd T.C. Doran, Therap. Drug Monit. 14 (1992) 408-415.

[175] G.L. Lensmeyer, B.E. Gidal and D.A. Wiebe, Therap. Drug Monit. 19 (1997) 292-300.

[176] H. Mita, R. Oosaki, Y. Mizushima, M. Kobayashi and K. Akiyama, J. Chromatogr. B 692 (1997) 461-466.

[177] M. Ishida, K. Kobayashi, N. Awata and F. Sakamoto, J. Chromatogr. B 727 (1999) 245-248.

[178] R.S. Plumb, R.D.M. Gray and C.M. Jones, J. Chromatogr. B 694 (1997) 123-133.

[179] M. Jemal, D. Teitz, Z. Ouyang and S. Khan, J. Chromatogr. B 732 (1999) 501-508.

[180] C. Souppart, M. Decherf, H. Humbert and G. Maurer, J. Chromatogr. B 762 (2001) 9-15.

[181] J.S. Janiszewski, M.C. Swyden and H.G. Fouda, J. Chrom. Sci. 38 (2000) 255-258.

[182] H. Zhang and J. Henion, Anal. Chem. 71 (1999) 3955-3964.

[183] M. Jemal, M. Huang, Y. Mao, D. Whigan and A. Schuster, Rapid Commun. Mass Spectrom. 14 (2000) 1023-1028.

[184] G. Rule and J. Henion, J. Am. Soc. Mass Spectrom. 10 (1999) 1322-1327.

[185] J. Janiszewski, R.P. Schneider, K. Hoffmaster, M. Swyden, D. Wells and H. Fouda, Rapid Commun. Mass Spectrom. 11 (1997) 1033-1037.

[186] H. Simpson, A. Berthemy, D. Buhrman, R. Burton, J. Newton, M. Kealy, D. Wells and D. Wu, Rapid Commun. Mass Spectrom. 12 (1998) 75-82.

[187] Y. Hsieh, J.-M. Brisson, K. Ng and W.A. Korfmacher, J. Pharm. Biomed. Anal. 27 (2002) 285-293.

[188] C.Z. Matthews, E.J. Woolf and B.K. Matuszewski, J. Chromatogr. A 949 (2002) 83-89.

[189] L. Yang, N. Wu and P.J. Rudewicz, J. Chromatogr. A 926 (2001) 43-55.

[190] C.Z. Matthews, E.J. Woolf, L. Lin, W. Fang, J. Hsieh, S. Ha, R. Simpson and B.K. Matuszewski, J. Chromatogr. B 751 (2001) 237-246.

[191] M.J. Rose, N. Agrawal, E.J. Woolf and B.K. Matuszewski, J. Pharm. Sci. 91 (2002) 405-416.

[192] J. Hempenius, R.J.J.M. Steenvoorden, F.M. Lagerwerf, J. Wieling and J.H.G. Jonkman, J. Pharm. Biomed. Anal. 20 (1999) 889-898.

[193] G. Hopfgartner, C. Husser and M. Zell,Therap. Drug Monit. 24 (2002) 134-143.

[194] Z. Liu, J. Short, A. Rose, S. Ren, N. Contel, S. Grossman and S. Unger, J. Pharm. Biomed. Anal. 26 (2001) 321-330.

[195] L. Yang, T.D. Mann, D. Little, N. Wu, R.P. Clement and P.J. Rudewicz, Anal. Chem. 73 (2001) 1740-1747.

[196] R. Bakhtiar, L. Khemani, M. Hayes, T. Bedman and F. Tse, J. Pharm. Biomed. Anal. 28 (2002) 1183-1194.

[197] R.C. Simpson, A. Skarbek and B.K. Matuszewski, J. Chromatogr. B 775 (2002) 133-142.

[198] D.A. Wells, G.L. Lensmeyer and D.A. Wiebe, J. Chrom. Sci. 33 (1995) 386-392.

[199] H. Lingeman and S.J.F. Hoekstra-Oussoren, J. Chromatogr. B 689 (1997) 221-237.

[200] C.F. Poole, S.K. Poole, D.S. Seibert and C.M. Chapman, J. Chromatogr. B 689 (1997) 245-259.

[201] K. Ensing, J.P. Franke, A. Temmink, X.-H. Chen and R.A. de Zeeuw, J. Forensic Sci. 37 (1992) 460-466.

[202] A. Koole, A.C. Jetten, Y. Luo, J.P. Franke and R.A. de Zeeuw, J. Anal. Toxicol. 23 (1999) 632-635.

[203] A.J. Tomlinson, S. Jameson and S. Naylor, J. Chromatogr. A 744 (1996) 273-278.

[204] A.J. Tomlinson, L.M. Benson, R.P. Oda, D. Braddock, B.L. Riggs, J.A. Katzmann and S. Naylor, J. Cap. Elec. 2 (1995) 97-104.

[205] A.J. Tomlinson and S. Naylor, J. Cap. Elec. 2 (1995) 225-233.

[206] A.J. Tomlinson, L.M. Benson, W.D. Braddock and R.P. Oda, J. High Resol. Chromatogr. 18 (1995) 381-383.

[207] L.M. Benson, A.J. Tomlinson, A.N. Mayeno and G.J. Gleich, J. High Resol. Chromatogr. 19 (1996) 291-294.

[208] A.J. Tomlinson, N.A. Guzman and S. Naylor, J. Cap. Elec. 6 (1995) 247-266.

[209] A.J. Tomlinson and S. Naylor, J. Liq. Chrom. & Rel. Technol. 18 (1995) 3591-3615.

[210] A.J. Tomlinson, L.M. Benson, S. Jameson, D.H. Johnson and S. Naylor, J. Amer. Soc. Mass. Spectrom. 8 (1997) 15-24.

[211] S. Naylor and A.J. Tomlinson, Talanta 45 (1998) 603-612.

[212] P.W.M. Reisinger, T. Kleinschmmidt and U. Welsch, Electrophoresis 13 (1992) 65-72.

[213] S.K. Poole and C.F. Poole, J. Planar Chrom. 2 (1989) 478-481.

[214] S.K. Poole, W.P.N. Fernando and C.F. Poole, J. Planar Chrom. 3 (1990) 331-335.

[215] H. J. Issaq, K.E. Seburn and J.R. Hightower, J. Liq. Chrom. & Rel. Technol. 14 (1991) 1511-1517.

[216] C. Regnault, P. Delvordre and E. Postaire, J. Chromatogr. 547 (1991) 403-409.

[217] P.W.M. Reisinger, T. Kleinschmmidt and U. Welsch, Electrophoresis 13 (1992) 65-72.

[218] C.G. Markell, D.F. Hagen and G. Schmitt, Anal. Chim. Acta 236 (1990) 157-164.

[219] E.R. Brouwer and H. Lingeman, Chromatographia 29 (1990) 415-418.

[220] L. Schmidt, J.J. Sun, J.S. Fritz, D.F. Hagen, C.G. Markell and E.E. Wisted, J. Chromatogr. 641 (1993) 57-61.

[221] K. Uchiyama, K. Ohsawa, Y. Yoshimura, J. Minowa, T. Watanbe and K. Imaeda, Anal. Sci. 8 (1992) 655-658.

[222] G. Durand, S. Chiron, V. Bouvot and D. Barcelo, Int. J. Environ. Anal. Chem. 49 (1992) 31-42.

[223] E.R. Brouwer, D.J. van Iperen, I. Liska, H. Lingeman and U.A.Th. Brinkman, Int. J. Environ. Anal. Chem. 47 (1992) 257-266.

[224] L.M. Davi, M. Baldi, L. Penazzi and M. Liboni, Pestic. Sci. 35 (1992) 63-67.

[225] L. Schmidt, J.J. Sun, J.S. Fritz, D.F. Hagen, C.G. Markell and E.E. Wisted, J. Chromatogr. 641 (1993) 57-61.

[226] I. Ferrer, D. Barcelo and E.M. Thurman, Anal. Chem. 71 (1999) 1009-1015.

[227] J.S. Fritz and J.J. Masso, J. Chromatogr. A 909 (2001) 79-85.

[228] L.A. Errede, G.B. Jefson, B.A. Langager, P.E. Olson, B.R. Ree, M.E. Reichert, R.A. Sinclair and J.J. Stofko, In: D.E. Leyden and W.T. Collin, Eds., Chemically Modified Surfaces, Volume 2: Chemically Modified Surfaces in Science and Industry, Gordon and Breach, New York (1987) 91-104.

[229] R. Saari-Nordhaus, L.M. Nair and J.M. Anderson, Jr., J. Chromatogr. A 671 (1994) 159-163.

[230] R. Saari-Nordhaus and J.M. Anderson Jr., J. Chromatogr. A 706 (1995) 563-569.

[231] A. Aistars, N.J.K. Simpson and D.C. Jones, Amer. Lab. 33 (2001) 50-54.

[232] D.A. Wells, LC-GC 17 (1999) 808-822.

[233] B. Kaye, W.J. Herron, P.V. Macrae, S. Robinson, D.A. Stopher, R.F. Venn and W. Wild, Anal. Chem. 68 (1996) 1658-1660.

[234] J.P. Allanson, R.A. Biddlecombe, A.E. Jones and S. Pleasance, Rapid Commun. Mass Spectrom. 10 (1996) 811-816.

[235] R.A. Biddlecombe and S. Pleasance, Proceedings International Symposium on Laboratory Automation and Robotics, Boston, MA USA (1996) 445-454.

[236] S. Pleasance and R.A. Biddlecombe, In: E. Reid, H.M. Hill and I.D. Wilson, Eds., Drug Development Assay Approaches, Including Molecular Imprinting and Biomarkers, The Royal Society of Chemistry, Cambridge (1998) 205-212; Methodological Surveys in Bioanalysis of Drugs, Volume 25, E. Reid, Ed.

[237] R.A. Biddlecombe and S. Pleasance, Proceedings International Symposium on Laboratory Automation and Robotics, Boston, MA USA (1998)

[238] I.D. Davies, J.P. Allanson and R.C. Causon, J. Chromatogr. B 732 (1999) 173-184.

[239] V. Cenacchi, S. Baratte, P. Cicioni, E. Frigerio, J. Long and C. James, J. Pharm. Biomed. Anal. 22 (2000) 451-460.

[240] W.Z. Shou, M. Pelzer, T. Addison, X. Jiang and W. Naidong, J. Pharm. Biomed. Anal. 27 (2002) 143-152.

[241] S.L. Callejas, R.A. Biddlecombe, A.E. Jones, K.B. Joyce, A.I. Pereira and S. Pleasance, J. Chromatogr. B 718 (1998) 243-250.

[242] I.D. Davies, J.P. Allanson and R.C. Causon, Chromatographia 52 (2000) S92-S97.

[243] A.E. Niggebrugge, E. Tessier, R. Guilbaud, L. DiDonato and R. Masse, Proceedings 48th American Society for Mass Spectrometry Conference, Long Beach, CA USA (2000).

Chapter 12

Solid-Phase Extraction: Strategies for Method Development and Optimization

Abstract

The fundamentals of solid-phase extraction (SPE) were presented in Chapter 11 along with procedures for using many of the versatile sorbent chemistries. An important step toward achieving success with SPE is learning how to develop new methods and improve existing ones. This chapter discusses rapid method development strategies in microplates by presenting two approaches—a generic methodology suitable for a wide range of structurally diverse analytes (used when *time* is most important) and a specific methodology designed to selectively extract a single analyte/internal standard pair or a small set of structurally similar analytes (used when *selectivity* is most important). It is essential when performing any type of method development that the analyst understands how to optimize SPE parameters. Rapid strategies for accomplishing this optimization are also presented. The ultimate goal in developing and using 96-well SPE methods is to achieve high throughput with time and labor savings. Efficient use of automation is the key to achieving this goal. Automation strategies for performing the method development approaches presented here are discussed in Chapter 13.

12.1 Method Development Objectives

12.1.1 Introduction

Method development for a new chemical entity establishes a process for its isolation from biological fluids and its analysis by LC-MS, LC-MS/MS or other analytical techniques. Method development is typically performed for the sample preparation step (off-line or on-line) as well as chromatographic separation. Usually the development of these tasks is performed independently, although in many cases valuable information that is useful in the sample preparation scheme can be learned from the liquid chromatography (LC) development. The method development process can also be performed for an

433

existing method when improvement is required in order to meet a particular need, such as addition of a metabolite or other analyte to an assay. Common goals for any method include higher sensitivity, greater selectivity and a reduced matrix effect. When a cleaner sample is presented to the analytical system by minimizing interferences from the sample matrix, LC column lifetime is increased and total system pressures are maintained.

The best approach to method development for sample preparation is to isolate and test one variable at a time. This approach can be performed rapidly by examining multiple variables in parallel. The historical association of method development with taking too much time has arisen in part from a history of hit or miss experiments which randomly looked at a number of variables simultaneously. Oftentimes, it is not known why a set of conditions worked or how much leeway exists for operator variation or changes in solution pH with each new batch. When the effect of variables is not completely understood, problems can result later when a method is transferred or another analyst performs the work.

This chapter discusses method development approaches specifically for solid-phase extraction performed off-line (independent of the analysis). Method development procedures for other sample preparation techniques such as protein precipitation, liquid-liquid extraction and on-line SPE are discussed in Chapters 6, 9 and 14, respectively.

12.1.2 Defining the Scope of the Project

Several important questions must be asked when beginning a method development project using solid-phase extraction as the method of choice:
- What are the requirements and goals for the method?
- What is known about the chemistry of the analytes?
- What investment in time can be made?
- What is known from earlier work with these or similar analytes?
- What resources are available to aid in this task?
- What is known about the sample matrix?

The answers to the above questions can help guide the selection of the SPE chemistry and product format to be evaluated. When method requirements conclude that elimination of the evaporation step is desirable, whether for improved efficiency or to eliminate instability on dry-down, the analyst is directed toward a low sorbent mass packed bed or disk product for the evaluation. A selective method required for a few structurally similar analytes may point toward a mixed mode chemistry that yields high recovery with low matrix effects. A more generic method for multiple analytes, as well as a straightforward on/off extraction method, may point toward a reversed phase polymer sorbent. An assessment of the range of analyte polarities to be encountered will help decide which specific sorbent chemistry should be examined [*e.g.*, analytes with strong polar character may benefit from use of FOCUS™ (Varian Inc., Harbor City, CA USA) while strongly hydrophobic character may benefit from a traditional styrene divinylbenzene polymer or one of the newer versions that have been modified with N-vinylpyrrolidone].

The time allotted for method development is important. The techniques presented here educate the reader in how to quickly evaluate a series of variables in one experiment. Following data analysis, the best one or two sets of extraction conditions can be selected for optimization of wash solvent and elution volume. When automation is available for this task, less hands-on time is required from the user. Table 12.1 provides a worksheet for the analyst to record information about the analytes and objectives for the method development project. This information provides a summary that will guide the analyst in selecting an SPE product format and defining the conditions to be evaluated.

12.2 Generic Approach to Method Development

12.2.1 Single Sorbent in a Universal Method

12.2.1.1. Polymer as a Generic Sorbent

12.2.1.1.1 Overall Objectives

Method development using a single sorbent chemistry involves a generic or universal set of extraction conditions which are expected to adsorb and desorb the majority of analytes encountered. When a range of analyte polarities is involved (*e.g.*, parent drug and several metabolites), a polymer sorbent has shown great utility (see Chapter 11, Section 11.2.1.2). The complexity of this generic method, in terms of the number of steps and solvents required, depends on the exact sorbent chemistry but generally consists of one methodology per sorbent type. The development process then consists simply of running the method using several analyte concentrations with multiple replicates and then assaying for recovery and performance (precision and matrix effect). This approach to method development is commonly used when little time is available to do the work and a method must be developed in 1–2 days.

Table 12.1
Worksheet for planning a solid-phase extraction method development project

Research program _____   Analyte _____

Sample matrix _____ Volume _____   Internal standard _____

Chemical Structure of Analyte	Any pKa information available?
	_____
	Dominant functional groups on analytes:
	_____

Solubility considerations: LC considerations and influence of pH:
Water? Methanol? Acetonitrile? DMSO?   Which mobile phases worked and didn't?

_____   _____

Any previous LLE experience?   Any previous SPE experience?
Sample pH; Organic solvents: Sorbents; Load pH; Wash; Elution; Vol.

_____   _____

SPE sorbents to evaluate: Bed mass and plate format to evaluate:

_____   _____

Sample pretreatment scheme: pH values to evaluate? Disrupt proteins?

_____   _____

Wash solvents/volumes to evaluate: Dry-down or direct inject eluate?

_____   _____

Elution solvents/volumes to evaluate: Sequential elution? Exhaustive elution?

_____   _____

Strategy and volumes for reconstitution −or− dilution followed by direct injection:

Once the generic method is run with a set of analytes and data are evaluated, optimization parameters can be examined in subsequent experiments, if desired. Most commonly, elution volume optimization is performed using the chosen extraction conditions. Using the smallest volume that yields consistent recoveries is beneficial since a smaller volume reduces the time required for evaporation. Likewise, if the volume is small enough, perhaps dilution followed by direct injection can be performed and time can be saved by eliminating the evaporation step. The goals of the elution will dictate the sorbent bed mass used for the first experiment; when dry-down elimination is the goal, a bed mass ≤ 25 mg should be used.

12.2.1.1.2 Comments on the Sample Load and Wash Steps

An example of the universal approach to solid-phase extraction is described in a report by Cheng *et al.* [1] for the extraction of verapamil and its metabolite in a plasma matrix. A hydrophilic lipophilic balanced (HLB) sorbent in a 96-well plate was used, Oasis® HLB (Waters Corporation, Milford, MA USA). The generic extraction method used is outlined in Figure 12.1. As executed, this method is a "catch all" approach that aims to capture or adsorb as many analytes as possible, each displaying different degrees of hydrophobicity and polarity.

The influence of pH on sample ionization and the extraction process is important to the success of any generic method. The Oasis HLB sorbent, and corresponding analogues from other manufacturers, relies on a reversed phase attraction of analyte to sorbent. The analyte should be loaded onto the sorbent bed at a pH in which it is neutral (unionized); acidic analytes are loaded under acidic conditions (2 pH units below their pKa) and basic analytes are loaded under basic conditions (2 pH units above their pKa). Should the analyte be ionized during loading, its recovery may be diminished since the lipophilic or weak van der Waals attraction of analyte to sorbent is not predominant.

The wash step in this procedure is 95% aqueous which is effective to displace proteins from the sorbent bed; lipophilic interfering components remain adsorbed, along with the analytes of interest. Elution with the polar solvent methanol is expected to disrupt the attraction of analyte to sorbent bed, eluting the analyte. However, since no selectivity is developed within this method, unwanted interfering components are also eluted [1]. Using a generic set of conditions is intended simply to extract the analytes of interest and the selectivity of the mass spectrometer is used to distinguish them from all other components in the injected sample.

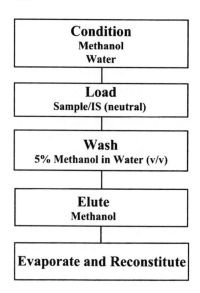

Figure 12.1. Typical generic extraction conditions using a reversed phase polymer sorbent. A modified set of conditions is suggested for use in method development, as shown in Figure 12.2.

12.2.1.1.3 Importance of pH Modification for Elution

Methanol has been shown to be successful in eluting analytes from a polymer sorbent as a component within a generic method. However, the author has experienced many instances in which pH modification of the methanol elution solvent was necessary. When the attraction of an analyte for a polymer sorbent is strong and is based on lipophilicity, the polar solvent methanol may not have enough solvent strength to disrupt the attraction and desorb the analyte; pH modification or another solvent may be needed. As an example, a proprietary analyte was loaded onto a 30 mg polymer sorbent under basic conditions to ensure that it was neutral (unionized) and the scheme in Figure 12.1 was followed. The sorbent was washed with 5% methanol in water and eluted with 100% methanol. The analyte was not recovered, even though the generic procedure was followed. In a follow-up experiment, samples were loaded onto the same sorbent bed in three parallel sets (n=4). Following loading, all wells received the same 5% methanol in water as the wash solvent. Equal volumes (1 mL) of three different elution solvents were applied in parallel and the results are shown in Figure 12.2.

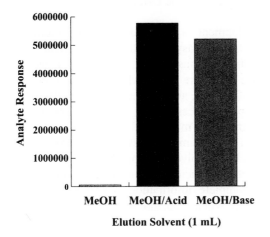

Elution Solvent (1 mL)

Figure 12.2. Comparison of analytical response for a proprietary drug candidate extracted from a polymer sorbent. Sample replicates (n=4) were extracted in parallel under identical conditions with three elution solvents: methanol, methanol/HCOOH (99:1, v/v) or methanol/NH$_4$OH (99:1, v/v).

The best recovery was obtained using acid modified methanol (containing 1% formic acid). The analyte in this example contains an amine group but at basic pH it is neutral; adsorption to the polymer sorbent is via a reversed phase mechanism. By eluting with methanol/formic acid, the amine is protonated and its nonpolar attraction to the sorbent is disrupted; both the positive charge and the nonpolar organic solvent facilitate the elution process. Base modification of methanol also provides an effective elution solvent by disrupting interactions at the sorbent surface and/or driving the equilibrium of ionization fully toward the unionized state.

Elution using methanol/acid in this example is more desirable than methanol/base when dilution followed by direct injection is the goal (using a low sorbent mass SPE product). Appropriate dilution of either solvent creates a mobile phase compatible solution suitable for injection. However, ammonium hydroxide containing solutions are not injected directly because of solvent incompatibility with the mobile phase and ionization source. Usually a base modified eluant solution must first be neutralized by addition of a small volume of acid or it can be evaporated, followed by reconstitution in a more desirable solvent.

12.2.1.1.4 Suggested Method Development Scheme Using a Polymer

An important goal in method development is to rapidly evaluate multiple variables. In the microplate format it is quite easy to use three sets of replicates and evaluate three elution schemes simultaneously since wells are arranged in an 8-row by 12-column grid. Since there are cases when methanol is not sufficient to fully elute the analytes of interest, an acid and a base modification of methanol should be processed in parallel. Essentially, three versions of the same experiment are run together within one microplate. Instead of using the generic procedure as listed in Figure 12.1 for a method development procedure, it is suggested to always use three elution solvents in parallel as outlined in Figure 12.3. By also examining the effect of acid and base modified methanol (or acetonitrile) on elution, more complete information is obtained in one experiment. The need to repeat a study when methanol alone did not work is eliminated, saving time.

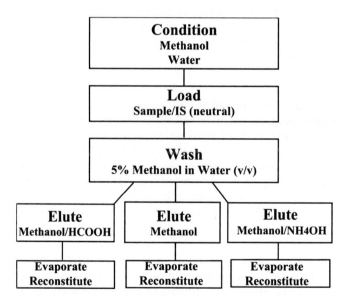

Figure 12.3. A method development scheme using a reversed phase polymer sorbent should always evaluate the influence of pH on elution. Three replicate sets of samples are extracted in parallel. The samples all receive the same load and wash treatment but separate elution conditions are used, as shown above. Acetonitrile can be substituted for methanol (or added to the scheme and run in parallel).

This generic approach as presented is performed when time is short and a method needs to be developed quickly. There is no desire to try multiple sorbents as the universal sorbent is expected to provide acceptable recovery for the majority of analytes encountered. The analyst with average skills in sample preparation is able to perform this method reliably. However, the sacrifice made with this approach is in selectivity. A more discriminating wash solution or series of solutions is the key to attaining better selectivity. A universal sorbent such as a polymer is able to be used in a more discriminating manner just as any other sorbent. Optimization of an extraction method in this manner is discussed in Section 12.3, "Selective and Rapid Approach to Method Development."

12.2.1.2 Mixed Mode as a Generic Sorbent

The sorbent chemistry chosen for a universal method is not required to be a polymer. Bonded silica sorbents can also be used effectively for this purpose. Recall that bonded silica sorbents display both primary and secondary interactions depending on solution pH, residual silanol content and solvents used, as described in Chapter 11, Section 11.2.1.1. Some laboratories use C2 or C8 bonded silica routinely and use the silanols to improve retention; elution is performed in acid modified organic solvent to disrupt these silanol interactions.

A more efficient manner in which to utilize the cation exchange ability of sorbents, whether bonded silica or polymer based, is to use a copolymeric mixed mode cation chemistry which has greater ion exchange capacity. Hydrophobic attraction through alkyl or aromatic groups is also imparted by the mixed mode sorbent. When mixed mode chemistry is used in a polymeric form, the silanol effect is removed which may result in more predictable results.

Mixed mode chemistry is used as a generic sorbent with the same objective as for a reversed phase sorbent—one method for all extractions regardless of analyte. When the analyte contains an amine group, protonation of the amine is exploited for cationic attraction; likewise, an acid group is made ionized for anionic attraction. When the analyte is neutral, the nonpolar affinity of the alkyl or aromatic groups on the sorbent is exploited and the ion exchange attraction may then be used to adsorb matrix interferences during the wash step.

The elution solvent shown to work optimally for mixed mode cation exchange is organic modified with base (*e.g.*, 2–5% NH_4OH); likewise, for mixed mode anion, the elution solvent is organic modified with acid (*e.g.*, 2–5% HCOOH).

Since methanol or acetonitrile is commonly used as the organic component of the elution solvent, during a method development experiment it is suggested that each be evaluated in parallel to look for relative differences in recovery. In the case when acetonitrile is found to elute analyte less completely than methanol, it may be a better choice for the wash solvent. Recall from Chapter 11, Section 11.2.2.4 that when analyte is strongly held by an ionic attraction, a 100% organic solvent can be used as a powerful wash to remove matrix interferences.

It is suggested to always use two elution solvent series in parallel for method development. More complete information on the elution characteristics of analytes is obtained in one experiment if the organic solvent (methanol or acetonitrile) is acid or base modified. Suggested schemes for generic method development using a mixed mode cation sorbent and a mixed mode anion sorbent are outlined together in Figure 12.4.

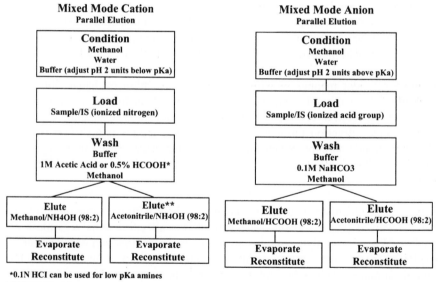

Figure 12.4. A method development scheme using a mixed mode sorbent can evaluate the influence of methanol vs. acetonitrile for elution. Two replicate sets of samples are extracted in parallel. The samples all receive the same load and wash treatment but separate elution conditions are used, as shown above. *Left*: Method for mixed mode cation sorbent; *Right*: Method for mixed mode anion sorbent.

A mixed mode ion sorbent, since it displays reversed phase functionality in addition to a strong cation or anion exchange, can be used in another way—as a pure reversed phase sorbent. When the method is designed appropriately (load analyte as neutral species, elute with 100% organic to disrupt van der Waals attraction or change pH to cause ionization of the analyte to break this nonpolar affinity), a typical reversed phase attraction scheme is in effect. Ideally, ionic interferences would be irreversibly trapped to the ionic functional groups of the sorbent as long as the analyte is not also charged at the same pH at which interferences adsorb.

12.2.2 Multiple Sorbents within a Method Development Plate

12.2.2.1 Considerations in Plate Configuration and Usage

The previous section discussed method development using a single sorbent chemistry with one generic set of extraction conditions. Two or three elution solvents are suggested for use in parallel to obtain more complete information within one experiment. When a wide range of analytes with great structural diversity is encountered, the inappropriate choice of the extraction conditions and/or the sorbent chemistry will lead to failure. The time invested in method development is best spent with a well designed experiment to maximize the data obtained. Thus, another valid approach uses multiple sorbent chemistries within a 96-well plate to simultaneously evaluate multiple sorbents within one experiment.

A method development plate can be obtained from a manufacturer as an off-the-shelf item or custom configured with the analyst's own arrangement of sorbent chemistries and/or sorbent bed masses. Likewise, the use of a modular plate allows the analyst to configure a custom plate as needed with the desired replacement wells. Advantages of the modular plate format include using only the exact number of wells needed and a straightforward transfer from individual column to microplate format when sorbent chemistries and brands match. An example describing the use of modular 96-well plates for method development is presented by Aistars *et al.* [2]. Svennberg *et al.* used modular plates (25 mg bed mass of C2, C4, C6, C8, ENV+ and 101 sorbents) with automation to optimize a solid-phase extraction method for an experimental drug in plasma [3].

Method development plates are commonly configured in two ways—with multiple chemistries all of the same bed mass or one or more chemistries in multiple bed masses. An evaluation of multiple chemistries directs the analyst to the best sorbent for extraction while multiple bed masses offer the ability to rapidly optimize the elution volume and confirm capacity for the extraction process.

The choice of sorbent chemistries to place into the method development plate is an important one. These chemistries can be chosen in a predictive sense, rationalizing which sorbents have been most useful in the past and which should apply in the future. For example, some of the most popular and useful chemistries are C18, C8, C2, mixed mode and polymeric. The short aliphatic C2 chain is attractive for the following reasons: allows easy steric access for organic molecules, analytes are held relatively well while matrix interferences often are not sufficiently held, and silanols on the silica surface assist with secondary interactions. C8 and C18 groups on silica allow better hydrophobic attraction; however, note that a long octadecyl chain can sometimes offer a restricted steric access for analytes. C18, while it is hydrophobic, tends to adsorb a large amount of unwanted matrix components in addition to the analyte. Attractive features of mixed mode and polymer sorbents have been mentioned previously.

Regarding the choice of sorbent chemistries, several questions are asked:
- Should mixed mode bonded silica be used or mixed mode polymer?
- Should unendcapped C18 be used for maximal silanol activity?
- Should two brands of C18 or C8 be compared?
- Should disks be used or packed particle beds?
- What sorbent bed mass should be evaluated?

Many variables should be considered when choosing the sorbents and bed masses for a method development plate. If 100 analysts were each asked to design a configuration, there likely would be at least 20 or more varieties specified. For this reason, manufacturers that offer method development plates in the single mold format frequently accept orders on a custom basis rather than stocking a dozen or more unique configurations.

Another factor that can potentially complicate the use of a method development plate, once configured, is the diversity in protocols for different sorbent chemistries. The protocol required for optimal extraction with a mixed mode cation sorbent involves pH adjustment of sample at two pH units below its pKa to ensure ionization, as well as unique wash (1M acetic acid) and elution solvents (organic modified by 2% NH_4OH). When the same solvents are delivered across the entire plate, the optimal conditions for mixed mode cation are not applicable to reversed phase extraction with direct injection into

LC-MS/MS systems; likewise, conditions typically used for reversed phase are not adequate to elute analytes from mixed mode cation since no base modifier is present. Mixing polymers with bonded silica sorbents can also require unique solutions and solvents for each. Disks and packed particle beds, if configured within the same plate, require different volumes of solution unless one larger volume is used (an excess volume for disks).

This diversity in protocols can be accommodated by some types of liquid handling workstations—the 4-/8-probe systems can utilize multiple reagents and even deliver different volumes to specific wells of the plate. The 96-tip workstation such as the Quadra® 96 (Tomtec Inc., Hamden, CT USA) can accommodate different solutions when a custom reagent reservoir is used in which the pattern matches the solvent delivery scheme; however, it cannot deliver different volumes (only one volume per solvent aspirate and dispense step). Automation for method development is discussed in more detail in Chapter 13.

A related note on the use of multiple sorbent chemistries for method development is that such an evaluation can be conveniently automated using the Prospekt™ on-line SPE system by Spark Holland (Emmen, The Netherlands; distributed in USA by LEAP Technologies, Carrboro, NC). Further information on method development using this on-line approach is found in Chapter 14, Section 14.6.3.

12.2.2.2 Empirical Determination of Optimal Sorbent Chemistries

Large numbers of structurally diverse compounds are screened for pharmacokinetic properties during the drug discovery process, as reviewed in Chapter 1. Janiszewski, Swyden and Fouda [4] described their application of a rational scheme for method development using automation as part of a standard protocol. Their approach utilized custom 96-well extraction disk plates to assess multiple sorbents and eluants simultaneously and empirically determine dominant extraction chemistries. Data were collected on over 100 discovery compounds using this approach.

The method development plates were configured with four bonded silica sorbents: C2, C8, C18 and mixed mode cation. These four sorbents were arranged sequentially in columns as three distinct sets (Figure 12.5). This arrangement is convenient as it also allows conducting experiments using partial plates of 32 samples at a time since the sorbent sequence is repeated in

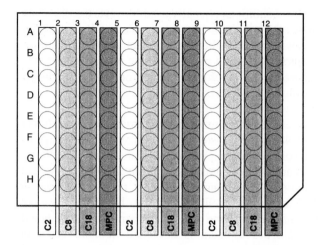

Figure 12.5. Configuration of a method development plate containing four bonded silica sorbents arranged sequentially in columns as three distinct sets of four sorbents.

columns 1–4, 5–8, and 9–12. Any unused wells are occluded with adhesive sealing film to maintain vacuum across the rest of the plate.

The semi-automated Quadra 96 was used for this study. Unique reagent delivery for elution was achieved by means of a custom-made reagent reservoir partitioned into twelve compartments in a 4-row by 3-column rectangular matrix. As the 96 tips aspirate liquids from the reservoir, each two rows contain a different elution solvent. When these liquids are dispensed into the development plate containing four sorbents repeating in three columns, two replicates per scheme (sorbent C2, C8, C18 and mixed mode cation with elution solvent A, B, C and D) are obtained.

Elution solvents used were:
  (A) Methanol
  (B) Methanol/1mM calcium acetate
  (C) Acetonitrile
  (D) Acetonitrile/5mM nitric acid

Using the disk plates, small eluant volumes (100 µL) were used and these were diluted with water (50–100 µL) in the collection device to achieve a 33–50% organic/aqueous concentration for direct injection into an LC-MS/MS system.

It was found that acetonitrile with an acidic additive (5mM nitric acid) yielded recoveries of >50% (deemed acceptable for discovery compounds) 80% of the time. Therefore, the SPE conditions having the broadest applicability were C8 or C18 sorbents with acetonitrile/5mM nitric acid as eluant. This approach was extended to a generic set of extraction conditions using single sorbent polymer plates (LMS, Varian Inc.). The polymer sorbent performed best using acetonitrile as eluant. The LMS polymer used in this manner showed >50% recovery for 60% of the compounds tested and >70% recovery for 38% of compounds [4].

An important lesson from this exemplary study by Janiszewski, Swyden and Fouda is that once the relationship among sorbent, eluant and analyte is understood, the choice of an optimal wash solvent can be deduced. For example, when secondary interactions with residual silanols on bonded silica sorbents are predominant, C2 using acidified methanol as eluant may provide optimal analyte recovery. Acetonitrile may yield poor recovery from a C2 sorbent. During data analysis from the method development experiment, a low recovery using acetonitrile eluant makes it clear that it can be diluted and used as an effective wash solvent because its ability to desorb analyte is poor.

For a relatively small investment in time (2.5 h for set up, extraction, data analysis and interpretation for about 40 total samples), a great deal of information is obtained. This theme is discussed further in the next section describing methodology to rapidly screen load and elution conditions in one experiment. A second experiment rapidly optimizes wash solvent composition to add selectivity to the method. A third experiment optimizes the volume of elution solvent to deduce the highest throughput.

12.3 Selective and Rapid Approach to Method Development

12.3.1 Single Sorbent and Bed Mass Evaluation—Strategy and Procedures

12.3.1.1 Overall Strategy

The microplate format for solid-phase extraction presents an 8-row by 12-column grid of 96 wells. Consider this arrangement of wells an empty canvas or template for designing a method development experiment that can simultaneously evaluate multiple variables in replicate sets. In one test the goal is to obtain as much information as possible about the affinity of analytes for a particular sorbent and how that attraction is modified by changes in pH.

When a defined strategy is followed, a selective bioanalytical method can be quickly developed and optimized. This scheme uses a single sorbent and bed mass as the extraction media. The following discussion presents a strategy from the author that has proved to be a rapid and successful approach to defining a selective bioanalytical SPE method. Experience has shown that when this practice is followed, the method developed for a compound in the discovery phase requires little modification as the compound progresses into preclinical and first time in man studies.

Rapid method development involves the following three experiments, usually in this order:

1. Determine the optimal pH conditions for loading analytes onto the sorbent and the most efficient solvents for eluting them. Three pH values are evaluated for the load step (acid, neutral and base); six elution solvents are evaluated (methanol and acetonitrile, alone and modified with acid and base), resulting in 18 total combinations (3 load pH by 6 elution solvents). The best performing combinations define the primary (and secondary, if present) mechanisms of attraction of analyte to sorbent.

2. Select one (or two) of the best load/elute combinations and perform a wash solvent optimization study. Determine the maximum percentage of organic in water that can be tolerated without significant analyte loss. The conditions for a method—load, wash and elute—have now been defined.

3. Select one (or two) of the most desirable load/wash/elute combinations and perform an elution volume optimization study. Determine the minimum volume required for consistent elution of analyte from the sorbent bed. Small elution volumes aid in throughput by reducing the dry-down time and offering the ability to eliminate the evaporation step in many cases.

Assuming satisfactory performance (recovery, cleanliness and matrix effect) has been obtained at this point, the analyst can proceed with a set of samples representing a standard curve, quality control and validation samples as directed by laboratory SOPs. Note that experiment #3 above can be optional when excess elution volume is used.

Appendix I contains different layouts of a microplate grid of 96 wells that the reader can photocopy and use to help design a method development experiment. The following grids are provided: samples 1–96 numbered by row and by column, and an empty plate of 96 wells.

12.3.1.2 Selection of Sorbent, Format and Bed Mass

The strategy presented for rapid and selective method development requires the proper choice of sorbent chemistry. This selection is made on the basis of the chemical structure of the analytes, their predicted mechanism of interaction with the sorbent (*e.g.*, reversed phase—hydrophobic, reversed phase—polar or mixed mode) and practical considerations such as the desired elution volume and elution solvent composition. A good compromise when choosing sorbent chemistry is to actually run the experiment twice using complementary chemistries, *e.g.*, a reversed phase polymer and bonded silica. The two extraction plates can be processed on the same day and two collection plates with eluate solutions are analyzed during an overnight LC-MS/MS run.

Multiple varieties of 96-well SPE plates, chemistries and formats are available, as discussed in Chapter 11, Section 11.4. General suggestions are to maintain an inventory of packed bed plates and disk plates from preferred vendors in only a few chemistries. Reversed phase in four varieties (a favorite polymer and C18, C8 and C2 bonded silica) as well as mixed mode are certainly the most used for bioanalysis, as determined from a review of the published literature. The choice of sorbent mass depends on the sample volume and matrix. Generally, drug discovery applications benefit from using disk plates and packed bed plates (2–25 mg) since sample matrix volumes do not exceed 100 µL. Preclinical and clinical applications with larger sample volumes (250–1000 µL) should also benefit from disks in many cases but packed particle bed masses up to 60 mg may be required more often.

12.3.1.3 Experiment #1: Load/Elute Screening

The first experiment evaluates the influence of pH on the sample load step and compares a methanol and an acetonitrile series of elution solvents. Each replicate set contains a defined load/elute combination. A practical number of combinations is 18—comprised of 3 load conditions (acid, neutral and base) crossed with 6 elution solvents (methanol/acid, methanol, methanol/base, acetonitrile/acid, acetonitrile and acetonitrile/base). The number of replicates for these 18 combinations can be a maximum of 5 (18 x 5 = 90) to fit within the 96-well matrix; however, for better visual representation 4 replicates are used here. This arrangement of buffers and solvents is shown in Figures 12.6 (samples arranged by rows) and 12.7 (samples arranged by columns).

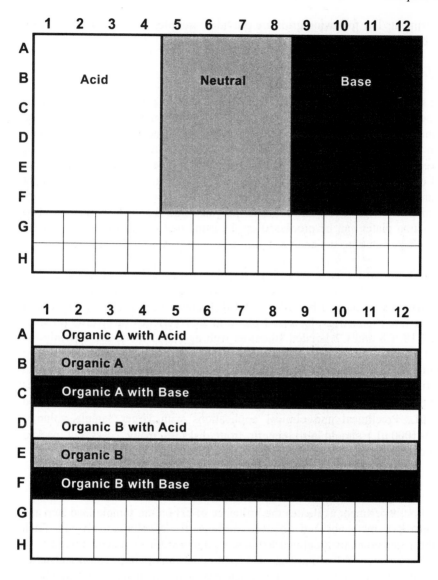

Figure 12.6. Samples are arranged within rows in method development experiment #1 (samples within columns are shown in Figure 12.7). *Top*: The influence of pH on the sample load step is evaluated using 3 aqueous buffers. *Bottom*: The effectiveness of 6 elution solvents is determined (*e.g.*, methanol and acetonitrile alone and each modified with acid or base). Recovery is determined for 18 load/elute combinations (n=4) as used in this single experiment.

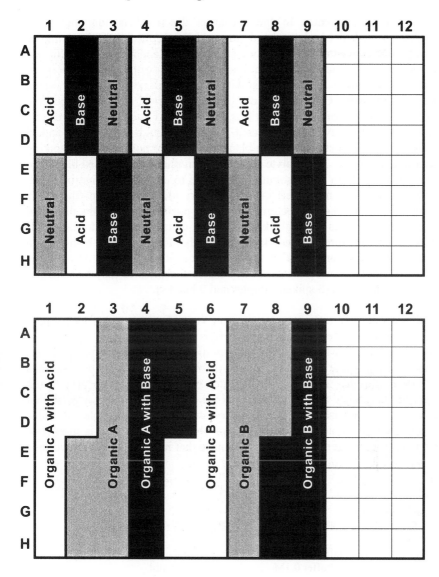

Figure 12.7. Samples are arranged within columns in method development experiment #1 (samples within rows are shown in Figure 12.6). *Top*: The influence of pH on the sample load step is evaluated using 3 aqueous buffers. *Bottom*: The effectiveness of 6 elution solvents is determined (*e.g.*, methanol and acetonitrile alone and each modified with acid or base). Recovery is determined for 18 load/elute combinations (n=4) as used in this single experiment.

The number of replicates in the scenarios described is four but what concentration should be chosen and how are the replicates to be utilized? Generally, one concentration of analyte and internal standard is chosen for this study. A practical choice is a concentration at the mid- or mid-to-high point of the calibration curve so that sensitivity limits are not challenged when determining recovery. Once a set of conditions is selected, recovery at multiple concentrations is determined but this first experiment is intended as a screening procedure. The four available replicate positions should consist of a spiked standard with internal standard and a blank intended for post-extraction spiking to determine recovery. The analyte response from this post-extracted blank can also be compared with a test solution of analyte to evaluate the matrix effect. The determination of matrix effect and recovery are discussed in Chapter 2, Sections 2.3.3 and 2.3.4, respectively. The four positions are commonly filled with 2 spiked standards and 2 blanks spiked post-extraction (after dry-down). If the analyst prefers, three spiked standards can be used with one blank, but a minimum of n=2 is preferred in the experience of the author. Also, if the number of samples can be increased to 90 then 5 replicates may be used per load/elute combination (3 spiked samples and 2 blanks).

The solutions needed to perform Experiment #1 are listed in Table 12.2. These solutions consist of the three prepared buffers (*e.g.*, pH 3, 7 and 12), the six elution solvents, and water used for the conditioning and the wash step. These buffers can all be made from the combination of two potassium phosphate salts (*e.g.*, mono- plus di-, di- plus triphosphate), with adjustment to pH 3 using phosphoric acid. Since the wash step is not yet optimized, water is used so that analyte loss from a wash solvent that was too strong is not a possible reason for low recovery. Also note that the volume of elution solvent used is in excess, determined from the sorbent bed mass, again so that insufficient elution volume is not a possible reason for low recovery.

Table 12.2
List of solutions used in each step of method development experiment #1

Condition	Load	Wash	Elute
Methanol	Acid buffer 0.1M	Water	Methanol/HCOOH (98:2)
Water	Neutral buffer 0.1M		Methanol
	Basic buffer 0.1M		Methanol/NH_4OH (98:2)
			Acetonitrile/HCOOH (98:2)
			Acetonitrile
			Acetonitrile/NH_4OH (98:2)

A major benefit of performing the experiment at acidic, neutral and basic pH is that the response of the analyte to these different conditions helps explain the mechanism of attraction to the sorbent. In a predictive sense, an experienced analyst should be able to describe what will happen but in a practical sense the primary and secondary interactions of bonded silica sorbents can be somewhat unpredictable. By evaluating the most frequent scenarios, the entire range of common pH values and solvents is covered and analyte responses are recorded. Also, in the event that the method needs further optimization during the drug candidate's lifetime, data are available that can help select an alternate or revised method without having to start over from the beginning.

12.3.1.4 Experiment #2: Wash Solvent Optimization

Once a set of load and elution conditions has been selected for use with a specific sorbent and the recovery and matrix effect are acceptable, a method is essentially defined. However, the use of 100% water as the only wash solvent does not provide the needed selectivity that is achievable with the use of solid-phase extraction. Water alone is recommended as the first wash solvent in a series since it displaces residual protein components from the sorbent bed. However, it should always be followed with another wash solvent, usually one that has some organic content. The goal is to use the maximum organic concentration that can be tolerated without significant analyte loss. The benefit is a cleaner eluate that is free of matrix interferences. Presenting a clean eluate to the LC system and mass spectrometer is very important to prolong column lifetimes and minimize source contamination from residual compacted components. As mentioned in the report by Janiszewski, Swyden and Fouda [4], a poor elution solvent identified from the first experiment can be useful as a wash solvent, usually more diluted.

A second SPE plate is used for the wash solvent optimization study. Since the full number of 96 wells is available for experimentation, the number of wash solvents and the number of replicates is decided by the analyst. Enough positions are available to optimize wash solvents for up to four load/elute combinations, assuming 4 replicates per solvent and six solvents per load/elute combination. Also, the analyst may choose to use 3 (or 2) replicates each and reserve 1 (or 2) well(s) for determination of matrix effect. The discussion here uses one load/elute combination for all the wells and twelve wash solvents are evaluated (n=4), as shown in Figure 12.8 for samples arranged within vertical columns. A thorough wash solvent series can be based on methanol (*e.g.*, 0, 5, 10, 20, 30 and 40% organic in water) and another based on acetonitrile (*e.g.*, 0, 5, 10, 20, 30 and 40% organic in water). Rather than have 0% organic (water)

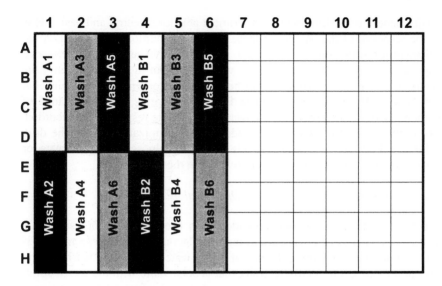

Figure 12.8. Method development experiment #2 optimizes the wash solvent composition (*e.g.*, %organic in water) to create a more selective solid-phase extraction method. Shown above is an evaluation using 12 wash solvents (n=4), although other numbers can be evaluated. The use of blank matrix and true blank wells that are spiked post-extraction can yield information on recovery and matrix effect, respectively. Samples are arranged within columns.

tested twice in this layout, one of the sets can be substituted with another solvent percentage if desired (*e.g.*, 15% organic). Note also that an adjustment in wash solvent pH can be evaluated instead of, or in addition to, the %organic in water. It is suggested to always include a high concentration of organic as the upper limit in the range of percentages, one that is expected to elute analyte, so that a maximum tolerated percentage can be reached. Failure to confirm this upper limit may require repeating the experiment because, when a higher percentage organic can be tolerated, its use ensures a more selective method.

12.3.1.5 Experiment #3: Elution Volume Optimization

After the completion of these two experiments using a single sorbent plate, optimal conditions are identified for the load, wash and elute steps to yield a selective solid-phase extraction method. The development of this method can end at this point but it is recommended to proceed and perform an elution volume optimization study. Knowledge of the elution characteristics can aid in determining whether or not the dry-down step can be eliminated. This approach consists of eluting analyte in a concentrated volume to allow for dilution, mixing and direct injection.

When dilution followed by injection is the goal, usually a basic organic eluate solution that uses NH_4OH is not chosen; rather, look for a second best elution solvent (organic or organic/formic acid) that, when diluted, is compatible with mobile phase for direct injection. If methanol and acetonitrile solvents are shown to be equivalent in the recovery of analyte from the sorbent, choose the solvent that more closely matches the mobile phase composition.

If the dry-down step cannot be eliminated due to sensitivity limits not being achieved, at least using a smaller elution volume should result in a shorter evaporation time. In this age of high throughout and great time constraints, it is not prudent to use a 1 mL elution volume when 400 µL is all that is necessary. Data obtained in this volume optimization study are used to confirm the minimum volume required for consistent recovery.

The numbers of elution volumes and replicates evaluated are again the decision of the analyst. The unused half of the plate from the wash solvent optimization study can be used for this volume optimization study or a new plate can be used. Generally, a complete elution profile is obtained using 8 different elution volumes in replicates of four; this arrangement is shown in Figure 12.9 for a plate in which samples are arranged within columns.

The volumes of elution solvent chosen for this study vary, depending on the sorbent bed mass in the SPE plate. Three general volume recommendations can be made based on the sorbent bed mass; these volumes are listed in Table 12.3. Note that at this point in the method development project, it can also be beneficial to use a modular plate having the same sorbent chemistry as used for the load/elute screening and the wash optimization study; volume optimization for different bed masses can then all be evaluated at once.

More information can be obtained when the elution volume optimization study is repeated with a second aliquot of solvent. Separate collection plates are used and the relative amount of analyte recovered in the first and second aliquots is determined to confirm the efficiency of analyte removal from the sorbent bed. When three to four sequential elutions are performed, this is known as an exhaustive elution study. It is wise to confirm that the analyte has been totally removed from the sorbent bed. Sometimes, an explanation for low recovery can be caused by an incomplete removal of analyte; it is only through an analysis of

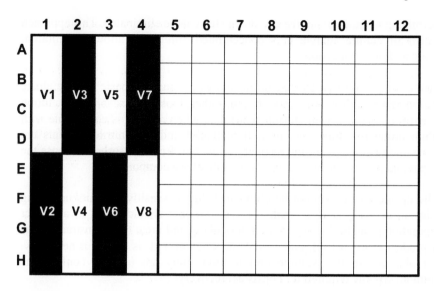

Figure 12.9. Method development experiment #3 optimizes the volume of elution solvent. The minimum volume that yields consistent recoveries should be used. When the volume is concentrated enough, it may be possible to dilute and then inject an aliquot directly into the MS, eliminating a dry-down step. Shown above is an evaluation using 8 volumes (n=4), although other numbers can be used. Samples are arranged within columns.

Table 12.3
Elution volume suggestions (µL) for an optimization study

Disk and Packed Bed Plate (5–25 mg)	Packed Bed Plate (30–50 mg)	Packed Bed Plate (60–100 mg)
75	250	300
100	300	400
125	350	500
150	400	600
200	500	700
250	600	800
300	700	900
400	800	1000

repeated elutions that this problem can be confirmed. An example of this occurrence is described in Section 12.4.2.6.

More complete removal of analyte from the sorbent bed has sometimes been demonstrated when analytes are eluted in two sequential half volume aliquots rather than one full volume aliquot. Usually, this occurrence is a result of the geometry below the extraction bed more than sorbent desorption efficiency, although the latter is observed on occasion. When a large surface area of plastic is encountered below the sorbent bed, a second aliquot helps to rinse residual analyte trapped on the plastic or in a groove and more complete recovery is obtained. Note that the volumes in Table 12.3 are chosen so that analyte response obtained using two sequential aliquots of 75 μL from a disk plate can be compared, *e.g.*, with a single 150 μL aliquot; likewise, 2 aliquots of 100, 125, 150 and 200 μL can each be compared with single aliquots of 200, 250, 300 and 400 μL, respectively.

12.3.2 Single Sorbent and Bed Mass Evaluation—Results

12.3.2.1 Experiment #1: Load/Elute Screening

The procedure for performing a method development experiment to evaluate the effect of pH modification using 3 sample load conditions and 6 elution solvents on the extraction of analyte from a single sorbent is discussed in Section 12.3.1.3. Figure 12.10 shows an example of the types of results that can be obtained. These data arise from the extraction of a proprietary analyte from rat plasma using a reversed phase polymer sorbent. Percent recovery is plotted for each of the 18 load/elute combinations, grouped first by methanol elution series and then by acetonitrile elution series.

Data shown in Figure 12.10 can be arranged in different ways (*e.g.*, by elution series, by load/elute combination with methanol and acetonitrile results side by side, or sorted in descending order from highest to lowest recovery). The important considerations to derive from the data focus on the following:

1. Which load/elute scenarios within each of the two solvent series show the highest recovery?
2. Explain how the analyte is interacting with the sorbent from analyzing recovery data under different pH conditions.
3. Which load/elute scenarios within each of the two solvent series show the highest recovery?
4. Explain how the analyte is interacting with the sorbent from analyzing recovery data under different pH conditions.

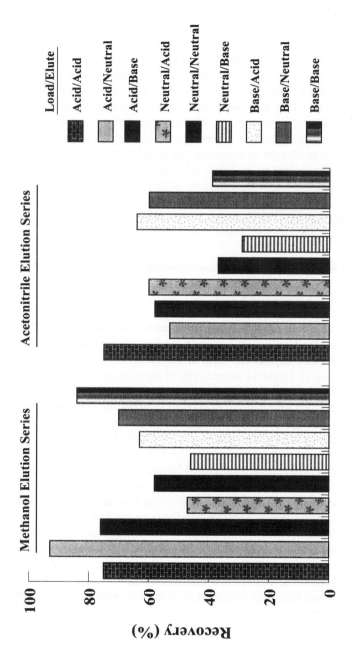

Figure 12.10. The results from a method development experiment generated in the microplate format are shown for a proprietary analyte extracted from rat plasma. A single sorbent 96-well SPE plate was used (a 30 mg polymer). Samples were selectively pretreated with buffer at three pH values prior to loading. Water was used as the wash solvent. Elution (excess volume) was performed to selected wells using a methanol and an acetonitrile series of solvents (unmodified and acid and base modified). A total of 18 load/elute combinations was evaluated in one screening experiment (n=2 per replicate set; averages shown). Recovery data yield information about the attraction mechanism of analyte to sorbent.

5. Which load/elute scenario within each of the two solvent series shows the highest recovery?
6. Explain how the analyte is interacting with the sorbent from analyzing recovery data under different pH conditions.
7. How do recoveries compare between the two solvent series?
8. What is the matrix effect for the best two or three combinations?
9. Which of the better performing elution solvents is most desirable to meet the objectives for this assay? (*e.g.*, compatible with mobile phase for dilution followed by direct injection).
10. Considering recovery, cleanliness of extract and elution solvent composition, is this sorbent a good choice to meet the objectives for this assay?
11. Should the method development continue with wash optimization using the same chemistry and bed mass or should the screening experiment be repeated with a different sorbent chemistry?

An analysis of these considerations follows. While knowing the chemical structure of the analyte in the examples in this section would be helpful in reinforcing the fundamentals of extraction chemistry, data for this approach using known marketed drugs were not available for the preparation of this chapter. Regardless, the approach presented is valid for any analyte. It is the author's intention to introduce the process for performing these experiments and the reasoning for data analysis and interpretation.

Within the methanol series, all three acid and base load combinations showed acceptable recovery; the three neutral load combinations were consistently lower. Neutral elution was best within the acid load series but base elution was best within the base load series. Analyte adsorbs to polymer quite well, overall; an acid load with methanol elution is quite acceptable for follow-up. Loading in acid medium may protonate some unwanted matrix interferences and wash these through the sorbent bed to waste. Neutral methanol is an effective solvent to elute adsorbed materials from the column so it is particularly important to find a selective wash step in the next experiment.

Within the acetonitrile elution series, recoveries are generally lower than with methanol. The same load/elute combinations as worked best for methanol were not the best performing using acetonitrile. It may be that some analyte is still adsorbed to the sorbent and not fully eluted, thus explaining the lack of correlation with data from the methanol series. Eluting these wells again with methanol may prove this hypothesis. Regardless, since higher recoveries are consistently obtained using methanol solvents, an acetonitrile elution would be abandoned from further development. However, data suggest that acetonitrile should be an effective wash solvent instead and the next experiment will examine this presumption.

12.3.2.2 Experiment #2: Wash Solvent Optimization

The procedure for performing a wash optimization study to evaluate the effect of organic/aqueous solvent composition on analyte recovery is discussed in Section 12.3.1.4. A thorough wash solvent series is commonly based on a methanol and an acetonitrile series (*e.g.*, 0, 5, 10, 20, 30 and 40% organic in water). Note that a modification of pH can be included in this evaluation as well. Figure 12.11 shows an example of the types of results that can be obtained. These data are from the extraction of another proprietary analyte from rat plasma using a C2 bonded silica sorbent. Analyte response is plotted for each of the wash solvents which are grouped first by methanol series and then by acetonitrile series.

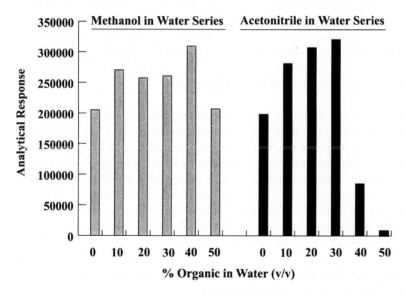

Figure 12.11. The results from a wash optimization experiment generated in the microplate format are shown for a proprietary analyte extracted from rat plasma. A single sorbent 96-well solid-phase extraction plate was used. A series of wash solvents (based on methanol in water and acetonitrile in water) was examined using this approach; identical load and elute conditions. Averages shown; n=4 per replicate set.

The goal of a wash optimization study is to determine the maximum tolerated organic composition before analyte response is reduced. All samples receive the same pretreatment for the load step and the same elution solvent; the only variable changed per replicate set is the wash solvent composition. While percent recovery can also be determined and plotted, Figure 12.11 uses analytical response for the y-axis simply because when time is of the essence a quick plotting of the data is usually made.

Note that analytical response shown in this graph also reflects ionization suppression in mass spectrometric analyses; the reduced response using 100% water for each solvent series may not reflect lower recovery as much as reduced response from the larger mass of matrix materials eluting with the analyte. As the method becomes more selective for analyte when a higher percent organic in water is used, response is improved as ionization suppression is reduced. Data show that this proprietary analyte can tolerate a 40% methanol/water wash but response is reduced at 50%. Only a 30% acetonitrile wash is tolerated before analyte response is reduced. Either choice would make a good wash solution; an examination of the matrix effect from 40% methanol versus 30% acetonitrile may be the deciding factor.

12.3.2.3 Experiment #3: Elution Volume Optimization

12.3.2.3.1 Elution Volume Profile—Multiple Volumes

The procedure for performing an elution volume optimization study to understand the profile of analyte desorption from the sorbent bed is discussed in Section 12.3.1.5. The goals of this study are to determine the optimal volume for elution (the minimum that yields consistent and acceptable recoveries) and, if the volume is small enough, to investigate whether dilution of eluate followed by direct injection is feasible (in a follow-up experiment). Figure 12.12 shows an example of the types of results that can be obtained. These data are from the extraction of another proprietary analyte from rat plasma using a C8 bonded silica sorbent in the disk format. Percent recovery is plotted against elution volume (range, 150–450 µL); sequential elution was performed three times and each aliquot was analyzed separately (after dry-down and reconstitution).

The data in Figure 12.12 show that an elution volume of 250 µL yields the best recovery (~96%). A larger volume could be selected, *e.g.*, 450 µL (99% recovery) but an increased gain in recovery is negligible and this larger volume

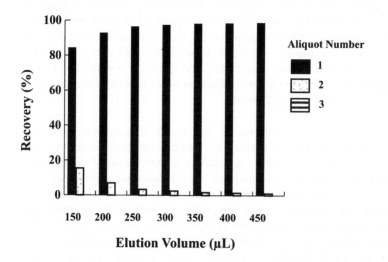

Figure 12.12. The results from an elution volume optimization study generated in the microplate format are shown for a proprietary analyte extracted from rat plasma. A single sorbent 96-well solid-phase extraction disk plate was used and 7 elution volumes were examined. Three sequential elutions were performed per volume; each fraction was isolated and determined separately; n=4 per replicate set (averages shown). The third aliquot for each volume did not elute measurable analyte.

can be predicted to take almost twice as much time for the dry-down step. In the interest of reducing total time required for sample preparation, a selection of 250 μL volume is valid. Note that for sensitive mass spectrometric detection, the maximal recovery is not always of paramount interest, especially for drug discovery support. An aliquot of 150 μL yields ~84% recovery; sufficient analyte mass is provided for analysis. It may be conjectured that precision is slightly more variable using 150 μL, although it should be within acceptable operating limits. A sufficient analyte recovery using only a 150 μL volume opens the possibility of eluting in this small volume, diluting with aqueous (*e.g.*, water or dilute formic acid), mixing and injecting an aliquot directly. This scenario is discussed in Section 12.3.2.3.3.

12.3.2.3.2 Elution Volume Profile—Repeated Aliquots of the Same Volume

The elution volume optimization experiment described in the previous section used a minimum volume of 150 μL and a maximum of 450 μL to elute analyte from a low sorbent bed mass disk plate. Data show that adequate recovery was achieved with this smallest volume. In some cases, when a particular objective is in mind, it may be informative to use repeated aliquots of the same volume to generate the elution profile. These objectives may include, *e.g.*, examining a smaller elution volume in more detail by looking at a narrower volume range (50–250 µL instead of 150–450 µL) and/or confirming recovery at small volumes with the intention of diluting the eluate and injecting an aliquot directly.

The results from a volume optimization experiment using repeated aliquots of 50 µL are shown in Figure 12.13. In this example, a C8 bonded silica disk plate was used for the extraction of a proprietary analyte from dog plasma and elution with 60% acetonitrile in water modified with formic acid. The solvent selected is one that is compatible with mobile phase for direct injection. It is seen that a volume of only 50 µL is enough to recover nearly 70% of analyte; two aliquots of 50 µL recover ~90%. This experiment confirms that a 100 µL volume can be used for elution.

12.3.2.3.3 Eliminating the Evaporation Step

There are several scenarios for eliminating the evaporation step. As discussed in the previous section, an elution solvent can be chosen that is compatible with

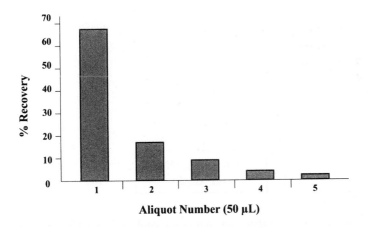

Figure 12.13. The results from a sequential recovery experiment using repeated 50 µL aliquots of elution solvent are shown for a proprietary analyte extracted from dog plasma using a C8 disk plate; n=4 per replicate set (averages shown).

mobile phase without any modification. Oftentimes, this solvent is a mixture of organic in aqueous, or organic in dilute formic acid. Another approach is to elute in pure organic solvent and then add an aqueous modifier into the collection well so that the resulting mixture is compatible with mobile phase for direct injection. Sometimes, the aqueous component is passed through the extraction bed and collected with the organic eluate. These scenarios are outlined in Figure 12.14.

Traditionally, maximal recovery is sought using solid-phase extraction methods which require dry-down of eluate followed by reconstitution. Simpson *et al.* [5] present an interesting and useful perspective on elution followed by direct injection. These authors propose that the elution volume should be optimized based on the concentration of analyte in the final elution fraction which determines the signal response of the LC-MS/MS. Three elution volumes were evaluated for analyte SR48968 (reprinted in Table 12.4). The internal standard was added post-extraction. The peak area ratio (SR48968/IS) was used to determine the signal response. A 75 µL elution volume yielded the highest response while the corresponding mass recovery was only slightly less than that obtained using 100 µL. Any volume over 75 µL diluted the final concentration to an extent not compensated by recovering additional mass.

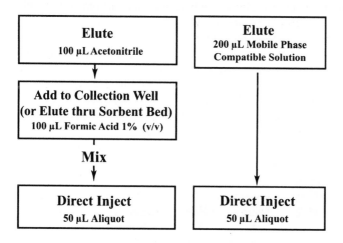

Figure 12.14. Elimination of the evaporation step is accomplished by eluting with a small volume of organic and adding an aqueous modifier (into each collection well or through the extraction bed). Alternately, a mobile phase compatible solution is used.

Table 12.4
Signal response of analyte SR48968 as a function of elution volume

Elution Volume (µL)	Average Peak Area Ratio (SR48968/IS)		
	Elution 1	Elution 2	Elution 3
50	5.10	1.41	0.326
75	5.16	2.79	0.355
100	3.97	2.08	0.160

n=4 replicates per elution volume; 3 sequential elutions
Reprinted with permission from [5]. Copyright 1998 John Wiley & Sons, Ltd.

Selected bioanalytical applications that detail these different approaches for eliminating the dry-down step in solid-phase extraction methods are presented in Table 12.5. Elution solvents, volumes and dilution schemes are listed to provide the reader with more practical examples of this time saving and valuable technique.

12.3.2.4 Building Common Scenarios into the Workflow Pattern

The procedures presented for rapid and selective method development using solid-phase extraction microplates are proven techniques. While the full approach may seem like a daunting task initially, especially for data analysis, the process can be streamlined in many ways. For example, as a workgroup, agree on the number of replicates to test and how they are to be arranged within the microplate. Be sure to allot for sample blanks as used for determination of analyte recovery and matrix effect. Once a protocol is decided, that procedure should be used routinely without modification. Each sample plate will be prepared in the same manner and a common set of reagents will be used. The collection plate for injection will have samples arranged in a common pattern and the sample queue, once created, will be used over and over.

Data analysis can be set up within a spreadsheet software program having the formulas for recovery already embedded. When data are acquired from the mass spectrometer system, they are saved as a text file and imported into the spreadsheet; the pertinent block of data is then copied and pasted into the reserved cells in the spreadsheet containing the formulas. Recalculation instantly derives all recoveries and averages. Data can then be graphed by selecting each series. This approach, including data analysis in a spreadsheet,

Table 12.5

Selected bioanalytical applications of solid-phase extraction in the microplate format that utilize direct injection of eluate*

Analytes	SPE Sorbent and Plate Type	Elution Solvent and Volume	Dilution Scheme	Injection Volume	Reference
Indinavir and L-756423	MPC Empore™ disk	ACN/water/NH₄OH (58:40:2) 225µL	1N Acetic acid 100µL	40µL	[6]
Chlorambucil and metabolite	C18 15mg Microlute™	0.1% HCOOH/0.1% HCOOH in ACN (3:7), 100µL	None	20µL	[7]
Everolimus and cyclosporine A	C18 AR disk SPEC•PLUS™	ACN 100µL	None	25µL	[8]
311C90 and metabolite	C18 50mg Microlute	50% ACN in 150mM ammonium formate pH 5 300µL	None	20µL	[9]
Ziprasidone	C2 Empore disk	5mM nitric acid/ACN, 100µL	Water, 50µL	50µL	[10]
SR46559	C8 Empore disk	Acetonitrile contg 0.1% trifluoroacetic acid, 200µL	Buffer, 200µL	100µL	[5]
Neurokinin-1 receptor antagonist and metabolite	Oasis® HLB 30mg	0.04% HCOOH in Methanol, 300µL	50mM ammonium formate buffer 300µL	100µL	[11]
Equilenin and progesterone	C8 Empore disk	ACN 100µL	0.2% HCOOH in water, 400µL	200µL	[12]

Fosinoprilat	C18 Empore disk	10mM Ammonium acetate in methanol 100µL, followed by 10mM Ammonium acetate pH 5.5, 100µL	None	n/a	[13]
Gleevec™ and metabolite	C8 Empore disk	1% 1N HCl in Methanol 200µL	None	10µL	[14]
Rofecoxib (Vioxx®)	C8 Empore disk	Acetonitrile 150µL	Eluates centrifuged into collection plate contg. 300µL water	135µL	[15]
Loratidine (Claritin®)	C18 Empore disk	150µL of (2mM ammonium acetate, 0.1% acetic acid, 0.1% HCOOH in acetonitrile/methanol 50:50)	300µL of (2mM ammonium acetate, 0.1% acetic acid, 0.1% HCOOH in water)	5µL	[16] [17]
Exemestane	C2 50mg	Acetonitrile contg 0.1% trifluoroacetic acid 300µL	None	80µL	[18]

*Analysis and detection by LC-MS/MS except for Reference [15] which was LC-Fluorescence
ACN=acetonitrile
n/a = information not available

lends itself nicely to n-in-1 determinations. A page is simply reserved within the spreadsheet file for each analyte. A summary page can be created to extract pertinent data from individual worksheet cells. Summary graphs with all analyte data can then be assembled as well.

The automation of this set of experiments for method development is the most desirable way to achieve rapid throughput. A 4-/8–probe liquid handling workstation is ideal for these tasks because it is fully programmable to deliver specific reagents into specified wells and also to use unique volumes per well or groups of wells. The 96-tip workstation will perform most of the method development tasks presented here except for the elution volume optimization; all 96 tips deliver the same volume of solvent each time to each well. Strategies for automation of solid-phase extraction are discussed in Chapter 13, including method development strategies.

12.3.2.5 Experiments on a Smaller Scale

There may be situations when the full 3 load by 6 elution study is not desirable whether due to time constraints, sample capacity on the mass spectrometer, number of available extraction wells or the need to examine a smaller subset of variables. The experimental outline as presented can be scaled down in any way imaginable but it is important to change only one variable within each set of conditions. Shown in Figure 12.15 are the results from a study that evaluated 3 load conditions with 3 elution conditions, using only an acetonitrile series. These data from the extraction of a proprietary analyte show clearly that acidic load conditions are unfavorable for analyte recovery from this C8 sorbent. Conditions when analyte is loaded in neutral buffer or base, and eluted in neutral or basic acetonitrile, are preferable. It is evident from this pattern that reversed phase attraction is predominant at neutral or basic pH and that this attraction is disrupted by acetonitrile alone or modified with base.

These data in Figure 12.15 can be further explored. The poor elution seen after loading in acid suggests that cationic attraction to silanols on the bonded silica surface is very strong, and acetonitrile does not display enough polar character to disrupt this interaction. A follow-up experiment was run in this example— the same wells were eluted again but with methanol. The methanol did elute analyte and 60% additional recovery was obtained (acid load). This finding confirms that the silanol attraction is predominant. Another conclusion from this example is that, since loading in acid and washing with acetonitrile did not elute much analyte, this scenario can make an effective load and wash

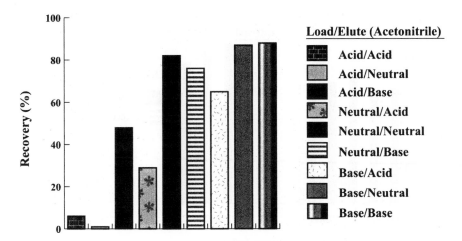

Figure 12.15. The results from a method development experiment generated in the microplate format are shown for a proprietary analyte extracted from plasma. A C8 96-well solid-phase extraction disk plate was used. Samples were selectively pretreated with buffer at three pH values prior to loading onto the sorbent. Water was used as the wash solvent. Elution (excess volume) was performed to selected wells using only an acetonitrile series of solvents (unmodified and with acid and base modified). A total of 9 load/elute combinations was evaluated in one screening experiment (n=2 per replicate set; averages shown). Acid/Acid refers to the pH for sample load and the pH for elution in an organic solvent.

combination. The elution solvent of choice is methanol in the revised procedure. Note that even failures in analyte recovery can aid in determining improved and more selective SPE methods.

One caveat in reducing the experimental variables examined in method development is that, when interpretation of the results is not straightforward, the analyst may have to repeat the experiment for the purpose of obtaining data on conditions that were omitted from the first experiment. Ironically, the savings in time from reducing the experimental variables can be nullified by having to repeat the work to include them. Perhaps a methanol elution should have been included in the study described above.

Other variables besides those presented here can be explored during method development. For example, specific questions might arise that need inquiry, such as "What is the influence of the volume of buffer used during the sample pretreatment step, or the volume used to wash the sorbent bed?"

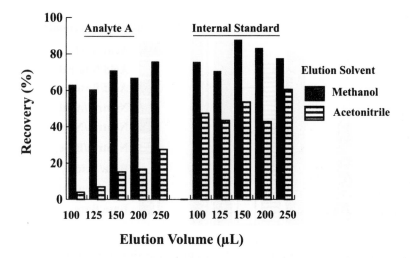

Figure 12.16. The results from a method development experiment in a reduced scale are generated in the microplate format and shown for a proprietary analyte extracted from plasma. A C8 96-well solid-phase extraction disk plate was used. Samples were eluted with methanol or with acetonitrile, each in varying volumes (n=2 per replicate set; averages shown).

Another example is to extract a number of wells to simultaneously compare methanol and acetonitrile as elution solvents for two analytes, and simultaneously perform an elution volume optimization from 100–250 µL. Data that illustrate this latter example are shown in Figure 12.16. In this single experiment much information is obtained for two analytes: methanol versus acetonitrile as elution solvents and elution volume optimization. The rapid approach to acquiring this data is based on using the 8-row by 12-column grid of available wells to design an efficient experiment. One variable at a time is examined per replicate set of samples; blanks are also extracted for recovery and matrix effect determinations.

12.3.2.6 Use of a Mixed Mode Sorbent

The protocol by which 3 load and 6 elution conditions are evaluated works best for reversed phase chemistries. This approach can also be used for a very pH specific sorbent, such as mixed mode cation or anion. If a nonpolar affinity is operating and is predominant, the experimental design should allow for its detection. However, note that for mixed mode cation when the analyte is a

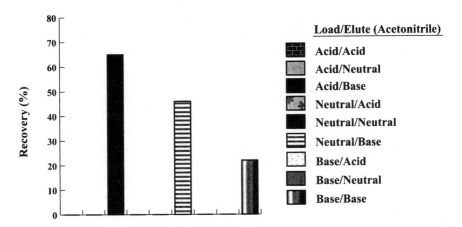

Figure 12.17. The results from a method development experiment generated in the microplate format are shown for a proprietary analyte extracted from rat plasma. A mixed mode cation SPE plate was used. Samples were selectively pretreated with buffer at three pH values prior to sorbent loading. Water was used as the wash. Elution (excess volume) was performed to selected wells using a methanol and an acetonitrile series of solvents (n=2 per replicate set; averages are shown).

basic amine, one load/elute combination will usually comprise the predominant set and the others may fail or give only moderate results, as shown in Figure 12.17. The usefulness of the 3 load by 6 elution solvent approach for mixed mode sorbents is then questionable. A more appropriate tactic may be to test different pH values that are more closely selected around the known or predicted pKa of the analyte, different wash solvents (*e.g.*, acetic acid versus formic acid; methanol versus acetonitrile), and/or different sources/percentages of base in organic solvent (*e.g.*, 2% vs. 5%; NH_4OH vs. triethylamine) for the load step. The method development approach using the microplate grid is flexible to meet each specific need.

12.4 Method Optimization and Troubleshooting

12.4.1 Introduction

Existing solid-phase extraction methods sometimes need adjustment or optimization for many reasons, such as the following: the method was not very selective in the first place (the first sorbent that worked reasonably well was chosen) and a change in matrix or the addition of a known metabolite to the assay revealed this deficiency; cleanliness requirements are not met, noted when the number of samples analyzed per day increased (LC column lifetime was found to be too short and/or high pressure system shutdown is encountered too frequently); a new clinical dosing mandate requires the assay of lower concentrations; analyte is found to be volatile on dry-down and a new elution scheme is required that eliminates the evaporation step; the method is being switched from an individual SPE column to the microplate because of higher throughput needs; the old automation that used only columns has been donated to a local university; or an existing method is too time consuming. The desired result of a method optimization is a robust, more selective procedure that is specific and reproducible from one analyst to another, as well as from one laboratory to another.

The steps involved in a typical solid-phase extraction method are shown in Figure 12.18. In order to optimize a design, the analyst must understand the different variables within each step and how a change may affect the final results. For a given analyte and detection technique, the most important variables are examined in the time allotted for method optimization. At the end of this section some common reasons for low recovery of analyte when using solid-phase extraction procedures are summarized.

12.4.2 Influence of Method Variables on the SPE Process

12.4.2.1 Sample Pretreatment

Two common reasons for failure of solid-phase extraction methods are that the pH conditions are not appropriate to promote maximal analyte binding and that the analyte is too strongly bound to matrix proteins or other components and cannot interact with the sorbent as it passes through. The importance of pH on the extraction process has been introduced in Chapter 2, Section 2.3.1. The reader is encouraged to review this material.

Many times, a buffer is not used to adjust pH but a basic or acidic solution is employed; while sometimes these solutions work, they do not possess the important buffering capacity that resists changes in pH when small amounts of acid or base are added (*i.e.,* the sample matrix itself). **Buffers should always be used during the sample pretreatment step**. Many solid-phase extraction methods require the analyte to be in either the ionic or neutral form for adsorption to occur; when pH is not properly controlled, variation can result and analytes may not adsorb.

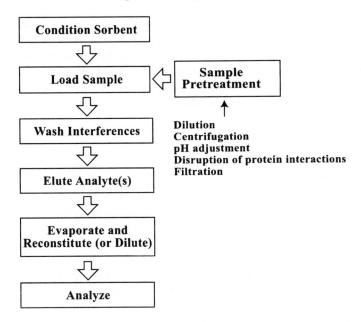

Figure 12.18. The general scheme for performing solid-phase extraction consists of several sequential steps. It is important in the method optimization process to understand the variables within each step and how changes may affect the final results.

Protein binding, when present, must be disrupted in the sample pretreatment step prior to sample loading onto the sorbent; otherwise, analyte will pass through the sorbent bed along with protein. Protein binding can be disrupted in many ways prior to the sample load step. Ke *et al.* [19] reported the best success using 40 μL 2% disodium EDTA per 100 μL mouse plasma; second best was 40 μL 2% formic acid in water (v/v). Other approaches include the addition of the following per 100 μL of matrix material: 40 μL 2% (w/v) trichloroacetic acid, 40 μL 2% acetic acid, 40 μL 2% trifluoroacetic acid, 40 μL 2% phosphoric acid or 200 μL acetonitrile. Pauwels *et al.* reported the best success among these similar techniques using formic acid [20]. Lensmeyer *et al.* [21] incubated serum, IS and 0.4M HCl at room temperature for 10 min to release steroids bound to proteins.

Sometimes a metabolic conjugate may need to be liberated before adsorption can occur and this hydrolysis is a part of the pretreatment step. Other times, dilution may be appropriate to reduce sample viscosity and improve flow characteristics. Centrifugation and filtration are also used in these situations.

The method of addition of internal standard to sample is sometimes a reason for poor method or extraction performance. Internal standards should be added directly to the sample matrix before the addition of diluent, buffer or protein denaturing agent so that the binding of the analyte can be mimicked. If internal standard is premixed with another solution (*e.g.*, within a buffer or an acetonitrile solution), inaccurate quantitation can result.

12.4.2.2 Sorbent Conditioning

It has been shown over the years and is generally accepted that conditioning the sorbent bed is required for bonded silica sorbents, especially for long chain alkyl sorbents such as octadecyl silane. Organic liquid (methanol or acetonitrile) is said to solvate the chains by extending them from their dry inwardly directed state so that they are ready to interact with aqueous analyte. Several of the newer polymer sorbents (*e.g.*, Oasis MCX, Waters Corporation) do not require organic solvent conditioning at all; sample is simply loaded onto the dry sorbent bed. The generally accepted protocol for conditioning is to use a volume of methanol followed by an equal or greater volume of water to displace the methanol from the sorbent bed. Typically, larger volumes than necessary are used for conditioning; these larger volumes have a detrimental effect on throughput but not on performance.

Variables that may, in some cases, affect method performance include the volume of solvent and the drying time between the methanol and water steps. These issues are much less of a concern with disk products and the lower sorbent bed mass products (*e.g.*, ≤50 mg), but are more prevalent with 100 mg sorbent bed masses and greater. Note that 3M Empore™ High Performance Extraction Disks can be conditioned with only 100 μL methanol [5], followed by water. They also perform well with only 50 μL methanol as conditioning solvent followed directly by sample matrix (no water conditioning in between) [10]. Methanol is used most commonly as the organic conditioning solvent but it is not the only choice; acetonitrile is frequently used and even dichloromethane is effective [20].

12.4.2.3 Sample Load Step

Most sorbent beds need to be moist and in the ready state before sample is applied during the load step. Also, the flow rate of the sample through the bed can sometimes be important since there must be enough residence time for analyte to interact with the sorbent as it flows through.

Acetonitrile or another organic solvent is sometimes added to the sample matrix as part of the addition of internal standard solution or to disrupt plasma protein binding. The precaution here is that analyte may precipitate out of solution along with the proteins as these are also forced out of solution by the organic solvent. The general guide is that the volume of organic loaded along with the total sample volume (matrix, buffer and internal standard solution) should not exceed 10% by volume. Thus, addition of 100 μL acetonitrile to 500 μL total sample and buffer solution results in a 17% organic concentration; dilution with 400 μL aqueous is required to reduce the total organic concentration to 10% or less.

12.4.2.4 Wash Step to Remove Interferences

During the wash step, overall analyte recovery must be balanced with the cleanliness of the final eluate. A water wash alone does not provide a clean eluate (residual protein and/or matrix components may coelute with analyte and be injected into the analytical system; over time this causes fouling of the LC column and the MS detection source). Water is recommended as the first of two (or three) sequential wash steps and should always be followed by a more selective wash solvent. Usually a small percentage of organic in water provides an effective wash; greater organic concentration is better as long as analyte is not removed. Sometimes, a change in pH may also be introduced and can aid in cleaning the interfering components from the sorbent bed prior to the elution step.

12.4.2.5 Incomplete Removal of Liquid from the Sorbent Bed

Oftentimes, long vacuum cycles and high vacuum pressures may leave residual aqueous wash solvent in the sorbent bed before elution. Vacuum is simply not the most efficient means to remove residual liquid; positive pressure is much better and is preferred. Following the vacuum step to remove wash liquid, it is recommended to mate the SPE plate with a collection plate and insert this combination into a centrifuge to remove residual liquids. A centrifugation time of 5 min at 3000 rpm is usually sufficient. Failure to remove this residual volume may adversely affect the completeness of final elution volumes. Residual aqueous in the elution solvent will also prolong the evaporation time. One report recommends rinsing the underside of the plate with methanol via a squirt bottle after the final wash step on the vacuum manifold, and then dry the tips by patting with a paper towel; the plate is then centrifuged [22]. A positive pressure manifold such as the model from Mallinckrodt Baker (Phillipsburg, NJ USA) is a very effective alternative to centrifugation.

An occasional problem that may result, particularly in clinical patient samples for bioanalysis, is that lipids may also adsorb onto the sorbent bed and elute with the analyte. This problem becomes greater with larger sample volumes. A strong wash solvent can aid in lipid removal from the sorbent but many times this is not possible due to a corresponding loss of analyte. One way to remove lipids, when contained in the eluate solution, is to add concentrated sulfuric acid, as reported by Pauwels *et al.* [20]. The eluate with sulfuric acid solution is mixed, centrifuged (*e.g.*, 5 min at 2000 x *g*), the organic layer removed and then evaporated to dryness. Reconstitution is followed by analysis.

12.4.2.6 Incomplete Analyte Elution

Incomplete elution of analyte can be a cause of low and inconsistent recoveries. The elution solvent may not have enough strength or affinity to desorb analyte from the sorbent bed and another choice may be more appropriate, or it may need pH modification in order to cause analyte release. Sometimes this solvent will elute analyte but in a slow gradual release rather than the rapid on/off desorption that is desired. The volume used may be insufficient as well for the mass of sorbent. Figure 12.19 demonstrates incomplete analyte elution; methanol only slowly desorbed analyte from the sorbent bed. A single aliquot of methanol showed low recovery but analyte was actually still retained on the sorbent bed. Elution with pH modified methanol was more effective.

A good practice is to always save the extracted 96-well SPE plate until data analysis and interpretation are complete. Then, if incomplete analyte elution is suspected, the wells can be eluted again using a different solvent. Analyte stability on the dry particle bed after a day is generally not a concern.

12.4.2.7 Concentration Step

It is sometimes observed that an analyte is volatile during an evaporation process that uses heat and nitrogen gas. This phenomenon is another reason for low recovery of analyte. Most often, this instability is noted for the first time in the method development process. It is important to remember for effective troubleshooting that this occurrence can be another reason for low analyte recovery. When it happens, the dry-down conditions may be modified (less heat with a prolonged evaporation time) or the step can be eliminated completely by using a lower sorbent bed mass product that yields a more concentrated analyte.

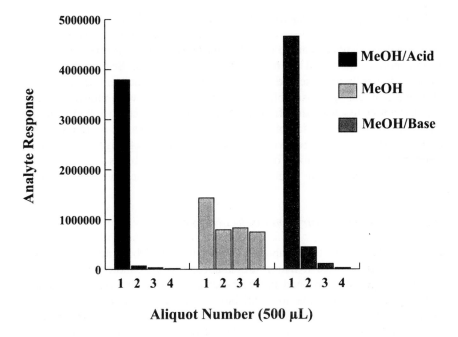

Figure 12.19. Results from an exhaustive elution study (4 sequential aliquots) demonstrate gradual release of analyte from C18 disk sorbent using methanol but a more efficient release using methanol modified with base or acid. Incomplete elution of analyte from one aliquot of methanol is the reason for low recovery in this example.

12.4.2.8 Reconstitution Step

The manner in which analytes are reconstituted following an evaporation step may be another reason for low recovery. This problem relates to solubility in the reconstitution solution, which is usually from 10–50% organic in aqueous (v/v). Solubility in organic is usually known for analytes but what about solubility in predominantly aqueous solutions? One approach to overcome this problem is that, after reconstitution, the collection plate is sealed and vortex mixed; sonication follows. Another vortex mix step is performed and the collection plate is taken to the autosampler for analysis [22]. It is also possible that residual analyte which had originally existed in perhaps 1 mL of eluate solution, after dry-down is adsorbed to the side of the tall collection vessel wall. The volume of reconstitution solution, on mixing, may be insufficient to reach up the sides of the wells to the original height.

12.4.2.9 Ion Suppression Due to Matrix Effect

A low analyte response can result from the presence of a matrix effect. This phenomenon is discussed in more detail in Chapter 6, Section 6.2.1. For example, it has been observed that a hemolyzed plasma sample, following extraction, can cause measurable ionization suppression with reduced analyte response compared with that from a regular plasma sample. Sometimes, the extraction of hemolyzed samples may result in nonlinearity of the standard curve and/or low analyte recovery. The use of a more selective sorbent (*e.g.*, mixed mode cation or anion), one which more strongly binds analyte but not matrix interferences, can often repair this assay to achieve the expected performance.

12.4.2.10 Sorbent Chemistry and Product Format

As presented early in this chapter, the selection of sorbent and the pH and solvent conditions have a major influence on extraction recovery. Existing methods should be questioned for their extraction efficiency with a particular sorbent; what previous data are available to explain why a particular sorbent was selected? Can another sorbent be more selective and yield a cleaner eluate? Can a smaller sorbent mass be used with success? Is there enough capacity with the sorbent mass chosen? When a method is developed from scratch, these issues can be appropriately addressed; however, when a method is transferred or passed on to a new analyst or laboratory and method improvement is requested, these and other questions should be asked.

12.4.2.11 Summary: Common Reasons for Low Recovery

The ability to troubleshoot solid-phase extraction methods requires a good understanding of potential problems and a keen attention to the details of exactly how a procedure is performed. Sometimes it is valuable to observe an analyst or colleague as the method is executed to watch for these details in technique. The manual preparation of the sample plate is a likely source for errors. When the procedure is entirely automated, attention is paid more toward extraction chemistry and physical placement of samples and solutions as well as the eluate collections. This chapter has introduced techniques for rapid method development and optimization. Through many examples the ability to troubleshoot methods has been introduced. Table 12.6 contains a checklist for method optimization to aid in identifying portions of a solid-phase extraction method that need attention or that are likely at fault to explain low recoveries.

Table 12.6
Checklist of issues to focus on for method optimization and troubleshooting

Method Step	Possible reasons for reduced method performance
Sample pretreatment	Sample pH is not adjusted with a buffer but a weak solution
	Analyte is not fully in the desired unionized (or ionized) form
	Analyte is too strongly bound to protein in sample matrix
	Method of addition of internal standard to sample is suboptimal
	Conjugate needs to be liberated to release analyte
Condition step	Volumes are not adequate to thoroughly wet sorbent bed
	Methanol is not displaced before sample is applied to sorbent
	Buffer may need to be passed through sorbent after water
Sample load step	Not enough bed mass for sample volume applied; low capacity
	Poor affinity of analyte for sorbent
	Flow rate too fast (not enough residence time with sorbent)
	Limit contact via reduced surface area on top of bed or disk
	Organic content of sample is >10%
Wash step	100% Water not used as first wash solution
	Insufficient % organic in wash solution to remove interferences
	Wash solvent is too strong; analyte is desorbed from sorbent
	Wash solution pH is not exploited for added selectivity
	Residual liquid remains after final wash step
Elution step	Elution solvent does not effectively desorb analyte
	Solvent pH not adjusted or is suboptimal
	Basic eluant solution not made fresh weekly
	Insufficient volume for elution
	Mismatched geometry of collection well to tips of SPE plate
	Vacuum too high; aerosolization or splashing of eluate
Concentration step	Analyte is volatile on dry-down and/or heat labile
Reconstitution step	Analyte is still bound to plastic high up in collection well
	Solubility of analyte in reconstitution solution is poor

12.5 Documentation of the Method Development Process

The lifetime of sample preparation methods has traditionally been very short, as they are constantly modified to meet changing assay requirements. Usually these changes result from needs for improved sensitivity and/or selectivity. The elucidation of a selective method early in the development cycle of a drug candidate ensures a longer method lifetime. Also, analysts have preferences in sample preparation techniques and product selection that play an important role in why methods are changed. It is well known that procedures also change hands quite often, from analyst to analyst and research laboratory to contract research organization. During these changes, methods can evolve in minor or major ways.

An observation by the author is that during these many processes of method improvement, method transfer and acceptance of methods, the original data concerning how and why that method was developed are not passed on. Each new analyst must rediscover certain key facets of how the analyte interacts with the sorbent and what constitutes efficiency in elution. In order to help the reader begin to document the results from the key experiments performed, Table 12.7 provides a summary worksheet for the method development process. This worksheet is ideally passed on with the written details of the method report.

12.6 Further Reading

The reader is encouraged to review several summaries of the method development process for further study. Two very general overviews are pertinent; one discusses the responsibilities that a developing laboratory should perform during the course of method development [23] and another introduces the SPE method development and troubleshooting process [24]. The construction of optimization charts is a very useful approach to evaluate and document each step of the extraction method. Two book chapters use optimization charts extensively in discussions of method development [25, 26]. A perspective on an overall approach to SPE method development is discussed by Simmonds, James and Wood [27].

Some of the experimental variables and procedural steps of solid-phase extraction are closely examined within various reports. Mayer and Poole [28] describe their investigation of the following variables: elution solvent and volume, drying time before elution, the conditioning step and sample flow rate. The effect of cartridge conditioning on the reversed phase extraction of basic

Table 12.7
SPE method development summary of variables examined

Research program _____   Analyte _____

Sample matrix _____ Volume _____   Internal standard _____

Sorbents   Notebook Reference
Polymer vs. bonded silica; mixed mode; brand names;
endcapping; silanols; bed mass

Sample Pretreatment   Notebook Reference
Effect of pH modification; protein binding disrupted;
volumes and ratios used

Wash Solvent Optimization   Notebook Reference
Wash solvents examined; percent organic tolerated;
changes in pH evaluated

Elution Solvent Optimization   Notebook Reference
Elution solvents examined; pH modifications made

Volume Optimization   Notebook Reference
Smallest volume yielding consistent recoveries

Elimination of the Dry-Down Step   Notebook Reference
Procedure used; volumes

Attraction Chemistry   Notebook Reference
Describe predominant attraction mechanism(s)

drugs is shared by Law [29]. The control and manipulation of the sample load and wash steps is described by Law and Weir [30]. The optimization of sample application conditions for basic drugs onto a cyano sorbent, including control of sample pH, is described by Hsu and Walters [31]. The influence of the evaporation step on loss of trace analyte from masking by solvent contaminants is illustrated by Crowley *et al.* [32].

Hennion provides a comprehensive review of the established and newer solid-phase extraction sorbents, their modes of interaction with analytes, on-line coupling with liquid chromatography and the method development process (including discussion of predictive models) [33]. Hennion also presents strategies for the prediction, rapid selection and optimization of SPE parameters with emphasis on the extraction of polar analytes [34]. Poole *et al.* review the kinetic and retention properties of solid-phase extraction devices from the perspective of method development [35]. A systematic approach based on orthogonal array designs for the optimization of SPE is described by Wan *et al.* [36]. Hughes and Gunton present a model to investigate selectivity as a function of the wash solvent composition [37].

A variety of reports from specific analytical methods contain useful information on aspects of the method development process. Fu *et al.* report how an observed assay error resulting from adsorption of analyte to the sample storage tube was identified and eliminated [38]. A book chapter discusses problems encountered during method development for the analysis of the calcium antagonist nisoldipine and how they were solved [39]. The method development process using a polymer sorbent in 96-well plates is described for verapamil and norverapamil [1], and methadone and a metabolite [40]. Naidong *et al.* share their approach for how sample preparation as well as mass spectrometric and chromatographic methods were optimized concurrently for six LC-MS/MS methods [41]. A comparative study of C18, C8 and polymer SPE sorbents, product formats (packed bed vs. disk), and retention and elution characteristics is reported by Samanidou *et al.* [42]. Svennberg *et al.* detail how they used factorial design to develop a 96-well plate method for an analyte in plasma with LC-MS analysis [3].

Acknowledgements

The author is appreciative to Gary Lensmeyer for his critical review of the manuscript and helpful·discussions. The line art illustrations were kindly provided by Pat Thompson, Woody Dells and Willy Lee.

References

[1] Y.-F. Cheng, U.D. Neue and L. Bean, J. Chromatogr. A 828 (1998) 273-281.

[2] A. Aistars, N.J.K. Simpson and D.C. Jones, Amer. Lab. 33 (2001) 50-54.

[3] H. Svennberg, S. Bergh and H. Stenhoff, J. Chromatogr. B (2002) in press.

[4] J.S. Janiszewski, M.C. Swyden and H.G. Fouda, J. Chrom. Sci. 38 (2000) 255-258.

[5] H. Simpson, A. Berthemy, D. Buhrman, R. Burton, J. Newton, M. Kealy, D. Wells and D. Wu, Rapid Commun. Mass Spectrom. 12 (1998) 75-82.

[6] M.J. Rose, S.A. Merschman, R. Eisenhandler, E.J. Woolf, K.C. Yeh, L. Lin, W. Fang, J. Hsieh, M.P. Braun, G.J. Gatto and B.K. Matuszewski, J. Pharm. Biomed. Anal. 24 (2000) 291-305.

[7] I.D. Davies, J.P. Allanson and R.C. Causon, J. Chromatogr. B 732 (1999) 173-184.

[8] L.M. McMahon, S. Luo, M. Hayes and F.L.S. Tse, Rapid Commun. Mass Spectrom. 14 (2000) 1965-1971.

[9] J.P. Allanson, R.A. Biddlecombe, A.E. Jones and S. Pleasance, Rapid Commun. Mass Spectrom. 10 (1996) 811-816.

[10] J. Janiszewski, R.P. Schneider, K. Hoffmaster, M. Swyden, D. Wells and H. Fouda, Rapid Commun. Mass Spectrom. 11 (1997) 1033-1037.

[11] D. Schütze, B. Boss and J. Schmid, J. Chromatogr. B 748 (2000) 55-64.

[12] G. Rule and J. Henion, J. Am. Soc. Mass Spectrom. 10 (1999) 1322-1327.

[13] M. Jemal, M. Huang, Y. Mao, D. Whigan and A. Schuster, Rapid Commun. Mass Spectrom. 14 (2000) 1023-1028.

[14] R. Bakhtiar, L. Khemani, M. Hayes, T. Bedman and F. Tse, J. Pharm. Biomed. Anal. 28 (2002) 1183-1194.

[15] C.Z. Matthews, E.J. Woolf and B.K. Matuszewski, J. Chromatogr. A 949 (2002) 83-89.

[16] L. Yang, N. Wu and P.J. Rudewicz, J. Chromatogr. A 926 (2001) 43-55.

[17] L. Yang, T.D. Mann, D. Little, N. Wu, R.P. Clement and P.J. Rudewicz, Anal. Chem. 73 (2001) 1740-1747.

[18] V. Cenacchi, S. Baratte, P. Cicioni, E. Frigerio, J. Long and C. James, J. Pharm. Biomed. Anal. 22 (2000) 451-460.

[19] J. Ke, M. Yancey, S. Zhang, S. Lowes and J. Henion, J. Chromatogr. B 742 (2000) 369-380.

[20] A. Pauwels, D.A. Wells, A. Covaci and P.J.C. Schepens, J. Chromatogr. B 723 (1999) 117-125.

[21] G.L. Lensmeyer, C. Onsager, I.H. Carlson and D.A. Wiebe, J. Chromatogr. A 691 (1995) 239-246.

[22] R.C. Simpson, A. Skarbek and B.K. Matuszewski, J. Chromatogr. B 775 (2002) 133-142.

[23] R.D. McDowall, LC-GC 17 (1999) 112-116.

[24] R.E. Majors, LC-GC 13 (1995) 852-858.

[25] G.L. Lensmeyer, In: S. Wong and I. Sunshine, Eds., Handbook of Analytical Therapeutic Drug Monitoring and Toxicology, CRC Press, Boca Raton (1997) 137-148.

[26] S.H. Ingwersen, In: N.J.K. Simpson, Ed., Solid-Phase Extraction: Principles, Techniques and Applications, Marcel Dekker, New York (2000) 307-330.

[27] R.J. Simmonds, C. James and S. Wood, In: D. Stevenson and I.D. Wilson, Eds., Sample Preparation for Biomedical and Environmental Analysis, Plenum Press, New York (1994) 79-85.

[28] M.L. Mayer and C.F. Poole, Anal. Chim. Acta 294 (1994) 113-126.

[29] B. Law, In: D. Stevenson and I.D. Wilson, Eds., Sample Preparation for Biomedical and Environmental Analysis, Plenum Press, New York (1994) 53-59.

[30] B. Law and S. Weir, J. Pharm. Biomed. Anal. 10 (1992) 487-493.

[31] C.-Y.L. Hsu and R.R. Walters, J. Chromatogr. 629 (1993) 61-65.

[32] X.W. Crowley, V. Murugaiah, A. Naim and R.W. Giese, J. Chromatogr. A 699 (1995) 395-402.

[33] M.-C. Hennion, J. Chromatogr. A 856 (1999) 3-54.

[34] M.-C. Hennion, C. Cau-Dit-Coumes and V. Pichon, J. Chromatogr. A 823 (1998) 147-161.

[35] C.F. Poole, A.D. Gunatilleka and R. Sethuraman, J. Chromatogr. A 885 (2000) 17-39.

[36] H.B. Wan, W.G. Lan, M.K. Wong, C.Y. Mok and Y.H. Poh, J. Chromatogr. A 677 (1994) 255-263.

[37] D.E. Hughes and K.E. Gunton, Anal. Chem. 67 (1995) 1191-1196.

[38] I. Fu, E.J. Woolf, B. Karanam, S. Vincent and B.K. Matuszewski, Chromatographia 55 *Supplement* (2002) S137-S144.

[39] S. Monkman, S. Cholerton and J. Idle, In: D. Stevenson and I.D. Wilson, Eds., Sample Preparation for Biomedical and Environmental Analysis, Plenum Press, New York (1994) 147-154.

[40] Y.-F. Cheng, U.D. Neue and L.L. Woods, J. Chromatogr. B 729 (1999) 19-31.

[41] W. Naidong, H. Bu, Y.-L. Chen, W.Z. Shou, X. Jiang and T.D.J. Halls, J. Pharm. Biomed. Anal. 28 (2002) 1115-1126.

[42] V.F. Samanidou, I.P. Imamidou and I.N. Papadoyannis, J. Liq. Chrom. & Rel. Technol. 25 (2002) 185-204.

Chapter 13

Solid-Phase Extraction: Automation Strategies

Abstract

The use of liquid handling workstations to automate the solid-phase extraction (SPE) procedure is an important step toward improving the throughput of this selective sample preparation technique. All liquid handling steps can be performed by a workstation—addition of internal standard and buffer to samples; conditioning of an SPE plate; sample transfer into the extraction plate; wash of the sorbent bed; analyte elution into a collection plate; and preparation for injection via dilution or reconstitution in a mobile phase compatible solvent. Also, sample reformatting from tubes into a microplate can be performed by a 4-/8-probe workstation. This chapter introduces procedures for the automation of the solid-phase extraction procedure in microplates using a 96-tip and a 4-/8-probe liquid handling workstation. Method development strategies for both types of workstations are also discussed.

13.1 Automation of SPE in Microplates Using a 96-Tip Workstation

13.1.1 Strategies

13.1.1.1 Introduction

A typical example of a 96-tip liquid handling workstation is the Quadra® 96 Model 320 (Tomtec Inc., Hamden, CT USA). The discussion of automation strategies here will focus on this instrument, although the general approach is similar for other 96-tip workstations and pipetting units. The Quadra 96 is a liquid handling system using 96 individual polypropylene tips for rapid throughput pipetting operations. The tips hold a maximum volume of 450 µL, which is the largest volume of all the semi-automated pipettors. It aspirates liquids from microplates or open vessels (solvent reservoirs). Each pipette tip is connected to its own independent piston. Pistons are moved up for liquid aspirating and down for liquid dispensing via a microprocessor controlled

485

stepper motor. As each piston moves up or down, it displaces an air column that causes liquid to be aspirated or dispensed. A double O-ring seals each piston to its pipette tip. The pipette tip head remains fixed in location.

The Quadra 96 has a six-position deck that is stepper motor driven to move left, right, forward and backward (Figure 13.1). All of the six positions or stages on the deck can be elevated to a user adjustable height. Elevating the stage is necessary because the tips, once in position, remain stationary. Liquids must be moved up to touch the tips for aspirating and, likewise, plates must be moved up for dispensing. The labware holding the tips is moved up for inserting tips into the head as well as for their removal.

13.1.1.2 Identify Tasks and Role for Automation

In designing an automation strategy for solid-phase extraction, the user must decide at what part of the method sequence the liquid handler first becomes involved. Preparation of the sample plate for extraction (sample plus internal standard addition and brief mixing) is performed off-line either manually or using a multiple probe liquid handling workstation. At this point, the prepared microplate can be placed onto the 96-tip workstation deck for all subsequent pipetting tasks.

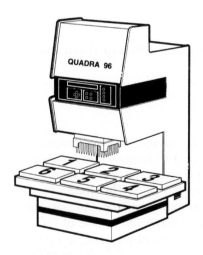

Figure 13.1. The Quadra 96 has a six-position deck that is stepper motor driven to move left, right, forward and backward. All of the six positions or stages on the deck can be elevated to a user adjustable height.

The following tasks are identified for the liquid handling workstation to perform during a solid-phase extraction procedure in microplates:

1. Condition the SPE plate with organic solvent and then water
2. Transfer sample from a microplate into the SPE plate
3. Wash SPE plate with water and any additional solutions
4. Transfer eluate solution to the SPE plate
5. Dilution of the eluate solution with an aqueous liquid for direct injection –or– reconstitution of the extract following evaporation

13.1.1.3 Plan the Deck Layout

The deck of the Quadra 96 has six positions that will accommodate the hardware (tips), extraction plate, sample plate and reagent reservoirs required to perform the procedure. Labware components sit on either a short or a tall nest, or placeholder, in each position. The actual nest type used influences the height that the stage is raised. A typical arrangement of these components on the Quadra 96 deck, as configured for SPE, is shown in Figure 13.2. Note that when more than three unique reservoirs are required for a procedure, one must be manually swapped out for another during a procedure.

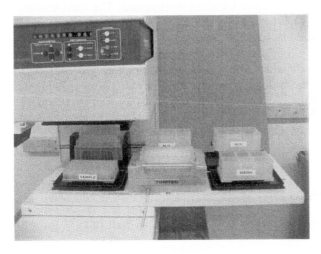

Figure 13.2. A deck layout for SPE commonly uses a tip jig in Position 1, three solvent reservoirs (Positions 2, 3 and 4), a vacuum manifold (Position 5) and a sample plate (Position 6). Photo reprinted with permission from Tomtec.

A consistent strategy for placing and using components on the deck of the Quadra 96 is recommended when using the short and tall nests (also known as reservoir and microplate nests, respectively). Each of the two nest types requires a different stage height setting, even though the same component may be placed on it. When both nest types are used within a procedure for the same component, *e.g.*, solvent reservoirs on both a short and a tall nest, it is necessary to determine and then recall a stage height setting for aspirating liquid when on a short nest and another setting when on a tall nest; errors can result in programming when a mistake is made by confusing the values for the short and tall nests. It is suggested that, for all reservoirs, standardize on one nest size (*e.g.*, tall nests); likewise, for all microplates, standardize the nest size used (*e.g.*, short nests), summarized in Table 13.1. When this strategy is followed, one common stage height setting is used for aspirating liquid from a reservoir and another for aspirating sample from a microplate.

13.1.1.4 Preparation of the Sample Plate

In order to use a 96-tip workstation, samples must be reformatted from individual source tubes into a microplate. This sample plate can be prepared off-line either manually or using a multiple probe liquid handler. Once the samples are arranged in this 8-row by 12-column array, all subsequent pipetting tasks can be performed by the liquid handling workstation. The ideal scenario is to have a 4-/8-probe liquid handling workstation prepare the full sample block (*e.g.*, plasma, internal standard and buffer), although not all laboratories are able to afford both a 96-tip and a 4-/8-probe liquid handling workstation. One way to introduce automation into a manual reformatting procedure is to deliver the samples into the wells of a sample plate but then let the Quadra 96 transfer internal standard and pH adjustment solution into each well; mixing can also be performed by aspirating and dispensing multiple times within the well. These steps can be added to the beginning of a program, if desired.

Table 13.1
A consistent strategy is recommended for placing components onto labware items

Component	Labware	Specific Labware Type
Reagent Reservoir	Microplate Nest	96-Well Tall Plate Nest
Assay Plate (*e.g.*, deep well plate)	Reservoir Nest	n/a

n/a, not applicable

13.1.1.5 Addition of Liquids and Vacuum Processing

Once the sample plate has been prepared and placed onto the deck of the Quadra 96, the first pipetting action by the instrument is to aspirate the solvents for conditioning the SPE plate—first methanol and then water. These solvents are sequentially aspirated from a reagent reservoir and dispensed into all the wells of the SPE plate; a manual vacuum step is performed in between. Note that the tip capacity of the Quadra 96 is only 450 μL (425 μL when using a 25 μL air gap). Volumes larger than this amount must be delivered using multiple pipetting steps.

A volume of sample matrix (already pretreated with internal standard and pH adjustment solution) is next aspirated from the sample plate and dispensed into the SPE plate; a manual vacuum step follows. Since the tips have now touched sample, they are shucked or ejected from the head assembly to waste and a new set of tips is loaded.

The wash solutions are next aspirated from reagent reservoirs; typically, 100% water is used as the first wash solution and after delivery it is processed through the SPE plate using vacuum. The same water reservoir as was used for the conditioning step is re-used for this wash step. A second wash solution follows from a different reagent reservoir, *e.g.*, 20% methanol in water (v/v) and then a vacuum step is performed to process the liquid through the SPE plate.

At this point in the procedure, all liquids have been collected to waste in a single compartment trough inside the manifold. This trough needs to be removed and emptied into a biological liquid waste container in preparation for the elution step. **Important Note: Vacuum is not strong enough to fully elute all residual liquid from an SPE plate. It is suggested to remove the SPE plate following the last wash step, mate it with a collection plate, and centrifuge (2000 rpm x 10 min) to remove all residual wash liquid.**

A clean collection plate is placed inside the vacuum manifold in preparation for the elution step. The dried SPE plate is repositioned on the manifold and the Quadra 96 aspirates and dispenses elution solvent into the SPE plate. The eluate is collected for further workup and analysis.

13.1.1.6 Evaporation and Reconstitution

The microplate containing the eluate is usually evaporated to dryness or, for small volume elutions, is diluted with aqueous for direct injection. Following the evaporation step, the plate containing dried residue is placed back onto a short nest on the deck of the workstation for the reconstitution procedure. A mobile phase compatible solution is placed into a reagent reservoir. The desired volume is aspirated from the reservoir and delivered into the destination wells of the microplate. The plate is removed from the deck; it is sealed, manually vortex mixed and then placed in an autosampler for injection of aliquots into a chromatographic system.

13.1.1.7 Throughput Considerations

The following example of throughput using a 96-tip workstation was previously mentioned in Chapter 11, Section 11.5.1.5, but is repeated here for completeness within this chapter and because it is a dramatic illustration of the potential of SPE automation. Janiszewski *et al.* [1] reported the first microplate SPE application on the Tomtec Quadra 96. A block of 96 samples was reported to be extracted in 10 min, which was about 30 times faster than manual single cartridge SPE methods. Elution in small volumes using an SPE disk plate allowed direct injection of eluate, saving additional time. Note that biological samples were reformatted off-line from tubes into a collection plate using a 4-probe workstation for subsequent processing on the Quadra 96. More detail about analytes, sample volumes and automation used for SPE in published applications can be found in Chapter 11, Table 11.8.

13.1.2 Method Example

A typical example of a 96-well solid-phase extraction method that is to be automated on the Quadra 96 is listed in Table 13.2. This SPE method must be translated into a series of sequential actions for the Quadra 96 to execute. A typical Quadra 96 program for performing this same method is detailed in Table 13.3. Each item, or numbered line in the program, involves an action to be performed at one of the six positions that is moved under the 96-tip aspirating/dispensing head. When pipetting, each line specifies a volume of liquid, an air gap volume, and the stage height setting as a number. Note that the stage height setting for each pipetting step depends on the specific nest type used and the preference of the analyst. Stage height settings are not included in this table as individual variations apply.

Table 13.2
Typical example of a microplate solid-phase extraction procedure

Prepare sample plate

- Aliquot 250 μL plasma from vials and transfer into 2 mL microplate

- Add IS 50 μL and briefly mix

- Add 300 μL buffer (50mM sodium phosphate, pH 4.0) and briefly mix

Condition plate: Methanol 150 μL; vacuum on/off

Condition plate: Water 250 μL; vacuum on/off

Load samples into SPE plate: 600 μL; vacuum on/off

Wash plate: Water 400 μL; vacuum on/off

Wash plate: 20% Methanol/water 400 μL; vacuum on/off

Remove SPE plate and centrifuge 2000 rpm x 10 min to remove residual liquid

Remove and empty waste trough inside vacuum manifold

Insert clean collection plate inside manifold

Place SPE plate back onto manifold

Elute: Acetonitrile 250 μL

Evaporation step (off-line)

Reconstitute dried extract in 100 μL mobile phase compatible solvent

The Quadra 96 program is run with a tip exchange step after sample delivery into the SPE plate. Some users prefer to also exchange tips after the last wash solution (right before the elution step). Pause steps are executed for tip exchanges and also for vacuum processing.

Note that four reagent reservoirs are needed to perform this procedure: methanol, water, 20% methanol in water and acetonitrile. However, only three positions are available on the deck of the instrument. Therefore, one reservoir will be swapped for another during the procedure. A schematic diagram of the components needed to perform this assay is shown in Figure 13.3.

Following the off-line evaporation step, a separate reconstitution program is run. The reconstitution procedure aspirates a fixed volume from a reagent reservoir containing mobile phase compatible solvent and delivers into the

Table 13.3
Typical Quadra 96 program for performing solid-phase extraction in the microplate format. See Figure 13.3 to identify the labware positions defined for this program. Stage height settings are not included as individual variations apply.

Condition Steps
1. Load Tips at 1
2. Mix 150 µL 3 times at Methanol in 2
3. Aspirate 150 µL from Methanol in 2, 25 µL Air Gap
4. Aspirate 10 µL [stage height 0] from 2, 0 µL Air Gap
5. Dispense 185 µL to SPE Plate in 5, 0 µL Blowout
6. Dispense 5 times at Methanol in 2 [Time Dispense of 0 µL]
7. Pause
8. Aspirate 250 µL from Water in 3, 25 µL Air Gap
9. Dispense 275 µL to SPE Plate in 5, 0 µL Blowout
10. Dispense 5 times at Water in 3 [Time Dispense of 0 µL]
11. Pause

Load Sample
12. Loop 2 times
13. Aspirate 300 µL from Sample Plate in 6, 25 µL Air Gap
14. Dispense 325 µL to SPE Plate in 5, 0 µL Blowout
15. End Loop
16. Shuck Tips at 6
17. Mix 0 µL 1 time at Water in 3, 0 µL Blowout
18. Pause

Wash Steps
19. Load Tips at 1
20. Aspirate 400 µL from Water in 3, 25 µL Air Gap
21. Dispense 425 µL to SPE Plate in 5, 0 µL Blowout
22. Dispense 5 times at Water in 3 [Time Dispense of 0 µL]
23. Pause
24. Aspirate 400 µL from Wash#2/Elution in 4, 25 µL Air Gap
25. Dispense 425 µL to SPE Plate in 5, 0 µL Blowout
26. Dispense 5 times at Wash#2/Elution in 4 [Time Dispense of 0 µL]
27. Pause

Elution Step
28. Mix 250 µL 3 times at Wash#2/Elution in 4
29. Aspirate 250µL from Wash#2/Elution in 4, 25 µL Air Gap
30. Aspirate 10 µL [stage height 0] from 4, 0 µL Air Gap
31. Dispense 285 µL to SPE Plate in 5, 0 µL Blowout
32. Dispense 5 times at Wash#2/Elution in 4 [Time Dispense of 0 µL]
33. Pause
34. Shuck Tips at 1
35. Quit at 2

TIPS in 1 *Tip Jig*	Methanol in 2 *Tall Nest*	Water in 3 *Tall Nest*
Sample Plate in 6 *Short Nest*	SPE Plate in 5 *Vacuum Manifold*	a) Wash #2 in 4 b) Elution in 4 *Tall Nest*

Figure 13.3. Typical deck layout of the Quadra 96 liquid handling workstation as configured for performing solid-phase extraction according to the method detailed in Table 13.2. Note that the reservoir in Position 4 is swapped during the procedure since this location must be shared.

plate containing dried extracts. If desired, this reconstitution step can be added to the end of the Quadra 96 program. However, it is suggested to maintain the reconstitution as a separate program so that the instrument is not occupied and unavailable for the length of time that it takes to dry down the eluate, typically 15–45 min. When the Quadra 96 is used daily by multiple analysts, there is often a waiting list or queue to perform other sample preparation assays. Closing the program at the end of the SPE procedure simply allows more efficient utilization of the instrument. The analyst who needs to reconstitute dried extracts then walks up to the instrument at the next available time opening, performs the procedure, and leaves the Quadra 96 available for the next procedure or analyst. Some tips and tricks in using the Quadra 96 for performing SPE in microplates are discussed next.

13.1.3 Tips and Tricks

13.1.3.1 Containing the Solvent Vapors

The most important factor to consider when using the Quadra 96 for SPE is that placing an organic solvent into an open reagent reservoir holding about 300 mL volume will allow volatile solvent vapors to rapidly escape and flood the surrounding atmosphere. In the interest of worker and laboratory safety, the Quadra 96 must be totally contained within a fume hood or surrounded with an efficient enclosure to contain and vent the solvent vapors.

13.1.3.2 Using the Polypropylene Tips

Many times, aspirating an organic solvent using the polypropylene tips in their dry state will result in leaking solvent due to weak surface tension. When pipetting organic solvents, several sequential steps are recommended:

1. Mix 3 times using the volume to be delivered and dispense back into the reagent reservoir
2. Aspirate the desired volume
3. Aspirate a 10 μL air gap at the tip ends (at a stage height setting of 0) and then move the deck to the desired position for the dispense step
4. After dispensing, remove solvent vapors from above the 96 piston head by using a time dispense of 0 μL five times

The tips hold a maximum volume of 450 μL. When a volume greater than 425 μL is to be dispensed (a 25 μL air gap in addition to the solvent volume is suggested), use multiple aspirate and dispense steps within a loop procedure. For example, when a volume of 1.5 mL is to be dispensed, specify a volume of 375 μL (plus a 25 μL air gap equals 400 μL) to be dispensed four times.

Tips are usually exchanged after any step that touches drug or analyte in solution to avoid contaminating a reagent reservoir in subsequent steps. The tips can be removed, or "shucked," into the "tip jig" but sometimes it may be convenient to shuck them into a used microplate. For example, after transfer of the samples to an SPE plate, the tips may be shucked into the empty sample plate in Position 6. Then, the plastic plate and tips can be removed in one step and disposed into a proper biohazardous waste receptacle. Shucking the tips into a different position than the tip jig provides a visual clue that new tips are to be loaded since that position remains empty.

Note that using the Quadra 96 to add internal standard to the sample plate delivers this solution to all 96 wells. In those instances where one or more double blanks (zero analyte level and no internal standard) are designated in the plate, the pipet tips corresponding to those exact well locations can simply be removed from the tip rack prior to pipetting. This selective tip removal approach can also be used for the reconstitution steps to keep selected wells free of solvent. System checks, blank mobile phase or other solution can be manually delivered into dry wells of the final analysis plate following reconstitution. Several tips may be removed manually from the tip rack but removal of too many will cause the assembly to become unbalanced and an error may occur on loading.

13.1.3.3 Centrifuging the Extraction Plate after the Final Wash Step

Vacuum is rarely strong enough to fully elute all residual liquid from an SPE plate. It is suggested to remove the SPE plate following the last wash step, mate it with a collection plate, and centrifuge (2000 rpm x 10 min) the combination to remove all residual wash liquid. When liquid is not completely removed from the sorbent bed, residual volume is added to the elution volume, resulting in longer evaporation time and/or inconsistent collected volumes.

13.1.3.4 Small Volume Elution

The use of low sorbent bed mass SPE plates (2–15 mg) allows elution in very small volumes (25–100 μL). The use of vacuum to process these small volumes through a disk plate or packed particle plate is often insufficient to fully isolate the total volume; it is more effective to use centrifugation. In this case, the SPE plate is mated directly with a collection plate and this combination is placed into the bottom half of a manifold (without the remaining components). Liquid is delivered directly into the plate and then the mated combination is carefully removed and taken to a centrifuge for processing. The manifold bottom provides the necessary stability to the mated plates to remain in position as liquid is delivered.

13.1.3.5 Vacuum Manifold Configurations

It is important to properly place the exit tips of the SPE plate slightly into or very near the top of the collection plate chosen for use inside the vacuum manifold. Shims should be used inside the manifold to fine-tune the height adjustment. When using the Tomtec manifold (#196-503), it is important to realize that the middle piece is used with short skirt SPE plates but is removed for use with tall skirt plates and modular plates such as the ISOLUTE® Array® (IST Ltd., Hengoed, United Kingdom, now an Argonaut Technologies Company). The modular VersaPlate™ (Varian Inc., Harbor City, CA USA) requires the use of the Varian manifold #1223-6101 due to its tall height.

13.1.3.6 Using Move Steps to Change Deck Position

Following a liquid transfer step into the SPE plate, the tips stay positioned right above the plate and manifold when it is time to activate the vacuum. It is then difficult to look on top of the plate to monitor liquid passage through the wells. It is advisable to move the tips out of the way by inserting a move step right before the pause step that cues the operator to run the vacuum. For example, execute a move step by performing a mix step with zero volume specified at Position 3, out of the way of the manifold, so that the SPE plate and manifold are clearly unobstructed for optimal viewing. A move step is also beneficial when loading tips; move the deck to Position 3 and then execute a pause step so that easy access to Position 1 is permitted to insert or remove tips.

13.1.3.7 Practice the Method before Processing Sample Unknowns

After a program has been written, it is advised to always run it once using blank solutions for practice. While the Simulation Mode on the Quadra 96 is useful for checking math errors in volumes used, it does not provide the needed movement to actually observe solvent leaking from tips or to identify the need to insert a move step. As the practice program is executed, take notes as to what steps to modify or add. The program cannot be edited while it is running; after it is finished, modify the program with the latest changes and then save the final version.

As an example, the following items were noted while observing a practice run of a newly created SPE program on the Quadra 96:

(1) Methanol drips and liquid remains on the tips after the methanol conditioning step. Insert 10 µL air gap after methanol aspiration and dispense the full volume at a stage height where tips contact the dispensed liquid.

(2) After the water conditioning step, the stage stays elevated and the tips remain resting in liquid for the vacuum step. Add a move step back to the water reservoir in Position 3 before the pause step for vacuum.

(3) After the sample load step, add a pause step so that the vacuum can be run. Without this pause step, vacuum is still pulling sample through the sorbent bed while the workstation starts the wash solvent addition.

(4) In the elution step with acetonitrile, the tips do not descend low enough into the SPE plate wells to top off the liquid. Adjust stage height setting to a greater value.

(5) After the program quits, it stops at Position 5. Add a move step to Position 2 so that the SPE manifold is forward and there is enough clearance to easily remove the SPE plate and eluate collection plate.

13.1.3.8 Use a Consistent Strategy for Deck Positions and Stage Height Settings

A consistent strategy aids the execution of SPE methods on the Quadra 96 liquid handling workstation, especially among multiple users:

- Microplates (sample plates and dried eluate plates for reconstitution) are placed on short nests
- Reagent reservoirs are placed on tall nests
- Tips are placed in Position 1
- A vacuum manifold, when used, is placed in Position 5
- The sample plate is placed in Position 6
- Shims are kept handy near the workstation for use inside the manifold
- A table listing stage height settings for each component and nest combination is used as a reference when writing methods

13.1.4 Applications Using a 96-Tip Workstation

Applications and methodology for the use of solid-phase extraction plates are described thoroughly in Chapter 11 and the reader is referred to Section 11.5 for details. A representative list of semi-automated applications performed using a 96-tip workstation with solid-phase extraction plates is provided in Chapter 5—Table 5.2 lists published references that specifically used the Quadra, Multimek or Personal Pipettor workstations. More detail about analytes, sample volumes, SPE plate chemistries and elution solvents used in reported applications is found in Chapter 11, Table 11.8. A published review by Rossi and Zhang provides an additional perspective on the automation of 96-well solid-phase extraction using a 96-tip workstation [2].

13.1.5 Automating Method Development Using a 96-Tip Workstation

Method development strategies for SPE have been described in Chapter 12. Most tasks can be automated using the Quadra 96 but keep in mind that selective solvent aspirating and dispensing to separate destination wells is not possible. When all 96 tips enter a reagent reservoir, all tips will deliver that reagent. Also note that variable volume delivery cannot be performed. In order to designate specific grids of tips to aspirate specific solvents (*e.g.*, an evaluation of six organic solvents for elution efficiency), the reagent reservoir needs to be customized. Instead of an open one compartment polypropylene reservoir, a multiple component reservoir can be used as shown in Figure 13.4.

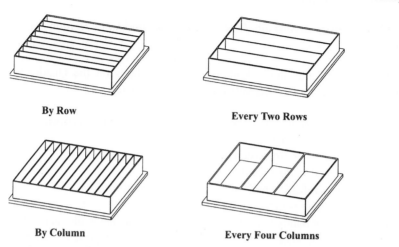

By Row

Every Two Rows

By Column

Every Four Columns

Figure 13.4. Custom reagent reservoirs for the Quadra 96 allow the placement of specific solutions or solvents into separate compartments.

These custom reservoirs are available from Tomtec in any configuration desired, *e.g.*, by row, by column, every two rows or every four columns. When one of these custom reservoirs is not available, a 2 mL microplate provides the ultimate custom reservoir; simply place different solutions into corresponding wells to achieve the pattern desired and use it as the reagent reservoir in the interim.

13.2 Automation of SPE in Microplates Using a 4-/8-Probe Workstation

13.2.1 Strategies

13.2.1.1 Introduction

The strategies for automating solid-phase extraction in the microplate format are thoroughly discussed in Section 13.1. Although this previous section focused on use of a 96-tip liquid handling workstation, the strategies presented are also pertinent for 4-/8-probe workstations. This section will focus only on the unique procedural aspects of performing automated solid-phase extraction using a 4-/8-probe workstation and is thus more concise. Typical examples of 4-/8-probe liquid handling workstations include the MultiPROBE II (Packard Instruments, Meriden, CT USA, now a part of PerkinElmer Life Sciences), the

Biomek 2000 (Beckman Coulter, Fullerton, CA USA) and the Genesis™ Series (Tecan US, Research Triangle Park, NC USA). This discussion of automation strategies will focus on the MultiPROBE II, although the general approach is similar for other workstations. Features, specifications and capabilities for these instruments as a group are described in Chapter 5, Sections 5.4.3 (semi-automated solutions) and 5.4.4 (fully automated solutions by addition of a gripper arm).

13.2.1.2 Deck Layout and Labware

A typical MultiPROBE deck configuration for solid-phase extraction in microplates (Figure 13.5) consists of the following labware components:

- Test tube or vial rack to contain the source samples
- Solvent troughs to contain internal standard solution, pH adjustment solution, methanol, water, 20% methanol in water, acetonitrile and the mobile phase compatible solvent (for reconstitution) or aqueous diluent
- Microplate that becomes the sample plate
- Vacuum manifold configured with SPE plate
- Disposable tip rack

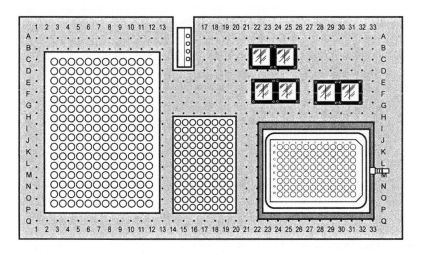

Figure 13.5. A deck layout for SPE commonly utilizes a test tube rack, a 96-well microplate, solvent troughs, and a solid-phase extraction plate on top of a vacuum manifold.

Disposable tips are used for all sample transfers and fixed probes are used for all solvent transfers. Typically, disposable tips having a 200 μL volume capacity are used; for larger sample volumes a 1 mL tip can be used. However, note that the 1 mL tips require the 4-panel extended deck platform, not the 2-panel regular deck.

13.2.1.3 Preparation of the Sample Plate

Since 4- and 8-probe liquid handling workstations accommodate tubes, the source samples can be placed directly onto the deck of the instrument in the appropriate rack. The probes are able to change their center-to-center spacing from tube width to microplate width (9 mm). Note that a recommended pretreatment step before using automation for pipetting is to first centrifuge the sample tubes to pellet any fibrin clots and particle debris at the bottom. The freeze/thaw cycle for plasma is known to introduce these clots and proper efforts such as centrifugation must be performed so that pipetting errors are not introduced. It is also possible to work with source plasma samples that are contained in a microtube rack rather than individual test tubes; in some studies the specification is made in the protocol to collect plasma in the microplate format and these individually prenumbered microtubes are a common choice.

An aliquot is aspirated, using unique disposable tips, from each set of sample tubes (4 or 8 at a time) and dispensed into the desired positions within the microplate wells. The well locations for the aspirating and dispensing steps are specified in the well maps for the source (test tube rack) and destination (microplate) labware. Once all the samples are transferred into the microplate, the internal standard (IS) is delivered to those same wells. The IS can be dispensed using fixed tips in multiple dispense mode but user preference is to use disposable tips for each transfer set. More accurate pipetting is achieved for small volumes of 50 μL or less using disposable tips. A recommended option after IS delivery is to perform a post-dispense mix to ensure adequate sample integration before the addition of buffer or pH adjustment solution to each well. Now that the sample plate is prepared, it can be taken to a 96-tip workstation for processing or used on the MultiPROBE. Note that, since the MultiPROBE utilizes tubes efficiently, samples can all be kept in tubes and then transferred to the SPE plate; user preference guides this decision.

Calibration standards and QC samples are usually placed into tubes and arranged within the sample tube rack so that they are delivered to a microplate in the same manner as the samples. However, a feature of a multiple probe liquid handling workstation is that the calibration standards and QC samples can actually be prepared by the instrument, rather than prepared off-line manually. Descriptions of this procedure for preparing standards and QC samples with a Biomek 2000 multiple probe liquid handling unit have been reported by Watt *et al.* [3] and Locker *et al.* [4]. More information on this application is described in Chapter 5, Section 5.6.4.2.

13.2.2 Tips and Tricks

13.2.2.1 Centrifuging the Extraction Plate after the Final Wash Step

Vacuum is rarely strong enough to fully elute all residual liquid from an SPE plate. It is suggested to remove the SPE plate following the last wash step, mate it with a collection plate and centrifuge (2000 rpm x 10 min) the combination to remove all residual wash liquid. When liquid is not completely removed from the sorbent bed, residual volume is added to the elution volume, resulting in longer evaporation time and/or inconsistent collected volumes.

13.2.2.2 Small Volume Elution

The use of low sorbent bed mass SPE plates (2–15 mg) allows elution in very small volumes (25–100 µL). The use of vacuum to process these small volumes through a disk plate or packed particle plate is often insufficient to fully isolate the total volume; it is more effective to use centrifugation. In this case, the SPE plate is mated directly with a collection plate and this combination is placed into the bottom half of a manifold (without the remaining components). Liquid is delivered directly into the plate and then the mated combination is carefully removed and taken to a centrifuge for processing. The manifold bottom provides the necessary stability to the mated plates to remain in position as liquid is delivered.

13.2.2.3 General Solvent Dispensing

The use of system liquid (deionized distilled water) as the source for the water conditioning and wash step greatly speeds solvent delivery as the probes do not need to refill from a solvent trough and go through a wash/rinse cycle. The use of the multiple dispensing feature also speeds the solvent delivery process; enough volume is aspirated from a methanol trough to deliver to 24 wells, *e.g.*, before refilling is required. However, ensure that the volume aspirated for multiple dispensing does not exceed the volume of the system tubing or else the excess is taken up by system liquid, not the intended solvent. A six-way valve can be added to the system as an alternative to using individual solvent troughs.

13.2.2.4 Use of Partial SPE Plates

Partial plates can effectively be used on a multiple probe workstation that can specify the destination wells; unused wells remain dry or used wells are bypassed. Vacuum strength across the SPE plate can be maintained by the occlusion of unused wells with adhesive backed tape.

13.2.3 Applications

Applications and methodology for the use of solid-phase extraction plates are described thoroughly in Chapter 11 and the reader is referred to Section 11.5 for details. A representative list of semi-automated applications performed using a 4-/8-probe workstation with solid-phase extraction plates is provided in Chapter 5—Table 5.2 lists published references that specifically used the Biomek 2000, Genesis, SPE 215 and MultiPROBE units. More detail about analytes, sample volumes, SPE plate chemistries and elution solvents used in reported applications is found in Chapter 11, Table 11.8.

Full automation for 96-well solid-phase extraction using a liquid handling workstation is achievable by the addition of a gripper arm to the processing configuration. This gripper arm, also called a robotic manipulator arm (RoMA) by Tecan, is able to assemble and disassemble a vacuum manifold, offering total walk away automation. A published application which used a gripper arm for SPE is reported by Schütze *et al.* [5]. Schütze described two main advantages of the Genesis workstation as (1) exact vacuum control from 30 mbar to about 700 mbar (approx. 1–21 in Hg), and (2) the additional capability offered by the robotic manipulator arm. The Biomek 2000 also has a gripper arm and two publications report its use for 96-well SPE [6, 7].

13.2.4 Automating Method Development Using a 4-/8-Probe Workstation

Method development strategies for SPE have been described in Chapter 12. The MultiPROBE II is able to perform all method development tasks, including the one experiment not possible with 96-tip workstations—elution volume optimization. Varying volumes can be delivered to specific destination wells in two ways: (1) unique well maps per replicate set with different program lines for execution and (2) a text file is linked that contains full information on the source solvent and the destination wells for dispensing.

When a laboratory is fortunate to have a 4-/8-probe and a 96-tip workstation, the 4-/8-probe instrument can be dedicated to sample reformatting to supply the 96-tip unit and to method development experiments as described in Chapter 12.

13.3 Full Robotics Integration with Multiple Task Modules

Strategies for the automation of solid-phase extraction procedures using traditional liquid handling workstations have been described. More complete automation of this application is achieved by utilizing a robotic or gripper arm to perform the task of disassembling and reassembling the vacuum manifold for the elution step, as described. However, additional off-line steps are still manually required, such as evaporation of eluate solution and centrifugation. The fully automated approach requires the ability to interface with multiple components and represents the ultimate in a custom configured application.

Biddlecombe and Pleasance have published their approach to 96-well solid-phase extraction that refines and further automates the traditional process using liquid handling workstations. They chose to interface this system with a Zymark robot (Zymark Corporation, Hopkinton, MA USA). The potential to analyze a complete clinical study (~800 h) in under 48 h using LC-MS/MS techniques was presented at the 1996 ISLAR conference [8] and summarized in a book chapter [9]. In order to automate the entire procedure, a customized SPE station was developed to condition the 96-well SPE plates (blocks), wash and elute into the appropriate collection plate. Conditioned blocks were transferred to the MultiPROBE for sample loading and then back to the SPE station for the wash and elution steps. Eluate in the collection plate was transferred to a cooled storage carousel in an autoinjector which could initiate 8 consecutive runs. Throughput of the developed system was reported as 12–16 blocks prepared in one working day, enough to supply two mass spectrometers. A refined perspective on this fully automated approach, including a discussion of enhancements made to the system, was presented by this same research group two years later at another ISLAR conference [10]. An application utilizing this custom built robotic system is described for the determination of salbutamol (albuterol) enantiomers in human plasma and urine [11].

When fast sample preparation in 96-well plates is combined with fast chromatographic analysis and detection, outstanding gains in throughput can be achieved. Deng *et al.* performed sample preparation using 96-well solid-phase extraction, followed by the use of four monolithic columns for parallel separation [12]. In this application, which also used multiplexed inlets on the mass spectrometer (discussed in Chapter 14, Section 14.4.4.2), overall throughput was reported as 30 s per plasma sample with gradient elution on a 10 cm long LC column; 1152 samples (twelve 96-well plates) were analyzed within 10 h. A Zymark track robot system interfaced with a Tecan Genesis liquid handling workstation was used for the simultaneous solid-phase extraction of four extraction plates in a fully automated manner.

Acknowledgments

The line art illustrations were kindly provided by Willy Lee.

References

[1] J. Janiszewski, R.P. Schneider, K. Hoffmaster, M. Swyden, D. Wells and H. Fouda, Rapid Commun. Mass Spectrom. 11 (1997) 1033-1037.
[2] D.T. Rossi and N. Zhang, J. Chromatogr. A 885 (2000) 97-113.
[3] A.P. Watt, D. Morrison, K.L. Locker and D.C. Evans, Anal. Chem. 72 (2000) 979-984.
[4] K.L. Locker, D. Morrison and A.P. Watt, J. Chromatogr. B 750 (2001) 13-23.
[5] D. Schütze, B. Boss and J. Schmid, J. Chromatogr. B 748 (2000) 55-64.
[6] S.H. Hoke II, J.D. Pinkston, R.E. Bailey, S.L. Tanguay and T.H. Eichhold, Anal. Chem. 72 (2000) 4235-4241.
[7] T.H. Eichhold, R.E. Bailey, S.L. Tanguay and S.H. Hoke II, J. Mass Spectrom. 35 (2000) 504-511.
[8] R.A. Biddlecombe and S. Pleasance, Proceedings International Symposium on Laboratory Automation and Robotics, Boston, MA USA (1996) 445-454.
[9] S. Pleasance and R.A. Biddlecombe, In: E. Reid, H.M. Hill and I.D. Wilson, Eds., Drug Development Assay Approaches, Including Molecular Imprinting and Biomarkers, The Royal Society of Chemistry, Cambridge (1998) 205-212; Methodological Surveys in Bioanalysis of Drugs, Volume 25, E. Reid, Ed.
[10] R.A. Biddlecombe and S. Pleasance, Proceedings International Symposium on Laboratory Automation and Robotics, Boston, MA USA (1998).
[11] K.B. Joyce, A.E. Jones, R.J. Scott, R.A. Biddlecombe and S. Pleasance, Rapid Commun. Mass Spectrom. 12 (1998) 1899-1910.
[12] Y. Deng, J.-T. Wu, T.L. Lloyd, C.L. Chi, T.V. Olah and S.E. Unger, Rapid Commun. Mass Spectrom. 16 (2002) 1116-1123.

Chapter 14

On-Line Sample Preparation: High Throughput Techniques and Strategies for Method Development

Jing-Tao Wu
Millennium Pharmaceuticals, Cambridge, MA USA

David A. Wells
Sample Prep Solutions Company, St. Paul, MN USA

Abstract

This chapter introduces some of the most popular and proven on-line sample preparation techniques used for pharmaceutical bioanalysis. While the conventional procedures are performed off-line in discrete steps before the analysis, on-line techniques combine the sample preparation and analysis steps into one integral process. The key component among the numerous formats of on-line sample preparation methods is a specially made extraction column. It is generally the characteristic nature of the extraction column that distinguishes one method from the other. Four types of extraction columns to be discussed in this chapter are: turbulent flow, restricted access media (RAM), monolithic and immunoaffinity. An additional column can be utilized after the extraction column to provide enhanced separation of components. Column switching is used to activate this separation function. Techniques and applications for turbulent flow, restricted access media and immunoaffinity columns will be discussed. On-line solid-phase extraction in a single use, disposable cartridge will also be discussed; an analytical column follows the extraction cartridge for additional separation. Comparisons of these different on-line choices will be made, as well as traditional off-line techniques for sample preparation. Strategies for method development will be introduced with the focus on the use of turbulent flow chromatography and disposable solid-phase extraction cartridges. Method optimization and troubleshooting of on-line methods are also presented.

14.1 Understanding the Technique

14.1.1 Introduction

Sample preparation and liquid chromatography-mass spectrometry (LC-MS) analysis are sequential processes within bioanalysis. The sample preparation techniques described in previous chapters (protein precipitation, PPT; liquid-liquid extraction, LLE; and solid-phase extraction, SPE) are commonly performed as separate steps preceding the analysis. Since these procedures are completed independently they are referred to as off-line techniques. In an effort to combine the two steps of sample preparation and LC-MS analysis into one integral process, on-line sample preparation techniques have been developed. In this approach, samples in biological fluids are injected directly onto a sample extraction column or cartridge which is part of a modified LC-MS system. Through various mechanisms, the analyte is retained while matrix components are diverted to waste. A change in flow path is made and the extracted analyte is then eluted for further separation and detection by mass spectrometric techniques. The samples injected on-line using this type of sample preparation system generally require minimal pretreatment or sometimes no pretreatment at all.

Numerous approaches for on-line sample preparation have been reported in the scientific literature. The aim of this chapter is to discuss those approaches that have been frequently used with success or have shown great promise.

14.1.2 Advantages

14.1.2.1 Greater Throughput

On-line sample preparation has become popular for bioanalysis in recent years due to its unique advantages compared with traditional sample preparation methods. By eliminating one or more off-line steps, the on-line methods provide a significant improvement in throughput by saving both time and labor. The hands-on analyst involvement with the sample preparation step is now considerably reduced. Although the on-line sample preparation and the LC-MS analysis are still fundamentally sequential steps, significant time savings can be realized by positioning the extraction and separation steps in parallel (dual-column mode). While the LC-MS system is detecting and analyzing one sample, the on-line sample preparation component can be programmed to extract the next sample and prepare it for injection as soon as the system is ready to accept it.

14.1.2.2 Fast Turnaround for Method Development and Sample Reanalysis

The improvement in sample throughput is even more evident in a non-GLP (Good Laboratory Practices) drug discovery environment. Here, sample batch sizes are relatively small (*e.g.*, 30–60 samples) and time is simply not available to develop selective sample preparation and analytical methods for each new series of analytes; new analogs within chemical drug classes can be encountered as often as weekly. Therefore, a rapid method development approach is essential in order to remove this potentially rate limiting step.

During the method development process, usually only a few samples need to be prepared and analyzed. Using off-line sample preparation methods, a reduction in sample number does not result in a proportional reduction in sample preparation time because this small number of samples still must be analyzed utilizing the complete multi-step procedure. However, using an on-line method, a much faster turnaround can be achieved; the analysis time is directly proportional to the number of samples.

Another related need for improved turnaround time is in the sample reanalysis procedure. When a sample concentration is found to be above the limit of quantitation another aliquot must be analyzed; using an on-line technique, only a few samples need to be prepared and again subjected to analysis.

14.1.2.3 Minimal Sample Loss by Reducing Nonspecific Adsorption

Sample loss due to nonspecific adsorption onto the surface of a sample tube or vial is a commonly encountered problem during off-line sample preparation for bioanalysis. The result is often poor quantitation and loss of sensitivity. Non-specific adsorption most often takes place during the steps after the analyte has been extracted from the biological matrix, since the analyte has lost the complex matrix for its binding. Using an on-line technique, the analyte stays within its biological matrix until it is injected into a chromatographic system. Sample loss due to nonspecific adsorption is minimized using this approach.

14.1.2.4 Little Investment in Additional Hardware

Most of the on-line sample preparation work involves only the use of conventional liquid chromatography equipment which is already in place within pharmaceutical research and development laboratories. A large capital purchase of new equipment is unnecessary to incorporate on-line sample preparation techniques; specialty extraction columns are acquired as consumable products from operating budgets. Scientists with basic skills in chromatography can comprehend and implement on-line sample preparation very quickly with minimal instruction. Many groups have set up their own on-line sample preparation systems and tailored them to unique applications. Some additional attractive features using on-line sample preparation methods are simplicity, flexibility and low cost relative to conventional off-line methods.

14.1.3 Disadvantages

Although many improvements have been made during the past few years, the ruggedness of an on-line system still cannot fully match that of the conventional systems using off-line sample preparation approaches; that is, unless a relatively more extensive sample pretreatment procedure such as protein precipitation is employed at the expense of throughput. Also, sample to sample carryover is generally more notable with on-line methods. Sometimes additional effort needs to be made during method development in order to reduce the carryover for on-line applications. With turbulent flow chromatography, one of the most frequently used on-line methods, solvent consumption and disposal may also be considered a disadvantage. In addition, if a high cost mass spectrometer is employed as the detector, its efficiency may be compromised since it is occupied with the on-line extraction process, particularly when an additional separation column is employed (dual-column mode). Some new approaches based on staggered or parallel sample introduction have been developed to address this issue and these will be discussed later in this chapter.

14.1.4 Usage

On-line sample preparation techniques, also referred to as direct sample analysis, have been used extensively for applications in clinical drug analysis, as reviewed by Wong [1]. However, as innovations in technology have occurred, and the need has arisen for greater throughput and rapid method development, the use of on-line sample preparation has now become widespread within the pharmaceutical industry. Indeed, on-line methods have been reported for bioanalysis in almost every stage of the drug discovery and development process, including *in vitro* microsomal stability screening, Caco-2 cell permeability studies, pharmacokinetic screening for bioanalytical support to drug discovery, preclinical GLP studies and clinical studies. In most of these functions, assay performance that is comparable to that obtained using conventional off-line sample preparation methods has been demonstrated. A resurgence of interest in column switching techniques, rapid chromatography and direct injection techniques has occurred in the support of bioanalytical drug discovery programs. The reader is referred to some general overviews that describe this rapid growth and recent popularity [2, 3].

14.2 High Throughput On-Line Sample Preparation Using a Single Column

14.2.1 Introduction

Among the numerous on-line sample preparation methods, the key component is a specially made extraction column or cartridge. It is generally the characteristic nature of the extraction column that distinguishes one method from the other. In some cases, this extraction column can serve two functions—on-line sample preparation and analysis. In most other cases an additional column is employed and column switching is used to activate the separation function (dual-column mode). Recently, the use of multiple columns in either parallel or staggered modes has been reported to improve the throughput for on-line sample preparation, as will be discussed in Section 14.4.

The simplest version of on-line sample preparation for bioanalysis is the single-column mode. Based on the extraction mechanism or column characteristics, four different primary approaches for direct injection are used: turbulent flow chromatography (TFC), restricted access media (RAM), monolithic columns and immunoaffinity extraction (IAE).

14.2.2 Turbulent Flow Chromatography (TFC)

14.2.2.1 Technology

TFC was first investigated by Pretorius and Smuts in 1966 as a fast analysis tool [4]. Their work showed that a turbulent flow would replace laminar flow as the dominant profile when a sufficiently high linear flow rate of mobile phase was attained (8–10 cm/sec). The generation of this turbulent flow allowed much more rapid mass transfer in the mobile phase, resulting in a significant reduction in plate height. A theoretical discussion of flow dispersion in chromatography, with reference to turbulent flow chromatography, is reported by Knox [5]. Further discussions of turbulent flow as a means for enhancing radial diffusion in open tubular columns are reported by Sumpter and Lee [6] and Bauer [7]. A general overview of the technology is presented by Oberhauser *et al.* [8].

Early reports of the use of TFC as a high throughput tool for direct injection in bioanalysis were published in 1997 and 1998 [9, 10]. Its performance has been described in general terms as best suited for repetitive analyses requiring modest separating power when throughput is critical [8]; these qualities fit well with the requirements for drug discovery bioanalytical support where it was first introduced. Within a few years, it quickly became a popular on-line sample preparation method for bioanalysis as its wide applicability for a diverse range of analytes in biological fluids was demonstrated. TFC has also been utilized successfully outside the drug discovery environment for preclinical and clinical applications. Note that TFC works well for analytes that are detected using LC-MS/MS [liquid chromatography (LC) with tandem mass spectrometry (MS/MS)] technology; it is not as appropriate for single quadrupole LC-MS detection using biofluids due to reduced selectivity from matrix interferences.

The U.S. Patent and Trademark Office has awarded a total of six patents as of July 2002 to Cohesive Technologies (Franklin, MA USA) for various aspects of the discipline of turbulent flow chromatography [11–16]. Cohesive has demonstrated a research investment and a sustained commercial effort in supporting and expanding the scope of this technology within the pharmaceutical and clinical research environments. While outside the scope of this text, note that TFC has also been applied to the determination of pesticides in water [17].

The key distinguishing feature in turbulent flow on-line extraction is the use of an ultrahigh linear flow rate that may possibly generate eddy currents and some turbulent profile in the solvent flow. When used for on-line extraction, the eddy currents are believed to play a significant role in improving ruggedness [18]. The generation of an ultrahigh flow rate is made possible by use of an extraction column having a small internal diameter (i.d.) and large particle size packing material (*e.g.*, 25–30 μm). The typical flow rate used is 4–5 mL/min with a typical column dimension of 1 x 50 mm.

The nature of the flow profile (laminar or turbulent) in this method is still under debate within the scientific community. "Turbulent flow" may not be the most accurate term for this method [19] but its use serves the purpose of clearly distinguishing this technique from other on-line sample preparation methods. Therefore, the term turbulent flow is adopted throughout this chapter to describe all on-line sample preparation work based on an ultrahigh flow rate.

14.2.2.2 System Configuration

The sorbent chemistry used in turbulent flow columns (typically 1 x 50 mm i.d.) is largely based on traditional reversed phase materials. Both spherical and irregular packing materials are in use. Typically the particles are 25 µm or larger in size, resulting in a reduced column backpressure at high flow rates. These large packing materials allow the use of end column frits having a large pore size (typically 10 µm or larger). The combination of the large pore size of the frits and the large particle size packing materials make it possible for large molecules in biological fluids to pass through without plugging the column.

A diagram of a single-column turbulent flow on-line extraction system is shown in Figure 14.1. The hardware configuration of the system is similar to a regular LC-MS/MS system except that (1) a turbulent flow extraction column is used instead of a regular analytical column and (2) a divert valve and a flow splitter are employed between the column and the mass spectrometer. The flow and gradient conditions and the programming aspects, however, are different from those used in conventional LC separation.

14.2.2.3 Procedure

When a sample in a biological fluid is injected into this system, the liquid chromatograph delivers a high flow rate (~5 mL/min) of a highly aqueous solvent (typically 100% water with pH or salt additives as needed). The divert

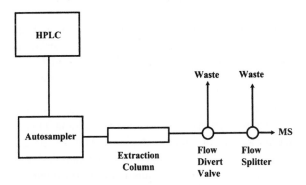

Figure 14.1. Schematic diagram of a single column turbulent flow on-line extraction system.

valve is set to waste for this load and wash period, which normally lasts for less than a minute. When a reversed phase sorbent is used in the extraction column, the analyte is retained via partitioning while hydrophilic components in the sample matrix are washed from the column and sent to waste. Most of the large molecules in the sample such as proteins are also washed from the column; two mechanisms are generally thought to be operating here—the physical size of proteins prevents them from entering the pores of the packing material to interact with the stationary phase and, since proteins are slowly diffusing molecules, the conditions of high flow rate keep them from accessing the inner surface area.

After the load and wash steps are completed, the composition is changed to a high organic solvent at the same flow rate. A divert valve is switched to direct the flow into a splitter, sending a portion of the flow into the mass spectrometer. Depending on the type of ion source and the ionization mode used, the flow is usually split from ~5 mL/min down to 0.3 to 1.0 mL/min. Now, the retained analyte is eluted from the column with some limited separation and is detected by the mass spectrometer. The elution step usually takes less than one minute. The composition is changed back to high aqueous to equilibrate the column in preparation for the next sample. Again, this equilibration step can be completed in less than one minute. Therefore, the entire on-line extraction and analysis procedure can usually be accomplished in about 3 min for each sample. This single-column mode provides the distinct advantage of high throughput and simplicity.

14.2.2.4 Applications

The single-column turbulent flow chromatography technique has been successfully reported for on-line sample preparation in many bioanalytical applications, as listed in Table 14.1. Ayrton *et al.* was one of the first groups to report the use of single-column TFC for the quantitation of pharmaceutical compounds in plasma [9]. An assay was validated in the range of 5–1000 ng/mL in plasma and used to support toxicokinetic studies. Similar applications were reported by different groups that used single-column TFC as a high throughput tool to quantify single components in plasma [10, 17, 20–22, 24–26]. TFC has also been used to quantify compounds in tissue samples. Van Eeckhout *et al.* reported the use of TFC for the determination of tetracyclines in bovine kidney [23]. Good accuracy and precision were achieved with recoveries from 71 to 91%; the limits of quantitation for four tetracyclines ranged from 36–47 µg/kg. Crean *et al.* [27] reported the use of TFC for the analysis of brain tissue and achieved much improved throughput.

Table 14.1
Selected bioanalytical applications using turbulent flow chromatography for on-line sample preparation (single-column mode) with LC-MS/MS detection

Analytes	Sample Matrix and Volumes	Extraction Column	Cycle Time (min)	Reference
Novel isoquinoline	Serum 100μL, IS 100μL, inject 50μL	Turbo-C18, 50μm, 1x50mm	2.5	[9]
Novel isoquinoline	Plasma 5μL, IS 5μL, inject 5μL	Oasis® 30μm, 0.18x50mm (custom made)	2.0	[20]
Proprietary analytes	Plasma 20μL, cfg, inject 20μL	Turbo-C18, 50μm, 1x50mm or Oasis 30μm, 1x50mm	2.5	[21]
Ro 46-2153 and Ro 18-5528	Human plasma 50μL, IS 50μL, cfg, inject 50μL	Turbo-C18, 50μm, 1x50mm	2	[22]
Proprietary analytes	Rat plasma 50μL, IS 50μL, cfg, inject 50μL	Oasis 30μm, 1x50mm	4	[10]
Novel Isoquinoline	Plasma 50μL, IS 50μL, inject 50μL	Oasis 30μm, 1x50mm or Prime C18, 1x50mm	1.2	[19]
Tetracyclines	Bovine kidney tissue homogenate (filtrate) 25μL	Oasis 30μm, 1x50mm	6.0	[23]

In this application, tissue was first homogenized with an 80% aqueous mixture and proteins were pelleted by centrifugation; supernatant was analyzed. TFC has been used with additional sample matrices, such as brain and liver homogenates, urine, intestinal perfusates, cerebrospinal fluid [28], plasma ultrafiltrate [29], bile, synovial fluid and whole blood [30].

Since no sample preparation is performed with turbulent flow chromatography, and thus no concentration effect, sample dilution can limit sensitivity. Ayrton *et al.* reported the use of LC on the capillary scale using a 0.18 x 50 mm Oasis® HLB extraction column (Waters Corporation, Milford, MA USA) packed with 30 μm particles instead of the traditional 1 x 50 mm dimensions with the same 30 μm particle size [20]. Great throughput (2 min per sample) was achieved using this technique and drugs were quantified in plasma at sub-ng/mL concentrations from a 2.5 μL plasma aliquot. The flow rate was reduced from 4 mL/min as is typical in TFC to 0.13 mL/min; elution of analytes in this reduced volume of mobile phase allowed greater concentration and then a reduced sample volume could be used. The applicability of this technique is for small volume studies, such as serial bleeding from mice.

The single-column mode of turbulent flow chromatography is the simplest and yields the shortest injection cycle. It is often the first approach tried in the development of a new assay. However, when the single-column mode does not present sufficient cleanup, a second column can be placed in series downstream from the first. The use of dual columns for turbulent flow chromatography is discussed in Section 14.3.3.

14.2.3 Restricted Access Media (RAM)

14.2.3.1 Technology

Another technique allowing the direct injection of plasma or serum on-line with the chromatographic system, using normal laminar flow liquid chromatography, involves the use of analytical columns containing restricted access media. The particles packed within RAM columns are designed to prevent or restrict large macromolecules from accessing the inner adsorption sites of the bonded phase. Methods based on the use of RAM extraction columns have been widely reported since early applications were demonstrated in the mid-1980s [31, 32]. The packing materials in a RAM column contain different chemistries at the external and internal surfaces. The pore size of this packing material is small (typically about 80 Angstroms); the inner surface is only accessible by molecules with a relatively small molecular size while the external surface is accessible by all.

Commercially available RAM columns, all silica based, include internal surface reversed phase (ISRP), semipermeable surface (SPS) and a hydrophobic shielded phase named Hisep™ (Supelco, Bellefonte, PA USA). The most popular column variety used in bioanalysis is the ISRP column. In this type, the internal surface is covered with a bonded reversed phase material and the external surface is covered with a nonadsorptive but hydrophilic material. This dual-phase column permits effective separation of the analyte of interest from macromolecules in the sample matrix; drugs and other small molecules enter the pores of the hydrophobic reversed phase to partition and be retained, while proteins and larger matrix components are excluded. Essentially, a combination of size exclusion chromatography and partition chromatography is observed. When serum or plasma is injected onto a RAM column, the proteins are excluded by the outer, hydrophilic layer and pass through to waste. Note that conventional polymers (*e.g.*, Oasis HLB) are subject to protein adsorption from nonspecific binding; column lifetimes are reduced ten-fold compared with RAM columns [33].

The method for preparation of an ISRP column follows. Silica particles (5 μm diameter, 80 Å pores) are derivatized with a glyceryl-propyl phase to yield a diol layer. A tripeptide (glycine-L-phenylalanine-L-phenyalanine; GFF) is attached to the glyceryl-propyl phases bound via amino groups. The phenylalanine moieties are cleaved using enzymatic treatment. Since this enzyme (carboxypeptidase A) cannot penetrate the inner pores of the silica, it leaves the GFF phase intact; a diol-glycine phase remains on the external surface [34, 35]. The exterior of the particle surface is hydrophilic and the interior is hydrophobic. This GFF phase is selective toward positively charged aromatic analytes and a range of neutral molecules. GFF II is a second generation phase with an improved bonding process; the GFF peptide is bonded to the silica surface through a monofunctional glycidoxypropyl linkage rather than a trifunctional linkage. The benefits of GFF II are reported by the manufacturer to include better sample retention, higher column efficiency and greater batch-to-batch reproducibility.

Additional ISRP support materials have been designed by research laboratories in an effort to overcome certain shortcomings of GFF and GFF II, and to improve performance for certain drug classes [35–37]. The BioTrap™ 500 column (ChromTech AB, Stockholm, Sweden) contains an outer coating of α1-acid glycoprotein (AGP). The use of AGP allows the column to tolerate high concentrations of organic in the mobile phase and to accept a large sample volume. Alkyl-diol silica (ADS) contains an outer diol layer and an inner alkyl layer of C4, C8 or C18 reversed phase material; it is available as LiChrospher® ADS (Merck KGaA, Darmstadt, Germany) and distributed in the USA by Iris Technologies (Lawrence, KS).

Alkyl-diol silica (25 μm particle size, pore diameter 60 Angstroms) is designed to be a precolumn packing for coupled column LC analysis. The outer particle surface is nonadsorptive to matrix components via its hydrophilic modification; its inner pore surface is accessible only by low MW compounds (<15,000

Dalton). The retention properties of ADS are due to its classic reversed phase interactions.

Similar to ISRP, the SPS phases consist of both hydrophilic outer and hydrophobic inner surfaces. However, in SPS the inner and outer surfaces are bonded separately so that each surface can be varied independently. The outer hydrophilic phase is a porous polyoxyethylene polymer covalently bonded to the silica surface and the inner phase is commonly C18, C8, phenyl or nitrile. These chemistries are more compatible with traditional buffered mobile phase solvents as used for reversed phase chromatography.

The Hisep phase is also based on silica. The inner hydrophobic phase consists of a disubstituted phenyl layer and the outer hydrophilic layer is a polyethylene glycol (PEG) network [38]. This hydrophilic PEG network shields the hydrophobic phenyl groups and prevents proteins from reaching the phenyl partitioning phase [34]. Size exclusion also plays a role in excluding proteins from the inner sites of the packing material.

The use of a flow path switching valve with RAM columns allows the direct injection of plasma samples. This on-line process involves four steps:
1. The extraction column is conditioned or regenerated
2. Analytes are loaded onto the column
3. A wash step removes proteins and other unwanted endogenous materials which flow to waste
4. Switching the mobile phase flow path then directs analytes in the flow stream into the detector

Just as with TFC, a dual-column approach can be used, whereby a more organic mobile phase is utilized to elute the analyte onto an analytical column for chromatofocusing. Dual-column RAM techniques are discussed in Section 14.3.4. A general overview of the use of restricted access materials in liquid chromatography has been published in two parts [39, 40].

14.2.3.2 Advantages and Disadvantages

The dual-phase nature of these RAM materials allows the direct injection of a large variety of biological sample matrices onto the column; little to no pretreatment is required. Pretreatment typically involves centrifugation of the sample to pellet particulates at the bottom of the sample container. Analytes are effectively separated from the macromolecules in the sample matrix; drugs and other small molecules enter the pores of the hydrophobic reversed phase to partition and be retained, while proteins and larger matrix components are excluded.

Using LiChrospher ADS as an example, this RAM column can be operated in either single-column or dual-column mode. In single-column mode, cleanup, extraction and separation occur simultaneously; there is no sample enrichment. Sample volumes for direct injection are small (<100 μL) in single-column mode; column overloading is a potential problem. There is limited peak capacity for analytes and the limit of detection is high. In contrast, for the dual-column mode of ADS, cleanup, extraction and separation occur sequentially; sample enrichment is realized. Sample volumes are large (100–500 μL) and limits of detection are much lower.

ADS column lifetime in single-column mode is short; from 100–500 injections of 10–20 μL sample volumes. Flow is only in one direction so the total mass injected goes toward the detector; regeneration requires additional time. In coupled column mode, about 2,000 injections of 50 μL can be made. A backflush mode cleans the precolumn; regeneration of the precolumn is included in the analytical cycle.

Some disadvantages regarding high throughput for RAM column usage are that retention times can be long (>10 min), washing of the column is required between injections, and the mobile phases are not always compatible with the ionization techniques used in LC-MS/MS. However, in the case of LiChrospher ADS, the extraction column conditioning step is part of the entire process so that while separation and quantification is occurring, the extraction column is being conditioned and prepared for the next injection. Other general disadvantages for RAM columns are low chromatographic efficiency, limited loading capacity and low extraction efficiency for highly (99.9%) protein-bound analytes [41]. Some possible solutions to lessen the influence of protein binding include: adding a dilute solution of trichloroacetic acid, diluting the sample with an aqueous solution, adjusting the solution pH or adding organic modifiers. The overall approach using RAM extraction columns is automated, however. There is a low capital outlay for automation equipment since this technique uses common HPLC hardware.

14.2.3.3 Applications

An application comparing three RAM columns (ISRP, SPS and Hisep) for the analysis of Non-Steroidal Anti-Inflammatory Drugs (NSAIDs) has been reported by Haque [42]. Note that these RAM columns are used within specific pH ranges and specific buffer concentrations, and that there is a maximum percentage organic (20–25%) that can be tolerated. Among the three columns, the Hisep is the only one that can be used below pH 5 without sacrificing column performance. The NSAIDs were the most retentive on the Hisep column at this lower pH. At its optimal pH of 6.5–7.5, the ISRP column retained the NSAIDs the least. The SPS column was found most desirable for these analytes based on retention time and peak shape. However, note that some users may consider the limited pH range and the low organic content tolerated by these columns to be disadvantages.

A comprehensive evaluation of the effects of altering pH and buffer concentration using ISRP columns was undertaken for 19 analytes spiked in serum [43] with analysis by UV. Quantitative recovery was seen for most analytes; however, the protein binding for phenelzine and tamoxifen in serum could not be disrupted and recovery was poor. A comparison of ISRP to C2 and C18 extraction columns showed that the ISRP material eliminated interferences as well as, or better than, the conventional reversed phase materials.

Although single-column RAM methods have been used primarily to analyze a single component in biological fluids, Rainbow *et al.* reported a method for the simultaneous determination of phenobarbital, carbamazepine, and phenytoin with satisfactory results [44]. The simultaneous determination of the carboxylate and lactone forms of 10-hydroxycamptothecin in human serum using a Hisep RAM column has been reported [45].

Restricted access materials have also been utilized as coatings for solid phase micro extraction (SPME) fibers. Although further discussion of SPME fibers is outside the scope of this chapter, note that ADS-coated SPME fibers were used for the SPME of benzodiazepines in urine [46]. The binding capacity, extraction efficiency and reproducibility of the fiber were found to be acceptable for ultraviolet determination of a wide concentration range of benzodiazepines in urine. Additional information about the SPME technology has been reviewed by Lord and Pawliszyn [47, 48] and some overviews of its use with biological samples have been reported [49–51].

14.2.4 Monolithic Columns

14.2.4.1 Technology

Monolithic columns have generated great interest as an alternative to particulate columns [52–58]. These monolithic columns consist of a one piece organic polymer or silica with flow-through pores and thus demonstrate a unique biporous structure. The smaller pores (mesopores, diameter about 13 nm) located on the silica skeleton provide the large surface area needed to achieve sufficient capacity. The larger pores (macropores, diameter about 2 µm) on the silica skeleton reduce flow resistance. Together, these specifications allow the use of high flow rates without generating high backpressure.

In addition to the clear implications for high speed separation, this biporous structure offers the unique advantage of direct injection of the sample matrix in bioanalysis. Small to medium sized analytes achieve good partitioning and separation through mesopores, while large molecules and particulates in the sample matrix pass directly through macropores. High separation efficiency can be maintained at greatly increased flow rates because the dependency of separation efficiency on flow rate is very small. An overview of the use of monoliths as stationary phases for separating biopolymers has been summarized by Iberer *et al.* [59]. An application for the use of monolithic columns for natural product isolation has been reported [60].

14.2.4.2 Applications

Dear *et al.* reported the use of a monolithic column for drug metabolite identification using filtered *in vitro* human liver microsome incubations as the sample matrix [61]. The column (4.6 x 50 mm) was operated at 4 mL/min and *in vitro* samples were injected directly. An adequate separation of six hydroxylated isomers of debrisoquine was achieved on a monolithic column in as little as 1 min. There was no loss in resolution compared with a typical packed particle LC column (2.1 x 150 mm C8 5 µm). Fast gradient short column LC (2.1 x 50 mm C18 5 µm) provided similar analysis times but the resolution was compromised compared with the silica rod or monolithic system. Overall, the total analysis time was reduced to 5 min from the original 30 min assay using a typical column.

Plumb *et al.* also examined the analytical throughput of the monolithic column system using 20 µL injection volumes (plasma diluted 1:1 with water and then centrifuged) [62]. Approximately 300 plasma samples were analyzed over a two day period using a simple and fast (1.5 min) linear gradient. Satisfactory separation was achieved but some degradation in separation performance was observed after 300 injections. Operation in gradient mode reduced the detrimental effect on separation performance observed in isocratic mode.

Note that turbulent flow chromatography has been combined with monolithic column separation for high throughput bioanalysis [63]. The total extraction and analysis time was reduced to 1.2 min by use of a 5 mL/min flow rate. Matrix suppression was noted but was eliminated by modifying the chromatographic conditions.

An application for monolithic columns as part of a routine bioanalysis strategy to support drug discovery programs has been reported by Wu [64]. The attractive features of monolithic columns for this application are high speed and good separation performance. A flow rate of 6 mL/min could be used without high backpressure. This flow rate is six-fold higher than commonly used and the run time was reduced by a factor of six. A run time of 1 min almost always provided sufficient chromatographic resolving power. A 12 h overnight run analyzed 600 samples that had been prepared by semi-automated 96-well solid-phase extraction. Retention time, peak response and sensitivity were all found to be reproducible throughout the run, demonstrating the ruggedness of this approach for high throughput bioanalysis.

Dear *et al.* investigated the potential of connecting up to 6 alkyl-bonded silica monolithic columns in series, producing a high resolution system with the benefit of both small particle performance and additional column length [65]. This system was demonstrated for metabolite characterization in biological fluids using both analytical and semi-preparative LC. The separations produced with the six-column configuration showed more resolving power than from a conventional packed column containing 3 μm particles. Various column configurations were examined with LC-UV, LC-NMR and LC-MS applications.

Deng *et al.* performed sample preparation using 96-well solid-phase extraction, followed by the use of four monolithic columns for parallel separations [66]. In this application, which also used multiplexed inlets on the mass spectrometer (discussed in Section 14.4.4.2), overall throughput was reported as 30 s per plasma sample with gradient elution on a 10 cm long LC column. In a related report by Hsieh, following acetonitrile precipitation of rat plasma for the simultaneous determination of a drug candidate and its metabolite, baseline separation was made with run times of 24 or 30 s under isocratic or gradient conditions, respectively [67].

The separation of methylphenidate and ritalinic acid was achieved in 15 s by using a 3.5 mL/min flow rate with a monolithic column; throughput was documented as 768 protein precipitated rat plasma samples (8 full microplates) analyzed within 3 h 45 min [68]. An additional bioanalytical assay using monolithic columns to analyze extracts from human plasma prepared by 96-well SPE has been reported for cyclooxygenase II inhibitors [69]; throughput was 5-fold greater than that achieved using typical LC columns.

14.2.5 Immunoaffinity Extraction (IAE)

14.2.5.1 Technology

Immunoaffinity extraction uses antibody-antigen interactions to provide a very high specificity for the molecules of interest. Antibodies are immobilized onto a pressure resistant solid support and then packed into LC columns for use in chromatography applications. These antibodies can remove a specific analyte or class of analytes from among all other materials in a sample. Their recognition is based on a particular chemical structure, rather than on a general reversed phase attraction such as occurs with solid-phase extraction. The great specificity of this technique can make it preferable to other approaches.

The binding of antibody to antigen is a reversible process and its equilibrium can be influenced by manipulating the solution pH and aqueous/organic composition. Note that an antibody can also bind one or more analytes that are structurally similar to the one of interest; this occurrence is called cross-reactivity. Cross-reactivity can be useful when a group of analytes is to be removed from solution, such as for the screening of small molecule combinatorial libraries [70].

14.2.5.2 Usage

IAE can be performed off-line, but it is time consuming; on-line extraction is preferred for higher throughput. This technique is also referred to as immunoaffinity solid-phase extraction and the particles as immunosorbents. A brief overview of chromatographic immunoassays, antibodies, and various binding immunoassays has been published [71]. Another overview introduces the concept of immunosorbents as applied to both off-line and on-line techniques [72]. Several reviews discuss advances in analytical applications of immunoaffinity chromatography [73, 74] and the characteristics and properties of immunosorbents with optimization techniques for their successful use [75, 76]. A broad perspective of the role of affinity based mass spectrometry techniques used in drug discovery is discussed [77]. Interestingly, in addition to isolating and quantitating analytes from complex matrices, it can be used to examine interactions and binding of drugs with serum proteins [78].

The coupling of immunoaffinity extraction with liquid chromatography allows a high degree of selectivity as extraction, concentration and isolation can be achieved simultaneously. The single-column mode can be utilized with immunoaffinity techniques but it is often not used because large volumes of aqueous buffer at low pH are commonly required for the elution step; the sample is in a dilute solution and a chromatofocusing, or concentration step at the head of an analytical column, is required to provide more sensitivity for detection. Also, passing totally aqueous solutions through the IAE column is desirable in prolonging its lifetime; in a dual-column mode this aqueous solution can be easily diverted to waste without adversely affecting the performance of the analytical column which prefers some organic concentration in the mobile phase. Flow diversion also permits a larger IAE column to be used as the larger volume passed through it also can be directed to waste.

A related approach to immunoaffinity extraction involves molecularly imprinted polymer (MIP) materials, also referred to as artificial antibodies. The molecular imprinting technique uses a highly selective binding area for a target analyte on a stationary phase. MIP materials have shown antibody-like affinities toward a target molecule and are used for selective extraction in an LC column, similar to immunosorbents [79]. Further discussion of MIP is outside the scope of this chapter and the reader is referred to the book edited by Sellergren [80], a review [81] and various applications of the technology as used to extract analytes from biological matrices [82–88].

14.3 High Throughput On-Line Sample Preparation Using Dual Columns

14.3.1 Configuration

Although the single-column mode previously described offers great simplicity for bioanalysis, the assay is often prone to matrix effects and interferences due to the limited chromatographic resolving power [10, 30]. In order to improve the separation capability a second column, which can be a regular analytical column used for separation, is added to the system. The primary technique used to connect the two columns is column switching.

Figure 14.2 shows one of the most popular configurations of a dual-column mode using column switching. In this configuration, an electrically actuated six-port valve is used to connect the two columns and perform the column switching function. Two LC pumps are necessary. LC Pump 1 is used for most of the sample extraction functions (load, wash, and equilibration steps) except elution. During these extraction steps, the switching valve is set to Position 1 (Figure 14.2A), which directs the effluent from the extraction column to waste and allows LC Pump 2 to perform separation on the analytical column. The valve is switched to Position 2 for the elution step (Figure 14.2B). LC Pump 2 now delivers flow through the extraction column eluting the analyte which passes on to the analytical column for further separation.

The dual-column mode provides the advantage of improved separation as well as detection sensitivity resulting from chromatographic focusing [30]. Some practical guidelines for on-line sample preparation using column switching with the use of dual columns have been summarized [89].

14.3.2 Advantages

This dual-column mode provides the advantages of improved separation performance as well as improved detection sensitivity as a result of chromato-graphic focusing in some cases and in other cases simply an additional resolution of analyte from the background matrix. Another advantage of dual-column configuration is that the extraction and separation processes

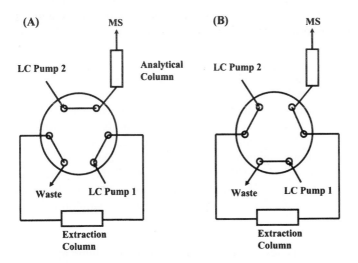

Figure 14.2. Schematic diagrams illustrating a column switching valve configuration for on-line sample preparation using dual columns: (A) Position 1 for the loading, extraction and equilibration steps and (B) Position 2 for the elution step.

can be performed simultaneously since they are now driven by two independent LC pumps. While the sample is running on the separation column, the extraction column can be used to extract the next sample. Therefore, all or part of the time that the system is used for on-line extraction can be buried in the run time of the separation column.

Most of the commonly used on-line sample preparation approaches are based on the principle of this dual-column configuration. These approaches include turbulent flow chromatography, restricted access media, solid-phase extraction using a disposable cartridge and immunoaffinity extraction. The major differences among these approaches are the type of the extraction column or cartridge, the solvent composition and the flow rate of the chromatographic system.

14.3.3 Turbulent Flow Chromatography with Column Switching

In the turbulent flow column switching approach, the same turbulent flow extraction column as described in Section 14.2.2 is used. LC Pump 1 operates in the same manner as described except that elution is performed using LC Pump 2 which operates under conventional flow conditions for the separation column. Usually a flow splitter is not needed before introducing the effluent into the mass spectrometer. This method is one of the most popular approaches for on-line sample preparation for bioanalysis because of the ruggedness of the turbulent flow extraction and the chromatographic resolution offered by the separation column.

This turbulent flow column switching approach has been used for the bioanalysis of a single analyte as well as for mixtures of multiple analytes. Selected applications using dual-column TFC are listed in Table 14.2. Jemal *et al.* used this scheme to effectively separate and quantify two positional isomers in plasma [10]. Multiple components in plasma were simultaneously quantified with good chromatographic separation and accuracy by Wu *et al.* [18] and Kollroser and Schober [90]. Lim *et al.* used this approach to simultaneously screen for metabolic stability and profiling [91]. With a total cycle time of 8 min, Lim's method was able to perform a separation of three isomeric metabolites.

While cycle time is typically several minutes using the dual-column approach, it has been reduced to 1.2 min using high flow rates with a monolithic column, as previously described [63]. Typically, for the same amount of sample injected into an LC-MS system, the use of high flow rates results in reduced sensitivity

Table 14.2
Selected bioanalytical applications using turbulent flow chromatography for on-line sample preparation (dual-column mode) with LC-MS/MS detection

Analytes	Sample Matrix and Volumes	Extraction Column	Analytical Column	Cycle Time (min)	Reference
Ketoconazole	Human plasma and IS mixture, supernatant following ACN PPT 250μL, inject 20μL	Turbo-C18 50μm, 1x50mm	Zorbax SB-C18 3.5μm, 4.6x50mm	4.5	[92]
Deprenyl, modafinil, haloperidol, nimodipine, K252a, CEP-1347	Rat plasma 100μL, ACN/IS 200μL, cfg, inject 25μL	Cyclone 50μm, 1x50mm	Eclipse XDB C18 3μm, 4.6x15mm	6.0	[28]
Proprietary analytes	Plasma 50μL, IS 50μL inject 10μL	Oasis® 30μm, 1x50mm	Capcell C18 5μm, 2.0x35mm	5.0	[93]
Proprietary analytes	Plasma 100μL, IS 100μL mix, cfg, inject 10μL	Oasis 30μm, 1x50mm	C18 5μm, 3.9x50mm	2.0	[94]
Proprietary analytes	Microsomal incubate 200μL after quenching with IS soln and cfg, inject 50μL	Turbo-C8 50μm, 1x50mm	Aquasil C18 3μm, 4x20mm	8.0	[91]
Lamisil® (terbinafine)	Human plasma and IS mixture inject 50μL	Turbo-C18, 50μm, 1x50mm	Eclipse XDB C18 3.5μm, 4.6x50mm	n/a	[95]
Proprietary analytes	Plasma 50μL, IS 50μL inject 10μL	Oasis 30μm, 1x50mm	Hypersil C18 5μm, 2x50mm	1.6	[96]

Analyte(s)	Sample Matrix and Volumes	Extraction Column	Analytical Column	Cycle Time (min)	Reference
Benzodiazepines, carbamazepine, haloperidol	Dog plasma 200µL, IS 20µL water 20µL, mix, inject 100µL or 200µL	Oasis 30µm, 1x50mm	Symmetry C18 or Develosil-MG C18, 2x150mm	n/a	[18]
Pravastatin	Human serum 200µL, IS 50µL cfg, inject 50µL	Oasis 30µm, 1x50mm	C18 5µm, 3.9x50mm	5.0	[10, 97]
Ondansetron	Human plasma 200µL IS 400µL, inject 50µL	Turbo-C18 50µm, 1x50mm	Reversed phase, unspecified, 4x20mm	2.8	[98]
Doxazosin, dofetilide, CP-122,288	Dog plasma 200µL, IS 200µL cfg, inject 200µL	Turbo-C18 50µm, 1x50mm	HTLC HiRes C18 10µm, 2.1x33mm	4.3	[99, 100]
Apomorphine, aminopterin, benzoylecgonine, carbamazepine, temazepam	Plasma 200µL IS 100µL, inject 60µL	Oasis 30µm, 1x50mm	YMC AQ 5µm, 2x100mm and 1x100mm	8.0	[30]
Diflunisal and clemastine	Plasma 100µL IS 100µL, inject 200µL	Oasis 25µm, 2.1x20mm	Symmetry C18 or Xterra™ MS C18 3.5µm, 2.1x30mm	3.0	[101]
Olanzapine, clozapine and metabolite	Human plasma and IS mixture inject 50µL	Oasis 30µm, 1x50mm	Symmetry® C18 5µm, 3x150mm	6.0	[90]

ACN, Acetonitrile; PPT, Protein precipitation; cfg., Centrifugation; IS, Internal Standard; n/a, not available since most electrospray ionization sources are concentration sensitive and the use of high flow dilutes the concentration in the elution band, regardless of the use of flow splitting. It has been reported that under conditions used in the Applied Biosystems/MDS Sciex Turbo V source, the detection is mass sensitive [102]. Therefore, the use of high flow rate does not result in any compromise in detection sensitivity if all column effluent is introduced into the source without splitting. The use of high temperatures and improved gas dynamics in the V source allow for full volatilization of the LC flow without splitting. Four 96-well plates filled with samples (plasma diluted 1:2 and centrifuged) were analyzed in 6 h using this method, resulting in a three-fold improvement in throughput compared with a typical 3 min gradient at lower flow rates.

14.3.4 Restricted Access Media with Column Switching

In the dual-column RAM method, a separation column is added to the system after the RAM column. Plasma or other matrix is injected directly onto the RAM column, proteins are washed from the column and sent to waste with mobile phase, and analytes are backflushed onto an analytical column using switching valves. The dual-column RAM approach has been widely reported to quantify compounds in plasma or serum; early reports used UV or fluorescence detection [33, 103–114].

An example of an LC-MS/MS application includes the work reported by Hogendoorn *et al.* [115]. Three different types of RAM columns (ISRP, SPS and ADS RP-18) were compared and a C18 analytical column was used in series after the RAM column. The ISRP column was found to have the best performance with a 100% aqueous buffer (pH 7.0); up to 200 μL sample volume could be injected. A lower limit of quantitation at 0.5 ng/mL was achieved for salbutamol and clenbuterol in plasma with a throughput of 5-7 samples per hour.

A report from van der Hoeven *et al.* details the direct injection of biological fluids (100–200 μL) for the determination of cortisol and prednisolone from plasma and arachidonic acid from urine [116]. Analytes were enriched on the ADS extraction column and eluted onto the analytical column. The limit of detection for prednisolone was 2 ng/mL. Jeanville utilized dual-column RAM extraction for 25 μL rat plasma using a BioTrap 500 column with analysis by time of flight (TOF) mass spectrometry; analytes were ecgonine methyl ester and cocaine [117]. Peng performed a drug stability study in plasma using an SPS extraction column (octadecyl silane inner phase) with a Symmetry® C18 analytical column and reported that this column switching arrangement had eliminated traditional protein precipitation, centrifugation, evaporation and reconstitution steps [118].

An interesting three-column configuration has been reported, in which a β-glucuronidase immobilized enzyme reactor column is positioned after an ISRP column but before the analytical C8 column [119]. These columns were connected with three six-port switching valves. Urine samples were injected onto the ISRP column and glucuronides were separated from the biological matrix; effluent from the ISRP column was directed to the β-glucuronidase column for deconjugation and then analytes were separated on the C8 column and detected at 280 nm.

Some restricted access materials were used as part of a novel system for the two-dimensional coupled column separation of complex biological samples for proteome characterization [120]. Two novel RAM extraction columns (research products of Merck KGaA, Darmstadt, Germany), based on cation and anion functionality, were placed in series. A column switching technique was developed using four parallel, short, reversed phase columns. The high resolution separation of small proteins <20 kDa from human hemofiltrate was demonstrated.

While the above reports utilized RAM columns as part of an existing LC system, another option for those users less comfortable and/or less familiar with the required hardware manipulations and valve switching schemes is to utilize a total system package configured for on-line RAM SPE coupled with LC. An example of this system is UNEXAS (Iris Technologies; available in Germany through Knauer GmbH, Berlin) which consists of a switching valve module, two LC pumps, one SPE pump, autosampler and software that controls the solid-phase extraction and chromatographic process. The system is fully automated and utilizes LiChrospher ADS SPE columns that allow >1,000 injections (50 μL) of undiluted plasma or urine that has undergone only a centrifugation or filtration step.

In addition to the many demonstrated applications of ADS using C4, C8 and C18 materials within the inner pores of the particles, a mixed mode column packing is available that also relies on the affinity and size exclusion principles of these RAM packing materials. The CAT-PBA™ column (Iris Technologies) is filled with a porous copolymer containing specially modified phenylboronic acid (PBA) as an affinity ligand. Low molecular weight analytes such as catecholamines (CAT; *e.g.*, epinephrine, norepinephrine and dopamine) reach the inner pore surface. Under slightly alkaline load conditions (pH 8.7), these catecholamines are extracted onto the PBA; selective retention occurs due to the formation of a cyclic ester between the boronic acid bound to the stationary phase and the 1,2-diol functional group on the catecholamines. Following a wash step to remove residual components, the pH is changed to acidic which causes the cyclic boronic acid ester to hydrolyze and the catecholamines are eluted onto an analytical column via a switching valve. Separation is achieved under reversed phase C18 conditions [121]. As the LC separation with electrochemical detection is taking place, the CAT-PBA™ SPE column is automatically regenerated and prepared for another extraction. The lifetime of the CAT-PBA™ SPE column has been reported by the manufacturer to exceed 2,000 untreated urine samples (100 µL each) and 1,000 untreated plasma samples (500 µL each).

14.3.5 Immunoaffinity Extraction with Column Switching

DNA adducts are promutagenic lesions normally present in tissues and these are believed to be important in the etiology of cancer. The analysis of specific etheno DNA adducts of deoxycytidine (etheno-dC) was performed using a reusable immunoaffinity extraction column with a graphitized carbon analytical column, an isotopically labeled internal standard and tandem MS detection [122]. The affinity column contained covalently bound monoclonal antibody specific for etheno-dC. The column bound etheno-dC from aqueous buffered solutions and released bound material using a mobile phase of methanol/water. Quantitative determination of etheno-dC was performed in calf thymus DNA, mouse liver and rat liver. Sensitivity using this technique was about 100-fold greater than that using traditional silica based sorbents.

Immunoaffinity extraction was coupled on-line with a packed capillary LC column and tandem MS in a report by Cai and Henion [123]. The small bore LC column aided in greater sensitivity by LC-MS. The IAE column (a 2.1 mm i.d. protein G column with immobilized antibody to analytes) operated at 2.5–4 mL/min while a coupled capillary trapping column and analytical column were run at 3.5 µL/min. The hallucinogen drug LSD (lysergic acid diethylamide) and its analogs were determined in spiked and positive human urine samples at low part-per-trillion levels. A related MS application again using capillary LC was demonstrated for five β-agonists (*e.g.*, clenbuterol) in bovine urine [124]. Another clenbuterol IAE has been reported using UV detection [125].

The steroids dexamethasone and flumethasone were determined from equine urine using an immunoaffinity extraction column coupled to an analytical column [126]; limits of detection were in the range 3–4 ng/mL using ion trap mass spectrometry. Additional applications reported for dual-column IAE include the determination of the mushroom toxins α- and β-amanitin from urine with MS analysis [127] and fluoroquinolone antibiotics from bovine serum with fluorescence detection [128].

14.3.6 Disposable Solid-Phase Extraction Cartridges with Column Switching

14.3.6.1 Introduction

Solid-phase extraction is an effective sample preparation technique in which analytes are selectively adsorbed onto sorbent particles, proteins and matrix materials pass to waste, and the analytes are desorbed (eluted) from the column prior to analysis. Typically the sorbent particles are bonded silica or polymeric. When the proper sorbent chemistry is chosen, such that affinity for the analytes of interest is obtained, SPE can yield high recoveries, selective separations and reproducible results. Further discussion of SPE is found in Chapter 11.

SPE has been performed off-line using individual cartridges or microplate wells. However, the technique is also automated using disposable extraction cartridges that are placed on-line with the chromatographic system. The use of individual disposable cartridges on-line in this manner removes the issue of sample carryover sometimes seen with on-line methods as discussed in detail in Section 14.7.2.

Typical examples of on-line SPE using disposable extraction cartridges are demonstrated by the Prospekt™ and Prospekt-2™ systems (Spark Holland, Emmen, The Netherlands) and the OSP-2 (Merck). The brand name PROSPEKT is derived from the terms "PRogrammable On-line Solid Phase ExtraKTion (sic)" while the name OSP is abbreviated from "On-line Sample Preparation." The Prospekt and the higher throughput version Prospekt-2 will be used as representative examples for on-line disposable extraction cartridges in this chapter. The Prospekt and Prospekt-2 systems are distributed in the USA by LEAP Technologies (Carrboro, NC).

14.3.6.2 Advantages

An important feature and advantage of on-line SPE, compared with off-line SPE, is direct elution of the analyte from the extraction cartridge into the LC system. The time consuming off-line steps of evaporation, reconstitution, and preparation for injection are eliminated, making on-line SPE more efficient and fully automated. Since the entire volume of eluate is analyzed, maximum sensitivity for detection is obtained.

Some other advantages of this on-line approach are that:
1. Samples and SPE cartridges are processed in a completely enclosed system protected against light and air
2. The operator is protected from working with hazardous and/or volatile organic solvents, and
3. There is less handling and manipulation involved with no transfer loss of analyte.

A unique feature of the Prospekt-2 is its ability to perform thermally assisted solid-phase extraction; controlling the temperature of the extraction cartridge may improve the performance of the overall extraction [129]. Temperatures for loading, washing and eluting (desorption) may be manipulated. Typically, raw sample matrix (combined with internal standard) is utilized for injection onto the SPE cartridge, although sometimes an additional sample preparation step may precede this on-line analysis, *e.g.*, protein precipitation [130, 131].

14.3.6.3 System Hardware and Configuration

The Prospekt-2 is an integrated sample cleanup and injection system consisting of an Endurance™ (Spark Holland) autosampler for use with microplates, LC pumps and the SPE cartridge system referred to as the ACE (Automated Cartridge Exchange) module, shown in Figure 14.3. Samples are introduced by the autosampler. Solvents are delivered to the cartridge by a high pressure dispenser. Cartridge exchange and valve switching are performed by the ACE module. The analytes, now purified, are eluted from the SPE cartridge using column switching. The flow path is directed to the LC analytical column and then to the detection system (*e.g.*, mass spectrometer). The entire arrangement is controlled from a computer using the SparkLink™ software with Easy Access™ system control (Spark Holland).

The Prospekt-2 uses two high pressure SPE cartridge clamps, two high pressure SPE solvent syringes and two SPE cartridge trays. In this configuration, elution is performed on one cartridge while the next cartridge is undergoing extraction. These short cycle times are ideal for keeping pace with LC-MS/MS analytical detection systems.

Figure 14.3. The Prospekt-2™ system for on-line solid-phase extraction consists of an autosampler, a cartridge exchange unit, and a pumping system. Photo reprinted with permission from Spark Holland.

The SPE cartridge is able to withstand LC system pressures to 300 bar and has a standard dimension of 10 x 2 mm; note that column dimensions of 10 x 3 mm and 10 x 1 mm are also available. A full range of sorbent chemistries (>38) is available in the cartridge format, including the versatile Oasis HLB polymer sorbent (Waters Corporation) as well as several other polymer chemistries. Typical particle sizes used are 40 µm, although the HySphere™ cartridges (Spark Holland) specifically use a smaller particle size <10 µm. Sorbent mass per cartridge is typically either 20 or 45 mg, depending on particle chemistry. The standard capacity of the ACE module is 192 cartridges (two trays containing 96 cartridges each); an optional feeder mechanism allows access up to 960 cartridges (10 trays).

14.3.6.4 Use with MS Systems for Staggered Parallel SPE and Analysis

The use of the Prospekt for single SPE with single LC analysis in a serial mode is straightforward. The original Prospekt configuration utilized this approach. With the advent of high throughput parallel and staggered parallel LC systems, the newer Prospekt-2 is more suitable for use with today's LC-MS/MS systems. Since staggered parallel extraction and elution can be performed, throughput of on-line SPE can be doubled; two analyses can be performed in the time taken for one analysis using the serial mode.

In order to achieve maximum throughput, the cycle time for on-line SPE must be faster than the LC run time. The flow path for parallel extraction and elution using the Prospekt-2 are shown in Figure 14.4. The type of LC-MS interface (APCI, atmospheric pressure chemical ionization or ESI, electrospray ionization) also influences time requirements. APCI interfaces are capable of handling conventional LC flow rates of 0.5 to 2.0 mL/min. Sample elution time can match this fast flow rate. However, with ESI interfaces, the flow is limited to about 0.1–0.2 mL/min. The time required to elute the analyte from the SPE cartridge now becomes the slowest step in the entire process and influences overall sample throughput. The use of two SPE cartridges in staggered parallel fashion, one being eluted while the other is performing extraction, now becomes an important feature in helping to meet throughput needs.

14.3.6.5 Applications

The original design of the Prospekt was reported in 1987 [132]; since that time, many applications have been reported in the literature that were largely developed for therapeutic drug monitoring as performed by clinical labs. The

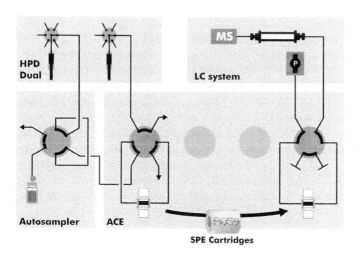

Figure 14.4. Schematic representations of parallel on-line solid-phase extraction and elution using dual cartridge clamps. Following extraction on the left cartridge, the cartridge is transferred to the right clamp for elution into the LC-MS/MS system. During analysis, extraction of the next sample is performed using a new cartridge in the left position. Illustration reprinted with permission from Spark Holland.

use of this disposable cartridge technique expanded into pharmaceutical applications, first using UV detection and then beginning in 1997 with reports using MS/MS detection. A list of some selected Prospekt applications is provided in Table 14.3. Note that most high throughput applications have been developed since the year 2000 with the introduction of the dual-column Prospekt-2.

The Prospekt-2 can be run in a generic mode, in which a universal sorbent chemistry (*e.g.*, HySphere Resin GP or Oasis HLB) is used with an LC column (*e.g.*, C18 4 x 30 mm, 3 μm) for a series of analytes; analytes may or may not be structurally similar. The LC run parameters usually involve the use of a fast gradient such as 5% acetonitrile to 95% acetonitrile and back to 5% within a 2 min window (flow rate 2.0 mL/min). This set of conditions is proposed as a first pass approach [133] and only if it fails (any quick modification proves unsuccessful) is the time taken to develop a more specific method. Note that the utilization of a fast gradient was found to contribute to matrix-induced ionization suppression, which was reduced by improving LC separation (increasing k') [134]. A related report further details the need for adequate chromatographic separation for quantitative bioanalysis using LC-MS/MS techniques [135]. Additional remedies for reducing the effects of matrix suppression on ionization have been reported as higher sample loading speed for the cartridge and lower sorbent hydrophobicity [136].

McLoughlin *et al.* reported the simultaneous determination of ten proprietary discovery compounds in plasma (cassette dosing) by APCI MS/MS using a cyano cartridge with the Prospekt; limits of quantitation ranged from 2.5–5 ng/mL for these compounds [137]. A simple C18 on-line extraction with LC-MS/MS detection was demonstrated as useful for the determination of 64 generic pharmaceutical compounds in support of drug discovery cassette dosing techniques [138]. This methodology is described in more detail in Chapter 1, Section 1.2.5. The extracted LOD (limit of detection) for 41 of these analytes was 25 pg and for 57 others was 250 pg, based on a 0.1 mL sample volume. Some additional generic approaches have been presented [139, 140].

While this generic approach meets the needs of little to no method development time, and no hardware changes, it does have some disadvantages. Gradient elution can sometimes be time consuming while waiting for the column to re-equilibrate to its starting mobile phase conditions. An SPE method developed to be selective for an analyte can often yield cleaner results than a generic method, as the specific method is customized for all steps of the process including load, wash and elution. However, it takes time (1–2 days) to develop

Table 14.3
Selected bioanalytical applications using on-line solid-phase extraction with disposable cartridges (Prospekt™ system)

Analytes(s)	Sample Matrix and Injection Volume	LOQ or LOD	SPE Sorbent	Cycle Time (min)	Analytical System	Reference
10 Proprietary compounds	Dog plasma 50µL	2.5–5 ng/mL LOQ	CN	8	LC-MS/MS	[137]
Clenbuterol	Urine 1mL	2 ng/mL LOD	HySphere C18	15	LC-MS/MS	[141]
Clenbuterol	Urine 1mL	50 ng/mL LOD 25 ng/mL LOD	C18 PVDB	11.5 8.5	LC-MS/MS	[142]
Eserine N-oxide	Human plasma 250µL	25 pg/mL LOQ	PLRP-S	7	LC-MS/MS	[143]
64 Generic drugs	Human plasma 100µL	250 pg/mL LOD	C18	4.5	LC-MS/MS	[138]
Paclitaxel	Human Serum 100µL	1 ng/mL LOQ	HySphere C18	1.5	MS/MS	[144]
Pranlukast and metabolites	Human plasma 100µL	10 ng/mL LOQ	Phenyl	5	LC-MS/MS	[145]
Rofecoxib	Human plasma 250µL	0.5 ng/mL LOQ	C8	10.5	LC-FLUOR	[146]
Ketorolac	Human plasma 1mL	5 ng/mL LOQ	C18	6	LC-UV	[147]
Irinotecan CPT-11 and metabolites	Rat plasma (diluted) 100µL	2.5–5 ng/mL LOQ	C18	9	LC-UV	[148]
Catecholamines	Urine (diluted) 200µL	1.3-3.0 ng/mL LOD	PLRP-S	12	LC-ECD	[149]

Analytes(s)	Sample Matrix and Injection Volume	LOQ or LOD	SPE Sorbent	Cycle Time (min)	Analytical System	Reference
Fluvastatin	Plasma (diluted) 200µL	0.5 nM LOQ	C2	14	LC-FLUOR	[150]
Anti-platelet activating factor BN-50727	Human plasma 500uL	3.75 ng/mL LOQ	CBA	6	LC-UV	[151]
Lesopriton and metabolite	Human plasma 100µL	1 ng/mL LOQ	C2	8.25	LC-FLUOR	[152]
Almokalant	Plasma (diluted) 650µL	2 nM LOQ	C2	10	LC-FLUOR	[153]
Codeine	Human plasma (diluted) 800µL	0.5 ng/mL LOD	C2	9	LC-UV	[154]
Nifedipine and nitrendipine	Human plasma 500µL	2 ng/mL LOQ	C2	11.5	LC-UV	[155]
Piroxicam	Human plasma 100µL	50 ng/mL LOQ	C8	12.5	LC-UV	[156]

LOQ, Limit of Quantitation; LOD, Limit of Detection
UV, Ultraviolet; FLUOR, Fluorescence; ECD, Electrochemical Detection

this more specific technique. Characteristic method development principles for on-line SPE are discussed in Section 14.6.3.

A full array of applications has been reported using the Prospekt on-line SPE technique. Its usefulness has been documented for bioanalytical applications providing drug discovery support [137, 138, 157, 158], developmental compound support [131, 143, 146, 148, 151, 159–162], drug screening for doping agents [141, 142, 163], therapeutic drug monitoring [129, 144, 145, 147, 150, 153–156, 164–168] and the analysis of endogenous materials [149, 169–175]. Now that the dual mode Prospekt-2 has demonstrated capability for higher throughput applications [130, 168, 176, 177], its utility is expected to increase for fast pharmaceutical analyses.

A general overview that highlights the use of the Prospekt-2 for high throughput analysis with MS techniques has been published by Koster and Ooms [133]. The applications described include a 2 min cycle time for 11 compounds extracted from serum using HySphere Resin GP cartridges; detection was by LC-MS/MS. Also, a 35 s cycle time is described for the extraction of carbamazepine from serum using HySphere C18 HD cartridges; detection by UV. An automated method development process is also introduced (see Section 14.6.3. for details).

It is possible to perform the on-line extractions and leave the analyte adsorbed to the cartridge for later analysis. This scenario may arise when policy states that biological samples cannot enter the mass spectrometer facility. Extractions can be performed in one location and the tray of cartridges, containing analyte but devoid of biological matrix, can be taken to another location and placed into an ACE unit coupled directly to an MS/MS for detection. There may even be a need to store analytes on extracted cartridges for analysis on another day. The stability of organophosphorous pesticides on C18 cartridges, following analysis, was evaluated in this manner over periods from 1–8 months at different temperatures [178].

The direct on-line coupling of SPE to MS/MS without an LC column has been reported for extraction of paclitaxel [144]; a cycle time of only 80 s was used for an LOQ (limit of quantitation) of 1 ng/mL. This approach requires the use of elution conditions that provide optimal ionization efficiency and analyte sensitivity. Another report established a number of low level assays in serum and plasma; quantitation limits were as low as 50 pg/mL using 200 μL sample volumes [179].

14.3.6.6 Special Related Techniques

14.3.6.6.1 On-Line SPE Using Traditional 96-Well Microplates

A special note is mentioned here of the capability for on-line solid-phase extraction offered by the SPE Twin PAL instrument (LEAP Technologies). Although the SPE Twin PAL does not technically match the Prospekt approach to fully enclosed on-line extraction and analysis, it does perform in a similar manner in an unattended, staggered parallel extraction mode. The SPE Twin PAL does not use individual SPE cartridges; rather, it processes standard 96-well solid-phase extraction microplates available from many vendors (see Chapter 11, Table 11.4 for a product listing). The system uses positive pressure for liquid processing through the extraction wells. The eluate is collected in a clean microplate that is moved into place below the extraction plate processing area at the designated time. The eluate from well #1 is injected as the extraction begins on well #2. One syringe (typically 1–5 mL) is used for SPE plate conditioning, loading and washing; a second syringe (typically 100 µL) is used for injecting small volumes of eluate into the chromatographic system. Essentially, this LEAP approach bridges the gap between off-line SPE in microplates and on-line fully enclosed SPE.

The SPE Twin PAL is controlled via the software Cycle Composer 4-2, which enables the user to control both PAL and the chromatographic system from one graphical interface. Method editing is possible with preprogrammed steps; more experienced users can use single step macros. This feature enables the user to add internal standards and perform dilutions of the sample, if desired. The flexibility of the SPE Twin PAL allows the user to perform any step which was previously done manually. The software is CFR 21 compliant.

LEAP Technologies has demonstrated full compatibility with major brands of SPE plates. Perhaps most appealing is the µElution Plate from Waters Corporation, which requires only a 25 µL elution volume (instead of 100–400 µL volumes from a variety of other products). A change in the well geometry of the plate allows the sorbent material to be stored in the lower end of the redesigned narrow tip instead of higher up within a larger diameter well bottom, thus achieving its lower elution volume. A great benefit of the system is that throughput can be increased when sample preparation times are longer than the analytical run time.

14.3.6.6.2 On-Line SPE Using Membrane Preconcentration Capillary Electrophoresis Mass Spectrometry (mPC-CE-MS)

A technique related to traditional on-line sample preparation methodologies involves the use of a preconcentration cartridge for analytical techniques in capillary electrophoresis-mass spectrometry (CE-MS). This preconcentration step yields improved limits of detection when using small sample volumes (typically <50 nL for a 50 μm internal diameter capillary). The use of a fabricated membrane cartridge for preconcentration and on-line coupling with CE-MS or CE-MS/MS has been reviewed [180]. Several additional papers discuss the development and details of this technique [181–184] as well as applications for its use in the bioanalysis of peptide mixtures [185–188]. On-line analyte concentration techniques for the determination of drugs, metabolic intermediates and biopolymers in biological fluids have been reviewed by Guzman *et al.* [189].

14.4 High Throughput On-Line Sample Preparation Using Multiple Columns

14.4.1 Introduction

Besides the popular dual-column mode, a new trend has emerged which involves the use of more than two columns for on-line bioanalysis. The objective of using this multiple column approach is to improve the overall throughput for bioanalysis by more efficiently utilizing the resources of the mass spectrometer. Since all on-line sample preparation methods are fundamentally chromatography based, they can be formatted into a parallel mode to significantly improve sample throughput. As the number of columns increases, so does the arrangement in which they can be placed. Some representative uses of multiple column separations are discussed next.

14.4.2 Dual-Column Staggered Extraction Coupled with Single-Column Separation

The purpose of the dual-column staggered extraction mode is to enhance the sample extraction throughput. A representative schematic diagram of this multiple-column mode, reported by Xia *et al.* [190], is shown in Figure 14.5. This system is very similar to the standard dual-column mode described in Section 14.3 except that two extraction columns are used and a ten-port valve replaces the six-port valve.

With this system as described, two on-line processes can be staggered on the two extraction columns. While one column is in the wash, load, or equilibration step, the other is in the elution step. The advantage of this system is that throughput can be doubled to feed the separation column. It is a very useful tool to increase the overall on-line analysis throughput when the run time on the separation column is considerably shorter than the extraction cycle time.

Dual-column staggered extraction for analysis of metabolic stability screening samples was reported by Ong *et al.* [191]. Two Turboflow columns were connected in parallel followed by a single analytical column operated under generic fast gradient chromatography conditions. Cycle time was 2 min per injection and carryover was compound dependent but did approach 0.5% for the worst case. Use of traditional dual-column mode in series (although not as fast) reduced carryover to <0.5%; this arrangement was preferred for plasma samples in which the limit of quantitation approached 0.1 ng/mL. Another application utilized on-line TFC with two parallel Oasis HLB columns for the *in vitro* analysis of drug metabolites following cytochrome P450 microsomal incubation [192]. Total run time was 1.5 min for all analytes; dynamic range (4–1000 ng/mL), accuracy and precision were comparable with those obtained using off-line protein precipitation.

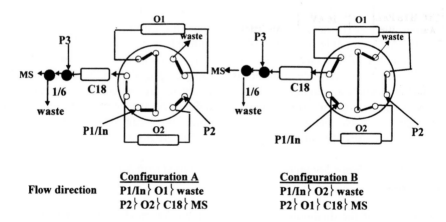

Figure 14.5. Schematic diagram of a staggered dual-column extraction coupled with single-column separation. O1 and O2 are the two extraction columns and P1 and P2 are the two extraction pumps. The injector (In) is part of an integral system for Pump 1. Pump3 (P3) is used for post-column infusion for the specific application and is not generally required. Reprinted with permission from [190]. Copyright 2000 John Wiley & Sons, Ltd.

LC-MS/MS applications using dual-column extraction with RAM coupled with single-column separation have also been reported. The analyte granisetron was determined on-line with an ISRP column and the calibration range was 0.1 to 50 ng/mL using an 80 μL plasma sample [193]. Needham [194] investigated the utility of BioTrap 500 RAM for use in providing bioanalytical support for drug discovery programs.

14.4.3 Single-Column Extraction Coupled with Dual-Column Staggered Separation

In some cases, the opposite situation can occur when the separation run time is considerably longer than the extraction cycle time. This is often the case when the separation of multiple components is involved. In order to improve the throughput of this application, single-column extraction coupled with dual-column separation has been developed. Figure 14.6 shows the configuration of

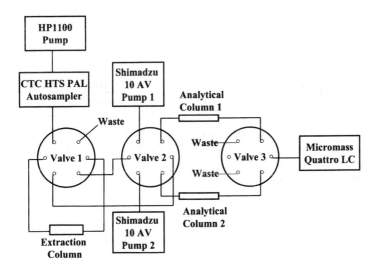

Figure 14.6. Instrument setup of the staggered separation LC-MS/MS system with on-line extraction. The system consists of two analytical pumps and columns, one extraction pump and extraction column, one autosampler, three electrically actuated 6-port valves, and a triple quadrupole mass spectrometer. Reprinted with permission from [195]. Copyright 2001 John Wiley & Sons, Ltd.

a system reported by Wu [195]. It has one turbulent flow extraction column coupled with two separation columns. The extraction column alternately loads extracted samples onto the two separation columns where staggered separations are performed. In this staggered fashion, the separation run time is reduced by about 50%, resulting in an overall improvement in throughput. The method was implemented for both single and multiple component assays and was found to be an effective tool for enhancing the throughput of screening drug candidates in biological fluids.

Another example of a staggered single-column extraction with dual-column separation is reported for the determination of vancomycin in serum and urine [196]. The on-line extraction step took 15 s followed by 90 s for the gradient chromatography. Serum and urine were vortex mixed with internal standard and then centrifuged before direct injection.

14.4.4 Multiple-Column Extraction Coupled with Multiple-Column Separation

14.4.4.1 Conventional Single Interface to Mass Spectrometer

Many literature reports detail the use of parallel LC columns (usually from 2 to 8) for analytical separation, *e.g.*, in high throughput microsomal stability screening of compound libraries [197] and in a multiple component bioanalytical assay [198]. On-line sample preparation was not utilized in these reports; rather, off-line sample preparation was employed. Further discussion is outside the scope of this chapter and the reader is referred to a review of parallel LC-MS and LC-MS/MS techniques [199]. In an effort to enhance sample throughput in bioanalysis, multiple extraction columns have been combined with multiple separation columns for on-line sample preparation and analyses.

One approach for multiple on-line extraction and separation uses two columns in parallel for the sample extraction and two analytical columns in parallel for separation and analysis [200]. This method allows the use of one of the extraction columns for analyte purification while the other column is being equilibrated in preparation for the next sample. Likewise, one analytical column performs the separation while the other column is being equilibrated. This dual mode permits a shorter run time because the time required for column reconditioning is not added to the total analysis time.

Instead of two parallel columns in a system configuration, four columns have been utilized for multiple analytical separations. Samples are injected sequentially onto the four separation columns to maximize use of the mass spectrometer. King *et al.* reported the use of a four-column staggered separation (without on-line extraction) [201]. This system is named Aria™ LX4 (Cohesive Technologies) and consists of four fully independent LC systems fed by two injection syringes on a twin arm HTS PAL autosampler (LEAP Technologies). Samples are introduced into the APCI-MS interface by use of a selection valve. A computer program completely controls all timing and triggering for the injection step, the gradient start time, and data collection. As with most other configurations described in this chapter, this system is based solely on column switching methods and does not require any modifications to the mass spectrometer interface.

The retention time windows that contain the analyte are detected one at a time by the mass spectrometer. For example, a single column method with an injection to injection cycle time of 6.0 min requires an acquisition time on the mass spectrometer of 1.5 min [28]. By staggering the injections among the four LC systems with an offset time of 1.5 min, a sample can be run using the Aria every 1.5 min injection to injection.

In addition to four parallel columns in an analytical system, four parallel extraction columns can be added for multiple on-line extraction and separation. Samples are injected onto the four extraction columns which are then eluted sequentially onto the four separation columns. Caulfield presented results using the Aria TX4 for the analysis of urinary free cortisol [202]; a sample result was obtained every 1.2 min and 150–200 samples can be analyzed within 8 h. McHugh also presented methodology for incorporating the Aria TX4 into a bioanalytical program supporting drug discovery [203]. The turbulent flow extraction column chemistry of choice was a Cyclone™ polymeric column (Cohesive Technologies).

14.4.4.2 Indexed Four-Sprayer Electrospray Interface to Mass Spectrometer

A slightly different approach for multiple-column extraction coupled with multiple-column separation is based on a commercially available multiplexed inlet mass spectrometer interface (MUX®, Micromass Inc, Beverly MA USA). This MUX interface utilizes a four-channel electrospray ion source on a triple quadrupole mass spectrometer for parallel analysis; an eight-channel version is also in use [204]. In the MUX technology variant the conventional electrospray probe and outer source assembly are replaced with a new source housing containing an array of four miniaturized, pneumatically assisted electrosprays. The position of the sampling rotor is monitored in real time enabling the four liquid inlets to be indexed.

Some previous reports of the use of this four channel multiplexed interface include use with a time-of-flight mass spectrometer for the qualitative analysis of a drug test mixture [205] and with triple quadrupole instruments. The simultaneous validation of LC-MS/MS methods for loratidine was made in four different preclinical matrices [206, 207]. Sample preparation for loratidine used 96-well solid-phase extraction rather than an on-line technique. An evaluation of this same interface was made using four pure analyte standards and diazepam in microsomal incubation media [208]. It was found that samples to be compared quantitatively must be analyzed through the same sprayer, and each channel must be independently calibrated. Deng *et al.* described the bioanalytical use of four monolithic columns for parallel separations and the MUX system with four multiplexed inlets [66], also following sample preparation using 96-well SPE. In this application, overall throughput was reported as 30 s per plasma sample with gradient elution on a 10 cm long LC column. Improvements in developing a high pressure gradient pumping system for the MUX, for the analysis of combinatorial libraries, have been communicated in a Letter to the Editor [209].

The first reports of a four-column parallel, on-line, turbulent flow extraction combined with a four-column parallel separation system and four-sprayer mass detection were reported by Bayliss *et al.* [210] and Deng *et al.* [211]. Both groups reported the implementation of the parallel system for high throughput quantitation of pharmaceutical compounds in biological fluids. In Bayliss's work, a four channel parallel single-column TFC system was interfaced to the MUX while in Deng's work, a four channel parallel dual-column TFC system was used with the MUX. Both systems were reported to be successfully used for direct plasma injection with an overall throughput increase by a factor of four.

Bayliss *et al.* [210] utilized the single-column turbulent flow chromatography mode with four sprayer mass detection, applied on both narrow bore (1 x 50 mm i.d. Oasis HLB 30 μm) and capillary scale (0.18 x 50 mm i.d. Oasis HLB 25 μm). Plasma was mixed with an aqueous internal standard (1:1) in a 96-well collection plate, and 20 μL or 5 μL injection volumes were made for narrow bore or capillary LC, respectively. Throughput was reported as 120 samples per hour (narrow bore LC) and 96 samples per hour (capillary LC).

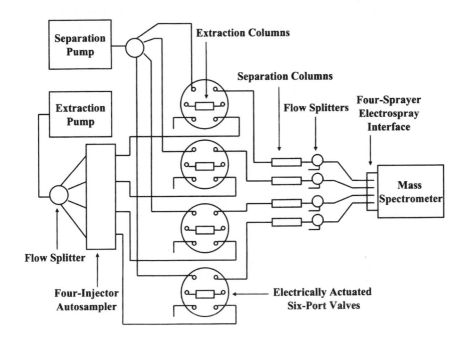

Figure 14.7. Instrument configuration for a multiple-column extraction coupled with multiple-column separation and four sprayer mass detection. Reprinted with permission from [211]. Copyright 2001 John Wiley & Sons, Ltd.

Deng *et al.* [211] used the dual-column turbulent flow chromatography mode with four extraction columns (1 x 50 mm i.d. Oasis HLB 30 μm), four separation columns and four sprayer mass detection. A schematic diagram of this system is depicted in Figure 14.7. Three processes were performed in parallel: on-line extraction, separation and mass spectrometric detection. The interchannel reproducibility of analyte response and sensitivity was found to be satisfactory, and interchannel crosstalk was found to be about 0.1%. Although this parallel system could not achieve the same LLOQ (Lower Limit of Quantitation) as on a conventional single channel system, a comparison of data generated on this system and a conventional system showed no more than 20% difference in measured concentrations. This parallel system was confirmed as a viable tool for high throughput quantitation of analytes in a fast paced drug discovery environment.

14.5 Comparisons of On-Line Versus Off-Line and Comparisons Among On-Line Methods

14.5.1 Introduction

Many comparisons among on-line sample preparation methods and conventional off-line methods have been made in an attempt to identify throughput gains and/or performance advantages. Similarly, comparisons have been made among several of the on-line sample preparation methods. This section will summarize some of the findings and comparisons reported from these evaluations.

14.5.2 Comparisons of On-Line Sample Preparation with Off-Line Sample Preparation

14.5.2.1 TFC vs. PPT, LLE or SPE

As an alternative approach to conventional sample preparation, on-line sample preparation has often been compared and cross-validated with off-line methods. Zimmer *et al.* performed studies using two proprietary compounds to compare the performance of single-column mode TFC with off-line SPE, liquid-liquid extraction (LLE), and protein precipitation (PPT) methods [21]. The TFC assays for both compounds passed full validation in compliance with Good Laboratory Practices (GLP). The quality of the data obtained with the TFC system, for one of the compounds analyzed from dog plasma, was similar to manual LLE and slightly better than semi-automated 96-well SPE in terms of accuracy and precision. The accuracy results for LLE, TFC, and SPE were −3.7 to 4.4%, −3.1 to 6.7%, and −4.1 to 10.6%, respectively. The precision results for the three methods were 3.1 to 5.1%, 3.1 to 6.8%, and 5.1 to 5.5%, respectively. The TFC assay achieved the same LOQ (1 ng/mL) as SPE; since an 8-fold larger plasma volume was used for LLE, its LOQ was 0.1 ng/mL. For the other compound, the accuracy and precision of the TFC assay were slightly better than data obtained using a validated PPT method. The accuracy and precision for TFC were −3.7 to 3.5% and 1.6 to 5.4%, respectively. For manual PPT, the accuracy and precision were 0 to 6.9% and 3.6 to 10.5%, respectively.

A comparison of TFC with LLE showed that the two techniques were equivalent in terms of total chromatographic run time, accuracy and precision [94]. However, the overall sample analysis throughput was doubled by using the TFC method. Good sensitivity with TFC was achieved using a small volume of biological matrix (equivalent to 5 µL).

The quantitative analysis of several development compounds from plasma was compared using on-line TFC, column switching after PPT, a micro column switching assay, and semi-automated 96-well SPE [22]. The LOQ for these analytes using TFC was found to be five-fold higher compared with the other methods examined. Disk SPE was fast and semi-automated, and minimized matrix effects. Column switching after PPT was fast, accurate and sensitive. Microbore LC allowed fast analysis with low sample consumption.

While comparing TFC with PPT for the quantitation of ketoconazole in human plasma, Ramos *et al.* concluded that the methods have very similar accuracy and precision [92]. Wu *et al.* compared the performance of dual-column TFC with SPE for ten structurally diverse drug compounds and found that the dynamic range, accuracy, and precision were very similar for the two methods [18]. Chassaing *et al.* cross-validated a dog pharmacokinetic study comparing dual-column TFC with off-line SPE and found that the half-lives obtained by TFC and SPE were 2.2 and 2.3 hr, respectively. The variation between the two sets of pharmacokinetic data was less than 12% [99, 100]. A report from Niggebrugge concluded that TFC used with an initial 96-well filtration of plasma provided a simple and cost effective assay that is as dependable as semi-automated 96-well SPE [212].

The total solvent consumption for TFC and SPE techniques was compared [9]. While more solvent is consumed using TFC than in conventional LC, the total volume of solvent used per analytical run for 96 samples is similar for TFC and for SPE followed by LC-MS/MS. About 1 L of mobile phase was used for a 4.5 h run using TFC; a total of 960 mL solvent volume was calculated for 96-well SPE (assuming 5 mL solvent per well for the condition, wash and elution steps) and analysis by conventional LC using 1 mL/min flow rate and a 5 min run time.

14.5.2.2 Prospekt vs. PPT, LLE or SPE

The performance of on-line SPE using a Prospekt has also been compared with off-line methods. Alexander *et al.* reported fully comparable results in accuracy and precision using the Prospekt and off-line semi-automated 96-well SPE [213]. Hedenmo *et al.* claimed good agreement between data obtained from a Prospekt assay and those from LLE [164]. Using 32 authentic plasma samples of clarithromycin and 28 samples of roxithromycin, they found the ratio of the Prospekt assay and LLE assay, at the 95% confidence interval (CI), to be 1.011 CI (0.977, 1.046) for clarithromycin and 1.007 CI (0.981, 1.034) for roxithromycin. Wang *et al.* compared Prospekt with semi-automated protein precipitation in microplates [157]. Both techniques extracted about 150 samples in 6–7 h and demonstrated similar performance, although the on-line Prospekt allowed a four-fold concentration of analyte. Borbridge *et al.* compared the on-line Prospekt approach to automated off-line SPE and reported that on-line SPE with its greater degree of automation obtained higher assay sensitivity and showed less matrix effects [162]. Yritia *et al.* compared on-line and off-line solid-phase extraction and reported slightly better detection limits with on-line SPE but overall equal performance for bioequivalence studies [156].

14.5.2.3 Immunoaffinity vs. SPE

In comparing on-line immunoaffinity extraction with off-line SPE, Cai and Henion reported a 20-fold improvement in the detection limit with the on-line immunoaffinity assay [123]. Ouyang *et al.* claimed up to a 474-fold improvement in detection limit by using on-line immunoaffinity methods [214].

14.5.2.4 Time Considerations

Most of the published work has reported a significant savings of time using on-line sample preparation methods compared with off-line manual methods. For example, Zimmer claimed a reduction in hands-on technician time from 8 h to 1 h by using TFC instead of manual LLE for each set of 100 samples [21]; total time for the extraction and analysis for TFC and LLE were 5.25 h and 16 h, respectively. Chassaing *et al.* reported a 50% time savings between TFC and manual SPE [99, 100]. However, with automated PPT and SPE, this may not always be the case.

Alexander *et al.* reported a total analysis time of 7.9 h using automated SPE and 15.1 h using the original Prospekt for the analysis of 130 samples [213]; however, note that the newer Prospekt-2 offers higher throughput via its two cartridge clamps for parallel SPE. Ramos *et al.* also claimed that automated PPT had a higher throughput than TFC [92]. In both cases, special liquid handlers and process optimization were employed.

TFC was compared with 96-well PPT in a collection plate and Quadra® 96 (Tomtec Inc., Hamden, CT USA) semi-automated transfer of the supernatant [92]. The injection to injection cycle time for TFC (4.5 min) was about 2.5 min longer than for 96-well PPT, so PPT was chosen as the assay of choice for the analysis of clinical samples.

Alexander reported that the extraction/analysis time using the Prospekt (4.5 min per sample) was longer than for 96-well SPE at <2 min per sample [213]. The bulk of the chromatographic cycle time for the Prospekt was associated with washes during the sample cleanup step. Throughput can be increased by running the assay on multiple instruments, using shorter LC columns or by operating these columns in parallel.

14.5.3 Comparisons Among Different On-Line Sample Preparation Methods

14.5.3.1 Introduction

All of the on-line sample preparation methods described in the previous section have been reported for bioanalysis and, therefore, meet some basic criteria for performance. However, because of the differences in hardware configuration and extraction mechanism, each of the on-line methods has its own advantages and disadvantages. No single approach works for all purposes. It is important to understand the capabilities of each system and then choose the best one that will satisfy the required application needs.

14.5.3.2 Single-Column and Dual-Column Turbulent Flow Chromatography

The single-column TFC mode offers the advantage of simplicity and throughput. However, in most cases, its performance is not comparable to that of the dual-column mode in terms of chromatographic separation and assay sensitivity. Shen *et al.* compared results from both single- and dual-column TFC and concluded that single-column TFC, although it lacks resolving power, offered satisfactory results with higher throughput for a dog pharmacokinetic study of a drug candidate [215]. Jemal *et al.* compared the use of single- and dual-column TFC and demonstrated additional separation and sensitivity achieved with the dual-column mode [10]. Mallet utilized both single- and dual-column modes with success, achieving relative standard deviation values less than 5% for both configurations; a basic and an acidic drug were analyzed [101]. Using 200 µL injection volumes, special attention had to be made to the washing solution (5% organic or less) to avoid protein precipitation in the waste lines.

Zeng *et al.* recently reported results from a comparison study of the single- and dual-column methods [216]. Four structurally diverse compounds (aminopterin, apomorphine, benzoylecgonine, and carbamazepine) in plasma were injected directly using the single- and dual-column modes, respectively. A significant improvement in chromatographic separation was observed.

Resolution data for adjacent peaks were 0, 0.09, and 0.37 using the single-column mode and 1.50, 1.06, and 6.24 using the dual-column mode. Also, because of the chromatographic focusing effect, the signal to noise ratios of these compounds in the dual-column mode showed enhancement by a factor of 6 to 14. Generally speaking, the single-column mode is most suitable for high level and single analyte assays such as microsomal incubations or single component pharmacokinetic studies. However, for trace level or multiple component bioanalysis, the dual-column mode should be considered.

14.5.3.3 Staggered and Parallel Sample Introduction

Both staggered and parallel sample introduction methods are effective ways to improve sample throughput. The staggered approach is relatively easier to set up because it does not involve any modification of the mass spectrometer hardware and the key technology used is column switching with timed events. The detection limit is the same as in a conventional assay [195, 201, 200]. The disadvantage of this method is that the extent of the throughput improvement depends on the blank chromatographic window available. If the retention times of analyte(s) and the internal standard spread out, this method will have less value. Also, fast separation using high flow rates and short columns is becoming more frequently used. This fast separation significantly reduces the portion of blank chromatographic window in the total run time, resulting in a limited enhancement for sample throughput in the staggered mode.

Parallel sample introduction using a multiple sprayer mass spectrometer, on the other hand, does not rely on the blank chromatographic window of a separation and therefore will always deliver a fixed multiple fold enhancement in throughput regardless of the type and nature of the separation. The disadvantage of the parallel sample introduction mode is the compromised detection sensitivity and the cost and complexity associated with modification of the mass spectrometer. Yang *et al.* [206] reported an increase in quantitation limit by a factor of three and Deng *et al.* [211] reported an increase by a factor of four.

14.6 Method Development Strategies

14.6.1 Introduction

This section describes some general method development procedures for on-line sample preparation. The focus of this discussion is on methods that are based on turbulent flow chromatography and on-line SPE using the Prospekt.

These two approaches represent the most widely used on-line sample preparation methods. Since there are many user adjustable parameters that can affect method performance, a better understanding of these operational parameters and how they relate to overall performance leads to a greater probability of success.

14.6.2 Method Development for Turbulent Flow Chromatography

14.6.2.1 Selection of Operation Mode

The first step in developing a turbulent flow chromatography method is deciding whether to use single-, dual- or multiple-column mode. As discussed in the previous section, this decision should be made based on the separation, sensitivity, and throughput requirements of the application. Generally speaking, if the assay requires the simultaneous quantitation of more than one analyte, or if the assay requires low or even a subnanomolar limit of quantitation, the dual-column mode is often the method of choice. Multiple-column staggered and parallel sample introduction are usually considered only when there is a strong demand for greater throughput (2 min per sample cycle time or faster). In the next section, method development procedures will be described for the dual-column mode only, but similar approaches can be applied to other modes of using turbulent flow chromatography.

14.6.2.2 Selection of Column

The second step in TFC method development is choosing the appropriate chemistry and type of columns for extraction and separation. Some of the more popular varieties of extraction columns are the Turbo C18 and Cyclone columns (1 x 50 mm) from Cohesive Technologies and the Oasis HLB column (1 x 50mm or 2 x 15 mm) from Waters Corporation. The Cyclone and Oasis columns use polymer based packing materials that are more sustainable to sharp backpressure changes during column switching techniques; therefore, they are more suitable for the dual-column mode. The selection of the analytical column is mostly based on the separation needs of the analyte(s). Generally, reversed phase columns with 2.1 or 4.6 mm i.d. are among the most frequently used. There is a recent trend to use short columns (20–30 mm in length) to improve throughput [28, 91, 217]. In order to avoid potential plugging of the analytical column, the use of a larger particle size packing material (10 μm) in the analytical column has been reported [99, 100].

14.6.2.3 Sample Preparation

Two sample preparation approaches for TFC have been widely reported in the literature. The first is a direct injection approach [9, 18, 99, 100, 190]. In a typical method, plasma samples are mixed with an equal volume of high aqueous internal standard solution and injected directly into the TFC system. Usually a centrifugation step is performed immediately before injection to remove particulates in the samples. This approach eliminates the majority of the sample handling steps performed in typical off-line methods and offers excellent throughput. The disadvantages include a relatively high risk of system failure due to plugging. Also, the internal standard must be dissolved in a predominantly aqueous solvent to avoid precipitating proteins within the sample matrix.

The second approach is used when internal standards need to be dissolved in a solution containing organic solvent. In this case, an indirect injection should be considered [28, 218, 219]. In a typical method, plasma samples are first precipitated using two volumes of acetonitrile containing the internal standard. The mixture is then centrifuged to pellet proteins at the bottom of the well or tube. The supernatant is injected into the TFC system. This indirect approach, besides dissolving the internal standard, generally provides better ruggedness and helps with highly protein-bound compounds since the precipitation step dissociates the binding. It is therefore a more generic method [28]. The disadvantages include the extra effort and time required in sample pretreatment and the compromised detection sensitivity. Sensitivity suffers because the samples are diluted with a high percentage of organic and the injection volume must be limited to avoid breakthrough. If the assay requires a large batch size (over 100 samples), or if the analyte structures change significantly from day to day, the indirect approach is usually the method of choice.

14.6.2.4 Solvent Selection and Flow Conditions for the On-Line Extraction Process

The on-line extraction process for TFC is relatively simple and several groups have reported the use of generic approaches for this step [19, 28, 99, 100]. Typically, an injection volume from 25 μL to 50 μL is used. The most common load and cleanup solvent is formic acid in concentrations from 0.05 to 0.1%. The use of formic acid helps to dissociate protein-bound analyte and match the pH with that of the mobile phase used in the elution step. The flow rate for the load and cleanup steps is normally 4 mL/min for a period of at least 30 s. It may be reduced to a lower value (0.3 mL/min) for a few seconds at the end of this step before the valve switches to the elution position. This approach should help to reduce potential damage to the extraction column and the pumping system due to the sudden pressure drop during column switching.

The elution step typically involves a fast gradient to backflush the analyte(s) from the extraction column and onto the separation column. The typical mobile phase solvents used for the separation step are mixtures of 0.05 to 0.1% formic acid and acetonitrile. The extraction column usually stays in loop with the analytical column for at least 1 min to ensure the complete transfer of the analyte onto the separation column. The extraction column is then switched back to the load and cleanup position for the flush and re-equilibration step. The flushing solvent is usually acetonitrile or a mixture of acetonitrile and methanol, which is delivered as solvent B on the extraction pump. The use of this flushing step reduces carryover.

The solvent and flow conditions used in a generic approach proposed by Herman are summarized in Table 14.4 [28]. In this table, there is an additional parameter for Pump 2 named the "Tee position" which is an approach to improve peak shape; use of this Tee is not common with other approaches and will be explained in Section 14.7.4. The valve configuration of the system used by Herman is shown in Figure 14.9

14.6.2.5 Generic Method for the On-Line Extraction of Drugs in Various Biological Matrices

In a drug discovery ADME laboratory providing bioanalytical support, new analytes and compound series are frequently encountered. The time required developing new methods for each analyte and/or series needs to be reduced in order to move toward true high throughput ADME screening. It is not possible to develop a unique method for each new compound being screened; rather, the approach taken is to develop one method that will work for >95% of compounds. Recent development of a polymeric stationary phase for the TFC extraction column has made this technique viable.

A sound and proven generic method that meets this objective was developed by Herman [28] utilizing turbulent flow chromatography with fast LC-MS/MS detection techniques. This method has been used with great success to screen over 1000 compounds in many different biological matrices. It utilizes a Cyclone polymeric extraction column with an Eclipse XDB C18 analytical column; extraction and analysis details have been reported [28].

Table 14.4
Liquid chromatography experimental conditions for a turbulent flow chromatography
dual-column method developed as a generic method for analytes

	Pump 1 - Cleanup Column	Pump 2 - Analytical Column
Column	Cyclone HTLC	Eclipse XDB C18
		4.6 x 15mm, 3μm, 120 A
Injection volume	25 μL	
Solvent A	0.05% formic acid in water	0.05% formic acid in water
Solvent B	0.05% formic acid in ACN	0.05% formic acid in ACN

Gradient for Pump 1

Time (min)	A%	B%	Valve Position	Flow (mL/min)
0.0	100	0	Load	4.0
0.5	100	0		4.0
0.6	100	0	Inject	0.3
2.1	60	40		
3.6	60	40	Load	4.0
4.1	100	0		4.0
6.0	100	0		4.0

Gradient for Pump 2

Time (min)	A%	B%	Tee Position	Flow (mL/min)
0.0	100	0	Out	1.2
0.6	100	0	In	1.2
2.1	100	0	Out	1.0
3.6	5	95		1.0
4.1	5	95		1.2
4.6	100	0		1.2

Reprinted with permission from [28]. Copyright 2002 John Wiley & Sons, Ltd.

14.6.3 Method Development for Disposable Solid-Phase Extraction Cartridges

14.6.3.1 Selection of Sorbent

The method development process for SPE using disposable cartridges, as
discussed for the Prospekt-2 system (see Section 14.3.6), involves the
examination of several . extraction variables with subsequent optimization.
These parameters include the sorbent chemistry, the composition of the load,
wash and elution solvents, sorbent particle size (40 μm or <10 μm), solvent

volumes and flow rates. A general overview of the method development process for on-line SPE coupled to LC and LC-MS has been published by Ooms *et al.* [220].

The sorbent chemistry is the first variable examined. A quick approach to evaluating sorbents is to examine their breakthrough curves. Once each SPE cartridge is equilibrated, a solution of the analyte is pumped directly through a UV detector to give an absorbance signal (the cartridge is bypassed). Then the cartridge is switched on-line and the UV signal drops to baseline as analyte is retained on the sorbent bed. When breakthrough occurs, the UV signal will rise back to its initial level. The volume of analyte solution that flows through the extraction cartridge before the signal rises is known as the breakthrough volume. The shape of the slope after breakthrough is an indication of desorption efficiency; the steeper the curve, the better the desorption efficiency. Ideally, quick desorption (steep slope) is desired so that the analyte is removed from the sorbent in a small volume. Together, the breakthrough volume and the slope of the curve are used to evaluate potential sorbents for a solid-phase extraction method. This information can all be gained within a single run. Examples of breakthrough curves are shown in Figure 14.8.

Some possible causes for low recovery of analyte from sorbent can be due to factors such as: poor affinity for the sorbent, lack of sufficient sorbent capacity for the mass of analyte loaded, strong adsorption to the sorbent (difficult or slow elution), slow mass transfer kinetics for the adsorption step, and adsorption to tubing. Note that method development trays are available, in which different sorbent chemistries in cartridges are contained within the 96-position tray. Screening of sorbent chemistries can then be easily performed in order to measure percent recovery, percent breakthrough and percent adsorption to system components for every combination of sorbent and cleanup parameters.

14.6.3.2 Optimization of Extraction Variables

Once a sorbent chemistry is selected, as described above, optimization experiments are performed. Recovery is measured in several experiments to evaluate analyte loss caused by adsorption, degradation and breakthrough. The dual cartridge system permits quick determination of the cause of analyte loss since two cartridges can be switched in series. One cartridge or both can be excluded from the solvent flow path.

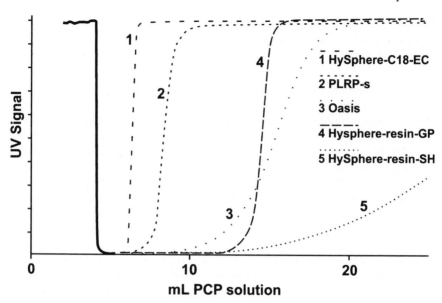

Figure 14.8. Examples of breakthrough curves for one analyte measured for different sorbent chemistries. The 100% level is measured at the point when the cartridge is switched in-line with the sample stream. The breakthrough volume is measured at the point when the detector signal has returned to 10% of its 100% level. The slope of the breakthrough curve is measured at the point when the signal has returned to 90% of its 100% level. Illustration reprinted with permission from Spark Holland.

First, two cartridges are placed in series to measure sorbent breakthrough. The analyte is introduced into the system by the autosampler and is transferred to the first cartridge. If that sorbent chemistry is optimal, the analyte will be completely retained; if not, analyte will break through the first cartridge and be trapped by the second cartridge. The extraction performance is monitored using three consecutive runs to look for analyte loss caused by: (a) adsorption, (b) recovery from the primary cartridge and (c) breakthrough. All tasks can be automated and run without operator involvement.

Analyte loss due to adsorption to the sample loop or tubing is monitored by directing the elution mobile phase through the loop and transfer tubing and to the analytical column, avoiding both SPE cartridges. Ways to reduce this adsorption include increasing the percentage of organic in the sample loading solvent or the wash solvent, and/or adding an ionic surfactant to the system mobile phase.

Recovery on the primary cartridge is determined by switching cartridge 1 in the flow path of the mobile phase for LC analysis; cartridge 2 is bypassed. Analyte recovery and matrix effects are determined. If recovery is lower than desired, a different sorbent chemistry can be tried or the percentage of organic in the sample transfer solvent can be manipulated.

Breakthrough is determined by taking the primary cartridge out of the flow path and eluting analyte retained on the second cartridge. The percentage breakthrough is calculated by comparison of the amount of analyte detected on the second cartridge with that detected on the first cartridge. If breakthrough is higher than desired, the analyte needs to be retained more strongly by the first cartridge and so another sorbent chemistry should be evaluated. Also, the percentage of organic within the mobile phase that transfer the sample can be manipulated.

The results from the experiments mentioned above will confirm the choice of sorbent chemistry by examining recovery, breakthrough and any loss due to adsorption in the system. If desired, the sorbent can be further evaluated by comparing the effect of particle size, reducing from 40 μm to 8 μm for those chemistries that are available in the smaller particle size. Additional studies can be performed to optimize flow rates and solvent compositions when further optimization is warranted. The chromatography also must be developed separately. The focus here is on the particular type of analytical column used, the mobile phase composition and the flow rate.

The rapid and straightforward approach that the Prospekt-2 employs for method development in solid-phase extraction is seen as a time savings compared with the traditional off-line approach. Determining the sorbent, sample load pH, wash and elution solvents and volumes are time consuming when done manually. Combining the two has been reported [177]; SPE methods can be developed using the Prospekt-2 and then transferred to off-line solid-phase extraction in the 96-well microplate format. Note that method development for off-line solid-phase extraction using microplates can be accomplished rapidly when combined with automation. Suggested further reading on this subject is found in Chapter 12, "Strategies for Method Development and Optimization" and Chapter 13, "Solid-Phase Extraction: Automation Strategies."

14.7 Method Optimization and Troubleshooting

14.7.1 Introduction

Some general method development procedures useful for on-line sample preparation have been described in the previous section. Now, some specific protocols and approaches to optimize and troubleshoot these on-line sample preparation methods are presented. Again, turbulent flow chromatography and SPE using disposable cartridges are used as the two primary models.

14.7.2 How to Reduce Carryover

Autosampler carryover is defined as the amount of analyte detected in a blank sample following the injection of a high standard; this subject has been reviewed by Dolan [221]. Carryover tends to be more of an issue in on-line sample preparation than in conventional off-line methods because the on-line system is directly in contact with the relatively large mass of proteins present in biological fluid. The proteins can adsorb onto the surface of the injector, transfer lines, column frits, and column packing material through various mechanisms. Since most of the drug-like compounds are highly protein-bound, a small portion of them can remain bound to the adsorbed proteins and remain in the system. A simple wash with organic solvent may not easily remove them. Wu *et al.* compared the carryover of ten compounds using both on-line and off-line sample preparation methods and noticed an increase in carryover with on-line methods by an average factor of two [18].

Several approaches can be used to reduce carryover in on-line sample preparation. The HTS PAL autosampler from LEAP Technologies is becoming known as the preferred autosampler for on-line sample preparation. It minimizes the use of transfer tubing by drawing samples directly into a regular syringe for injection. Another feature of this autosampler is its capability to use two wash solvents to clean the syringe, needle and the injector. Therefore, an acidic and a basic, or a highly aqueous and a highly organic solvent can be used for washing [18].

Takarewski *et al.* reported several approaches to reduce carryover in TFC [222]. The most significant improvement claimed was the modification of a Gilson autosampler using a PEEK tee which allowed for cleaning of the injection port. Chassaing *et al.* reported the use of an oversized loop, a three step column wash with acid, base and a high percentage organic solution, and 1% acetic acid in methanol/water (50/50, v/v) as the autosampler wash. It was reported that no carryover was observed after more than 2000 injections [99, 100].

14.7.3 How to Improve Ruggedness

System ruggedness is also a very important criterion for on-line sample preparation. Most of the on-line work reported to date has claimed good ruggedness for the injection of hundreds of plasma samples a day. However, because of the complexity of the biological matrices, there is always a high risk of system failure. For dual-column TFC, the failure can occur as a result of the fouling of the analytical column, plugging of the extraction column or plugging of the injector. The fouling of the analytical column results from column exposure to macromolecules such as proteins that have not been completely removed from the sample matrices during the on-line extraction step. A simple approach for reducing this problem is placing a frit or guard column in-line.

The clogging of the extraction column or injector is often due to the particles precipitated from sample matrices. Generally, high speed centrifugation before injection is helpful to pellet particulates at the bottom of the well and reduce the chances of clogging, particularly if the internal standard is prepared in a solution containing some organic solvent. Lachance *et al.* compared the effects of dilution, filtration and centrifugation on column lifetime in a TFC single-column method [223]. Compared with single aqueous dilution, a centrifugation step at 13,000 rpm x 5 min increased column lifetime by a factor of 2 to 3. Sample dilution followed by filtration through a 0.45 μm 96-well filter plate allowed from 800–1000 injections to be made on the extraction column. Injection of supernatant from protein precipitated plasma was documented to increase column lifetime to 1000 samples compared with 200–500 samples for injection of neat plasma [224]. Another report identified that cooling of the plasma samples kept extraction column pressure steady through 300 injections, while pressure slowly accumulated starting at about 125 samples injected when the samples were at ambient temperature during a long overnight run [225]. The cleaning of the injector should also be performed as a routine maintenance task as it will aid in reducing carryover.

The use of an indirect approach greatly improves ruggedness. As described in Section 14.6.2.3, the most commonly used indirect approach is an off-line protein precipitation step before on-line analysis. This step helps to reduce the risk of system failure from each of the three aspects discussed above. In a slightly different approach, Niggebrugge *et al.* reported the use of a plasma filtration method before injection [212]. After adding the internal standard, the plasma mixture was filtered through a 1.2 μm 96-well filtration plate (Millipore, Bedford, MA USA) to remove small particulates. Unlike the protein precipitated samples which contain a high percentage organic solvent, the organic content of the filtrate is little to none and a larger volume (100 μL) of the filtrate may be injected into the system to achieve good sensitivity.

On-line sample preparation using the Prospekt-2 usually provides satisfactory ruggedness since the extraction cartridge is only used once. Two additional factors contribute to ruggedness—the full volume of eluate solvent is analyzed and there are no transfer steps which can introduce variability in volume retrieval or utilization.

14.7.4 How to Improve Peak Shape

Good peak shape helps to achieve higher sensitivity and more reliable quantitative results. Using TFC in dual-column mode with a backflush will almost always give better peak shape than the single-column mode due to the chromatofocusing effect provided by the second column. Gradient elution is often used to improve peak shape. Ideally, in a dual-column gradient elution mode, the extraction column should have less retention of the analyte than the separation column so that the analyte eluted from the extraction column can be refocused at the head of the separation column. However, this is often not the case. The extraction column usually contains a less selective and more retentive sorbent in order to make it more applicable for analytes having more diverse structures. When the elution solvent reaches the organic content needed to elute the analyte from the extraction column, it is often not weak enough to refocus the analyte on the separation column. Herman recently reported a dual-column TFC setup developed by Cohesive to improve the peak shape [28]. The configuration of the system is shown in Figure 14.9.

This system from Herman utilized a 200 μL loop stored with the elution solvent having a certain organic content. The system also used a modified valve (Valve 2 in Figure 14.9) which incorporated a Tee. During the elution step, this solvent is used to desorb the analyte from the extraction column. Meanwhile, the separation pump delivers a highly aqueous solvent to mix with the elution solvent and dilute the organic content; the analyte that was just eluted from the extraction column can be refocused onto the head of the separation column. Good peak shapes were achieved for most of the compounds analyzed with this system.

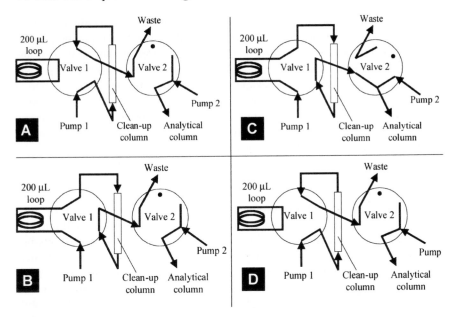

Figure 14.9. Schematic diagrams of valve configurations for a dual-column turbulent flow chromatography system utilized for generic extractions. Reprinted with permission from [28]. Copyright 2002 John Wiley & Sons, Ltd.

14.7.5 How to Achieve a Cleaner Extraction

When an on-line sample preparation method is used to support studies in a screening mode, the three aspects discussed above are usually the most important ones. However, when an on-line method is used to support studies for compounds that are moving into the late discovery stage or development stage, it is always worthwhile to make the effort to further optimize the on-line method. One of the important goals in method optimization is to yield cleaner extractions which help to improve assay reliability by reducing ion suppression and/or matrix interferences.

As observed with conventional off-line sample preparation methods, where some simple chemistry can optimize the extraction process, on-line methods can be fine-tuned using similar approaches. Ding and Neue developed a procedure to perform cleaner on-line extractions [226]. It was based on manipulating both pH and organic content of the load and cleanup solvents. More effective separation of matrix components from the analyte of interest was achieved. For a basic analyte, a cleanup solvent having a pH higher than its pKa is used. The analyte is neutral at this pH and shows strong retention on the sorbent. Therefore, a certain percentage of organic content can be used in the cleanup solvent to wash off the mildly retentive and nonretentive ionic interferences, resulting in cleaner extracts. Likewise, for an acidic analyte, a cleanup solvent having a pH lower than its pKa is used. The analyte is neutral at this low pH and shows strong retention on the sorbent. Again, a certain percentage of organic can be used in the cleanup solvent to wash off the mildly retentive and nonretentive ionic interferences, resulting in cleaner extracts.

14.7.6 Various Troubleshooting Techniques

Compounds that do not extract using turbulent flow chromatography in a generic screening approach, as reported by Herman [28], generally fall into two categories. First, they may be very hydrophilic compounds that are not retained by the chemistry of the reversed phase extraction column or the analytical column operated with a low organic concentration in the mobile phase (*e.g.*, 8% organic was reported). The compound can be transferred from the extraction column to the analytical column with less organic in the mobile phase or the chemistry of the analytical column can be changed to one that more strongly retains hydrophilic compounds. Second, compounds having high hydrophobic character may not be completely transferred to the analytical column at 40% organic, therefore reducing sensitivity. In this case, simply increase the percentage organic in the loop that is used for this transfer step.

Acknowledgments

The collaboration of Jing-Tao Wu as coauthor for this chapter is gratefully acknowledged. The authors thank Mohammed Jemal and Hiroyuki Kataoka for their careful review of the full manuscript. The authors also thank Ahmed Aced (Iris Technologies), Ingo Christ (LEAP Technologies), Bert Ooms (Spark Holland) and Joe Takarewski (Cohesive Technologies) for reviewing and contributing to portions of the chapter. The line art illustrations were kindly provided by Pat Thompson and Woody Dells.

References

[1] S.H.Y. Wong, In: S. Wong and I. Sunshine, Eds., Handbook of Analytical Therapeutic Drug Monitoring and Toxicology, CRC Press, Boca Raton (1997) 137-148.

[2] B.L. Ackermann, A.T. Murphy and M.J. Berna, Amer. Pharm. Rev. 5 (2002) 54-63.

[3] M.L. Powell and M. Jemal, Amer. Pharm. Rev. 4 (2001) 63-69.

[4] V. Pretorius and T.W. Smuts, Anal. Chem. 38 (1966) 274-281.

[5] J.H. Knox, J. Chromatogr. A 831 (1999) 3-15.

[6] S.R. Sumpter and M.L. Lee, J. Microcolumn Sep. 3 (1991) 91-113.

[7] H. Bauer, Chromatographia 27 (1989) 238-242.

[8] C.J. Oberhauser, A.E. Niggebrugge, D. Lachance, J.J. Takarewski, M.M. Pegram and H.M Quinn, LC-GC 18 (2000) 716-724.

[9] J. Ayrton, G.J. Dear, W.J. Leavens, D.N. Mallett and R.S. Plumb, Rapid Commun. Mass Spectrom. 11 (1997) 1953-1958.

[10] M. Jemal, Y.-Q. Xia and D.B. Whigan, Rapid Commun. Mass Spectrom. 12 (1998) 1389-1399.

[11] H.M. Quinn and J.J. Takarewski, U.S. Pat. No. 5,772,874 (30 June 1998).

[12] H.M. Quinn, R.A. Menapace and C.J. Oberhauser, U.S. Pat. No. 5,795,469 (30 August 1998).

[13] H.M. Quinn and J.J. Takarewski, U.S. Pat. No. 5,919,368 (6 July 1999).

[14] H.M. Quinn, R.A. Menapace and C.J. Oberhauser, U.S. Pat. No. 5,968,367 (19 October 1999).

[15] H.M. Quinn and J.E. Brann III, U.S. Pat. No. 6,110,362 (21 Nov 2000).

[16] H.M. Quinn and J.E. Brann III, U.S. Pat. No. 6,149,816 (21 Nov 2000).

[17] A. Asperger, J. Efer, T. Koal and W. Engewald, Biologische Abwasserreinigung 16 (2001) 123-143.

[18] J.-T. Wu, H. Zeng, M. Qian, B.L. Brogdon and S.E. Unger, Anal. Chem. 72 (2000) 61-67.

[19] J. Ayrton, G.J. Dear, W.J. Leavens, D.N. Mallett and R.S. Plumb, J. Chromatogr. A 828 (1998) 199-207.

[20] J. Ayrton, R.A. Clare, G.J. Dear, D.N. Mallett and R.S. Plumb, Rapid Commun. Mass Spectrom. 13 (1999) 1657-1662.

[21] D. Zimmer, V. Pickard, W. Czembor and C. Müller, J. Chromatogr. A 854 (1999) 23-35.

[22] G. Hopfgartner, C. Husser and M. Zell, Therap. Drug Monit. 24 (2002) 134-143.

[23] N. Van Eeckhout, J.C. Perez, J. Claereboudt, R Vandeputte and C. Van Peteghem, Rapid Commun. Mass Spectrom. 14 (2000) 280-285.

[24] C. D'Arienzo, J. Wang, D. Wang-Iverson and P.J. Gale, Proceedings 46th American Society for Mass Spectrometry Conference, Orlando, FL USA (1998).

[25] E.M. Shobe, M.G. Williams, J. Palandra, P.A. Bombardt and T.G. Heath, Proceedings 47th American Society for Mass Spectrometry Conference, Dallas, TX USA (1999).

[26] J.A. Townsend, J. Pav and J.J. Takarewski, Proceedings 47th American Society for Mass Spectrometry Conference, Dallas, TX USA (1999).

[27] C.S. Crean and S.C. Werness, Proceedings 49th American Society for Mass Spectrometry Conference, Chicago, IL USA (2001).

[28] J.L. Herman, Rapid Commun. Mass Spectrom. 16 (2002) 421-426.

[29] A.E. Niggebrugge, C. MacLauchlin, D. Dai, L.A. Ford, A.T. Menendez and A.S. Chilton, Proceedings 50th American Society for Mass Spectrometry Conference, Orlando, FL USA (2002).

[30] H. Zeng, J.-T. Wu and S.E. Unger, J. Pharm. Biomed. Anal. 27 (2002) 967-982.

[31] I. H. Hagestam and T. C. Pinkerton, Anal. Chem. 57 (1985) 1757-1763.

[32] S.E. Cook and T.C. Pinkerton, J. Chromatogr. 368 (1986) 233-248.

[33] C. Schafer and D. Lubda, J. Chromatogr. A 909 (2001) 73-78.

[34] T.C. Pinkerton, J. Chromatogr. 544 (1991) 13-23.

[35] J. Haginaka, Chromatography 23 (2002) 1-12.

[36] E. Yamamoto, K. Murata, Y. Ishihama and N. Asakawa, Anal. Sci. 17 (2001) 1155.

[37] J.A. Perry, B. Invergo, H. Wagner, T.J. Szczerba and J.D. Rateike, J. Liq. Chrom. & Rel. Technol. 15 (1992) 3343-3352.

[38] D.J. Gisch, B.T. Hunter and B. Feibush, J. Chromatogr. 433 (1988) 264 - 268.

[39] K.-S. Boos and A. Rudolphi, LC-GC 15 (1997) 602-611.

[40] A. Rudolphi and K.-S. Boos, LC-GC 15 (1997) 814-823.

[41] T.E. Gundersen and R. Blomhoff, J. Chromatogr. A 935 (2001) 13-43.

[42] A. Haque and J.T. Stewart, Biomed. Chromatogr. 13 (1999) 51-56.

[43] G.D. George and J.T. Stewart, J. Liq. Chrom. 13 (1990) 3861-3889.

[44] S.J. Rainbow, C.M. Dawson, T.R. Tickner, J. Chromatogr. 527 (1990) 389-396.

[45] J. Ma, C.-L. Liu, P.-L. Zhu, Z.-P. Jia, L.-T. Xu and R. Wang, J. Chromatogr. B 772 (2002) 197-204.

[46] W.M. Mullett and J. Pawliszyn, Anal. Chem. 74 (2002) 1081-1087.

[47] H.L. Lord and J. Pawliszyn, LC-GC 16 (1998) S41-S46.

[48] H. Lord and J. Pawliszyn, J. Chromatogr. A 885 (2000) 153-193.

[49] G. Theodoridis, E.H.M. Koster and G.J. de Jong, J. Chromatogr. B 745 (2000) 49-82.

[50] H. Lord and J. Pawliszyn, J. Chromatogr. A 902 (2000) 17-63.

[51] S. Ulrich, J. Chromatogr. A 902 (2000) 167-194.

[52] H. Minakuchi K. Nakanishi, N. Soga, N. Ishizuka and N. Tanaka, Anal. Chem. 68 (1996) 3498-3501.

[53] H. Minakuchi K. Nakanishi, N. Soga, N. Ishizuka and N. Tanaka, J. Chromatogr. A 762 (1997) 135-146.

[54] B. Bidlingmaier, K.K. Unger and N. Von Doehren, J. Chromatogr. A 832 (1999) 11-16.

[55] R. Asiaie, X. Huang, D. Farnan and C. Horvath, J. Chromatogr. A 806 (1998) 251-263.

[56] A. Podgornik, M. Barut, A. Strancar, D. Josic and T. Koloini, Anal. Chem. 72 (2000) 5693-5699.

[57] K. Cabrera D. Lubda H.-M. Eggenweiler, H. Minakuchi and K. Nakanishi, J. High Resol. Chromatogr. 23 (2000) 93-99.

[58] D. Lubda K. Cabrera W. Kraas, C. Schaefer and D. Cunningham, LC-GC Europe 14 (2001) 1-4.

[59] G. Iberer, R. Hahn and A. Jungbauer, LC-GC 17 (1999) 998-1005.

[60] P. Zollner, A. Leitner, D. Lubda K. Cabrera and W. Lindner, Chromatographia 52 (2000) 818-820.

[61] G. Dear, R. Plumb and D. Mallett, Rapid Commun. Mass Spectrom. 15 (2001) 152-158.

[62] R. Plumb, G. Dear, D. Mallett and J. Ayrton, Rapid Commun. Mass Spectrom. 15 (2001) 986-993.

[63] S. Zhou, M.J. Larson, X. Jiang and W. Naidong, Proceedings 50th American Society for Mass Spectrometry Conference, Orlando, FL USA (2002).

[64] J.-T. Wu, H. Zeng, Y. Deng and S.E. Unger, Rapid Commun. Mass Spectrom. 15 (2001) 1113-1119.

[65] G.J. Dear, D.N. Mallett, D.M. Higton, A.D. Roberts, S.A. Bird, H. Young, R.S. Plumb and I.M. Ismail, Chromatographia 55 (2002) 177-184.

[66] Y. Deng, J.-T. Wu, T.L. Lloyd, C.L. Chi, T.V. Olah and S.E. Unger, Rapid Commun. Mass Spectrom. 16 (2002) 1116-1123.

[67] Y. Hsieh, G. Wang, Y. Wang, S. Chackalamannil, J.-M. Brisson, K. Ng and W.A Korfmacher, Rapid Commun. Mass Spectrom. 16 (2002) 944-950.

[68] N. Barbarin, D.B. Mawhinney, R. Black and J. Henion, J. Chromatogr. B (2002) in press.

[69] P.T. Vallano, R.S. Mazenko, E.J. Woolf and B.K. Matuszewski, J. Chromatogr. B 779 (2002) 249-257.

[70] R. Wieboldt, J. Zweigenbaum and J. Henion, Anal. Chem. 69 (1997) 1683-1691.

[71] D.S. Hage and M.A. Nelson, Anal. Chem. 73(7) (2001) 199A-205A.

[72] N. Delaunay-Bertoncini, V. Pichon and M.-C. Hennion, LC-GC Europe 14 (2001) 162-172.

[73] D.S. Hage, J. Chromatogr. B 715 (1998) 3-28.

[74] M. de Frutos and F.E. Regnier, Anal. Chem. 65(1) (1993) 17A-25A.

[75] N. Delaunayi, V. Pichon and M.-C. Hennion, J. Chromatogr. B 745 (2000) 15-37.

[76] D. Stevenson, J. Chromatogr. B 745 (2000) 39-48.

[77] M.A. Kelly, T.J. McLellan and P.J. Rosner, Anal. Chem. 74 (2002) 1-9.

[78] D.S. Hage, J. Chromatogr. B 768 (2002) 3-30.

[79] B. Bjarnason, L. Chimuka and O. Ramstrom, Anal. Chem. 71 (1999) 2152-2156.

[80] B. Sellergren, Ed., Molecularly Imprinted Polymers: Man-made Mimics of Antibodies and their Applications in Analytical Chemistry, Elsevier, Amsterdam (2001); Techniques and Instrumentation in Analytical Chemistry, Volume 23.

[81] K. Ensing, C. Berggren and R.E. Majors, LC-GC 19 (2001) 942-954.

[82] W.M. Mullett and E.P.C. Lai, Anal. Chem. 70 (1998) 3636-3641.

[83] C. Crescenzi, S. Bayoudh, P.A.G. Cormack, T. Klein and K. Ensing, Analyst 125 (2000) 1515-1517.

[84] C. Berggren, S. Bayoudh, D. Sherrington and K. Ensing, J. Chromatogr. A 889 (2000) 105-110.

[85] L.I. Andersson, J. Chromatogr. B 739 (2000) 163-173.

[86] W.M. Mullett and E.P.C. Lai, J. Pharm. Biomed. Anal. 21 (1999) 835-843.

[87] R.F. Venn and R.J. Goody, Chromatographia 50 (1999) 407-414.

[88] P. Martin, I.D. Wilson, D.E. Morgan, G.R. Jones and K. Jones, Anal. Commun. 34 (1997) 45-47.

[89] R.E. Majors, K.-S. Boos, C.-H. Grimm, D. Lubda and G. Wieland, LC-GC 14 (1996) 554-560.

[90] M. Kollroser and C. Schober, Rapid Commun. Mass Spectrom. 16 (2002) 1266-1272.

[91] H.K. Lim, K.W. Chan, S. Sisenwine and J.A. Scatina, Anal. Chem. 73 (2001) 2140-2146.

[92] L. Ramos, N. Brignol, R. Bakhtiar, T. Ray, L.M. McMahon and F.L.S. Tse, Rapid Commun. Mass Spectrom. 14 (2000) 2282-2293.

[93] Y. Hsieh, M.S. Bryant, G. Gruela J.-M. Brisson and W.A Korfmacher, Rapid Commun. Mass Spectrom. 14 (2000) 1384-1390.

[94] M. Jemal, M. Huang, X Jiang, Y. Mao and M.L. Powell, Rapid Commun. Mass Spectrom. 13 (1999) 2125-2132.

[95] N. Brignol, R. Bakhtiar, L. Dou, T. Majumdar and F.L.S. Tse, Rapid Commun. Mass Spectrom. 14 (2000) 141-149.

[96] M. Jemal, Z. Ouyang, Y.-Q. Xia and M.L. Powell, Rapid Commun. Mass Spectrom. 13 (1999) 1462-1471.

[97] M. Jemal and Y.-Q. Xia, J. Pharm. Biomed. Anal. 22 (2000) 813-827.

[98] D. Lachance, C. Grandmaison and L. DiDonato, Proceedings 48th American Society for Mass Spectrometry Conference, Long Beach, CA USA (2000).

[99] C. Chassaing, J. Luckwell, P. Macrae, K. Saunders, P. Wright and R. Venn, Chromatographia 53 (2001) 122-130.

[100] C. Chassaing, P. Macrae, P. Wright, A. Harper, J. Luckwell, K. Saunders and R. Venn, Presented at 11th International Symposium on Pharmaceutical and Biomedical Analysis, Basel, Switzerland (2000).

[101] C.R. Mallet, J.R. Mazzeo and U. Neue, Rapid Commun. Mass Spectrom. 15 (2001) 1075-1083.

[102] E.B. Jones, M. Kadkhodayan and B. Pitamah, Proceedings 50th American Society for Mass Spectrometry Conference, Orlando, FL USA (2002).

[103] W.R.G. Baeyens, G. Van der Weken, J. Haustraete, H.Y. Aboul-Enein, S. Corveleyn, J.P. Remon, A.M. Garcia-Campana and P. Deprez, J. Chromatogr. A 871 (2000) 153-161.

[104] R. Oertel, K. Richter, T. Gramatte and W. Kirch, J. Chromatogr. A 797 (1998) 203-209.

[105] T. Gordi, E. Nielsen, Z. Yu, D. Westerlund and M. Ashton, J. Chromatogr. B 742 (2000) 155-162.

[106] R.M. Mader, B. Rizovski and G.G. Steger, J. Chromatogr. B 769 (2002) 357-361.

[107] W.M. Mullett and J. Pawliszyn, J. Pharm. Biomed. Anal. 26 (2001) 899-908.

[108] P. Chiap, A. Ceccato, R. Gora Ph. Hubert, J. Geczy and J. Crommen, J. Pharm. Biomed. Anal. 27 (2002) 447-455.

[109] G. Friedrich, T. Rose and K. Rissler, J. Chromatogr. B 766 (2002) 295-305.

[110] A. El Mahjoub and C. Staub, J. Chromatogr. B 742 (2000) 381-390.

[111] P. Kubalec and E. Brandsteterova, J. Chromatogr. B 726 (1999) 211-218.

[112] S. Vielhauer, A. Rudolphi, K.-S. Boos and D. Seidel, J. Chromatogr. B 666 (1995) 315-322.

[113] T. Ohta S. Niida and H. Nakamura, J. Chromatogr. B 675 (1996) 168-173.

[114] D. Song, W. Guillaume Wientijes and J.L.-S. Au, J. Chromatogr. B 690 (1997) 289-294.

[115] E.A. Hogendoorn, P. van Zoonen, A. Polettini, G.M. Bouland and M. Montagna, Anal. Chem. 70 (1998) 1362-1368.

[116] R.A.M. van der Hoeven, A.J.P. Hofte, M. Frenay, H. Irth, U.R. Tjaden, J. van der Greef, A. Rudolphi, K.-S. Boos, G. Marko Varga and L.E. Edholm, J. Chromatogr. A 762 (1997) 193-200.

[117] P.M. Jeanville, J.H. Woods, T.J. Baird III and E.S. Estapé, J. Pharm. Biomed. Anal. 23 (2000) 897-907.

[118] S.X. Peng, M.J. Strojnowski and D.M. Bornes, J. Pharm. Biomed. Anal. 25 (1999) 343-349.

[119] M. Pasternyk (Di Marco), M.P. Ducharme, V. Descorps, G. Felix and I.W. Wainer, J. Chromatogr. A 828 (1998) 135-140.

[120] K. Wagner, T. Miliotis, G. Marko-Varga R. Bischoff and K.K. Unger, Anal. Chem. 74 (2002) 809-820.

[121] R.E. Majors, LC-GC 20 (2002) 332-344.

[122] D.W. Roberts, M.I. Churchwell, F.A. Beland, J.-L. Fang and D.R. Doerge, Anal. Chem. 73 (2001) 303-309.

[123] J. Cai and J. Henion, Anal. Chem. 68 (1996) 72-78.

[124] J. Cai and J. Henion, J. Chromatogr. B 691 (1997) 357-370.

[125] B.A. Rashid, P. Kwasowski and S. Stevenson, J. Pharm. Biomed. Anal. 21 (1999) 635-639.

[126] C.S. Creaser, S.J. Feely, E. Houghton and M. Seymour, J. Chromatogr. A 794 (1998) 37-43.

[127] H.H. Maurer, C.J. Schmitt, A.A. Weber and T. Kraemer, J. Chromatogr. B 748 (2000) 125-135.

[128] C.K. Holtzapple, S.A. Buckley and L.H. Stanker, J. Chromatogr. B 754 (2001) 1-9.

[129] B. Ooms and E. Koster, Proceedings 48th American Society for Mass Spectrometry Conference, Long Beach, CA USA (2000).

[130] C.D. James, J.A. Dunn and O. Halmingh, Proceedings 48th American Society for Mass Spectrometry Conference, Long Beach, CA USA (2000).

[131] M.C. Woodward, G. Bowers, J. Chism, L. St. John-Williams and G. Smith, Proceedings 48th American Society for Mass Spectrometry Conference, Long Beach, CA USA (2000).

[132] M.W.F. Nielen, A.J. Valk, R.W. Frei, U.A.Th. Brinkman, Ph. Mussche, R. de Nijs, B. Ooms and W. Smink, J. Chromatogr. A 393 (1987) 69-83.

[133] E. Koster and B. Ooms, LC-GC Europe 14 (2001) 55-57.

[134] F.E. Wolf, K. Xu and K. Miller, Proceedings 48th American Society for Mass Spectrometry Conference, Long Beach, CA USA (2000).

[135] M. Jemal and Y.-Q. Xia, Rapid Commun. Mass Spectrom. 13 (1999) 97-106.

[136] B. Ooms, E. Koster and P. Ringeling, Proceedings 50th American Society for Mass Spectrometry Conference, Orlando, FL USA (2002).

[137] D.A. McLoughlin, T.V. Olah and J.D. Gilbert, J. Pharm. Biomed. Anal. 15 (1997) 1893-1901.

[138] F. Beaudry, J.C.Y. Le Blanc, M. Coutu and N.K. Brown, Rapid Commun. Mass Spectrom. 12 (1998) 1216-1222.

[139] W.D. van Dongen, D. van de Lagemaat, F. Schoutsen, R.J. Vreeken, E.R. Verheij, J. van der Greef and K. Ensing, Proceedings 48th American Society for Mass Spectrometry Conference, Long Beach, CA USA (2000).

[140] G. Haak, A. Schellen, B. Ooms, D. van de Lagermaat, W. van Dongen, R. Vreeken and E. Verheij, Proceedings 48th American Society for Mass Spectrometry Conference, Long Beach, CA USA (2000).

[141] C.H.P. Bruins, C.M. Jeronimus-Stratingh, K. Ensing, W.D. van Dongen and G.J. de Jong, J. Chromatogr. A 863 (1999) 115-122.

[142] M.W.J. van Hout, C.M. Hofland, H.A.G. Niederlander and G.J. de Jong, Rapid Commun. Mass Spectrom. 14 (2000) 2103-2111.

[143] A. Pruvost, I. Ragueneau, A. Ferry, P Jaillon, J.-M. Grognet and H. Benech, J. Mass Spectrom. 35 (2000) 625-633.

[144] A. Schellen, B. Ooms, M. van Gils, O. Halmingh, E. van der Vlis, D. van de Lagemaat and E. Verheij, Rapid Commun. Mass Spectrom. 14 (2000) 230-233.

[145] A. Marchese, C. McHugh, J. Kehler and H. Bi, J. Mass Spectrom. 33 (1998) 1071-1079.

[146] J.Y.-K. Hsieh, L. Lin and B.K. Matuszewski, J. Liq Chrom. & Rel. Technol. 24 (2001) 799-812.

[147] J. Sola J. Prunonosa H. Colom, C. Peraire and R. Obach, J. Liq Chrom. & Rel. Technol. 19 (1996) 89-99.

[148] A. Kurita and N. Kaneda, J. Chromatogr. B 724 (1999) 335-344.

[149] A. Pastoris, L. Cerutti, R. Sacco, L. De Vecchi and A. Sbaffi, J. Chromatogr. B 664 (1995) 287-293.

[150] H. Toreson and B.-M. Eriksson, Chromatographia 45 (1997) 29-34.

[151] J. Prunonosa L. Parera C. Peraire, F. Pla O. Lavergne and R. Obach, J. Chromatogr. B 668 (1995) 281-290.

[152] G. Garcia-Encina, R. Farran, S. Puig, M.T. Serafini and L. Martinez, J. Chromatogr. B 670 (1995) 103-110.

[153] H. Svennberg and P.-O. Lagerstrom, J. Chromatogr. B 689 (1997) 371-377.

[154] J.A. Pascual and J. Sanagustin, J. Chromatogr. B 724 (1999) 295-302.

[155] M. Yritia P. Parra E. Iglesias and J.M. Barbanoj, J. Chromatogr. A 870 (2000) 115-119.

[156] M. Yritia P. Parra J.M. Fernandez and J.M. Barbanoj, J. Chromatogr. A 846 (1999) 199-205.

[157] J. Wang, S.-Y. Chang, C. D'Arienzo and D. Wang-Iverson, Proceedings 48th American Society for Mass Spectrometry Conference, Long Beach, CA USA (2000).

[158] H. Ghobarah, J.C. Flynn, J.D. Laycock and K.J. Miller, Proceedings 48th American Society for Mass Spectrometry Conference, Long Beach, CA USA (2000).

[159] J. Prunonosa J. Sola C. Peraire, F. Pla O. Lavergne and R. Obach, J. Chromatogr. B 677 (1996) 388-392.

[160] G.J. de Jong, M. Jeronimus, C.H.P. Bruins, W.D. van Dongen and K. Ensing, Chromatographia *Supplement* 52 (2000) S25.

[161] C. Nieto, J. Ramis, L. Conte, J.M. Fernandez and J. Forn, J. Chromatogr. B 661 (1994) 319-325.

[162] L. Borbridge, D. Lourenco and A. Acheampong, Proceedings 48th American Society for Mass Spectrometry Conference, Long Beach, CA USA (2000).

[163] B. Starcevic, E. Di Stefano and D.H. Catlin, Proceedings 48th American Society for Mass Spectrometry Conference, Long Beach, CA USA (2000).

[164] M. Hedenmo and B.-M. Eriksson, J. Chromatogr. A 692 (1995) 161-166.

[165] O.V. Olesen and B. Poulsen, J. Chromatogr. 622 (1993) 39-46.

[166] E.A. Martin, R.T. Heydon, K. Brown, J.E. Brown, C.K. Lim, I.N.H. White and L.L. Smith, Carcinogenesis 19 (1998) 1061-1069.

[167] A. Desroches, M. Vranderick, E. Federov, M. Mancini and M. Allard, Proceedings 48th American Society for Mass Spectrometry Conference, Long Beach, CA USA (2000).

[168] E.H.M. Koster, B.A. Ooms and H.A.G. Niederlander, Proceedings 50th American Society for Mass Spectrometry Conference, Orlando, FL USA (2002).

[169] C. Opper, W. Wesemann, G. Barka H. Kerkdijk and G. Haak, LC-GC 12 (1994) 684-698.

[170] G. Hotter, I. Ramis, G. Bioque, C. Sarmiento, J.M. Fernandez, J. Rosello-Catafau and E. Gelpf, Chromatographia 36 (1993) 33-38.

[171] I. Ramis, G. Hotter, J. Rosello-Catafau, O. Bulbena C. Picado and E. Gelpi, J. Pharm. Biomed. Anal. 11 (1993) 1135-1139.

[172] M. Hedenmo and B.-M. Eriksson, J. Chromatogr. A 661 (1994) 153-159.

[173] D. Barron, J. Barbosa J.A. Pascual and J. Segura, J. Mass Spectrom. 31 (1996) 309-319.

[174] O. Halmingh, M. Van Gils and B. Ooms, Amer. Clin. Lab. 18 (1999) 6-7.

[175] A.P. Bruins, D.B. Robb, H.A.M. Peters and P.L. Jacobs, Proceedings 48th American Society for Mass Spectrometry Conference, Long Beach, CA USA (2000).

[176] H. Bi, R.N. Hayes, R. Castien, O. Halmingh and M. van Gils, Proceedings 48th American Society for Mass Spectrometry Conference, Long Beach, CA USA (2000).

[177] D. Tang, P. Gerry and O. Kavetskaia, Proceedings 50th American Society for Mass Spectrometry Conference, Orlando, FL USA (2002).

[178] S. Lacorte, N. Ehresmann and D. Barcelo, Environ. Sci. Technol. 29 (1995) 2834-2841.

[179] G.D. Bowers, C.P. Clegg, S.C. Hughes, A.J. Harker and S. Lambert, LC-GC 15 (1997) 48-53.

[180] A.J. Tomlinson, N.A. Guzman and S. Naylor, J. Cap. Elec. 6 (1995) 247-266.

[181] A.J. Tomlinson, L.M. Benson, R.P. Oda D. Braddock, B.L. Riggs, J.A. Katzmann and S. Naylor, J. Cap. Elec. 2 (1995) 97-104.

[182] A.J. Tomlinson and S. Naylor, J. Cap. Elec. 2 (1995) 225-233.

[183] A.J. Tomlinson, L.M. Benson, W.D. Braddock and R.P. Oda, J. High Resol. Chromatogr. 18 (1995) 381-383.

[184] A.J. Tomlinson, L.M. Benson, S. Jameson, D.H. Johnson and S. Naylor, J. Amer. Soc. Mass. Spectrom. 8 (1997) 15-24.

[185] A.J. Tomlinson, S. Jameson and S. Naylor, J. Chromatogr. A 744 (1996) 273-278.

[186] L.M. Benson, A.J. Tomlinson, A.N. Mayeno and G.J. Gleich, J. High Resol. Chromatogr. 19 (1996) 291-294.

[187] A.J. Tomlinson and S. Naylor, J. Liq. Chrom. 18 (1995) 3591-3615.

[188] S. Naylor and A.J. Tomlinson, Talanta 45 (1998) 603-612.

[189] N.A. Guzman, S.S. Park, D. Schaufelberger, L. Hernandez, X. Paez, P. Rada A.J. Tomlinson and S. Naylor, J. Chromatogr. B 697 (1997) 37-66.

[190] Y.-Q. Xia D.B. Whigan, M.L. Powell and M. Jemal, Rapid Commun. Mass Spectrom. 14 (2000) 105-111.

[191] V.S. Ong, K. Cook, J. Bernstein and W. Brubaker, Proceedings 50th American Society for Mass Spectrometry Conference, Orlando, FL USA (2002).

[192] M.R. Anari, K. Chan, H.K. Lim, E.A. Dierks, K. Stams, S.E. Ball, C. Tio, J. Kao, J.A. Scatina F.S. Abbott and D. Kwok, Proceedings 48th American Society for Mass Spectrometry Conference, Long Beach, CA USA (2000).

[193] V.K. Boppana C. Miller-Stein and W.H. Schaefer, J. Chromatogr. B 678 (1996) 227-236.

[194] S.R. Needham, M.J. Cole and H.G. Fouda, J. Chromatogr. B 718 (1998) 87-94.

[195] J.-T. Wu, Rapid Commun. Mass Spectrom. 15 (2001) 73-81.

[196] R.T. Cass, J.S. Villa D.E. Karr and D.E Schmidt Jr, Rapid Commun. Mass Spectrom. 15 (2001) 406-412.

[197] R. Xu, C. Nemes, K.M. Jenkins, R.A. Rourick, D.B. Kassel and C.Z.C. Liu, J. Amer. Soc. Mass. Spectrom. 13 (2002) 155-165.

[198] C.K. Van Pelt, T.N. Corso, G.A. Schultz, S. Lowes and J. Henion, Anal. Chem. 73 (2001) 582-588.

[199] A.B. Sage, D. Little and K. Giles, LC-GC 18 (2000) S20-S29.

[200] Y.-Q. Xia C.E.C.A. Hop, D.Q. Liu, S.H. Vincent and S.-H.L. Chiu, Rapid Commun. Mass Spectrom. 15 (2001) 2135-2144.

[201] R.C. King, C. Miller-Stein, D.J. Magiera and J. Brann, Rapid Commun. Mass Spectrom. 16 (2002) 43-52.

[202] M. Caulfield, Presented at North Jersey American Chemical Society Mass Spectrometry Discussion Group Meeting, Somerset, NJ USA (January 2002).

[203] C. McHugh, Presented at Cohesive Technologies User Group Meeting at 50th American Society for Mass Spectrometry Conference, Orlando, FL USA (2002).

[204] S.M. Chesson, J.P. Collins and J.L. Dage, Proceedings 48th American Society for Mass Spectrometry Conference, Long Beach, CA USA (2000).

[205] V. de Biasi, N. Haskins, A. Organ, R. Batemen, K. Giles and S. Jarvis, Rapid Commun. Mass Spectrom. 13 (1999) 1165-1168.

[206] L. Yang, T.D. Mann, D. Little, N. Wu, R.P. Clement and P.J. Rudewicz, Anal. Chem. 73 (2001) 1740-1747.

[207] L. Yang, N. Wu and P.J. Rudewicz, J. Chromatogr. A 926 (2001) 43-55.

[208] D. Morrison, A.E. Davies and A.P. Watt, Anal. Chem. 74 (2002) 1896-1902.

[209] D. Tolson, A. Organ and A. Shah, Rapid Commun. Mass Spectrom. 15 (2001) 1244-1249.

[210] M.K. Bayliss, D. Little, D.N. Mallett and R.S. Plumb, Rapid Commun. Mass Spectrom. 14 (2000) 2039-2045.

[211] Y. Deng, H. Zeng, S.E. Unger and J.-T. Wu, Rapid Commun. Mass Spectrom. 15 (2001) 1634-1640.

[212] A.E. Niggebrugge, E. Tessier, R. Guilbaud, L. DiDonato and R. Masse, Proceedings 48th American Society for Mass Spectrometry Conference, Long Beach, CA USA (2000).

[213] T.C. Alexander, B.T. Hoffman, K.L. Wheeler, J.R. Perkins and J. Henion, Proceedings of the American Association for Pharmaceutical Scientists Annual Meeting, Denver, CO USA (2001)

[214] S. Ouyang, Y. Xu and Y.H. Chen, Anal. Chem. 70 (1998) 931-935.

[215] J. Shen, J-L Tseng, M. Lam and B. Subramanyam, Proceedings 49th American Society for Mass Spectrometry Conference, Chicago, IL USA (2001).

[216] J-T Wu, H. Zeng, A. Deng and S.E. Unger, Proceedings 2nd North American Bioanalytical Forum, Kansas City, MO USA (2002).

[217] A.C. Hogenboom, W.M.A. Niessen and U.A.Th. Brinkman, J. Chromatogr. A 841 (1999) 33-44.

[218] V.C.X. Gao, W.C. Luo, Q. Ye and M. Thoolen, J. Chromatogr. A 828 (1998) 141-148.

[219] G.I. Kirchner, C. Vidal, W. Jacobsen, A. Franzke, K. Hallensleben, U. Christians and K.-F. Sewing, J. Chromatogr. B 721 (1999) 285-294.

[220] J.A.B. Ooms, G.J.M. Van Gils, A.R. Duinkerken and O. Halmingh, Amer. Lab 32 (2000) 52-57.

[221] J..W. Dolan, LC-GC Europe 14 (2001) 148-154.

[222] J.J. Takarewski, M.M. Pegram, J.P. Kiplinger, M.J. Dameron and P.M. Lefebvre, Proceedings 48th American Society for Mass Spectrometry Conference, Long Beach, CA USA (2000).

[223] D. Lachance, Y.G. Leblanc and C. Grandmaison,Proceedings of the American Society for Mass Spectrometry, (2002)

[224] S. Clarke, J. Cook, G. Imrie, I. Smith and T. Noctor, Proceedings 50th American Society for Mass Spectrometry Conference, Orlando, FL USA (2002).

[225] J. Takarewski, D. Magiera and H. Quinn, Proceedings 47th American Society for Mass Spectrometry Conference, Dallas, TX USA (1999).

[226] J. Ding and U.D. Neue, Rapid Commun. Mass Spectrom. 13 (1999) 2151-2159.

Chapter 15

Advances in Sample Preparation for Bioanalysis

Abstract

The drug development process within the pharmaceutical industry is dynamic and many scientific advances often occur within different disciplines at similar times. As the demands, needs and resources within scientific research change, there is a constant give and take relationship and a shifting of the rate limiting step in the overall drug development process. At one time, this rate limiting step was clearly in sample preparation but in the more advanced research environments it has shifted to data analysis, reporting and efficient sharing of information among multiple research sites around the world. This chapter presents two recent advances in sample preparation for bioanalysis—solid-phase extraction in a novel card format and in a higher density 384-well plate. Some additional information is briefly introduced on other techniques that have shown great potential to positively impact how sample preparation and analysis are performed in drug development research. Common themes for these processes include a rethinking of how sample preparation can be performed, achievements in higher throughput and realizations of greater selectivity.

15.1 Solid-Phase Extraction in a Card Format

15.1.1 Introduction

The use of animals to assess ADME (Absorption, Distribution, Metabolism and Elimination) characteristics is a costly and time consuming process. While animals are used to assess drug toxicity, particularly in toxicokinetic studies, current trends use *in vitro* screens which have matured in recent years and been shown to be fairly predictive. These *in vitro* methodologies use enzymes, tissues and cell cultures to allow researchers to screen for drug characteristics such as cell absorption, metabolic stability, drug-drug interactions, clearance, bioavailability and toxicity [1–3]. Instrumentation to accommodate cell maintenance has matured in recent years to the point where high throughput

575

testing using these *in vitro* screens is now a viable approach to investigate the absorption and metabolism of drugs.

Various metabolic reactions are performed *in vitro* within the microplate format. A reaction is stopped at selected time points by the addition of cold acetonitrile and then centrifugation is performed to pellet the proteins at the bottom of the wells. The supernatants are collected but often require further preparation to desalt the samples, isolate and/or concentrate the analytes of interest (including metabolites), and reduce potential matrix effects. This sample preparation step is commonly performed on-line (*e.g.*, turbulent flow chromatography) [4] or off-line (*e.g.*, solid-phase extraction in the 96-well format) [5]. In the constant effort to improve sample throughput, Olech *et al.* have reported the use of a novel format, simultaneous solid-phase extraction (SPE) of 96-wells in a disposable card, with serial elution directly into the mass spectrometer, therefore making the need for a chromatography step optional [6–8]. The specifications of this format and reports of its application for sample cleanup of *in vitro* metabolic reactions will now be detailed.

15.1.2 SPE Card Specifications and Usage

The SPE card (Figure 15.1) is constructed by securing a PTFE particle-loaded membrane (Empore™, 3M Corporation, St. Paul, MN USA) within a plastic frame matching the outer dimensions of a microplate. The membrane itself (~0.5 mm thickness, ~10 μm C18 bonded silica or polymer sorbent particles) is protected above and below by a sheet of blown microfiber material (for details about the Empore particle-loaded membrane technology see Chapter 11, Section 11.4.3.4). Welded into the sheet are 96 discrete elution zones (7 mm diameter, 9 mm center-to-center spacing).

The Empore SPE Card was designed to precisely fit into a modified cell harvester instrument called Harvex (Tomtec Inc., Hamden, CT USA) where it is securely clamped into place using a roller locking device (Figure 15.2, *left*). A unique double O-ring seal prevents crosstalk among the defined extraction zones within the membrane. The instrument performs the necessary steps for sorbent activation, load and wash as listed in Table 15.1.

Using vacuum, methanol is aspirated from a solvent reservoir placed near the base of the instrument and flows vertically through PTFE-coated stainless steel tips (connected above to 96 individual fluid lines) and up through each well of the SPE Card. A solvent reservoir of water is manually swapped to complete a conditioning procedure. A microplate containing the *in vitro* metabolic reaction

Figure 15.1. The Empore™ SPE Card consists of a PTFE particle-loaded membrane secured within a plastic frame matching the outer dimensions of a microplate. Photo reprinted with permission from 3M Corporation.

mixture is then manually inserted and aspiration proceeds; analytes are adsorbed to the sorbent particles within the membrane and liquid flows through to waste. A solvent reservoir containing a suitable wash solution (water or a small percentage organic in water) is then aspirated and passed through the SPE card, followed by drying with air. All 96 samples are now extracted onto the particle-loaded membrane.

The elution procedure is performed by placing the dry SPE card into another instrument, the Elutrix (Tomtec), shown in Figure 15.2 (*right*). SPE cards are inserted into a storage tower on the Elutrix and a shuttle mechanism securely loads the bottom card from the stack into the unit. A unique double O-ring seal

Table 15.1
Typical solid-phase extraction conditions used with an SPE card

Step	Solvent and Volume	Time (s)
Condition	Methanol 200μL	2
	Water 100μL	2
Load	Sample 50–100μL	10 (pulse at 1 s 10-times)
Wash	Water 250μL	2
Dry	Air	20

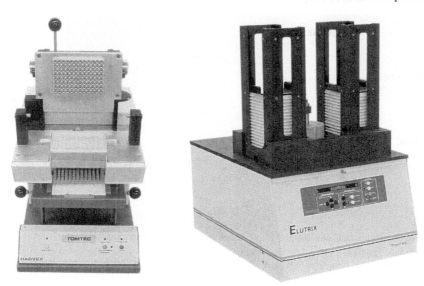

Figure 15.2. *Left*: The Harvex instrument is used for sample loading onto an SPE card. *Right*: The Elutrix instrument is used for eluting sample directly into a mass spectrometer or other detection device. Together these two units comprise a commercial system for utilizing an SPE card. Photos reprinted with permission from Tomtec.

prevents crosstalk among the defined elution zones within the membrane. The plumbing of the lines is configured so that a single isocratic eluent solution is pumped through each well of the SPE card within the defined zone and into an API-ESI mass spectrometer; this path can also be bypassed via a switching valve placed in series.

The direction of the fluid path is through the bottom of the well and up, just as in the sample loading step. Eluent solution is pumped through the lines and directly into the mass spectrometer. A typical composition of eluting solvent is 2mM ammonium acetate combined with 50–90% methanol or acetonitrile. Typical conditions reported are 1–3 mL/min flow rate, elution time of ~12–15 s per well and an MS split ratio of 5:1 [6, 7]. The small particle size used within the Empore membrane mounted into the card actually mimics the performance of a discrete LC column [7] and therefore allows elution to be performed in this manner. An illustration of the card secured into the Elutrix for this procedure is shown in Figure 15.3.

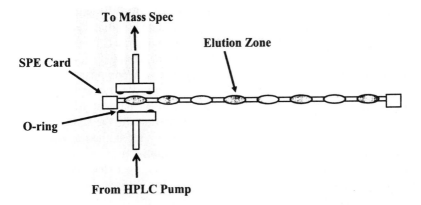

Figure 15.3. Schematic diagram showing an SPE card secured into an Elutrix instrument (Tomtec) for analyte elution. Illustration reprinted with permission from 3M Corporation.

15.1.3 Applications

The SPE card and Harvex protocol were demonstrated to permit rapid off-line sample cleanup followed by direct elution into a mass spectrometer in an initial report presented by Olech *et al.* [6–8]. Hepatic extraction ratios obtained from metabolic stability studies of buspirone and propranolol in human hepatocytes were similar between samples analyzed by conventional methods and those analyzed using the SPE card protocol. Precision across the SPE plate was tested using a stock solution of buspirone (1 µM) and was reported to be less than 7%. The time required to load samples onto the SPE card was 2–5 min. Total time to elute all 96 samples was 27 min, averaging ~17 s per sample.

Several interesting aspects to solid-phase extraction arise using this SPE card approach. It is conceivable that when multiple *in vitro* incubation reactions are performed within a given day that they can all be processed on the SPE card and stored in the dry state for future elution and assay. Used in this manner, a novel and compact storage mechanism is now available. Since the Elutrix has the capacity to hold about 24 stacked microplates or about 30 of the thin SPE cards, overnight runs of up to 30 cards are feasible. Note that any cellular debris that is aspirated along with liquid volume is never eluted as it is trapped on the underside of the membrane wells; elution proceeds in the same direction so the debris stays fixed.

Although the Empore SPE Card was designed to mate directly with existing Tomtec equipment, its format is versatile and open to alternate processing methods. Researchers at Cornell University have demonstrated its use with a Quadra® 96 (Tomtec) liquid handling workstation for sample loading under vacuum using a modified setup. Elution of analytes is accomplished with a pneumatically assisted electrospray device. The sprayer extracts sample directly from the card by infusion of the elution solvent into the SPE media coaxially with the inlet of the sprayer tube which penetrates to the center of the card. The solvent is aspirated into the sprayer tube by the combined suction created with the nebulizing gas and the electrospray process itself. By balancing the flow rate against the suction of the sprayer, the extraction solvent is prevented from spreading into adjacent modules. The SPE card is held on a robotic platform that can be moved in the X-Y-Z axes under control of a Labview program [9]. An application was presented by these researchers for the extraction of methylphenidate from urine samples spiked with standards between 6.6 ng/mL and 3300 ng/mL. Acceptable quantitation results with good sensitivity were shown to be possible without a subsequent chromatographic separation step.

The SPE card presents a novel format for processing relatively clean matrices such as ADME *in vitro* samples. Although its use is not fully characterized at the time of publication of this book, other sample matrices such as plasma are under investigation for their applicability with the SPE card. Also, preliminary studies have not shown crosstalk among wells within an individual plate. Card-to-card carryover depends on analyte characteristics and concentration but has been shown to vary between 0.01–0.05%. Whether or not elution from the card can be combined with an LC column to afford separation is another area to be investigated. The open format of the card certainly lends itself to alternate processing mechanisms and higher throughput configurations for elution, such as multiple heads or sprayers.

15.2 Solid-Phase Extraction in a 384-Well Plate

15.2.1 Introduction

The use of solid-phase extraction in a 96-well plate has been rapidly adopted since its introduction in late 1996. Several dozen applications of the format have been demonstrated in the published literature and novel strategies for its automation have been conceived and introduced. The combination of 96-tip liquid processing with 96-well extraction is a model for high throughput bioanalysis. Full automation of the process using a robotic arm provides around-the-clock operation for multiple assays. High throughput applications for 96-well SPE are reviewed in Chapter 11, Section 11.5.

As the number of samples required to be analyzed by LC-MS/MS techniques increase and as timelines become shorter, inevitably the issue of raising high throughput to the next level becomes of greater interest. Higher density plate formats, such as 384- and 1536-well, are established and used in automated high throughput screening applications with success. These higher density formats allow greater numbers of samples to be processed but also reduce sample volume and solvent consumption, resulting in cost savings. The feasibility of performing bioanalysis in a 384-well SPE plate (16 rows x 24 columns) has been investigated by two research groups and some details of their published applications are presented next.

15.2.2 Applications

15.2.2.1 Sumatriptan in Human Plasma

A 384-well SPE method for the 5-HT agonist sumatriptan in human plasma was reported by Biddlecombe, Benevides and Pleasance [10]. Since these 384-well SPE plates are not commercially mass produced, prototypes machined from polyethylene were used for this work. Several sorbent mass loadings were evaluated to select an optimal bed mass; a 5 mg mass was chosen. The sorbent evaluated was Oasis® HLB (Waters Corporation, Milford MA USA). Details about this sorbent chemistry are found in Chapter 11, Section 11.2.1.2. The dimensions of the extraction well are reported as 3.54 mm i.d. with a 180 µL volume capacity; total well height is ~29.2 mm (tapered at the bottom) and the diameter of the narrow exit tip is ~0.5 mm. Samples were loaded into a 384-well plate by transfer from individual tubes into four 96-well plates using a MultiPROBE II (Packard Instruments, Meriden, CT USA, now a part of PerkinElmer Life Sciences); a RapidPlate™ 96/384 (Zymark Corporation, Hopkinton, MA USA) was used for the 96-well to 384-well transfer steps. The RapidPlate was also used for all liquid additions to the extraction plate.

This report by Biddlecombe *et al.* details several potential difficulties in adapting to the 384-well SPE format. A liquid handling workstation capable of using small volumes with good accuracy and precision was noted as a requirement. Today, several models are available that meet this requirement. Elution volumes of 25–75 µL are expected to be used with 384-well SPE plates. Typical volumes for the condition, load and wash steps are 100–150 µL. The issue of sample reformatting is also an important one; this task is potentially time consuming but can be aided by having samples in a microtube rack or microplate from the study collection site. The availability of disposable tips having the required narrow dimensions to fit into the wells was noted as another potential issue. Liquid processing through the plate using vacuum is also of concern, as it is difficult to collect small volumes without splattering or nebulizing the liquids spraying from the small diameter exit hole, and these exiting liquids easily may adsorb to the wells of the collection plate. Centrifugation was deemed the best option until such point as vacuum manifolds with finer control and optimal geometries become available. An evaporation unit capable of 384-well dry-down was not available so a concentration step was not able to be used in the SPE method.

Performance of the 384-well SPE method for sumatriptan in human plasma was equivalent to that of the 96-well SPE method. Recoveries across the 384-well plate were lower (68% ± 15%, n=119) than the >90% achieved on 96-well plates but the specificity was still adequate to achieve the limit of quantitation (LOQ). Intra-assay precision (%CV) and accuracy (%bias) did not exceed ± 20%. The overall sample preparation time was about half that needed to prepare 4 x 96-well plates on an existing automated system; the time for sample reformatting from tubes to 384-well plate slowed any dramatic potential gain in throughput. Note that there was no real difference in LC-MS/MS analysis time, requiring a 5.5 min run time per sample (38 h total analysis time), regardless of which method was used to prepare the samples.

Biddlecombe *et al.* raise some questions regarding feasibility of this technique. One important issue is that the stability of the LC-MS/MS response must be maintained over the time period required to assay 384 samples. In terms of sample numbers, the need for a 384-well plate exists for clinical studies but each analyte typically is analyzed by its own optimized procedure and samples from individual studies are analyzed separately. The number of total samples *per study* is an important criterion in justifying the need for a 384-well plate. The issue of repeat analysis is a valid one; the cost effectiveness of using a partial plate is of concern. Also, the range of various accessory products in the market capable of working with the 384-well format needs to be expanded, *e.g.*, an evaporation unit and a vacuum manifold unit capable of working with this format.

15.2.2.2 Methotrexate and Metabolite in Human Plasma and Urine

An assay for the antineoplastic drug methotrexate and its 7-hydroxy metabolite was developed in the 384-well format using a C18 5 μm glass fiber disk [11]

(Orochem Technologies, Westmont, IL USA). A Personal Pipettor 550-3S (Apricot Designs, Monrovia, CA USA) benchtop liquid handling instrument was used. An analytical run of 384 samples was performed with a calibration curve range of 1 to 50 μg/mL methotrexate and 50 to 1000 ng/mL 7-hydroxymethotrexate in urine, and 5 to 250 ng/mL and 5 to 100 ng/mL in plasma, respectively.

Experiments were conducted to determine the optimal sorbent particle size (5, 20 or 40–50 μm) in the glass fiber disk (equal packing mass per disk) and elution volume (60–240 μL). A particle size of 5 μm and an elution volume of 60 μL were chosen for the assay. The injection volume used was 40 μL (a dry-down step was not utilized). A modified vacuum manifold was used for the 384-well SPE procedure with urine, while centrifugation was used for plasma due to the potential for foaming or bubble formation. Sample volumes extracted were 30 μL. Overall, excellent precision (%CV) and accuracy (%bias) were obtained at three QC (Quality Control) levels for each analyte using a single calibration series at the beginning and at the end of the run. Recovery of analytes from urine was >95%. The analytical run time was 14 h for this assay. Experiences in transferring a 96-well SPE method for amlodipine in human plasma to a 384-well plate are detailed in a presentation from this same research group [12]. The method transfer was problematic since the identical sorbent chemistry was not available in the 384-well version and a new assay had to be developed.

15.3 Future Directions

15.3.1 Molecularly Imprinted Polymers

Solid-phase extraction is an attractive and well established technique but it can suffer from generic selectivity and nonspecific interactions [13]. In an effort to improve SPE, molecular recognition mechanisms offering specificity and selectivity have been coupled with such a separation procedure to provide high resolving power and ruggedness. Immunoaffinity SPE [14] is a step in the right direction but a selective antibody must be developed for each analyte. This process requires animals and is lengthy and time consuming (months) [15]. Research into molecularly imprinted polymers (MIPs) has been an active area for many years and has been applied toward several analytical techniques in many disciplines other than bioanalysis, such as environmental analysis, biosensors, liquid chromatography and various binding assays. MIP solid-phase extraction involves preparing a sorbent with a predetermined selectivity for a particular analyte or group of structural analogues, *e.g.*, caffeine [13], clenbuterol [16–18], theophylline [19, 20], nicotine [21], darifenacin [22] or propranolol [23]. That prepared sorbent is then put into the familiar column or cartridge format for solid-phase extraction.

The reader is referred to other sources for comprehensive information about molecularly imprinted polymers. A basic overview of the principles of molecular imprinting, advantages and disadvantages, and some examples of their use in various routine applications is discussed by Ensing *et al.* [15, 24]. Developments and applications of MIP in analytical chemistry [25] and drug bioanalysis [26] are reviewed by Andersson. An excellent reference book on the subject has been edited by Sellergren [27].

15.3.2 Chip Based Sample Preparation

Interest in micro fabrication technologies has spread to new areas beyond their original application in the semiconductor industry. Microfluidic devices have been made that transport tiny volumes of liquids through microchannels that interact with chemicals; these chemicals activate indicators and/or perform a reaction. This microfabricated device is often referred to by the name laboratory-on-a-chip [28]. Such systems offer the integration of multiple steps in a complex analytical procedure and the ability to operate in parallel mode. Special needs for sample preparation are necessary in these systems, particularly when the fluid moving through a microchannel is of biological origin.

Separation of nucleic acids and proteins are two of the most promising applications of microchip analysis. These separations are carried out in etched channels in either capillary electrophoresis (CE), capillary electro-chromatography (CEC) or pressure-assisted electrochromatography mode and detection is made by either laser-induced fluorescence or mass spectrometry [29]. Among several advantages of the chip format for bioanalysis is the possibility for eliminating carryover by avoidance of a redundant fluid path [30]. The future holds great potential for these systems in clinical diagnostics and certain portable on-the-spot testing procedures.

An application of chip based capillary electrophoresis/mass spectrometry for the on-chip separation and electrospray detection of small drug molecules is described by Deng, Zhang and Henion [31]. Also reported by this group is development of a microsprayer (a pneumatically assisted electrospray interface) that can be used to couple glass or plastic chip based separation devices to a mass spectrometer operating with an atmospheric pressure ionization interface [32, 33]. This sprayer can also sample from various 96-, 384- and 1536-well microplates, as well as the SPE card mentioned in Section 15.1. Advion Biosciences (Ithaca, NY USA) has continued this research in a commercial effort, developing a monolithic nanoelectrospray nozzle micro fabricated in a silicon wafer for coupling with nanoelectrospray mass spectrometry. The aim of this effort is to develop microfabricated devices for combined sample preparation and electrospray mass spectrometry analysis of samples [34].

The reader is referred to other sources for comprehensive information about chip based sample preparation. A review of the latest developments in bioanalysis using microfluidic devices is presented by Khandurina and Guttman [28]. A brief summary of microchips, microarrays, biochips and nanochips is provided by Kricka [35]. Nanotechnology in bio/clinical analysis is discussed by Guetens *et al.* [36]. Chromatography and electrophoresis on chips is discussed by Regnier *et al.* [37].

15.3.3 Solid-Phase Extraction

Sample preparation will likely remain an essential requirement prior to LC-MS/MS analysis. Lower sorbent bed masses for solid-phase extraction will be used more frequently to eliminate the evaporation and reconstitution steps, which add precious time to a procedure. Smaller elution volumes overall will be the goal and, as demonstrated with the SPE card in Section 15.1, direct injection of the entire volume or a greater proportion of the eluate volume aids in achieving gains in sensitivity. This direct injection concept becomes especially important with the use of narrow bore and capillary LC coupled to mass spectrometry detection, as greater sensitivity and reduced solvent consumption are realized. Sorbent chemistries which are selective yet easy to use (such as the mixed mode chemistries) will be employed more often. These copolymerized phases can be used as a typical reversed phase sorbent or, by changing the solvents and pH, can be used as an ion exchanger. Certainly a smaller number of sorbents will be used for drug bioanalysis, reserving the specialty phases for specialty cases. Just as filtration and SPE are accomplished together by adding a prefilter on top of a sorbent bed, other combinations of sample preparation techniques may be joined together.

The on-line sample preparation techniques are evolving to the point of greater acceptance via their demonstrated high throughput gains, greater ease of method development and ease of use. Manufacturers are packaging complete on-line systems using components that are integrated and optimized out of the box. Since the extraction column is the key component in an on-line sample preparation system, it will likely remain the focus for future development. Among the various column types for on-line extraction, monolithic columns show the greatest potential due to their unique biporous structure, potentially low cost mass production, and their flexibility in column frits (not required). Possible new improvements in sample preparation methodology may also come from the combination of some of these technologies: turbulent flow chromatography, restricted access media and/or monolithic columns.

Another future direction for on-line techniques may be the development of pre-fabricated column arrays for parallel extraction. Here, multiple parallel extraction columns can be installed or replaced simultaneously with controlled intercolumn variability. This approach would allow higher throughput on-line extraction without the complexity.

Besides column technology, other areas that may see future improvements include chemical or mechanical approaches to reduce carryover and high throughput sample pretreatment to improve method ruggedness. Carryover and robustness are currently two of the key concerns for on-line sample preparation. Some generic approaches to address these problems greatly support the growth of this promising technique.

Acknowledgments

The author appreciates the contribution by Jing-Tao Wu to the Future Directions section of this chapter and the review of portions of the manuscript by Rich Matner (3M Corporation) and Robert Speziale (Tomtec).

References

[1] D.A. Smith and H. van de Waterbeemd, Curr. Opin. Chem. Biol. 3 (1999) 373-378.

[2] P.R. Chaturvedi, C.J. Decker and A. Odinecs, Curr. Opin. Chem. Biol. 5 (2001) 452-463.

[3] C.K. Atterwill and M.G. Wing, Alternatives to Laboratory Animals 28 (2000) 857-867.

[4] H.K. Lim, K.W. Chan, S. Sisenwine and J.A. Scatina, Anal. Chem. 73 (2001) 2140-2146.

[5] J.J. Zheng, E.D. Lynch and S.E. Unger, J. Pharm. Biomed. Anal. 28 (2002) 279-285.

[6] R.M. Olech, R.A. Pranis, J.R. Jacobson, C.A. Perman, B.A. Bowman, J. Soldo, R. Speziale, T.W. Astle, M.J. Cole, J.S. Janiszewski and K.M. Whalen, Proceedings 49th American Society for Mass Spectrometry Conference, Chicago, IL USA (2001).

[7] J. Janiszewski, R.M. Olech, R.A. Pranis, J.R. Jacobson, C.A. Perman, B.A. Bowman, J. Soldo, R. Speziale, T.W. Astle, M.J. Cole and K.W. Whalen, Proceedings International Symposium on Laboratory Automation and Robotics, Boston, MA USA (2001).

[8] R.M. Olech, R.A. Pranis, M.J. Cole and J.S. Janiszewski, Proceedings 50th American Society for Mass Spectrometry Conference, Orlando, FL USA (2002).

[9] T. Wachs and J. Henion, Proceedings 50th American Society for Mass Spectrometry Conference, Orlando, FL USA (2002).

[10] R.A. Biddlecombe, C. Benevides and S. Pleasance, Rapid Commun. Mass Spectrom. 15 (2001) 33-40.

[11] G. Rule, M. Chapple and J. Henion, Anal. Chem. 73 (2001) 439-443.

[12] D.A. Campbell, T.J. Ordway, K.T.M. Dillon, L.M. Irwin, J.R. Perkins and J. Henion, Proceedings 49th American Society for Mass Spectrometry Conference, Chicago, IL USA (2001).

[13] G. Theodoridis and P. Manesiotis, J. Chromatogr. A 948 (2002) 163-169.

[14] D. Stevenson, J. Chromatogr. B 745 (2000) 39-48.

[15] K. Ensing, C. Berggren and R.E. Majors, LC-GC Europe 15 (2002) 2-8.

[16] C. Crescenzi, S. Bayoudh, P.A.G. Cormack, T. Klein and K. Ensing, Anal. Chem. 73 (2001) 2171-2177.

[17] C. Crescenzi, S. Bayoudh, P.A.G. Cormack, T. Klein and K. Ensing, Analyst 125 (2000) 1515-1517.

[18] C. Berggren, S. Bayoudh, D. Sherrington and K. Ensing, J. Chromatogr. A 889 (2000) 105-110.

[19] W.M. Mullett and E.P.C. Lai, Anal. Chem. 70 (1998) 3636-3641.

[20] W.M. Mullett and E.P.C. Lai, J. Pharm. Biomed. Anal. 21 (1999) 835-843.

[21] A. Zander, P. Findlay, T. Renner, B. Sellergren and A. Swietlow, Anal. Chem. 70 (1998) 3304-3314.

[22] R.F. Venn and R.J. Goody, Chromatographia 50 (1999) 407-414.

[23] P. Martin, I.D. Wilson, D.E. Morgan, G.R. Jones and K. Jones, Anal. Commun. 34 (1997) 45-47.

[24] K. Ensing, C. Berggren and R.E. Majors, LC-GC 19 (2001) 942-954.

[25] L.I. Andersson, J. Chromatogr. B 745 (2000) 3-13.

[26] L.I. Andersson, J. Chromatogr. B 739 (2000) 163-173.

[27] B. Sellergren, Ed., Techniques and Instrumentation in Analytical Chemistry, Volume 23: Molecularly Imprinted Polymers: Man-made Mimics of Antibodies and their Applications in Analytical Chemistry, Elsevier, Amsterdam (2001).

[28] J. Khandurina and A. Guttman, J. Chromatogr. A 943 (2002) 159-183.

[29] R.S. Plumb, G.J. Dear, D.N. Mallett, D.M. Higton, S. Pleasance and R.A. Biddlecombe, Xenobiotica 31 (2001) 599-617.

[30] B.L. Ackermann, M.J. Berna and A.T. Murphy, Curr. Topics Med. Chem. 2 (2002) 53-66.

[31] Y. Deng, H. Zhang and J. Henion, Anal. Chem. 73 (2001) 1432-1439.

[32] T. Wachs, Y. Deng and J. Henion, Proceedings 48th American Society for Mass Spectrometry Conference, Long Beach, CA USA (2000).

[33] G.A. Schultz, T.N. Corso, S.J. Prosser and S. Zhang, Anal. Chem. 72 (2000) 4058-4063.

[34] J. Henion, S.J. Prosser, T.N. Corso and G.A. Schultz, Amer. Pharm. Rev. 3 (2000) 19-29.

[35] L.J. Kricka, Clin .Chim. Acta 307 (2001) 219-223.

[36] G. Guetens, K. Van Cauwenberghe, G. De Boeck, R. Maes, U.R. Tjaden, J. van der Greef, M. Highley, A.T. van Oosterom and E.A de Bruijn, J. Chromatogr. B 739 (2000) 139-150.

[37] F.E. Regnier, B. He, S. Lin and J. Busse, Trends Biotechnol. 17 (1999) 101-106.

Appendix I: Microplate Worksheets

	1	2	3	4	5	6	7	8	9	10	11	12
A	1	2	3	4	5	6	7	8	9	10	11	12
B	13	14	15	16	17	18	19	20	21	22	23	24
C	25	26	27	28	29	30	31	32	33	34	35	36
D	37	38	39	40	41	42	43	44	45	46	47	48
E	49	50	51	52	53	54	55	56	57	58	59	60
F	61	62	63	64	65	66	67	68	69	70	71	72
G	73	74	75	76	77	78	79	80	81	82	83	84
H	85	86	87	88	89	90	91	92	93	94	95	96

	1	2	3	4	5	6	7	8	9	10	11	12
A	1	9	17	25	33	41	49	57	65	73	81	89
B	2	10	18	26	34	42	50	58	66	74	82	90
C	3	11	19	27	35	43	51	59	67	75	83	91
D	4	12	20	28	36	44	52	60	68	76	84	92
E	5	13	21	29	37	45	53	61	69	77	85	93
F	6	14	22	30	38	46	54	62	70	78	86	94
G	7	15	23	31	39	47	55	63	71	79	87	95
H	8	16	24	32	40	48	56	64	72	80	88	96

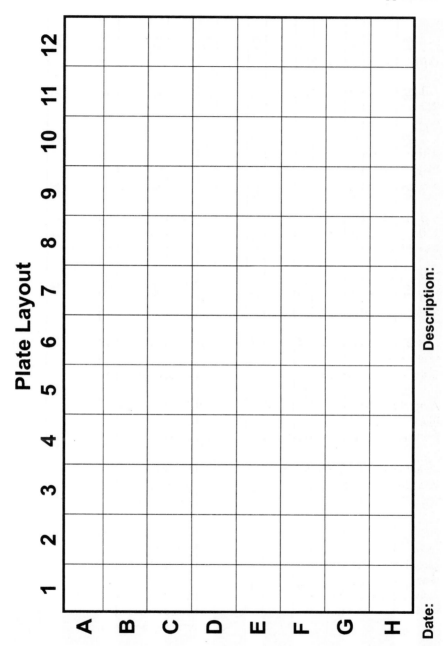

Plate Layout

	1	2	3	4	5	6	7	8	9	10	11	12
A												
B												
C												
D												
E												
F												
G												
H												

Date: **Description:**

Appendix II: List of Vendors

ABgene North America
Rochester, NY USA
www.abgene.com

Advion Biosciences
Ithaca, NY USA
www.advion.com

Agilent Technologies
Wilmington, DE USA
www.agilent.com

Alcott Chromatography Inc.
Norcross, GA USA
www.alcottchromatography.com

Alltech Associates Inc.
Deerfield, IL USA
www.alltechweb.com

Amersham Bioscience Corporation
Piscataway, NJ USA
www.amershambiosciences.com

Analytical Sales & Service Inc
Pompton Plains, NJ USA
www.analytical-sales.com

Ansys Technologies Inc.
(now a part of Varian Inc.)
Lake Forest, CA USA
www.ansysinc.com
www.metachem.com
www.varianinc.com

Apogent Discoveries
Hudson, NH USA
www.apogentdiscoveries.com

Applied Separations Inc.
Allentown, PA USA
www.appliedseparations.com

Apricot Designs Inc.
Monrovia, CA USA
www.apricotdesigns.com

Argonaut Technologies Inc.
Foster City, CA USA
www.argotech.com

Axygen Scientific Inc.
Union City, CA USA
www.axygen.com

BD Biosciences
Bedford, MA USA
www.bdbiosciences.com

Beckman Coulter Inc.
Fullerton, CA USA
www.beckman.com

Bio-Tek Instruments Inc.
Winooski,VT USA
www.biotek.com

Brinkmann Instruments Inc
Westbury, NY USA
www.brinkmann.com

CERA Inc.
Baldwin Park, CA USA
+1 626 814 2688

Chemical Separation Corporation
Phoenixville, PA USA
+1 610 935 6066

Chromacol Ltd.
Hertfordshire, United Kingdom
www.chromacol.com

ChromTech AB
Stockholm, Sweden
www.chromtech.se

Cohesive Technologies Inc.
Franklin, MA USA
www.cohesivetech.com

Corning Inc
Acton, MA USA
www.corning.com/lifesciences

Diazem Corporation
Midland, MI USA
www.diazem.com

Diversified Biotech Inc.
Boston, MA USA
www.divbio.com

Fisher Scientific International
Pittsburgh, PA USA
www.fishersci.com

GeneVac Inc.
Valley Cottage, NY USA
www.genevacusa.com
www.genevac.co.uk

Gilson Inc.
Middleton, WI USA
www.gilson.com

Glas-Col
Terre Haute, IN USA
www.glascol.com

Greiner Bio-One Inc.
Longwood, FL USA
www.greinerbiooneinc.com

Hamilton Company
Reno, NV USA
www.hamiltoncompany.com

Harvard Apparatus Inc.
Harvard Bioscience Inc.
Holliston, MA USA
www.harvardapparatus.com
www.harvardbioscience.com

Hitachi High Technologies
San Jose, CA USA
www.hii-hitachi.com

Horizon Instrument Inc.
Horizon Specialty Inc.
King of Prussia, PA USA
www.horizon-specialty.com

IKA Works Inc
Wilmington, NC USA
www.ika.net

Innovative Microplate
Chicopee, MA USA
www.innovativemicroplate.com

IRIS Technologies LLC
Lawrence, KS USA
www.iristechnologies.net

IST, Ltd.
(now Argonaut Technologies)
Hengoed, United Kingdom
www.ist-spe.com
www.argotech.com

J-KEM Scientific Inc.
St. Louis, MO USA
www.jkem.com

Jouan Inc.
Winchester, VA USA
www.jouan.com

Kendro Laboratory Products
Newtown, CT USA
www.kendro.com

Kimble / Kontes
Vineland, NJ USA
www.kimble-kontes.com

Knauer GmbH
Berlin, Germany
www.knauer.net

Labconco Corporation
Kansas City, MO USA
www.labconco.com

LabCyte
Union City, CA USA
www.labcyte.com

Labnet International
Woodbridge, NJ USA
www.labnetlink.com

LEAP Technologies Inc.
Carrboro, NC USA
www.leaptec.com

Macherey-Nagel Inc.
Easton, PA USA
www.mn-net.com

Mallinckrodt Baker Inc.
Phillipsburg, NJ USA
www.mallbaker.com
www.jtbaker.com

Matrix Technologies Corporation
Hudson, NH USA
www.apogentdiscoveries.com

MDS Sciex
Concord, Ontario Canada
www.mdssciex.com

Merck KGaA
Darmstadt, Germany
www.merck.de/chromatography

Mettler-Toledo Autochem
Vernon Hills, IL USA
www.mtautochem.com
www.bohdan.com

MicroLiter Analytical Supplies Inc.
Suwanee, GA USA
www.microliter.com

Micromass Inc.
Beverly, MA USA
www.micromass.co.uk

Micronic BV
Lelystad, The Netherlands
www.micronic.com

Millipore Corporation
Bedford, MA USA
www.millipore.com

Misonix Inc.
Farmingdale, NY USA
www.misonix.com

Molecular Probes Inc.
Eugene, OR USA
www.probes.com

Nalge Nunc International
Rochester, NY USA
www.nalgenunc.com

Nichiryo America Inc.
Flanders, NJ USA
www.nichiryo.com
www.nichiryo.co.jp

Organomation Associates
Berlin, MA USA
www.organomation.com

Orochem Technologies Inc.
Westmont, IL USA
www.orochem.com

Packard Instrument Company
(now PerkinElmer Life Sciences)
Meriden, CT USA
www.packardinstrument.com
www.perkinelmer.com

Pall Corporation Life Sciences
Ann Arbor, MI USA
www.pall.com

Perkin Elmer Life Sciences
Boston, MA USA
www.perkinelmer.com

PGC Scientifics Corporation
Gaithersburg, MD USA
www.pgcsci.com
www.pgcscientifics.com

Phenomenex Inc.
Torrance, CA USA
www.phenomenex.com

Porvair Sciences Ltd.
Middlesex, United Kingdom
www.porvair-sciences.com

Qiagen Inc.
Valencia, CA USA
www.qiagen.com

Robbins Scientific Corporation
Sunnyvale, CA USA
www.robsci.com
www.apogentdiscoveries.com

Shimadzu Scientific Instruments
Columbia, MD USA
www.ssi.shimadzu.com

Shiseido Co. Ltd.
Ginza, Chuo-ku, Japan
www.shiseido.co.jp/hplc

Simport
Beloeil, Quebec Canada
www.simport.com

SLR Systems
Vancouver, WA USA
www.slrsystems.com

Spark Holland
Emmen, The Netherlands
www.spark-holland.nl

Spectrum Laboratories Inc.
Rancho Dominguez, CA USA
www.spectrumlabs.com
www.spectrapor.com

SPEware Corporation
San Pedro, CA USA
www.speware.com

SPEX CertiPrep
Metuchen, NJ USA
www.spexcsp.com

SSP Companies Inc.
Ballston Spa, NY USA
www.sspinc.com

Sun International
Wilmington, NC USA
www.sun-sri.com

Supelco Inc.
Bellefonte, PA USA
www.supelco.com
www.sigmaaldrich.com

Tecan US
Research Triangle Park, NC USA
www.tecan-us.com
www.tecan.com

Techne Inc.
Princeton, NJ USA
www.techneusa.com
www.techneuk.co.uk

Thermo CRS
Burlington, Ontario Canada
www.thermocrs.com
www.crsrobotics.com

Thermo Forma
Marietta, OH USA
www.thermoforma.com

Thermo IEC
Needham Heights, MA USA
www.thermoiec.com
www.iec-centrifuge.com

Thermo Labsystems
Franklin, MA USA
www.thermo.com/labsystems
www.labsystems.fi

Thermo Orion
Beverly, MA USA
www.thermoorion.com

Thermo Savant
Holbrook, NY USA
www.thermosavant.com

3M Corporation
St. Paul, MN USA
www.mmm.com/empore

Tomtec Inc.
Hamden, CT USA
www.tomtec.com

Trade Winds Direct Inc.
Gurnee, IL USA
www.tradewindsdirect.com

United Chemical Technologies Inc.
Bristol, PA USA
www.unitedchem.com

Varian Inc.
Harbor City, CA USA
www.varianinc.com

Velocity 11
Palo Alto, CA USA
www.velocity11.com

VWR International
Chicago, IL USA
www.vwr.com

Waters Corporation
Milford, MA USA
www.waters.com

Whatman Inc.
Clifton, NJ USA
www.whatman.com

Zinsser Analytic GmbH
Frankfurt, Germany
www.zinsser-analytic.com

Zymark Corporation
Hopkinton, MA USA
www.zymark.com

Subject Index

Note: Figures and Tables are indicated by *italic page numbers*

599